Electricity transmission, distribution and storage systems

Related titles:

High temperature superconductors (HTS) for energy applications
(ISBN 978-0-85709-012-6)

Metropolitan sustainability
(ISBN 978-0-85709-046-1)

The coal handbook: Towards cleaner production Volume 1: Coal production
(ISBN 978-0-85709-422-3)

Details of these books and a complete list of titles from Woodhead Publishing can be obtained by:

- visiting our web site at www.woodheadpublishing.com
- contacting Customer Services (e-mail: sales@woodheadpublishing.com; fax: +44 (0) 1223 832819; tel.: +44 (0) 1223 499140 ext. 130; address: Woodhead Publishing Limited, 80 High Street, Sawston, Cambridge CB22 3HJ, UK)
- in North America, contacting our US office (e-mail: usmarketing@woodheadpublishing.com; tel.: (215) 928 9112; address: Woodhead Publishing, 1518 Walnut Street, Suite 1100, Philadelphia, PA 19102–3406, USA)

If you would like e-versions of our content, please visit our online platform: www.woodheadpublishingonline.com. Please recommend it to your librarian so that everyone in your institution can benefit from the wealth of content on the site.

We are always happy to receive suggestions for new books from potential editors. To enquire about contributing to our Energy series, please send your name, contact address and details of the topic/s you are interested in to sarah.hughes@woodheadpublishing.com. We look forward to hearing from you.

The team responsible for publishing this book:

Commissioning Editor: Sarah Hughes
Publications Coordinator: Steven Mathews
Project Editor: Anneka Hess
Editorial and Production Manager: Mary Campbell
Production Editor: Adam Hooper
Project Manager: Newgen Knowledge Works Pvt Ltd
Copyeditor: Newgen Knowledge Works Pvt Ltd
Proofreader: Newgen Knowledge Works Pvt Ltd
Cover Designer: Terry Callanan

Woodhead Publishing Series in Energy: Number 38

Electricity transmission, distribution and storage systems

Edited by
Ziad Melhem

WOODHEAD PUBLISHING

Oxford Cambridge Philadelphia New Delhi

Published by Woodhead Publishing Limited,
80 High Street, Sawston, Cambridge CB22 3HJ, UK
www.woodheadpublishing.com
www.woodheadpublishingonline.com

Woodhead Publishing, 1518 Walnut Street, Suite 1100, Philadelphia,
PA 19102–3406, USA

Woodhead Publishing India Private Limited, 303 Vardaan House, 7/28 Ansari Road,
Daryaganj, New Delhi – 110002, India
www.woodheadpublishingindia.com

First published 2013, Woodhead Publishing Limited

British Library Cataloguing in Publication Data
A catalogue record for this book is available from the British Library.

Library of Congress Control Number: 2013947638

ISBN 978-1-84569-784-6 (print)
ISBN 978-0-85709-737-8 (online)
ISSN 2044-9364 Woodhead Publishing Series in Energy (print)
ISSN 2044-9372 Woodhead Publishing Series in Energy (online)

The publisher's policy is to use permanent paper from mills that operate a sustainable forestry policy, and which has been manufactured from pulp which is processed using acid-free and elemental chlorine-free practices. Furthermore, the publisher ensures that the text paper and cover board used have met acceptable environmental accreditation standards.

Typeset by Newgen Knowledge Works Pvt Ltd
Printed by Lightning Source

Contents

Contributor contact details

(* = main contact)

Editor

Z. Melhem
Oxford Instruments NanoScience
Tubney Woods
Abingdon, Oxfordshire,
OX13 5QX, UK

E-mail: Ziad.MELHEM@oxinst.com

Chapter 1

H. L. Willis PE
Quanta Technology
4020 Westchase Boulevard
Raleigh, NC 27518, USA

E-mail: lwillis@quanta-technology.com

Chapters 2 and 3

K. Bell* and C. Booth
University of Strathclyde
Department of Electronic and Electrical Engineering
204 George Street
Glasgow, G1 1XW, UK

E-mail: keith.bell@strath.ac.uk; campbell.d.booth@strath.ac.uk

Chapter 4

K. Kauhaniemi
University of Vaasa
P.O.Box 700 FI-65101, Vaasa, Finland

E-mail: kimmo.kauhaniemi@uva.fi

Chapter 5

J-L. Bessède
Schneider Electric
37 quai Paul-Louis Merlin
38050 GRENOBLE, Cedex 09, France

E-mail: jean-luc.bessede@schneider-electric.com

Chapter 6

D. Van Hertem*
University of Leuven (KU Leuven)
Department of Electrical Engineering, Division ELECTA
Kasteelpark Arenberg 10 (PB2445)
3001, Heverlee, Belgium

E-mail: dirk.vanhertem@esat.kuleuven.be

M. Delimar
University of Zagreb
Faculty of Electrical Engineering and Computing (FER)
Department of Energy and Power Systems
Unska 3, 10000, Zagreb, Croatia

E-mail: marko.delimar@fer.hr

Chapter 7

K. Wang and M. L. Crow*
Department of Electrical and Computer Engineering
Missouri University of Science and Technology
1870 Miner Circle
Rolla, Missouri, 65409, USA

E-mail: crow@mst.edu

Chapter 8

G. C. Stevens*
GnoSys Global Ltd and University of Surrey
Surrey Research Park
Guildford, Surrey, GU2 7YD, UK

E-mail: g.stevens@gnosysgroup.com

A. Vaughan
Head of Electrical Power Engineering Research Group
University of Southampton
Southampton, SO17 1BJ, UK

Chapter 9

W. Hassenzahl
Advanced Energy Analysis
11015 Haven St.
Las Vegas, NV 89183, USA

E-mail: advenergy1@aol.com

Chapter 10

J. Oberschmidt*, M. Klobasa and F. Genoese
Competence Center Energy Policy and Energy Systems
Fraunhofer Institute for Systems and Innovation Research
BreslauerStraße 48
76139, Karlsruhe, Germany

E-mail: julia.oberschmidt@gmx.de; marian.klobasa@isi.fraunhofer.de; fabio.genoese@isi.fraunhofer.de

Chapter 11

P-J. Tsai and S. L. I. Chan*
School of Materials Science and Engineering
University of New South Wales
Sydney, NSW 2052, Australia

E-mail: pj.ben.tsai@gmail.com; sli.chan@unsw.edu.au

Chapter 12

M. Skyllas-Kazacos* and C. Menictas
School of Chemical Engineering
University of New South Wales
Sydney, NSW 2052, Australia

E-mail: m.kazacos@unsw.edu.au; c.menictas@unsw.edu.au

T. Lim
School of Life Sciences and Chemical Technology
Ngee Ann Polytechnic
Singapore

E-mail: TMLim@ntu.edu.sg

Chapter 13

P. Tixador
Grenoble INP/Institut Néel – G2Elab
CNRS Bâtiment E-B.P. 166
38 042 Grenoble, Cedex 09, France

E-mail: pascal.tixador@grenoble-inp.fr

Woodhead Publishing Series in Energy

1 **Generating power at high efficiency: Combined cycle technology for sustainable energy production**
Eric Jeffs

2 **Advanced separation techniques for nuclear fuel reprocessing and radioactive waste treatment**
Edited by Kenneth L. Nash and Gregg J. Lumetta

3 **Bioalcohol production: Biochemical conversion of lignocellulosic biomass**
Edited by Keith W. Waldron

4 **Understanding and mitigating ageing in nuclear power plants: Materials and operational aspects of plant life management (PLiM)**
Edited by Philip G. Tipping

5 **Advanced power plant materials, design and technology**
Edited by Dermot Roddy

6 **Stand-alone and hybrid wind energy systems: Technology, energy storage and applications**
Edited by John K. Kaldellis

7 **Biodiesel science and technology: From soil to oil**
Jan C. J. Bart, Natale Palmeri and Stefano Cavallaro

8 **Developments and innovation in carbon dioxide (CO_2) capture and storage technology Volume 1: Carbon dioxide (CO_2) capture, transport and industrial applications**
Edited by M. Mercedes Maroto-Valer

9 **Geological repository systems for safe disposal of spent nuclear fuels and radioactive waste**
Edited by Joonhong Ahn and Michael J. Apted

10 **Wind energy systems: Optimising design and construction for safe and reliable operation**
Edited by John D. Sørensen and Jens N. Sørensen

11 **Solid oxide fuel cell technology: Principles, performance and operations**
Kevin Huang and John Bannister Goodenough

12 **Handbook of advanced radioactive waste conditioning technologies**
Edited by Michael I. Ojovan
13 **Membranes for clean and renewable power applications**
Edited by Annarosa Gugliuzza and Angelo Basile
14 **Materials for energy efficiency and thermal comfort in buildings**
Edited by Matthew R. Hall
15 **Handbook of biofuels production: Processes and technologies**
Edited by Rafael Luque, Juan Campelo and James Clark
16 **Developments and innovation in carbon dioxide (CO_2) capture and storage technology Volume 2: Carbon dioxide (CO_2) storage and utilisation**
Edited by M. Mercedes Maroto-Valer
17 **Oxy-fuel combustion for power generation and carbon dioxide (CO_2) capture**
Edited by Ligang Zheng
18 **Small and micro combined heat and power (CHP) systems: Advanced design, performance, materials and applications**
Edited by Robert Beith
19 **Advances in clean hydrocarbon fuel processing: Science and technology**
Edited by M. Rashid Khan
20 **Modern gas turbine systems: High efficiency, low emission, fuel flexible power generation**
Edited by Peter Jansohn
21 **Concentrating solar power technology: Principles, developments and applications**
Edited by Keith Lovegrove and Wes Stein
22 **Nuclear corrosion science and engineering**
Edited by Damien Féron
23 **Power plant life management and performance improvement**
Edited by John E. Oakey
24 **Electrical drives for direct drive renewable energy systems**
Edited by Markus Mueller and Henk Polinder
25 **Advanced membrane science and technology for sustainable energy and environmental applications**
Edited by Angelo Basile and Suzana Pereira Nunes
26 **Irradiation embrittlement of reactor pressure vessels (RPVs) in nuclear power plants**
Edited by Naoki Soneda
27 **High temperature superconductors (HTS) for energy applications**
Edited by Ziad Melhem
28 **Infrastructure and methodologies for the justification of nuclear power programmes**
Edited by Agustín Alonso

29 **Waste to energy conversion technology**
Edited by Naomi B. Klinghoffer and Marco J. Castaldi

30 **Polymer electrolyte membrane and direct methanol fuel cell technology Volume 1: Fundamentals and performance of low temperature fuel cells**
Edited by Christoph Hartnig and Christina Roth

31 **Polymer electrolyte membrane and direct methanol fuel cell technology Volume 2: *In situ* characterization techniques for low temperature fuel cells**
Edited by Christoph Hartnig and Christina Roth

32 **Combined cycle systems for near-zero emission power generation**
Edited by Ashok D. Rao

33 **Modern earth buildings: Materials, engineering, construction and applications**
Edited by Matthew R. Hall, Rick Lindsay and Meror Krayenhoff

34 **Metropolitan sustainability: Understanding and improving the urban environment**
Edited by Frank Zeman

35 **Functional materials for sustainable energy applications**
Edited by John A. Kilner, Stephen J. Skinner, Stuart J. C. Irvine and Peter P. Edwards

36 **Nuclear decommissioning: Planning, execution and international experience**
Edited by Michele Laraia

37 **Nuclear fuel cycle science and engineering**
Edited by Ian Crossland

38 **Electricity transmission, distribution and storage systems**
Edited by Ziad Melhem

39 **Advances in biodiesel production: Processes and technologies**
Edited by Rafael Luque and Juan A. Melero

40 **Biomass combustion science, technology and engineering**
Edited by Lasse Rosendahl

41 **Ultra-supercritical coal power plants: Materials, technologies and optimisation**
Edited by Dongke Zhang

42 **Radionuclide behaviour in the natural environment: Science, implications and lessons for the nuclear industry**
Edited by Christophe Poinssot and Horst Geckeis

43 **Calcium and chemical looping technology for power generation and carbon dioxide (CO_2) capture: Solid oxygen- and CO_2-carriers**
Paul Fennell and E. J. Anthony

44 **Materials' ageing and degradation in light water reactors: Mechanisms, and management**
Edited by K. L. Murty

45 **Structural alloys for power plants: Operational challenges and high-temperature materials**
Edited by Amir Shirzadi, Rob Wallach and Susan Jackson
46 **Biolubricants: Science and technology**
Jan C. J. Bart, Emanuele Gucciardi and Stefano Cavallaro
47 **Advances in wind turbine blade design and materials**
Edited by Povl Brøndsted and Rogier P. L. Nijssen
48 **Radioactive waste management and contaminated site clean-up: Processes, technologies and international experience**
Edited by William E. Lee, Michael I. Ojovan, Carol M. Jantzen
49 **Probabilistic safety assessment for optimum nuclear power plant life management (PLiM): Theory and application of reliability analysis methods for major power plant components**
Gennadij V. Arkadov, Alexander F. Getman and Andrei N. Rodionov
50 **The coal handbook: Towards cleaner production Volume 1: Coal production**
Edited by Dave Osborne
51 **The coal handbook: Towards cleaner production Volume 2: Coal utilisation**
Edited by Dave Osborne
52 **The biogas handbook: Science, production and applications**
Edited by Arthur Wellinger, Jerry Murphy and David Baxter
53 **Advances in biorefineries: Biomass and waste supply chain exploitation**
Edited by Keith W. Waldron
54 **Geological storage of carbon dioxide (CO_2): Geoscience, technologies, environmental aspects and legal frameworks**
Edited by Jon Gluyas and Simon Mathias
55 **Handbook of membrane reactors Volume 1: Fundamental materials science, design and optimisation**
Edited by Angelo Basile
56 **Handbook of membrane reactors Volume 2: Reactor types and industrial applications**
Edited by Angelo Basile
57 **Alternative fuels and advanced vehicle technologies: Towards zero carbon transportation**
Edited by Richard Folkson
58 **Handbook of microalgal bioprocess engineering**
Christopher Lan and Bei Wang
59 **Fluidized bed technologies for near-zero emission combustion and gasification**
Edited by Fabrizio Scala
60 **Managing nuclear projects: A comprehensive management resource**
Edited by Jas Devgun

61 **Handbook of Process Integration (PI): Minimisation of energy and water use, waste and emissions**
Edited by Jiří J. Klemeš
62 **Renewable heating and cooling system**
Edited by Gerhard Stryi-Hipp
63 **Environmental remediation and restoration of contaminated nuclear sites**
Edited by Leo Van Velzen

For my wife Salam and my sons Nabil and Tariq.

Preface

For over a century, a significant number of the world's population has enjoyed the benefits of electric power – a clean, economical and manageable energy source that leads to material improvements in quality of life and industrial efficiency and enhanced prosperity. At present, there are few places of social and economic importance on this planet that do not have utility-supplied electric power. Along with roads and bridges, telephone, the World Wide Web, water and sewer, electricity has become part of the very foundation upon which first-world countries have built their quality of life and economic prosperity and this is essential to the development and prosperity of the rest of the world. Stable and secure supply of electric power is essential for societies and is critical for national security, public health and economic prosperity. It is an integral part of daily life that people often take for granted.

In the past few decades, electric power has been produced and delivered to electric energy consumers in all nations over electric power systems which are almost all based on the same 'central station' paradigm: power is produced in bulk at relatively few places but consumed at many locations. The power system serving a city or region is dominated by a few large central generating stations, each consisting of from one to perhaps half a dozen industrial-scale power production machines (generators) along with the ancillary equipment needed to operate and maintain them in good working order. Transmission lines carried the power to points throughout the region, where it was passed to smaller-capacity lines (distribution) on which it was routed through neighbourhoods and eventually to individual homes, businesses, and other energy users that each use only a tiny fraction of the power produced by the average-size generator.

I undertook this book project with the aim of providing a reference source for electricity transmission, distribution and storage (TDE) systems. The book provides a structured account of the fundamental principles of TDE and current status of development of primary components. The book is a comprehensive review of the materials, architecture and performance of electricity transmission and distribution networks, and the application and integration electricity storage systems. The intended audience for the book are scientists, researchers and engineers in the power industry, research laboratories, and students at the senior/ graduate level in universities.

The book is divided into three parts: Part I reviews in Chapter 1, the fundamentals of transmission and distribution (T&D) of electric power and provides an introduction to T&D networks, infrastructure, reliability and engineering challenges as well as the different regulation and planning. In Chapter 2, a description of the principal technical limits required for satisfactory operation of T&D networks and the corresponding controlled elements required for safe and continuous supply of power is discussed. It also covers the impact of renewable generation and the potential contribution of energy storage. The protection of T&D networks during a fault or abnormal conditions is presented in Chapter 3, whereas technical issues of successful integration of distributed energy resources (DER) is presented in Chapter 4.

In Part II, the emphasis is on the development and application of advanced materials for electricity transmission and distribution (T&D) networks. Chapter 5 reviews the development of the state-of-the-art of switch gear materials for high voltage (HV) applications with particular focus on circuit breakers. This is followed in Chapter 6 by covering the technology required for long distance power transfers via high voltage direct current (HVDC) and the impact on renewable and the rapid increase in energy consumption on reviving the technology of HVDC technology. In Chapter 7, the description of Modern Flexible AC Transmission System Controllers and future trends is presented, with particular focus on the three primary modern FACTS devices: the static synchronous compensator (STATCOM), the static synchronous series compensator (SSSC) and the unified power flow controller (UPFC). Effective power transmission is dependent on dielectric materials; Chapter 8 presents the critical elements used in power transmission devices and the challenges in enabling reliable transmission that will require development of new materials like Nanodielectrics.

Innovation in superconducting power applications has the potential to become a leading twenty-first century technology for enhancing the capacity of power equipment and improving efficiency and reliability. Superconducting materials is another revolutionary material with huge potential impact on T&D. The past two decades saw many prototyping activities using superconducting materials from TDE of energy. Superconducting applications in the long term, especially those working at high temperatures will lead to a step change and revolutionise our energy generation, delivery and management. In Chapter 9 an elaborate overview of the technology of superconducting power cables and superconducting fault current limiters is presented. With a compelling discussion of the historical development of superconducting cables and its current status, this is followed by a description of fault current limiters and their developments.

The final part, Part III, is reserved for a review of the critical elements of electricity storage techniques. In Chapter 10 the economic and environmental assessment of electricity storage systems is reviewed and discussed. This is primarily driven by the increased share of intermittent renewable energies in power generation, in the future, electricity storage could be a key technology of strategic importance with regard to grid stability and security of supply. In this context, this chapter analyses the economic feasibility of applying electricity storage technologies in energy markets. A particular focus is on stationary storage technologies for management of power fluctuations due to intermittent renewable energies.

There are different storage technologies such as pumped hydro energy storage and lead acid batteries as well as innovative alternatives such as advanced adiabatic compressed air energy storage and redox flow batteries. Other storage technologies are under consideration like hydrogen storage, lithium ion batteries and sodium sulphur batteries.

In Chapter 11, materials and chemistry for nickel-based batteries are presented. Nickel-based batteries are similar in a way to nickel hydroxide electrodes' utlisation as positive plates in the systems. Ni-based batteries have been, and still are the most important power sources for a wide range of electronic devices. A comprehensive review is presented on the past and present available Ni-based battery systems and their performance. The current popular and novel materials that have potential in the applications of Ni-based batteries is discussed, followed by the challenges and the future trend of the batteries.

Chapter 12 reviews another technology for medium- and long-term storage utilising the novel redox flow batteries for grid stabilisation, needed for example, when integrating renewable energy sources into the electricity grids. Flow batteries offer high energy efficiencies, very long cycle life and good cost structures for applications requiring more than two hours of storage capacity.

In the final chapter (13), the promising technology of Superconducting Magnetic Energy Storage (SMES) is reviewed. SMES is one of the very few direct electric energy storage systems particularly promising for high-power and short-time applications (pulse current power sources especially). An SMES unit can release its energy very quickly and may have energy efficiency greater than 95 %.

I hope that the contents presented in this book will help the reader and all interested parties in TDE engineering and technology to have a better understanding of the recent and future developments in this critical aspect of power delivery. It is important that scientists and engineers continue developing the various TDE technologies, improve materials performance, and demonstrate cost effectiveness as it has the potential for enabling a step change in many applications that will affect our quality of life.

This book idea originated following a workshop on Transmission, Distribution and Storage sponsored by the UK Materials. I would like to thank all chapter authors who provided an excellent contribution on their respective area of expertise.

Ziad Melhem
Oxford, UK

Part I
Fundamentals of electricity transmission and distribution

1

Introduction to transmission and distribution (T&D) networks: T&D infrastructure, reliability and engineering, regulation and planning

L. WILLIS, Quanta Technology, USA

DOI: 10.1533/9780857097378.1.3

Abstract: Power transmission and distribution systems have evolved over the past twelve decades into vast interconnected systems of equipment built around large centralized generation plants. In the twenty-first century the evolution will begin to include significant portions of de-centralized, distributed generation and storage, much of it based on renewable and 'micro-grid' technologies. Power utilities will not evolve their systems completely to this type of configuration, but instead pragmatically combine their old central-station systems with additions and augmentation based on distributed generation and micro-grids when and where most suitable to their needs. Future power systems will be an amalgamation of old and new types providing considerably more flexibility as to reliability, economy and fit to extreme aesthetic and user demand requirements than traditional systems, but often requiring considerably more context-responsive and dynamic control than in the past.

Key words: power systems, power transmission, power distribution, micro-grids, distributed generation, renewable generation, distributed storage, future of power systems, electric utilities, electric load, electric demand.

1.1 Introduction

For over a century, an increasing portion of the world's population has enjoyed the benefits of electric power – a clean, controllable, and economical energy source that leads to material improvements in quality of life and industrial efficiency. At present, there are few places of economic importance on this planet that do not have utility-supplied electric power. Along with roads and bridges, telephone, and water and sewer, electricity has become part of the very foundation upon which first-world countries have built their quality of life and economic prosperity. During the past century, electric power has been produced and delivered to electrical energy consumers in all nations over electrical power systems that are almost all based on the same 'central-station' paradigm: power is produced in bulk at relatively few places but consumed at many. The power system serving a city or region is dominated by a few large central generating stations, each consisting of from one to perhaps half a dozen industrial-scale power production machines (generators) along with the ancillary equipment needed to operate and maintain them in good working order. Transmission lines carry the power in bulk quantities to points throughout the region, where it is passed to smaller-capacity lines (distribution) on which it is routed through neighborhoods and eventually to individual homes, businesses, and other energy users (Fig. 1.1), which each use only a tiny fraction of the power produced by the average-size generator. Typically, there are several orders of magnitude more points of consumption – perhaps 100 000 times as many – as power generation points. Engineering standards, which here will be taken to mean the institutionalized and documented 'way of doing things,' vary, sometimes significantly, from one continent or region of the world to others, but the vast majority of utility and industrial power systems on earth have been built to be, and continue to be, operated within this overall central-station system concept.

A *power T&D system* is that portion of the power system that moves power from where it is produced to where it is consumed: basically, it is the entire power system sans generators. The T&D system interconnects all the disparate parts of the power system and thus to a great extent determines the character of that system.

Beginning in the late twentieth century and continuing into the twenty-first century, significant changes – advances if perhaps not true breakthroughs – began to occur in several of the technologies that made up electric power systems. It became possible to build electric power systems fed by many more, but individually smaller, generating sites (Fig. 1.2). No longer would there be thousands of times fewer generating stations than energy consumers: conceivably, the ratio could be one to one. Such *distributed power systems,* in which power production is dispersed widely throughout the energy consumer base rather than concentrated at a few generating stations as in a traditional power system, had different reliability, maintainability, and operability characteristics, as well as different economies of scale, etc., which shaped their use differently from traditional power systems. Neither type of system, traditional or distributed, is necessarily better. They are merely different. What seems clear is that the power system of the future will be neither one, but instead a hybrid mix of both.

This chapter focuses on the delivery part of the system, the T&D systems. It will review and summarize the overall structure, function, design, and performance of modern electric power T&D systems. At times it will present and discuss at an almost elementary level, basic concepts behind system

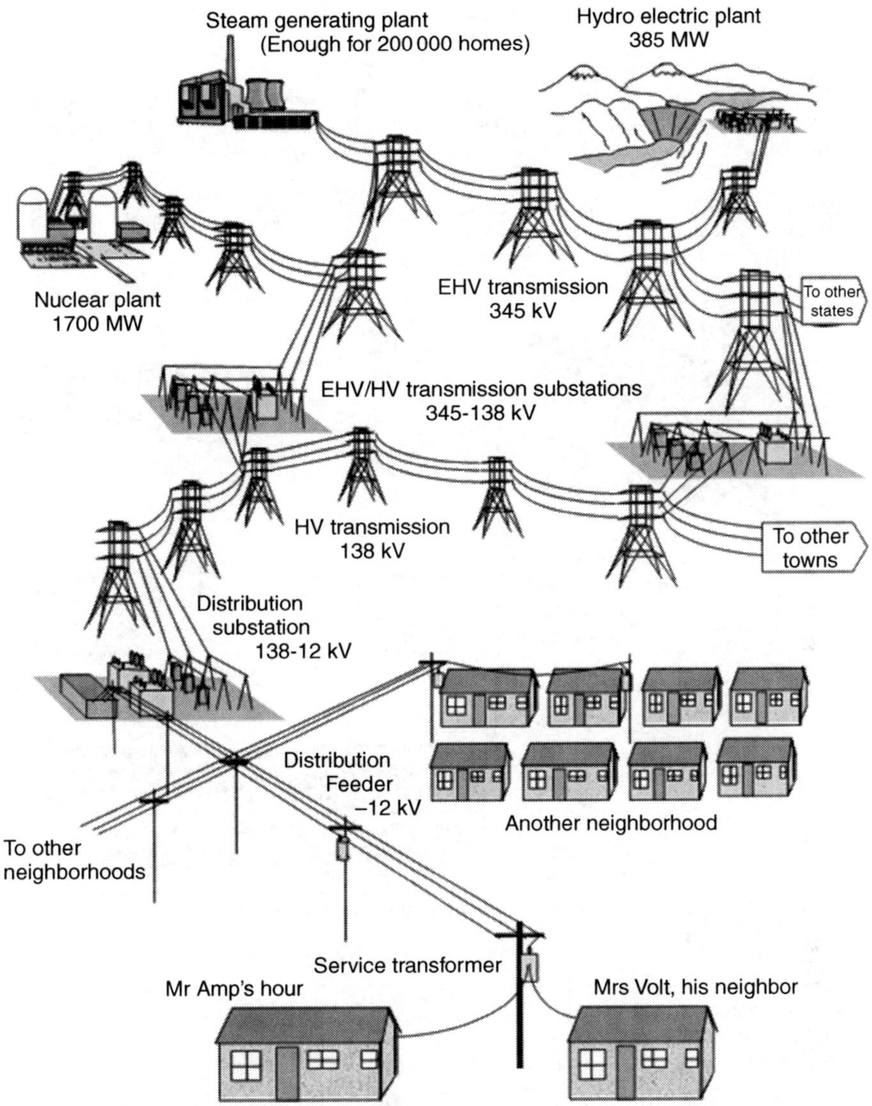

1.1 The structure of a traditional power system, dominated by several large central-station generator plants and the bulk transmission system connecting them. That transmission not only moved the bulk power around the system but in many ways determined the character of the power system itself.

design and operation, when these basics play into the differences between traditional and modern distributed power systems. These basics are among the factors that once required the traditional type of design solution, but can now be accommodated by something different, and thus they are a key to

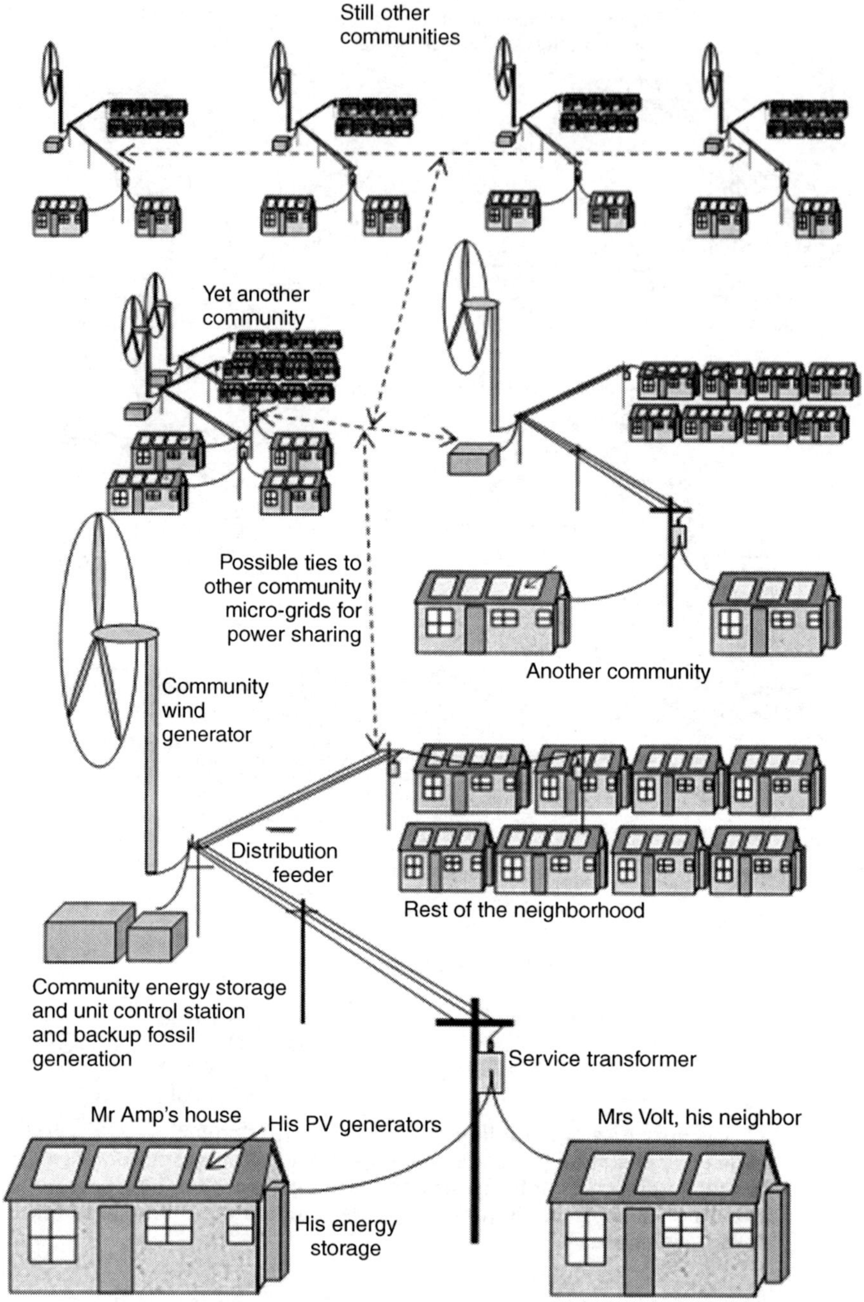

1.2 The overall structure of a distributed power system, in which individual customers may have generation and the 'power system' may only be a local micro-grid connecting a number of local consumers together for power and reliability sharing.

understanding how the transition from one to the other, or the hybrid melding of the two, can be accomplished.

This chapter begins in Section 1.2 with a look at the traditional power system, a power delivery paradigm dominated by large central-station generators and composed of a network of high-voltage transmission lines, medium voltage primary feeder systems, and low-voltage service lines to customers. Section 1.3 then takes a look at three technology changes and the smart distributed power systems they enable, and discusses how and why the distribution power systems that they enable differ from traditional designs, and what that can mean in terms of cost and performance.

Regardless of type, power systems must be planned, engineered, designed, built, and operated. Their parts and subsystems must be regularly maintained, repaired when broken and replaced when they fail or wear out. Someone must pay for all of that, and that is usually done by charging consumers for the power they according to well-established principles that basically price it per unit of use. Finally, the entire system must be managed, which is the job of utility companies, either public or private as the case may be, or the owner in the case of large industrial power systems. Section 1.4 summarizes these aspects of power systems.

Finally, whatever the advantages of distributed power systems, most developed nations around the world have traditional power systems in place, woven into the fabric of every city and town and region, and without which the local society and economy could not function. While there are often serious concerns related to reliability, pricing, or environmental issues, for the most part these traditional systems function well, or at least well enough. Most tellingly, they are in place and paid for – almost always an overwhelming reason to keep them despite any problems they may present. But distributed systems offer considerable advantages that are important to modern society and increasing individuals, companies, and governments are turned to them to meet their expanding power needs. Section 1.5 looks at electric infrastructures and the forces and factors in favor of retaining the traditional power system, as against moving to the newer distributed paradigm.

1.2 Characteristics of traditional and nontraditional power systems

Whether traditional or distributed, an electrical power system consists of equipment interconnected and operated in a coordinated fashion in order to route power from where it is produced to where it is consumed. Strictly speaking this simple definition could describe a flashlight, which includes a power source (battery), conducting 'power transmission' pathways (often the body of the flashlight itself) to route power to the electrical load: in this case a light bulb that consumes the power, and a control system (switch) in command of the operation. However, as normally used, *power system* is reserved for more powerful and expensive assemblages of equipment and generally does not include the end-use equipment (the light bulb in the case of the flashlight) but only the production, transmission, control, and hand-off points to the demands. With exceptions too rare and specialized to go into here, all power systems anywhere on the planet, and of either traditional or distributed types, have the following characteristics discussed below:

1.2.1 Constant supply voltage

For whatever reason, mankind has chosen to build constant-voltage power systems. The perfect constant-voltage power system would provide an unvarying, identical voltage at each electric consumer's location, not varying *at all* regardless of time, system operating conditions, or the amount of usage. In practice, voltage is permitted to vary with time and conditions by small amounts – for example up to 3% during short periods of time. But despite this, the concept of use is: vary the amount of power drawn from the system by varying the current drawn from the system. For example, a device that needs 12 watts – a small light bulb – is designed to draw 0.1 amps from a source of 120 V. By contrast a device needing 120 watts, say a small motor, would be designed to draw 1 amp from that same 120 V, and a 1200 watts hairdryer would be designed to draw 10 amps at 120 V. This is accomplished by varying the electrical impedance of the device using the power as respectively 10, 1, and 0.1 ohms in the three preceding examples. All power systems throughout the world operate on this principle. All consumer equipment is designed for this constant-voltage/variable current environment. Power system design, operation, protection, and safety principles are built upon this basic concept. It is worth noting that it would have been possible to build an electric infrastructure around constant current systems, in which a relatively constant current flowed through lines and variable usage need was controlled by changing the voltage at the point of use. Arguably, such systems could have been made to work, and work well. But while interesting, discussion of the path not taken is not productive for this discussion and will not be pursued here.

1.2.2 Alternating current

Power systems work with alternating current, in which voltage and current flow throughout the system cycle, or reverse their direction, either 50 or 60 times a second. This is in contrast to DC (direct current) power, in which the voltage does not vary in direction. Alternating current (AC) systems have one overwhelming advantage with respect to traditional power system design: they permit the use of transformers. Electric engineers view a transformer as a device that changes the voltage-current combination of an otherwise identical amount of power: a transformer with a 10:1 turns ratio takes a kilowatt of power at 100 V and 10 amps and transforms it into a kilowatt of power at 1000 V and 0.1 amp. Except for a very small amount of electric power consumed in the operation of the transformer, the amount of power does not change significantly. But the *economy of scale of power transmission* changes drastically with voltage. Transformers permit the use of high voltage transmission lines that can economically move power long distances in bulk quantities *and* low-voltage lines that move conveniently small amounts of power into a neighborhood of homes, in the same power system. Transformers only work with AC power, so all power systems use only AC power.

1.2.3 Natural stability

If all equipment in a system itself is functioning and operating as intended, a traditional power system will continue to operate as it is at the moment, providing electric power of good voltage, quality, and continuity, even if disturbed by an unexpected equipment failure or a reasonable change in load. The system is naturally stable – it tends to return to its current operating mode even if disturbed, as long as no controls on equipment or system operational settings are changed. While this might seem to be an unremarkable characteristic, it is actually a quality that can be difficult to assure, and one that can consume a great deal of engineering effort. Many types of systems are naturally unstable, to the point that not only will a minor disturbance upset their operating behavior, but a natural characteristic of their normal operation is to try to move to an unacceptable operating mode, even without provocation, thus requiring constant correction and change of control inputs, etc. But in a traditional utility or industrial power system, the system is designed to have a reasonable natural stability at any moment, so that it is stable enough that no human interaction is required to keep it running for the next few seconds, even if some disturbances were to occur. Large modern systems depend on automatic control to assure this: the system is not inherently unstable, but has so much operating complexity that it is likely a human alone would inadvertently go back into a situation in which it was, at least locally, in jeopardy. Automation assures that control actions to keep it operating are taken quickly, without human interaction, should they be needed. The added volatility of renewable energy and other changes accompanying distributed power systems are one reason why improved, 'smart' control systems will be needed in the future, even if one foregoes the economy and performance advantages smart equipment may convey.

1.3 Customer requirements and demand

The traditional power system is one in which a relatively few central generating stations provide power to a very large number of consumption points, through a T&D system that uses lines and equipment at several voltage levels to affect economical transmission of the energy from production sites to consumer sites. Some fundamental 'truths' about the demand shape the design of the traditional power system are:

1.3.1 Geographic spread of consumption points

Points of consumption are scattered throughout a sizable geographic region. The consumers of electric power are sometimes close together geographically, but often scattered widely. A large metropolitan utility system may have to deliver power to 200 000 separate locations distributed over 5000 square miles (Fig. 1.3). The power system has to run a wire to every one of these points, building a network of lines that connects to all the energy consumers over a wide region.

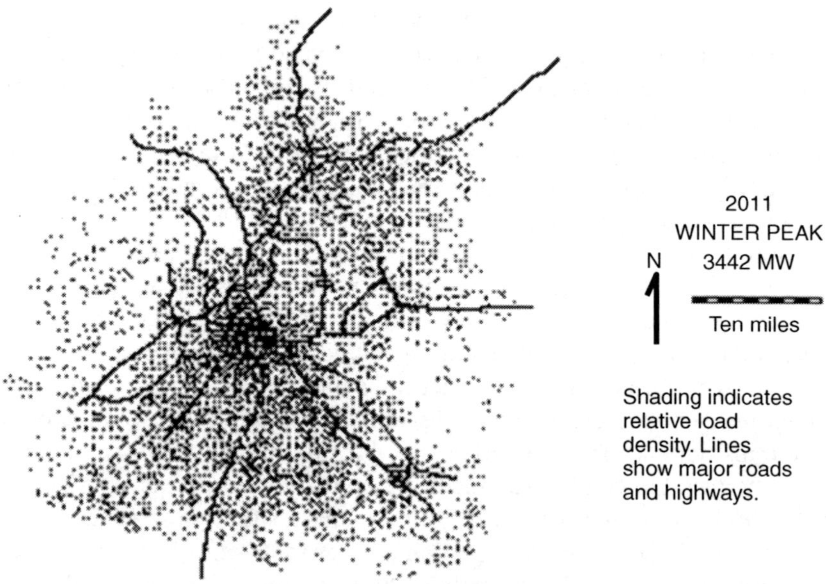

1.3 A map showing the locations of nearly 3500 MW of peak demand for a city. Only eight power stations provide power to the entire city. The T&D system's job is to route power from those eight stations to all 385 000 energy demand sites scattered throughout the region.

1.3.2 Demand for low-voltage (service voltage) power

The demand was and still is for low-voltage (service voltage) power. The 'ready-to-use form' requirement for power delivery means that the power must be of rather constant voltage, as discussed previously, and of relatively low voltage. Over 90% of the power consumed in most power systems is at the lowest service voltage offered – as low as 100 V line to ground in portions of Japan, 120 V or 240 V line to ground in the USA, and 250 V in the UK and Europe. Most household appliances and business equipment are designed for this voltage, and substantially higher voltages are not practical or economical for building wiring, or deemed safe.

1.3.3 Immediacy of power delivery

Electric power has to be delivered to appliances and equipment at the instant of consumption. A reality for traditional power system designers was that electric power could not be efficiently or economically stored. It had to be delivered to each point of consumption, in ready-to-use form, at the very instant of use. This situation has changed somewhat in the early twenty-first century. One can argue about whether modern batteries and energy storage technologies are 'efficient and economical'. There are situations where a positive business case can be made for their use as stationary energy storage units, but there are many more situations where a case cannot currently be made. This is one reason some distributed power system concepts work well enough to be viable alternatives to more traditional designs, and others do not.

1.3.4 Reliability

Reliability of service must be outstanding. Generally, no more than 2 h without power per year, (99.98% availability) is considered a reasonable level of power availability. Many electrical utilities are striving to limit the time that an average customer is without power to no more than 1 h per year.

1.3.5 Economy

While power systems cumulatively cost billions upon billions of dollars, they provide millions of consumers with thousands of kilowatt hours of power every year. The unit cost of electric energy, delivered in reliable and ready-to-use form, is quite economical in most locales around the globe: a major reason for electrical power's widespread use.

1.3.6 Safety

Safety is a major consideration. Electrical equipment can present a hazard to energy consumers, innocent bystanders and, particularly, to those who have to operate and service it.

1.3.7 Ease of use

The no-hassle use of power is recognized as a major market factor by electric utility pioneers of the nineteenth and early twentieth centuries. This means the consumer does not have to be involved in any way with the management and operation of the power system: they have to do nothing more than decide to use the power and be willing to pay for it. Once established, the 'mindless ease' with which power could be used by the average homeowner or business was often neglected as an important factor behind its popularity. In the early twenty-first century, this is again a major factor of consideration for smart and distributed power systems: whatever advantages they offer, they cannot require the homeowner or small businessperson to be involved in their operation, or complicate the process of easy usage. This is proving to be an important factor, as will be discussed later.

1.3.8 Equity and price subsidization

Issues of equity and price subsidization often dictate policy at government and corporate levels, with a very democratic view of electricity usage. It is so enabling of economic prosperity and improved quality of life that system design, operation, and price policies must provide at least basic access to all. These considerations often heavily shape the conditions and constraints under which utilities and users alike must operate.

1.3.9 Environmental and aesthetic issues

Environmental and aesthetic issues occasionally dominate local decisions about the details of power system construction and nature, and are often a part of national government policy. No system can be built without some 'side effects,' and power systems are no exception. The power production is not without its emissions or environmental issues. T&D equipment is unattractive in most cases and takes up room. Everyone wants the benefits of power without the pollution and other side effects of energy production or the aesthetic and 'societal fit' issues that come with the need for a large T&D system. At times finding a balance is difficult. Distributed power systems do not eliminate these concerns, but have different characteristics and interactions, and thus fit certain situations with unique characteristics that make them much more a solution to some local balance needs. This is one reason that may drive their future popularity.

1.4 Principles and natural laws governing T&D system design

The layout of traditional power systems was heavily influenced by engineering decisions aimed at accommodating and working around several 'truths' about T&D systems and the laws of physics that determine their performance and cost.

1.4.1 Economy of scale in power generation

Without exception, there is a very large positive economy of scale in all types of traditional power generation. Pick *any* traditional power generation method and technology, and a bigger generating unit or station (one capable of producing more power) is potentially more cost-effective and economical than a smaller unit of the same type and technology. For many types of traditional power generation, such as coal- and natural-gas-fired steam power plants, diesel powered generating plants, and nuclear power, the generator is basically a Carnot cycle engine, and the major reason for this economy of scale is that 'physics is on the side of the larger unit': it can be designed to have lower percentage thermal losses. A 500 MW natural-gas-fired combined cycle power plant will be slightly more efficient than a 50 MW unit of the same type and technology. This natural physical advantage applies also to fuel cells, which are Carnot cycle devices in a real sense, oxidizing rather than combusting fuels and producing heat as a byproduct of electrical product: a 500 kW solid oxide fuel cell might be impressively efficient, but a 500 MW fuel cell power plant designed with the same technology will be noticeably more efficient. Similarly, solar thermal and solar tower generation technologies also have a noticeable economy of scale in unit and station size.

With exceptions too fine to discuss here, wind and photovoltaic (PV) power production differ from other forms of generation in having almost no physical economy of scale. Their physical efficiency does *not* materially improve with size. A 100 MW photovoltaic power plant is composed of perhaps 100 000 PV panels, a 1 MW PV plant only 1000. All the panels are exactly as efficient, so the two plants are potentially equally efficient. The same goes for a 150 MW wind park composed of 50 3.0 MW turbines as compared to one composed of only five: each has the same natural physical efficiency. But despite this, the larger PV plant and the larger wind park will both be slightly more cost-effective in practice than the smaller ones, due to *concentration*. A wind park with 50 wind turbines can be built and managed for less money than ten wind parks each with five wind turbines, which can be operated with slightly less labor and maintenance cost, and often has less total environmental and societal esthetic impact. This is a factor to consider, but not an overwhelming factor favoring large size, as is a combination of this factor *and* the thermal efficiency advantages of scale discussed earlier.

But since wind and PV power have far lower total (sum of practical and physical) economies of scale, modern power systems tend to distribute them more. Factors having to do with location, ownership, unique customers wants and needs, and similar issues often mean that small installations – what is called distributed generation – is selected even though it might be less than optimally efficient. This is a key point with regard to modern (distributed) power plants: these generation types have economies of scale, but often not substantial enough to outbalance other important factors. That was not the case throughout most of the twentieth century.

Thus, the practical advantages of concentration, added to the physical efficiency advantages brought by larger size, meant there was a *significant* business case for concentrating the power generation of a system in a few very large central-station power plants, even if that power was to be consumed by hundreds of thousands or even millions of consumers scattered over a large geographic region. Designers of traditional power systems worked to include the largest possible generating units and stations in their systems. A number of factors militated against building only one or a very few power plants. Chief among these was reliability: the utility did not want to place all its eggs in one basket, so to speak. Thus, typically, a large regional power system might have one or two dozen power plants, all large but none larger than, say, 8% of the total. The result was power systems where the power was generated at a few 'bulk' sites but distributed to perhaps several hundred thousand consumer sites. The T&D system is the network of lines, facilities, and equipment that links these few big generating sites to the myriad smaller consumption points in a way that provides for the safe, dependable, and economical flow of satisfactory amounts of power to all consumption points.

1.4.2 Economical movement of power

Service voltages are useless for moving power any great distance. The 120/240 V single-phase utilization voltage used in the United States, or the 250 V/416 V three-phase used in 'European systems' are not equal to the task of economically moving power more than a few hundred yards. The application of these lower voltages for anything more than very local distribution at the neighborhood level results in unacceptably high electrical losses, severe voltage drops, and astronomical equipment cost.

It is most economical to move power in bulk at high voltage. The higher the voltage, the lower the cost per kilowatt to move power any distance, and the greater the distance that power can be moved with any particular level of efficiency. But the higher the voltage, the greater is the capacity and cost of transmission lines. A very high voltage line is potentially more economical in the movement of power, no matter how economy is measured, than a lower voltage line. However, one must realize that it is the 'giant economy size,' and while always giant, it will only achieve its economy of scale if applied to move large amounts of power. Therefore, for any specific amount of power and distance, there is a voltage level that is best from the standpoint of overall materials, labor, and lifetime operating cost.

1.4.3 Cost of changing voltage levels

It is costly to change voltage levels but not prohibitively so, for it is done throughout a power system (that is what transformers do) – but voltage transformation is a major expense that does nothing to move the power any distance in and of itself.

The overall concept of a power delivery system layout that has evolved to best handle the needs, constraints, and factors discussed earlier is a hierarchical system of decreasing voltage levels each with increasing numbers of components, as shown in Fig. 1.4. As power is dispersed throughout the

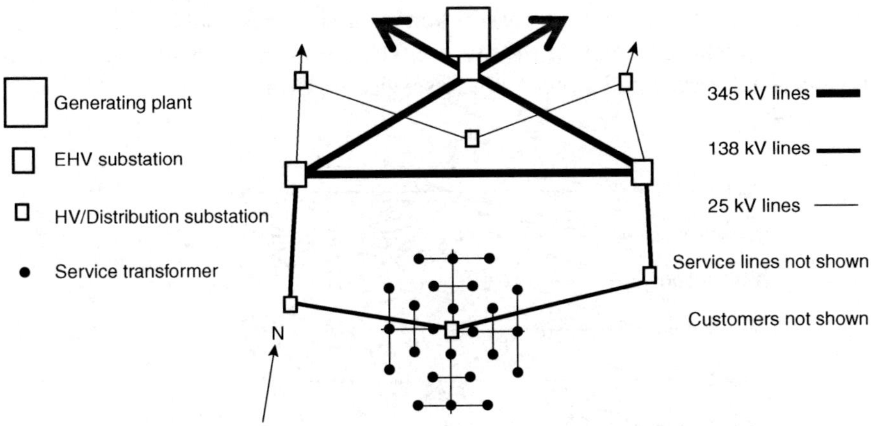

1.4 Traditional power system design evolved into a hierarchical structure with several distinct voltage levels. A key concept is 'lower voltage and split,' which is done from three to five times during the course of power flow from generation to customer.

service territory, it is gradually moved down to lower voltage levels, where it is moved in ever smaller amounts (along more separate paths) on lower capacity equipment until it reaches the customers. The key element is a 'lower voltage and split' concept, as exemplified at a distribution substation: incoming lines at a voltage such as 138 kV may number two to four, but outgoing primary feeders at a voltage of between 4 and 34 kV.

1.5 Layers or levels of the traditional T&D system

One useful consequence of the hierarchical structure of the traditional power delivery system is that it can be thought of as composed of several distinct *levels* or *layers* of power equipment, as illustrated in Fig. 1.5. Each level consists of many units of fundamentally similar equipment – same nominal voltage, roughly the same capacity, doing roughly the same job – located in different parts of the utility service territory in order to 'cover' the entire utility service territory, and interconnecting at that level or the next highest. For example, all of the distribution substations are planned and laid out in approximately the same manner and do roughly the same job. All are composed of roughly similar equipment doing the same job: large quantities of power are brought into the substation, lowered in voltage via the substation's transformers, and routed out to nearby neighborhoods and communities on lower voltage lines than those on which the power entered the substation. Some may be 'larger' than others in both physical and equipment terms – one could have 450 MVA transformers and another two, etc., but, fundamentally, all perform the same function in the same way for the same reasons, and hopefully, with the same result (reliable local power delivery with good economy and safety). Taken as a set, these substations constitute the 'substation level' of the system. Their service areas fit together in a mosaic, each covering its piece of the service territory.

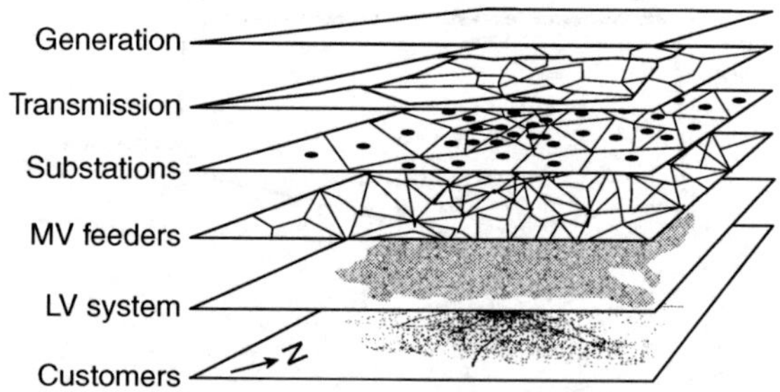

1.5 A traditional power system lends itself to being thought of as composed of layers, each consisting of all the equipment at a certain voltage level and function, with equipment at that layer serving all of the utility territory.

Likewise, the feeders the substations route power into are all similar in equipment type, layout, and mission, and all service transformers to which those feeders route power are similarly serving the same basic mission and are designed with similar planning goals and to similar engineering standards.

Thus, power can be thought of as flowing 'down' through these various levels, on its way from power production and the wholesale grid to the energy consumers. As it moves from the generation plants (system-level) to the energy consumers, the power travels through the transmission level, to the sub-transmission level, to the substation level, onto and through the primary feeder level, and onto the secondary service level, where it finally reaches the customer. Each level takes power from the next higher level in the system and delivers it to the next lower level in the system. In almost all cases, each flow of power is split into several paths at or shortly after transition down to the next level.

Each level is fed power by the one above it, in the sense that the next higher level is electrically closer to the generation.

The nominal voltage level and the average capacity of equipment drop from level to level, as one moves from generation to customer. Transmission lines operate at voltages of between 69 and 1100 kV and have capacities between 50 and 2000 MW. By contrast, distribution feeders operate between 2.2 and 34.5 kV and have capacities somewhere between 2 and 35 MW.

Each level has many more pieces of equipment in it than the one above. A system with several hundred thousand customers might have 50 transmission lines, 100 substations, 600 feeders, and 40 000 service transformers.

As a result, the net capacity of each level (number of units times average size) increases as one moves toward the customer. A power system might have 4500 MVA of substation capacity but 6200 MVA of feeder capacity and 9000 MVA of service transformer capacity installed. This greater-capacity-at-every-lower-level characteristic is a deliberate design feature of most power systems, and is required both for reliability reasons and to accommodate diversity of local peak demands.

Reliability drops as one moves closer to the customer because there is a larger amount of equipment that could fail between the sources of power and that point. A majority of service interruptions are a result of failure (either due to aging or to damage from severe weather) of equipment relatively near the customer (Table 1.1).

Table 1.2 gives statistics for a typical system. The net effect of the changes in average size and number of units is that each level contains a greater total capacity than the level above it – the service transformer level in any utility system has considerably more installed capacity (number of units times average capacity) than the feeder system or the substation system. Total capacity increases as one heads toward the customer because of non-coincidence of peak load and equipment guidelines made to accommodate that.

Table 1.1 Equipment/power flow unexpected outage data for a traditional power system in good condition

Level	Number of times/year	Total hours/year
Generation	0.01	0.10
EHV	0.03	0.10
HV	0.10	0.30
Distribution bus	0.15	0.15
Feeder	0.50	0.50
Service	0.50	0.35
Customer	0.02	0.15
Totals	1.30	1.65

Table 1.2 Equipment level statistics for a medium-sized electric system

Level of system	Voltage (kV)	Number of units	Avg. Cap. (MVA)	Total Cap (MVA)
Transmission	345, 138	53	250	13 250
Sub-transmission	138, 69	192	85	16 320
Substations	138/25, 69/12	825	22	18 150
Primary Feeders	23.9, 13.8	2550	8	20 400
Service Trans.	0.12, 0..24	346 500	0.077	26 681
Secondary/Service	0.12, 0..24	1 000 000	0.025	25 000
Customer	0.12	1 000 000	0.021	21 000

1.5.1 The transmission level

The transmission system is a network of three-phase lines operating at voltages generally between 115 and 765 kV. Capacity of each line is between 50 and 2000 MVA. The term 'network' means that there is more than one electrical path between any two points in the system. Networks are laid out in this manner for reasons of reliability and operating flow – if any one element (line) fails, there is an alternate route and power flow is (hopefully) not interrupted.

In addition to their function in moving power, portions of the transmission system – the largest elements, namely its major power delivery lines – are designed, at least in part, in order to meet system stability needs. The transmission grid provides a strong electrical tie between generators, so that each can stay synchronized with the system and with the other generators. This arrangement allows the system to operate and to function evenly as the load fluctuates and to pick up load smoothly if any generator fails – what is called stability of operation. (A good deal of the equipment put into transmission system design, and much of its cost, is for these stability reasons, not solely or even mainly for moving power.)

1.5.2 The sub-transmission level

The sub-transmission lines in a system take power from the transmission switching stations or generation plants and deliver it to substations along their routes. A typical sub-transmission line may feed power to three or more substations. Often, portions of the transmission system – bulk power delivery lines, lines designed at least in part for stability as well as power delivery needs – do this too, and the distinction between transmission and sub-transmission lines becomes rather blurred.

Normally, sub-transmission lines are in the range of capacity of 30 MVA up to perhaps 250 MVA, operating at voltages from 34.5 kV to as high as 230 kV. With occasional exceptions, sub-transmission lines are part of a network grid – they are part of a system in which there is more than one route between any two points. Usually, at least two sub-transmission routes flow into any one substation, so that feed can be maintained if one fails.[1]

1.5.3 The substation level

Substations, the meeting points between the transmission grid and the distribution feeder system, are where a fundamental change takes place within most T&D systems. The transmission and sub-transmission systems above the substation level usually form a network, as discussed above, with more than one power flow path between any two parts. But from the substation on to the customer, arranging a network configuration would simply be prohibitively expensive. Thus, most distribution systems are radial – there is only one path through the other levels of the system.

Typically, a substation occupies an acre or more of land, on which the various necessary substation equipment is located. Substation equipment consists of high- and low-voltage racks and busses for the power flow, circuit breakers for the T&D level, metering equipment, and the 'control house,' where the relaying, measurement, and control equipment are located. But the most important equipment – what gives this substation its capacity rating, are the substation *transformers*, which convert the incoming power from transmission voltage levels to the lower primary voltage for distribution.

Individual substation transformers vary in capacity, from less than 10 MVA to as much as 150 MVA. They are often equipped with tap-changing mechanisms and control equipment to vary their windings ratio so that they maintain the distribution voltage within a very narrow range, regardless of larger fluctuations on the transmission side. The transmission voltage can

[1] Radial feed – only one line – is used in isolated, expensive, or difficult transmission situations, but for reliability reasons it is not recommended.

swing by as much as 5%, but the distribution voltage provided on the low side of the transformer stays within a narrow band, perhaps only ± 0.5%.

Very often, a substation will have more than one transformer. Two is a common number, four is not uncommon, and occasionally six or more are located at one site. Having more than one transformer increases reliability – in an emergency, a transformer can handle a load much over its rated load for a brief period (e.g., perhaps up to 140% of rating for up to 4 h). Thus, the T&D system can pick up the load of the outaged portions during brief repairs and in emergencies.

Equipped with from one to six transformers, substations range in 'size' or capacity from as little as 5 MVA for a small, single-transformer substation, serving a sparsely populated rural area, to more than 400 MVA for a truly large six-transformer station, serving a very dense area within a large city.

Often T&D planners will speak of a *transformer unit,* which includes the transformer and all the equipment necessary to support its use – 'one-fourth of the equipment in a four-transformer substation.' This is a much better way of thinking about and estimating cost for equipment in T&D plans. For while a transformer itself is expensive (between $50 000 and $1 000 000), the busswork, control, breakers, and other equipment required to support its use can double or triple that cost. Since that equipment is needed in direct proportion to the transformer's capacity and voltage, and since it is needed *only* because a transformer is being added, it is normal to associate it with the transformer as a single planning unit – add the transformer, and the other equipment along with it.

Substations consist of more equipment, and involve more costs, than just the electrical equipment. The land (the site) has to be purchased and prepared. Preparation is nontrivial. The site must be excavated, a grounding mat (wires running under the substation to protect against an inadvertent flow during emergencies) laid down, and foundations and control ducting for equipment must be installed. Transmission towers to terminate incoming transmission must be built. Feeder getaways – ducts or lines to bring power out to the distribution system – must be added.

1.5.4 The primary feeder or Medium Voltage (MV) level

Feeders, typically either overhead distribution lines mounted on wooden poles or underground buried or ducted cable sets, route the power from the substation throughout its service area. Feeders operate at the primary distribution voltage. The most common primary distribution voltage in use throughout North America is 12.47 kV, although anywhere from 4.2 to 34.5 kV is widely used. Worldwide, there are primary distribution voltages as low as 1.1 kV and as high as 66 kV. Many distribution systems use several primary voltages – for example 23.9, 13.8, and 4.16 kV.

A primary feeder distributes between 2 MVA and more than 30 MVA, depending on the conductor size and the nominal voltage level and the utility's use of it to serve load. European and American practice differs slightly in how primary or MV feeders are laid out. European systems are most often built as loops, while American practice is typically a *dendritic* or *dendrillic* configuration – repeated branching into smaller branches as the feeder moves out from the substation toward the customers. In combination, all the feeders in a power system constitute the *primary feeder system*. There are typically from two to twelve emanating from one substation – again, one purpose of the substation it to reduce voltage and therefore one should expect to see the 'lower voltage and split' function in a well-designed substation. However, there are some substations with only one feeder. The most the author has ever seen from a single substation is 84 (Commonwealth Edison's Northwest substation on the edge of downtown Chicago).

The main, three-phase trunk of a feeder is called the *primary trunk* or *primary loop* and may branch or have switching options toward several branches or alternative main routes. These main branches end at open points where the feeder meets the ends of other feeders – points at which a *normally open switch* serves as an emergency tie between two feeders. In addition, each feeder will be divided, by normally closed switches, into several switchable elements. During emergencies, segments can be re-switched to isolate damaged sections and route power around outaged equipment to customers who would otherwise have to remain out of service until repairs were made.

By definition, the feeder consists of all primary voltage level segments between the substations and an open point (switch). Any part of the distribution level voltage lines – three-phase, two-phase, or single-phase – that is switchable is considered part of the primary feeder. The primary trunks and switchable segments are usually built using three phases, with the largest size of distribution conductor (typically this is about 500–600 MCM conductor, but conductors over 1000 MCM is not uncommon, and the author has designed and built feeders for special situations with up to 2000 MCM conductor) justified for reasons other than maximum capacity (e.g., contingency switching). Often a feeder has excess capacity because it needs to provide back-up for other feeders during emergencies.

The vast majority of distribution feeders worldwide and within first-world nations like the United States are overhead construction: wooden pole with wooden cross-arm or post insulator. Only in dense urban areas, or in situations where esthetics are particularly important, can the higher cost of underground construction be justified. In this case, the primary feeder is built from insulated cable, which is pulled through concrete ducts that are first buried in the ground. Underground feeder costs from three to ten times what overhead does.

Many times, however, the first several hundred yards of an overhead primary feeder are built underground even if the system is overhead. This underground portion is used as the *feeder getaway*. Particularly at large substations, the underground getaway is dictated by practical necessity, as well as by reliability and esthetics. At a large substation, 10 or 12 three-phase, overhead feeders leaving the substation means from 40 to 48 wires hanging in midair around the substation site, with each feeder needing the proper spacing for electrical insulation, safety, and maintenance. At a large-capacity substation in a tight location, there is simply not enough overhead space for so many feeders. Even if there is, the resulting tangle of wires looks unsightly and, perhaps most important, is potentially unreliable – one broken wire falling in the wrong place can disable a lot of power delivery capability.

The solution to this dilemma is the underground feeder getaway, usually consisting of several hundred yards of buried, ducted cable that takes the feeder out to a riser pole, where it is routed above ground and connected to overhead wires. Very often, this initial underground link sets the capacity limit for the entire feeder – the underground cable ampacity is the limiting factor for the feeder's power transmission.

1.5.5 The service transformers

Service transformers lower voltage from the primary voltage to the utilization or customer voltage, normally around 240–250 V in European systems and 120/240 V two-leg service in most power systems throughout North America. In overhead construction, service transformers are pole-mounted and single-phase or three-phase, between 5 and 166 kVA capacity. Since power can travel economically only up to about 200 feet at utilization voltages, there must be at least one service transformer located reasonably close to every customer; hence, there may be a dozen or more on a European primary circuit, or several hundred scattered along the trunk and laterals of any given feeder in an American system.

Underground service, as opposed to overhead pole-mounted service, is provided by padmount or vault type service transformers, again either single- or three-phase. The concept is identical to overhead construction, with the transformer and its associated equipment changed to accommodate incoming and outgoing lines that are underground.

While passing through these transformers, power is reduced in voltage to the final utilization voltage (120/240 V in the United States, 240 V in Europe) and routed onto the secondary system or directly to customers. In cases where the system is supplying power to large commercial or industrial customers, or the customer requires three-phase power, between two and three transformers may be located together in a transformer bank and interconnected in such a way as to provide multiphase power. Several different connection schemes are possible for varying situations.

1.5.6 The secondary and service level

Secondary circuits, fed by the service transformers, route power at utilization voltage directly to the customer, usually in an arrangement in which each transformer serves a small radial or loop circuit serving from one to several dozen homes and businesses, leading directly to the meters of customers in the immediate vicinity. At most utilities, the layout and design of the secondary level is handled through a set of standardized guidelines and tables, which are used by engineering technicians and clerks to produce work orders for the utilization voltage level equipment. In the United States, the vast majority of this system is single-phase.

1.5.7 European vs American distribution systems

T&D systems used in Europe and North American are broadly similar: central-station-oriented, with high-voltage transmission networks and a series of three to five voltage layers including the customer service level. A wide range of specific standards on voltages and equipment types are used, for example 400 vs 345 kV, or grounded Y distribution vs delta. European systems are normally operated at 50 Hz alternating current, while American systems operate at 60 Hz. Such differences are important but not fundamental. Overall, the structures of systems in both continents, and in places around the world that adhere to the standards of either, are as depicted earlier in Fig. 1.1.

The major difference between American and European systems is at the distribution level. Service voltage in Europe is almost exclusively around 240 V (around 416 V phase-to-phase), whereas in the US it is 120 V (208 V phase-to-phase). This is a two-to-one ratio and it makes a substantial difference in the design limitations that shape the service and primary feeder levels. Generally, twice the voltage allows power to be moved twice as far, which means distributed at service voltage can cover four times the ground area. Thus, service transformers in European systems are, for this reason alone, roughly four times the size, and one-fourth the number, of those in American systems. Also, this situation makes the use of three-phase service level distribution (0.416 kV phase-to-phase) lines practical and economical, whereas 208 volt phase-to-phase service circuits and wiring are used in American systems only in cases where three-phase 208 volt service is desired by the customer. Three-phase lines can move power roughly twice as far, with equivalent losses and voltage drop, as single-phase, so that times two distance becomes times four, and the ratio or ground area that can be efficiently covered is *sixteen* times more. As a result, European systems have far fewer, and far larger, service transformers. In European systems, service transformers are typically 250–1500 kVA in size, whereas in the US they are 15–100 kVA.

1.5.8 The lateral level in American systems

One consequence of the small service transformers and limited distance that 120 V service lines can route power to customers is that American systems require primary voltage laterals. Laterals, short stubs or line segments that branch off the primary feeder, represent the final primary voltage part of the power's journey from the substation to the customer in an American system. A lateral is directly connected to the primary trunk and operates at the same nominal voltage. A series of laterals tap off the primary feeder as it passes through a community, each lateral routing power to a few dozen homes.

Normally, laterals do not have branches, and many laterals are only one- or two-phase – all three phases are used only if a relatively substantial amount of power is required, or if three-phase service must be provided to some of the customers. Normally, single- and two-phase laterals are arranged to tap alternately different phases on the primary feeder, as shown below, in an attempt by the distribution planning engineer to balance the loads as closely as possible.

Typically, laterals deliver from as little as 10 kVA for a small single-phase lateral to as much as 2 MVA. In general, even the largest laterals use small conductors (relative to the primary size). When a lateral needs to deliver a great deal of power, the planner will normally use all three phases, with a relatively small conductor for each, rather than employ a single-phase and use a large conductor. This approach avoids creating a significant imbalance in loading at the point where the lateral taps into the primary feeder. Power flow, loadings, and voltage are maintained in a more balanced state if the power demands of a 'large lateral' are distributed over all three phases.

Laterals (wooden poles) are built overhead or underground. Unlike primary feeders and transmission lines, single-phase laterals are sometimes buried directly. In this case, the cable is placed inside a plastic sheath (that looks and feels much like a vacuum cleaner hose), a trench is dug, and the sheathed cable is unrolled into the trench and buried. Directly buried laterals are no more expensive than underground construction in many cases.

1.5.9 Dendrillic versus loop systems

Another difference between US and European systems is in the layout of the primary distribution circuits. US practice is very much to lay out circuits in a branching fashion, often called a dendrillic or dendritic layout. A circuit leaves the substation and branches and re-branches, so that overall it has a sole starting point and many ending points. A few, perhaps as many as six, of these end points might be connected to other circuit end points via normally open switches, for contingency back-up purposes and field switching. By contrast, European practice is often to use loop feeders, in which a 'loop' is fed at two points (both ends) and operated either as a closed loop or with an open tie point. In practices, neither paradigm is absolute. There are loop feeders used occasionally in American systems, and dendritic feeder topography can be found in many European systems, at least in rural areas.

To some extent, the larger size and fewer number of service transformers in a European system leads to this approach. But partly, the practices of distribution circuit layout in American and European type systems are just a matter of practice: it has been done that way in the past, it works, so why change?

1.6 Modern smart distributed power distribution systems

A confluence of three major technological development trends has created a new type of power system capability that differs noticeably from the traditional central-station centric system depicted above. The three technological trends are: distributed resources, energy storage, and smart systems.

1.6.1 Technology trend 1: Improving cost effectiveness of distributed resources

The LV and MV levels of the traditional power system are, in a very real sense, their own distributed resources. The LV utility network in particular is distributed over the service territory, reaching every single customer and sized locally in direct proportion to local customer energy needs. However, the term 'distributed resources' in modern power systems is used solely to refer to power systems in which the electric energy itself is produced by machinery, facilities, or systems that are distributed throughout the service area rather than concentrated in a few large central-station generating plants, as was depicted earlier in Fig. 1.2. Distributed resources include small generators that might include:

- low head of small hydro-power generation,
- wind energy generation,
- micro-turbine powered generation,
- high- and medium-speed diesel generation,
- photovoltaic power generation,
- small solar thermal and tower generation.

These small generators are distributed throughout the utility service area, although not necessarily in direct proportion to the customer demand. For example, in a rural area a three-turbine wind generator facility of 1 MVA capability might be located in a tilled field two miles from the farmhouse and harvest processing/drying facilities that create the demand for most of that power. In an urban area, a 2 MVA PV panel set located on office rooftops might produce power that at times is moved several miles to serve nearby residential demand. But invariably these distributed generating sources are, on average:

- Closer to the energy consumers than is central-station generation. Hence, power delivery costs are potentially lower, reliability of power transmission is higher and esthetic and environmental impacts of the power transmission lines are all lower than for power delivery in a traditional system (all three because the pathway for power delivery is, ultimately, shorter than in a traditional system).
- Less efficient in overall unit cost of power than larger central-station generating plants, as was discussed earlier. The margin might be small or large depending on technology and characteristics specific to each situation.
- The usefulness and popularity of distributed resources rests on the economic, service quality, social, and market advantages that being closer to the customer has, as opposed to any disadvantages created by the lower potential efficiency of per-unit power production.

Demand response as a distributed resource

In some instances, distributed resources (DR) also include non-generating resources that can be dispatched much like generation. An example is demand response, or load control, in which certain loads can be switched off for a time to keep system resources and demand in balance.[2] From the standpoint of many system operating decisions aimed

[2] Demand response is also abbreviated 'DR', in some cases, leading to confusion.

at achieving and maintaining that balance, it makes little difference if generation is increased or demand reduced. Dispatchable load control, whether directly (the appliances and equipment themselves are shut down) or indirectly (voltage on a feeder is reduced slightly lowering the load of connected loads) the result is the same: at the system operator's request, a change has been made to help control the ratio of generation to load.

In some cases, utility planners, managers, and regulators will restrict 'demand response' to methods that are directly dispatchable (controllable in near-real time) by the utility or system operator – methods such as direct load control or active demand limiters. In other cases, however, the term demand response includes programs and load-influencing methods that rely on customer or automatic (customer programmed) price-sensitive responses from the demand, such as real-time-pricing (RTP) methods. These methods do not have the temporal immediacy of direct load control – the demand reduction may take seconds or minutes to affect, or it might not happen at all – customers may override these methods in critical situations, etc. Thus they are less certain in their effect and left out of DR considerations by some, but sometimes included as 'demand response' and 'distributed resources' by others. It is best to inquire in each instance to avoid ambiguity and confusion.

1.6.2 Technology trend 2: Effective and economically justifiable energy storage

It has always been possible to store electric energy, including storing it in what is effectively in alternating current form. Even in the first half of the twentieth century, alternating current energy could be 'stored' overnight or for a longer period of times, using lead–acid/rectifier-inverter sets, compressed air storage, or pumped hydro-power plants. Into the late twentieth century, all three technologies improved in efficiency and performance/price. However, these energy storage methods typically did not always have a good economic performance on a small scale (distributed) basis. Pumped hydro- and compressed-air storage have a very substantial economy of scale due to the 'concentration' effect discussed earlier as well as for several natural physical reasons, and those technologies were rather more widely used in traditional power systems, but almost exclusively in central-station roles. The technology for battery systems and other small energy storage devices such as flywheels, etc., even into the early 1990s, just did not have a positive business case for widespread use.

The only exception to this rule was the very widespread adoption of very small lead–acid and carbon–zinc based UPSs as back-up for critical loads and energy needs. The use of such very distributed systems skyrocketed in the last quarter of the twentieth century. Two key points to the business case success of these units are illuminating as to the larger issue of energy storage in general. These are:

1. The growing number of appliances and equipment that need absolute continuity of service, such as digital devices and robotic machinery. Overall, the value of reliability and continuity of service as opposed to energy itself is increasing.
2. The duty cycle of UPS devices typically has them fully discharging fewer than ten times a decade: the unit's purpose is to standby with available power, not provide it on a routine basis. As such UPS batteries do not 'fatigue' or wear out due to daily cycling, as do lead–acid batteries, after only a few hundred charge–discharge cycles.

In the last decade of the twentieth century, and into the twenty-first, improvements in chemical and energy storage control technologies improved electrical storage (super- and ultra-capacitors), chemical (battery and mixture systems) and mechanical storage (flywheels, etc.) in nearly every important performance category. Energy density improved substantially, to the point where batteries and, in some cases, flywheels became light and compact enough to permit practical electrical vehicles for personal and light commercial use. From the standpoint of power systems, where weight and size are far less important criteria, the most important improvement was the lifetime cycle counts. Late twentieth century lead–acid batteries could perhaps go through 500 charge–discharge cycles before 'wearing out.' Modern lithium ion batteries can go through five to seven times as many, which makes for a much better business case for non-UPS applications. In addition, changes in storage control technology were also important. Brought about by digitization and considerable research for the electric vehicle industry, these led to a reduction in the economy of scale of power systems energy storage units and an ability to control storage quickly enough to make it instantly dispatchable and in many cases fast and accurate enough in control to be a system stability resource. Increasingly, positive business cases could be made for energy storage attached to primary (MV), LV, and even customer facilities.

1.6.3 Technology trend 3: 'Smart' systems

Smart systems and smart grid have almost as many definitions and interpretations as there are people in the power industry. It is not much of an exaggeration to say that 'smart grid' is always defined as the use of a particular company's or person's latest technology or idea, in a way that maximizes the business case for the purchase of those products and services. But throughout all 'smart' technologies, there are two general areas of improved capability that combine to make smart power equipment 'smart,' and that lead to smart grids, whatever they may be. These are discussed as follows.

Equipment-to-equipment communication

Largely due to improved bandwidth-cost-performance of digital communications, individual and small units of a power system can communicate in near-real time with both a central system if needed and, more important, with other nearby equipment. An end-of-feeder power monitor can inform the utility's distribution management system whenever it senses no power. It can also inform a nearby switch that there is no power at its site. A recloser on one circuit can know the status (open–closed) and loading of a recloser on a nearby circuit that forms the alternate route to the loads it protects. Such communication is not only possible, but becoming routine, made with commodity equipment.

Sensors and monitoring equipment

Technological advance is widening the range of characteristics in a power system that can be measured and tracked, a good example being phasor measurement units – PMUs. In addition, almost across the board, the cost of remote sensing of equipment status and power has been greatly improved as has the periodicity (frequency) of how often readings can be or are taken.

System-level anticipation and control

These two technology improvements have led to substantial improvements in the control and performance of traditional power systems, using schemes in which remote monitoring is used to inform a central control (either automated or human), which will respond via remote control of generation and T&D equipment. This has and will continue to make a difference in the performance of traditional power systems.

But these advances have made an even larger extension to the potential for *coordinated* control of DR and the surrounding local distribution system. The character of this change is subtle but fundamental. In traditional power systems, there were many units of equipment that were *automatic*. Reclosers and sectionalizers performed rather complicated actions – as they were built to do. Capacitor banks could be built with switches to turn them on or off depending on voltage, power factor, loading, or a combination. Voltage regulators and line-drop compensators varied voltage according to their 'programming.' All this equipment was automatic, equipment taking actions based on what voltage, current, power factor, or other conditions (temperature, time of day) it measured at *its* location.

But the two capabilities above, along with the use of cheap digital computation, now permit equipment to monitor conditions nearby, or elsewhere, and respond to that, not just the conditions at its location alone. Thus, *groups of equipment* can be programmed to work together to behave in a similarly automatic manner, in a coordinated way. Distribution and customer-site equipment can be built so that it will 'understand' the interactions and dependencies it has with neighboring equipment, and essentially reprogram its automatic actions in response to local conditions and needs. For example, what were reclosers or sectionalizers in the past become 'smart switches,' aware of the network configuration, loading, and outages in nearby circuits and able to determine how to respond in various contingency and operation situations, should they develop. DR systems can vary response and priorities based on local conditions, automatically. The type of control and control topology does not matter.[3]

Potentially, this permits an isolated or local portion of the power system to be autonomous in some sense, at least for periods of time and/or under certain conditions (such as an outage of equipment upstream). This alone would not lead to any substantial change in the status quo of power system design, except that when combined with DR discussed earlier, it creates an *independent micro-grid.* If provided with enough local generation and energy storage, the local power distribution system in a neighborhood can fend for itself: it can provide for its own energy need, and operate on its own. The extreme case for this is the isolated micro-grid: a power system covering only a small area and a few customers, and not connected to the larger regional power system at all. An electric utility might build such a system – in fact is obligated to do so in most regulatory venues – if that micro-grid can economically and efficiently serve a group of customers with equivalent service quality to what others in the utility territory are receiving. But 'micro-grid' also can include situations where a group of individually independent energy consumers (each has sufficient on-site generation, energy storage, and demand response control to meet 100% of their own needs) intertie their private systems for purposes of mutual reliability and efficiency improvement.

[3] The end product is essentially the same regardless of the details of how this control is affected: whether this capability is executed through a central hub, or whether each device figures it out for itself in an independent manner, or if a hierarchical control consisting of many local hubs operates the system, from a big picture result, the end product is the same.

But in addition, one can talk about a *virtual micro-grid,* an area of distribution within a larger power system that manages the local balance of generation, storage, demand, and circuit operation in its neighborhood, so that the power transfer across its boundaries is nil. This local distribution system is tied to the larger, traditional power system, but does not normally exchange power with it due to its generation and storage resources, and manner of local control.[4] Figure 1.2 illustrated this concept, an entire utility system covered by small neighborhood systems that might or might not be inter-tied.

1.7 Factors affecting the T&D system of the future

This section will discuss trends and factors that influence the decisions utilities and consumers will make about electric service and power systems of the future. It will then conclude the chapter with a discussion of how these might shape the power systems of the future.

1.7.1 Economic benefit of utilizing the existing system

The existing system has already been paid for. This might not be strictly, completely true, but from a practical sense it is sufficiently true to be an overwhelming factor in many decisions. The utility is 'stuck' with its current system. Much of the equipment and facilities *are* in place and *are* paid for: old enough to be fully depreciated and amortized, but still hopefully not so old as to be worn out. Given this, the existing system can and should be used to the best effect possible, and so the traditional power system will be employed as long as it lasts.

The capital investment in a power system is significant. Equipment is expensive because it is designed for long life and high levels of safety, and thus is very robust. The labor required for construction and installation is considerable, and expensive (much if it requires special skills and training). A lot of it is neither utilities, consumers, nor society as a whole can afford to make wholesale replacements of existing systems just because they are obsolete in some ways compared to newer technologies, or because they are getting old. Practically speaking, all parties can only make incremental investments and changes. The exception is the customer end. Often, customers can afford to make investments in UPS, power quality, and on-site

[4] Exceptions would be during times of emergencies such as when local generation has an unexpected outage, when power would flow from the system into the neighborhood, or when local generation owners wish to sell power into the regional grid, in which case it would flow out and up onto the transmission system.

generation that provide individually justifiable site- or appliance-specific benefits to their electric service.

1.7.2 Increasing need for service reliability

Viewed over the very long term – since the beginning of the electric era in the late nineteenth century, there has been a continuous growth in the need for absolute continuity of electric service and improved power quality (voltage regulation, etc.), as measured by the value that reliable and quality service has and the willingness of consumers to bear the cost of improved reliability and power quality. The reliability of modern first-world utility power systems is quite good – an average of around .75 SAIFI (System Average Interruption Frequency Index) and 100 min SAIDI (System Average Interruption Duration Index), non-inclusive of natural disasters and major storms, a delivery performance of better than 99.98%. Voltage regulation is generally within 3%. However, the widespread use of digital controls, robotic machinery, and smart systems for many critical infrastructures means that electric service interruptions, voltage sags, and high harmonic contents, even if infrequent and/or of short duration, can have noticeable, sometimes significant, costs and consequences that electric energy consumers wish to avoid.

That said, it is not clear that any long-term societal or consumer need for even higher levels of electric service reliability and quality will be satisfied by a general improvement in the performance of utility power systems. The aforementioned widespread use of UPS systems shows that appliance- and installation-specific reliability augmentation is a viable option in many cases. Many of the needs can be met by such means. And at some point, the cost of improving the reliability of power systems, which serve all consumers and all demands, is not justified by the value that reliability provides to those few consumers and demands that require improved service reliability. The power industry may be near that point. In many cases the marginal cost of reliability and power quality per kW is lower when using such equipment that when provided it through improvements in the electric grid. The only widespread exception is likely to be an increasing regulatory expectation for improved storm and energy response: better planning for major storms, quicker and more optimally managed damage repair and restoration. This will be enabled by smarter distribution management systems (DMS), outage management systems (OMS), and field resource management systems (FRMS).

Regardless, area-, site-, installation-, and appliance-specific UPS and power quality equipment options will certainly be an option often, probably increasingly, used in the future, whether owned and/or operated by the utility, the consumer, or third party service providers. With regard to future and evolving power systems, the important points with respect to reliability and power quality are:

- Demand for high levels of reliability of service and power quality will continue to increase, at least in some industries and some areas of society.
- What matters is the reliability at the consumer/appliance level. If a homeowner's lights and power do not go out, it is irrelevant to them if the distribution feeder circuit serving their neighborhood is out of service. Similarly, if the lights do go out, the effect is the same, whether due to the utility circuit having a problem or because of a failure in the consumer's UPS. It is the combination of service quality and reliability given by the utility and any special service equipment installed locally that is what matters to each consumer.
- Localized 'solutions' to reliability and power quality, such as whole-building or even neighborhood-scale UPS and power regulation systems, or high-reliability local virtual mini-grids, provide the ability for utilities and consumers to vary reliability where and as needed. High-reliability service can be arranged only for those areas/consumers/loads that need it and are willing to pay the price. In fact, it is not necessary for the utility to be involved and many regulatory venues may decide that, beyond providing a satisfactory standard level of reliability, utilities are not required or even allowed to address the market for superior levels of reliability. However, it can be difficult to separate reliability from efficiency in some ways.[5] For this reason, many regulatory venues may permit or even encourage the utility to offer different types of service availability at different pricing structures.

1.7.3 Growth of demand for electric power

Historically, electricity demand has grown steadily from the late nineteenth century through to the early twenty-first century, in the total amount of power consumed, in the number and variety of devices that use electricity, and in the number of people and businesses that wish to use these devices. It is worth noting that no one individual, homeowner, or business wants electric power. They want only the products that its use creates: cold milk and beer in a refrigerator, a home heated comfortably and safely in winter, sheet steel rolled into pipe, air liquefied and separated into its constitute elemental gases, garage doors raised and lowered, illumination when and where

[5] A load that is 'controlled' – turned off for a period of time – in the interests of system or energy efficiency or due to a resource constraint has been denied service and thus is without power. Since RTP demand response systems work on a price–signal basis, one can view that all interruptions in service have an economic price and should be viewed through this one lens, in which case the utility offering very high levels of reliability for high prices is only consistent with its demand response program.

needed, 80 diagonal inches of dazzling three-dimensional moving imagery of the evening news, and transportation to and from work each day in a zero-emissions car or electric mass transit system. There is every expectation that these trends will continue unabated for the foreseeable future.

1.7.4 Aging infrastructure

A very real issue for utilities is aged infrastructure. Many existing power systems have considerable amounts of equipment in them that are old and nearly worn out, and there are large areas of nearly any major power system where most equipment is old. Old equipment breaks down and fails (end of life) more often than new equipment: the utility sees an increase in service reliability problems and operating costs. To prevent further deterioration of performance and reliability, the aged equipment requires more frequent maintenance and more extensive service, further raising operating costs. And, older equipment may not be as efficient or competitive (no digital controls, no vacuum interrupters, etc.) as modern equipment, leading to a service, cost, and net business disadvantage.

The challenge for a utility or industrial power system owner is that of the proportionality of their problem to the viable business solutions they have. Electric power equipment is designed for harsh service and long lifetimes, so in almost all cases it is incredibly robust. The average lifetime of most categories of power system equipment is longer than might be thought – perhaps 50–70 years (not allowing for storm damage of non-failure related causes of replacement). But another consequence of that design is that the standard deviation of lifetime is also very long, too. Average equipment lifetime for a particular type of equipment in one system might be 70 years, but some of that equipment will fail with routine problems at only 30 years, while other units will provide good service past 100 years. This variation in lifetimes between 'good' and 'bad' equipment may be significant portion of its average lifetime. Add to this the fact that the equipment in a utility system was installed over a long period of time – decades – and failures and equipment service problems created by wear and tear can appear, both perceptually and statistically, to be very random in nature: not predictable or manageable.

Furthermore, 'high' levels of aging equipment failure are not that high. The failure (end of life) rate for equipment in a power system in very good condition is well below, .1% in just about every equipment category – less than 1 in 1000 units failure per year. A failure rate so high that it is 'completely unacceptable' to all – unaffordable to the utility, unacceptable from a service standpoint for consumers and regulators – would be only 0.5% annually. Finally, a replacement cost of even 1% of equipment annually, whether because it actually failed and had to be replaced, or the utility proactively replaced it in a preventative program, is currently unaffordable to most utilities. It would raise the average utility's capital spending by a noticeable amount – usually between 20% and 80%. The industry as a whole has adjusted to a cost-basis and spending expectation that does not include any monies for significant rates of replacement/renewal. Thus, overall, managing equipment aging and its effects is very challenging, and replacement rates that would permit a utility to renew the type of system or convert to distributed systems or new technologies are low.

1.7.5 Aged metropolitan core systems

This situation is likely to become a particularly urgent problem for electric utilities in the central portions of many large American and European cities, where despite decades of development to a point that electricity demand is quite dense, that demand is steadily but slowly growing. In such metropolitan core areas, most T&D facilities and structural elements for the power system are often *much* more than half a century old. Further, urban crowding means expanded substation sites, new ROW, or even more and wider underground cable duct space is practically unavailable. The combination of aged equipment and facilities with such restrictions on expansion often leaves no viable options within the traditional central-station paradigm that the utility can take to adequately meet future power delivery needs in the area.

There is an expectation among those outside the industry, and some within, that this situation is actually fortuitous, that utilities will have to replace these aged urban power systems with new, distributed resource smart systems in the next few decades. Certainly, those technologies will contribute, but the load densities and reliability needs of electric consumers in metropolitan cores are extreme, often beyond levels that local distributed and renewable resources and smart technologies have been tested against, and in some cases beyond what they have even conceptually been planned to meet. Finally, the cost of any new traditional, nontraditional, or hybrid power system that can meet those needs will be very significant: right now most utility executives do not see how they can be paid for. And finally there is a practical challenge: how does one make the transition from one to the other, while keeping the lights on and working in a crowded urban area where disruptions or traffic and metro core activity have a 'cost' of their own? Renewal/replacement/expansion of power systems in these areas is likely to be one of the biggest challenges facing retail delivery utilities in the twenty-first century.

1.7.6 Consumer control and free retail markets

The wholesale power market is de-regulated in many parts of the world, and the retail market in some areas. Again, historically, there has been a long-term trend toward de-regulation of utility markets, most notably telephone. One promise of smart technology is that it will permit customized and user-specified choices or the quality–quantity–timing–price combination that each energy consumer wants. At present there is no proof that a viable societal and market mechanism that satisfies consumers, regulators, utilities, and political considerations has been found, but there is an expectation that in the future, energy consumers will have more choice about, and more control over, their own energy supply than only controlling its usage.

1.7.7 New technology

The noted American power engineer Jim Burke often observed that 'If technology makes it possible to do something people will expect you to do it.' As new materials, inventions, systems, and communications capabilities become available, individuals and society alike will expect them to be utilized in the electric power industry. For example, 'everyone' has been told that smart systems will lead to reduced costs, more choice, better service, etc. There is an expectation both that smart systems, whatever that means, will be used, and that benefits, often not fully identified but recognizable when they occur, will flow from that.

The issue for many utilities is in developing a technology-use plan that can meet those expectations on the one hand, while not creating a significant technology-based business risk on the other. *Investment stranded by obsolescence* is increasingly recognized by utility executives as a major source of business risk, leading to reluctance to plunge fully into any new technology without a plan for recourse and 'technology diversity.' They do not want to bind their companies too tightly to the long-term use and cost of equipment or system technologies that might be eclipsed or to long-term business commitments that may become obsolete or not preferred as new technologies become available. The chief problem is that the traditional power industry period of equipment usage and financial deprecation is much longer than the technological half-life of modern systems.

An example is the very great investment some utilities have made in providing service to 'data centers' – huge digital communications control and internet server warehouses which have a nearly 24/7 demand of over 40 or 50 MVA. Many utilities made sizable investments in new circuits and reliability/power quality equipment to serve these data centers, planning to pay back that investment with proceeds from the sale of that large amount of power over the next many years. However, it is possible that advances in optical computing will lead to a new generation of digital equipment and servers with higher computing and switching speeds combined with power/cooling needs reduced by an order of magnitude. Server companies would install such equipment and reduce their demand, and the utility's revenues from that site, before the original investment is paid back.

1.8 Conclusion

The majority of power systems on this planet are central-station oriented systems with layered transmission–substation–distribution-service levels of a traditional design. These systems are in place, largely paid for, and too expensive to replace quickly. Many are old but replacement rates due to wear and tear mean renewal or evolution half-lives are on the order of two to three decades at the least. New technologies permit the widespread distributed generation and demand response systems, energy storage on a central and distributed manner, and very responsive monitoring, control, and operational flexibility of both traditional and local mini-grid and off-grid power systems.

Meanwhile, the demand for electric power is expected to continue to grow, as will the need for reliability of service and power quality. Societal expectation as well as technology capability will be that customers can have choice and control over their power supply, and a role in the management of the electric service. DR and smart systems mean that many can, if they wish, afford to take over complete ownership and operation of equipment to satisfy their energy needs, and go 'off the grid.' There is little indication that most consumers will: most homeowners can grow food in the backyard, face few if any regulatory or legal hurdles in doing so, yet few choose to do so. It will likely be the same with home and local power generation.

Thus, the power systems of the future will very likely be one that is an evolution of the traditional central-station systems in place now, 'renewed' through good maintenance and repair, life-extension and replacement of equipment, and upgrades of control equipment, combined with selective use by the utility of DR and storage. Smart equipment and systems will be used to coordinate and get the performance needed both globally and locally out of all equipment, and to monitor for and solve problems created by aging, high demand, and increasing reliability ends. Certain parts of the traditional power system will be augmented, added to, and expanded, while other parts will be removed or recast in distributed applications, perhaps. Traditional power systems will evolve into hybrid central and distributed systems, utilizing their traditional resources plus their DR, energy storage, and smart technology so that they are, in places, mini-grids or virtual mini-grids, while retaining portions of the high-voltage T&D grids and the large-scale efficient power generation facilities that have been a staple of the power industry for more than a century.

1.9 Sources of further information and advice

Burke J. J., *Power Distribution Engineering: Fundamentals and Applications*, Marcel Dekker, New York, 1994, 376 pages.

Lakervi E. and E. J. Homes, *Electricity Distribution Network Design,* Peter Peregrinus Ltd, London, 1995.

Philipson L. and H. Lee Willis, *Understanding Electric Utilities and De-Regulation,* Marcel Dekker, New York, 2006, 499 pages.

Santacana E., editor, *Electrical Transmission and Distribution Reference Book*, ABB Electric Systems Technology Institute, Raleigh, 1997, 860 pages.

Willis H. L., *Power Distribution Planning Reference Book*, Second Edition, Marcel Dekker, New York, 2004, 1184 pages.

2

Transmission and distribution (T&D) network monitoring and control

K. BELL and C. BOOTH, University of Strathclyde, UK

DOI: 10.1533/9780857097378.1.39

Abstract: This chapter describes the main technical limits that must be satisfied in the operation of T&D networks and how various system elements are controlled to ensure safe, continuous supply of electricity. Some discussion is given on the impact of renewable generation and the potential contribution of energy storage, along with some other developments that promise to provide greater flexibility to system operators.

Key words: power system operation, control, power system stability.

2.1 Introduction

A power system consists of a very large number of interconnected elements, as illustrated in Fig. 2.1, each of which must be adequately controlled, protected and managed if the final supply of power to users of electricity is to be safely and continuously achieved. Inevitably, disturbances to the system do occur and, when they do, the system's ability to continue to operate safely is a key measure of its success.

In this chapter, we review the main requirements for control and management. In so doing, we consider the monitoring and measurements that are required. Finally, we outline some newer developments that are beginning to facilitate the meeting of the newer challenges outlined elsewhere in the book, in particular the accommodation of highly variable and uncertain power from renewables, notably wind and solar. Chapter 3 reviews the related topic of the protection of T&D networks.

2.1.1 Requirements for network monitoring, control and protection

In this section, the general control requirements for electric power systems and the means by which they are achieved are described. Although each aspect interacts with the others, they can be thought of as falling under a number of headings:

- control of system frequency;
- control of voltages around the system;
- control of currents flowing through different components of the system;
- ensuring stability of the system;
- enabling short-circuit faults to be cleared safely.

Each of the above aspects makes use of dedicated control equipment or particular conventions in respect of operation of the system as a whole. They will now be briefly described in turn.

2.2 Control of system frequency

One of the key restrictions on the operation of electric power systems has been the lack, to date, of some means of storing electrical energy on any large scale in an economical way. At the time of writing, the best that is available is 'pumped storage', where surplus electrical energy is used to pump water in a lower reservoir up a hill where it is stored in a higher reservoir as potential energy until it is needed, at which time it is allowed to fall, the kinetic energy being converted to electrical energy by turbines. However, the typical power rating of such facilities is of the order of a few hundreds of MW with a storage capacity of a few thousands of MWh and requires a significant capital outlay and the availability of suitable sites for the reservoirs. (As can be seen from Part III of this book, the pursuit of cheap and reliable storage of electrical energy for use in different contexts is the subject of much work.)

The current relative lack of storage means that the rate of production of electrical energy must match the rate of consumption; in other words, generation of power must match demand for it. Because, depending on the technology, it takes time to get individual generating units into the condition of generating power, the demand for electricity must be forecast and generation scheduled to start, ramp up its output, ramp down and stop accordingly. As will become evident from the discussion of reserve below, whatever the market structure within which electricity supply takes place, at some point the scheduling of generation must be coordinated to ensure that, in 'real time', there is neither too much nor too little. Generally, this will entail scheduling of power production from the cheapest units (taking into account limitations on how often individual units can start and stop and how quickly their output can be ramped) along with sufficient 'response' and 'reserve'. Given forecasts of demand, the available generation, a set of costs or prices of generation and the technical limits, an optimal schedule can be derived using 'unit commitment' and 'economic dispatch' software, and the results communicated to the operators of each unit. (See, for example, Wood and Wollenberg, 1994.)

All power systems of any reasonable size in the world at present use alternating current. This is for historical reasons and because of the convenience by which transmitted and distributed power can be transformed between different voltage levels through exploitation of electromagnetic effects in transformers. (Power is transmitted with lower losses at higher voltages and transformed back to lower voltages with lower insulation requirements for final consumption.) The main consequence of an excess of generation relative to demand will be an increase in the frequency at which voltages and currents on the system alternate – in effect, excess energy entering the system produces an increase in the speed of rotating 'synchronous generators' and of the kinetic energy of the system. On the other hand, a deficit of generation relative to demand will lead to kinetic energy being extracted to meet the demand for electrical and hence system frequency falling.

Variations in system frequency would not be a problem were it not for the fact that much of the equipment connected to the system has been designed to operate only within certain frequency limits. This is particularly true of the large turbogenerators that continue to provide most of the electrical energy in almost all large power systems around the world. Consistency between what a power system operator provides and what designers and operators of plant connecting to the system can assume about system frequency is ensured by the setting of statutory limits on system frequency, typically ±1% of nominal. Because of the importance of reliable operation of individual generating units and the reality of disturbances to the system leading to variations in system frequency (see discussions on disturbances and responses later in this section), generation equipment connected to transmission systems is generally required to be capable of operating at system frequencies outside of those statutory limits, typically ±3% of nominal. Such requirements are generally defined in system specific 'grid codes' with which grid users such as generators must comply, usually enforced by the system operator.

2.2.1 Contributions to control of system frequency

As noted, a system frequency above nominal is indicative of an excess of energy entering the system; a system frequency below nominal is indicative of a deficit. Since, in the timescales relevant to control of system frequency, that frequency is the same on all parts of the system, it can be readily measured anywhere on the system. Control of system frequency is most readily achieved by appropriate variation of the output of generators; at least in

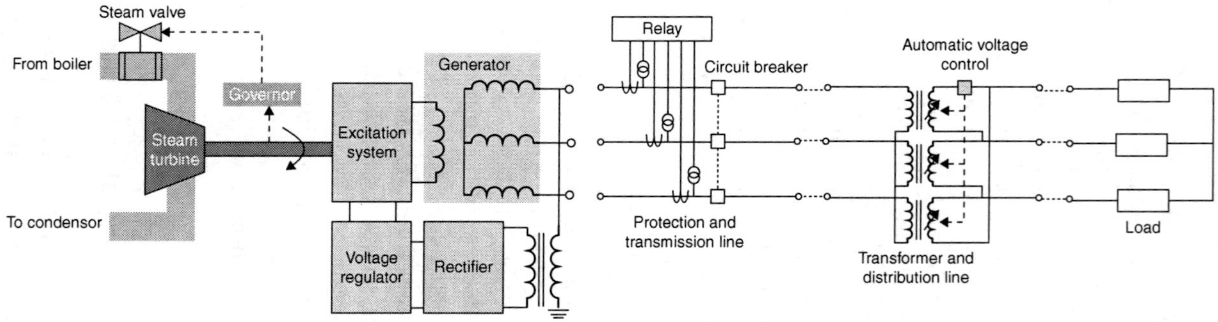

2.1 Schematic representation of some main components of a three-phase power system (based on a diagram in Glover *et al.*, 2007).

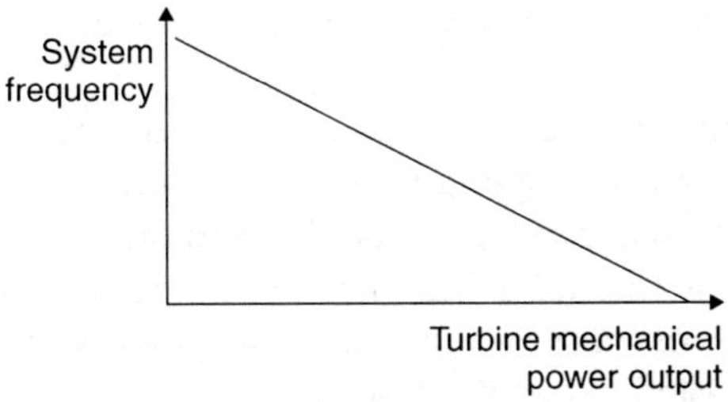

2.2 An idealised typical frequency droop characteristic. (The mechanical output power from a generator's prime mover is regulated in response to measured values of system frequency.)

response to small variations, this can be achieved automatically by the action of governors on appropriately equipped generators that respond to variations in system frequency measured locally to the generator by increasing the rate of input of energy when the frequency is low and decreasing it when it is high. The change in input power relative to a deviation of frequency is known as 'droop'; it is typically set so that there would be a 4% drop in the generator's speed between no load and full load (see Fig. 2.2.). The use of droop characteristics on a number of generators allows the response to frequency changes to be shared between them.

The physical effect of governor action on plant that relies on steam pressure to drive the turbogenerator is to open a valve to let more steam through the turbine (in order to increase the electrical energy produced), or close it to let less through. Similarly, for hydro plant, more or less water will be allowed to pass and, for wind turbines, the pitch of the blades will be appropriately adjusted to change the amount of kinetic energy in the wind that is converted to the kinetic energy of the turbine and hence to electricity.

Response to a high system frequency – the reduction of electrical output – is generally always available. Even if particular generators can only operate stably above certain levels of output, in the worst case, they can be tripped. In addition, other control systems within the power station should respond to the governor action by ensuring that steam pressures back through the generation system are not excessive.

Any response to a low system frequency depends on it being possible to increase the electrical power output of at least some generators. This means that they should not already be operating at their maximum power output, i.e. some 'headroom' should have been made available in advance. Normally, generators are most efficiently operated at their maximum output; the owner of the generator would seek recompense for the loss of earnings associated with a reduced output in order to make some 'headroom' available. (This is one reason why, although many grid codes require wind farms to have the capability to provide frequency response, they are not often used – reductions in output, and loss of earnings from renewables incentives or subsidies are particularly expensive.)

For steam driven generators, the automatic low frequency response described above is soon exhausted; opening the valves to let more steam through will lead to a reduction in pressure back through the generator system if more steam is not generated. For the increased electrical output to be maintained, a higher rate of input of fuel will be required which entails another action, one that is not always automatic.

2.2.2 Response and reserve

The above considerations in avoiding an excessive fall in system frequency following a disturbance lead to a distinction between 'response' – a relatively short-lived but fast increase in generation facilitated by an automatic control system action, known in Europe as 'primary' control – and 'reserve', a slower increase in generation output that is not available immediately but which can be sustained for a longer period. In Britain, a distinction is made between 'primary response' – 'the increase in generation or decrease in demand in response to a fall in system frequency that will be released increasingly with time over the first 0–10 s and sustainable for a further 20 s' – and 'secondary response' – 'the increase in generation or decrease in demand in response to a fall in system frequency that will be fully available by 30 s from the time of the start of the Frequency fall and be sustainable for at least a further 30 min' (National Grid, 2012).

The amount of response and reserve that are required, and how quickly it should be delivered, depends on the rate of change of system frequency. That, in turn, depends on the total equivalent inertia of the system, which is made up of all the rotating mechanical plant that is coupled electrically. A large power system serving a very large electrical load will have a large number of generators operating. Especially when these are conventional steam plant, these have large rotating masses, all of which interact through the electromagnetic forces acting across the air gaps within the turbogenerators and transmitted through the electrical network. Similarly, motor loads connected to the system are rotating masses that have inertia. Thus, a large system operating with a high demand will have a lot of generation operating and a much higher inertia than a small system. Moreover, demand on any one system varies through the day and hence the amount of generation that is operating varies, and so does the inertia.

Typically, the biggest challenge to control of system frequency is presented by a sudden reduction in the generation that is operating on the system and a consequential reduction in system frequency due to the imbalance of total generation and demand. This is usually caused by a failure leading to the trip of one or more generating units. Loss of single generator units is not uncommon and might be caused by loss of the electrical connection between the generator and the network but, more often, is due to some fault within the power station. Where there is one prime mover powering more than one generating unit, it is clear that failure of that prime mover will cause multiple generating units to cease generating. While that is the most likely single cause of a multiple loss, multiple events can occur within a short span of time. To carry sufficient reserve to cover all the possible combinations would be extremely expensive. System operators make a judgement on how much reserve – in the case of safeguarding against under-frequency, the provision of 'headroom' – to schedule. This decision is generally taken centrally on behalf of the system as a whole by human operators, albeit informed by analytical software tools which, in turn, depend on input of real-time or historical measurement data on demand and performance of individual generators. As is described in for example Dent *et al.* (2010), a system operator's judgement is informed by an assessment of risk: what is the probability of different amounts of generation becoming unavailable within different time horizons, and what would be the consequences of those losses if there was insufficient reserve? This permits a cost-benefit analysis in which the cost of making headroom available (which obliges the generators providing reserve to not be operating at their most efficient level) is compared with the probability and impact of not being able to meet demand.

The longer the time horizon, the greater the uncertainty and the greater the risk – there is simply more time for unplanned, 'bad things' to happen; the shorter the time horizon, the more certain the system operator can be about what will be going on. On the other hand, it must be considered that much additional generation – reserve – takes some minimum time to reach a certain level of output. Generators that are already operating but part-loaded – 'spinning reserve' – have a certain ramp rate limited by the properties of the prime mover. Other generators take time to get ready for synchronisation with the system. For example, a cold coal power station might take 24 h to warm up and generate enough steam to synchronise with the system and start inputting electrical energy to it. Other generators, such as open-cycle gas turbines (OCGT) and hydro, can start quite quickly but, in the case of OCGT, are very expensive to operate.

To help rationalise the above considerations, system operators will tend to think of different categories of reserve that need to be made available within different periods ahead of time and different quantities of power in each; as conditions change on the system, the reserve made available can then be either utilised or stood down. The categories defined by the system operator in Great Britain help to manage a disturbance such as that

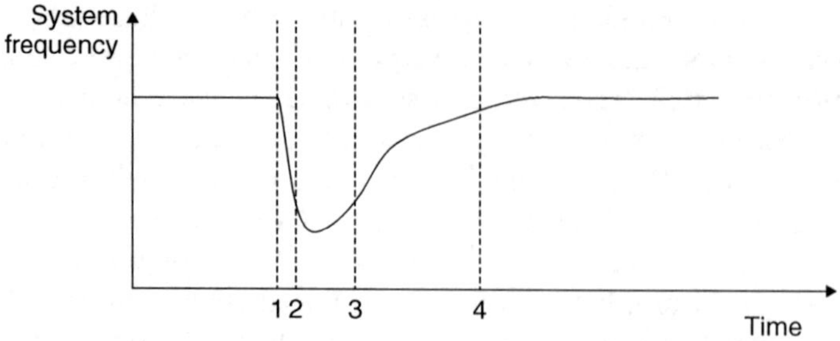

2.3 Classifications of reserve in Britain. At time 1, a loss of some generation has occurred. Between times 2 and 3, 'response' operates automatically within a few seconds to arrest the fall in system frequency. Between times 3 and 4 (a few minutes after the disturbance),other fast spinning reserve and, later, 'short-term operating reserve' are called upon to restore system frequency to normal levels.

shown in Fig. 2.3. 'Response' (referred to in parts of Europe as 'primary reserve') takes place automatically through governor action. 'Fast reserve' and 'short-term operating reserve' (often referred to in Europe as 'secondary' and 'tertiary' reserve) are called upon manually by the system operator, the former to cover short-term variations within a couple of minutes of them occurring, and the latter to deal with recovery of system frequency after loss of a large infeed of power to the system, e.g. a large generator or an interconnector that imports power from another system which, in Britain, might cause the system frequency to drop from around 50 Hz to around 49.5 Hz. In addition, 'contingency reserve' is what is made available some hours ahead of time to cover uncertainties around the actual level of demand and generators that break down or fail to start.

2.2.3 The effect of renewable generation

One of the issues raised in Chapter 4 on renewable generation is its variability and uncertainty. Wind generation is especially challenging in those respects: wind speeds are not constant and are difficult to predict and, as a consequence, so is the output from wind farms. If system frequency is always to be kept within limits, sufficient reserve must be made available to cover the variations, subject to a risk and cost-benefit analysis. However, flexible demand – electricity use that consumers are willing to adapt to the prevailing circumstances, for example by reducing or delaying their consumption – may be expected to become a significant contributor to the regulation of system frequency as the necessary communication, control and information processing become available to deal with the very large number of potential individual participants.

Clearly, good forecasting is not only of demand for electricity but also available wind power is a prerequisite for optimal scheduling of reserve. Just as the error in a forecast is greater, the further ahead is the time for which the forecast has been made, so too is the range of generation technologies that can be scheduled, as illustrated in Fig. 2.4. Different timescales should also be taken into account when utilising flexible demand to provide low frequency reserve since, at some point, that demand for electricity can be expected to return, i.e. interrupted loads will be restored.

An overview of wind forecasting techniques can be found in Ernst (2012) while, at the time of writing, integration of wind forecasts and quantification of the associated risk over different time horizons with unit commitment is the subject of considerable research (see, for example Gubina *et al.*, 2009). Good wind forecasting depends on a good view of 'real-time' wind outputs being provided for the system operator that can be used along with outputs from within the previous few hours and some characterisation of trends to produce a forecast. In some countries (such as, at the time of writing, Ireland and Spain), the provision of real-time information on wind farm outputs by their operators to the system operator is an obligation.

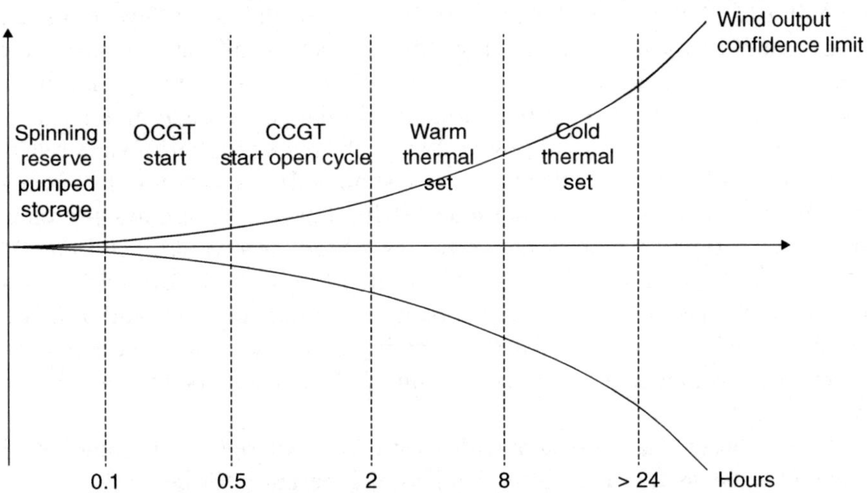

2.4 Reserve requirements in different periods ahead of time and the notice periods required for use of different generators. (The need for reserve is directly related to the uncertainty or variability of the balance between available generation and the forecast demand.) (CIGRE WG C1.3, 2006).

Stored energy does provide an alternative to part-loading of generation to provide reserve. For example, this is largely how pumped storage facilities are used – when electrical energy is relatively cheap, e.g. when it is windy, energy can be stored ready to be released when there is a sudden shortage of generation relative to demand. As storage technologies develop, including distributed forms of storage, they should play an increasingly important part in overall power system balancing.

2.2.4 Managing large disturbances

It was observed above that trips of multiple generators are rare but can – and do – occur. They can lead to excursions of system frequency outside not only of statutory limits but also outside the stable operating ranges of generators. Where a low frequency excursion has occurred due to a shortage of generation relative to demand, this can be disastrous: more generation trips making the situation even worse, possibly leading to system frequency instability and its collapse; in other words, a blackout. As described in, for example IEEE Task Force (2007) and CIGRE WG C1.17 (2010), this has happened on a number of occasions in different parts of the world.

To safeguard against frequency instability, 'under-frequency load shedding' (UFLS) (IEEE Task Force, 2007) has been installed in many countries. This constitutes a number of relays that will act to disconnect particular feeders on the distribution system when the system frequency measured locally is below a particular threshold. While the control action is local, in order to avoid creating local power flow problems a number of such relays will generally be distributed around the system and set at different frequency thresholds. Typically, the number and distribution of UFLS relays and their frequency settings will be designed for the system as a whole such that 5% of load will be shed at a first value of system frequency, a further 5% at a lower frequency and so on, the idea being to reduce total demand to being closer to total generation and arrest a fall in frequency without shedding too much, inconveniencing an excessive number of consumers and causing an over-correction.

As previously noted, a system with a low inertia will see a more rapid fall in system frequency following a sudden loss of generation (or increase of load) than one with a higher inertia. Large continental synchronous areas with high inertia, such as those in Western Europe or regions of the United States, tend not to experience problems with over- or under-frequency. (The exception, as observed in Europe in 2008, for example, is when the system splits into separate electrical islands for other reasons and, where the system as a whole had a good balance between generation and demand, the individual islands – which had been exporting power to neighbouring areas or importing it – might not.) However, operators on island power systems, either electrically isolated from neighbouring areas or connected to them only via high voltage direct current links (HVDC – see Chapter 6), must pay a lot of attention to control of system frequency and, even if rational reserve scheduling policies are in place, can sometimes expect to see frequency excursions.

An example of such an event took place in Great Britain in May 2008. Within 2 min on one particular day, two large generating units tripped, due to events within the respective power stations and independently of each other. This led to a dramatic fall in system frequency, so much so that the first stage of UFLS operated.

While it might appear that the UFLS succeeded in saving the system and preventing a further fall in frequency, subsequent investigation using simulations suggested that the system would, on this occasion, have survived without it. However, two unexpected things were noticed: there was a further fall in frequency some time after the initial events; and the response in terms of increases in output from generators on the transmission system was better than expected. The latter was due to more generators than expected having their governors in operation; the former was concluded to be due to a number of small generators connected within the distribution system (so not visible to the transmission system operator) tripping as frequency decreased.

It has long been anticipated that, on power systems in the industrialised world, the proportion of generation that is connected to the distribution system, i.e. 'distributed' or 'embedded' generation of small scale and connected at lower voltages, would increase. This was the case for the initial growth in wind power through the 1990s. However, one of the reasons why the lights have generally stayed on following electricity supply industry liberalisation and the disaggregation of ownership of generation has been that individual, large power stations have continued to contribute to operation of the system as whole, through frequency regulation and, as will be described in Section 2.4, voltage regulation, driven partly by statutory requirements – through grid codes – for control capability and, partly by commercial 'ancillary service' arrangements for provision of control. Distributed generation has, until now, faced no such obligations or incentives. Furthermore, because of safety concerns about continued operation of generation on portions of distribution networks that have become isolated from the main grid, they have been deliberately tripped when a connection to the main part of the system seems to have been lost.

Up to time of writing, in most large power systems control of system frequency remains a task undertaken at a transmission level with distribution connected generation operating simply at the maximum available output. This is now changing for a number of reasons: distribution connected generation is growing such that it begins to have a significant influence on the behaviour of the system as a whole; and if, as some authors have suggested, islanded operation of distribution networks is to be viable following faults on the connections back to the main grid, it *must* regulate frequency. Also, as will be addressed below, it is becoming necessary to restrict operation of some distribution connected generation for reasons of network thermal or voltage constraints.

High rate of change of frequency, so high that there is risk of system frequency going outside limits before corrective action can be taken, is a concern with growing use of wind generation regardless of the voltage at which it is connected, especially on island systems such as those in Britain, Ireland or Mediterranean islands. The older wind turbines still in operation at the time of writing have quite low inertia; the newer ones that use synchronous machines connected to the main AC system via back-to-back voltage source converters rated for full output of the turbine, i.e. 'fully rated converters' (FRC) make no 'natural' contribution to the inertia of the system at all as the power electronic converter makes a complete decoupling between the electro-mechanical systems on either side. While it has been suggested that a response similar to inertia can be synthesised by appropriate control of the converter, concern has been raised that this might still be too slow on island systems, especially under low load conditions (CIGRE C1.3, 2006), or that excessive extraction of kinetic energy from the rotors would lead to the wind turbines shutting down.

2.3 Ensuring system stability

The IEEE/CIGRE Joint Task Force on Stability Terms and Definitions (Kundur *et al.*, 2004) noted that 'The power system is a highly nonlinear system that operates in a constantly changing environment'. As noted above, and elsewhere in this book, disturbances to the system do happen. These include outages of generators or network equipment, as well as continual variations in the demand for electricity.

The IEEE/CIGRE Joint Task Force (Kundur *et al.*, 2004) defined power system stability as: 'the ability of an electric power system, for a given initial operating condition, to regain a state of operating equilibrium after being subjected to a physical disturbance'. However, it also noted that, because the means of analysing the effects of disturbances and the ways of dealing with them might be different, a number of categories of power system stability might also usefully be defined. These are depicted in Fig. 2.5.

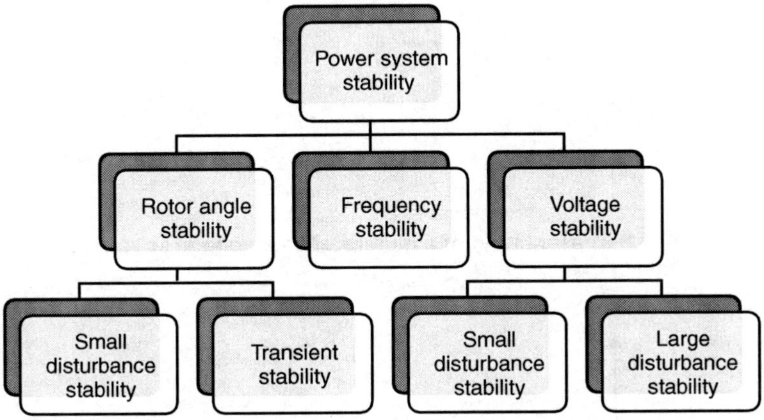

2.5 Classification of power system stability (Kundur *et al.*, 2004).

As can be seen from Fig. 2.5, one of the aspects is 'frequency stability'. This concerns whether the frequency of the system remains within acceptable limits and is what was discussed in the preceding section of this chapter.

The two other main aspects are voltage stability and rotor angle stability. The former is closely related to control of voltage and reactive power and will be discussed in Section 2.4. The latter is often the first thing people think of in the context of 'power system stability', and concerns interactions of the mechanical and electrical sub-systems of the power system.

The transient aspect of rotor angle stability concerns the system's immediate response to a large disturbance, such as a short-circuit fault on a branch of the network. The immediate consequences of this are perhaps most easily understood in terms of a fault near to a particular synchronous machine (most of the generators on a modern power system are synchronous machines though many new generators, such as wind turbines, are connected via power electronic converters which lead to quite different behaviours) and the effect on the operation of that machine. In a first approximation, the machine can be represented as a voltage source behind a reactance that is in turn connected to the grid. The voltage behind the reactance can be referred to as the excitation voltage, E, while the voltage at the point of connection to the grid is referred to as V (see Fig. 2.6.).

The electric power supplied by the generator to the grid can be found to be

$$P_e = \frac{EV}{X} \sin \delta \qquad [2.1]$$

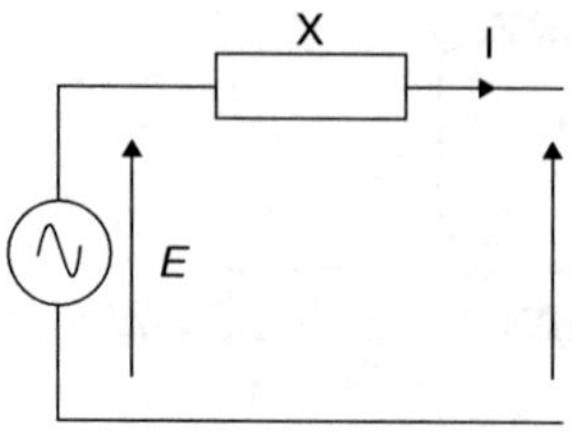

2.6 Representation of a generator as a voltage source behind a reactance.

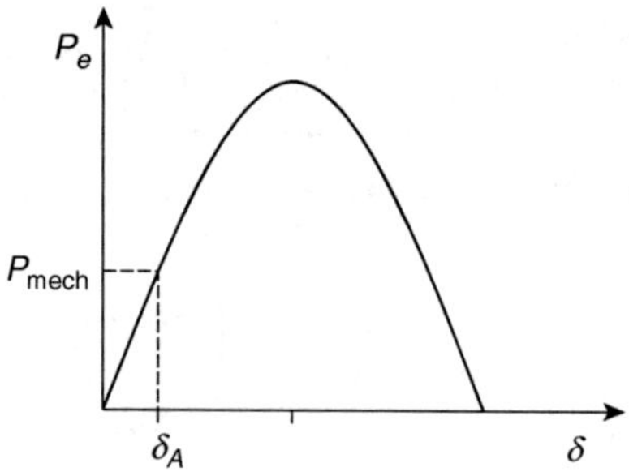

2.7 Relationship of electrical power output with rotor angle δ.

where X is the reactance of the generator, δ is the machine rotor angle, i.e. the angle between the terminal voltage V and the excitation voltage E (see, for example, Glover *et al.*, 2007[1]). The relationship of P_e with δ for fixed values of E, V and X is as shown in Fig. 2.7.

Under steady state conditions, the input mechanical power to the machine, P_{mech}, that enables the rotor to turn at synchronous speed will be approximately equal to P_e. (Actually, P_e will be a bit less than P_{mech} due to losses within the machine. However, for the purposes of simplifying the discussion below, we will assume that losses can be neglected and, under steady state, $P_e = P_{mech}$.) For given values of E, V and X, the value of P_{mech} determines the value of δ.

A generator's terminals will typically be connected to the rest of the grid via one or more network branches. When a short-circuit fault occurs on the network near the generator's connection with the grid, the voltage V seen at the generator terminals will fall to zero or near zero (depending on the nature and location of the fault). Hence, the electrical output power from the generator to the grid is also zero or near zero. Under this condition, and assuming the input mechanical power has been unchanged, $P_{mech} > P_e$. This

[1] Strictly, V and E are 'root mean squared' (RMS) voltages rather than the amplitudes of the sinusoidal waveforms. See Section 2.4 for further discussion.

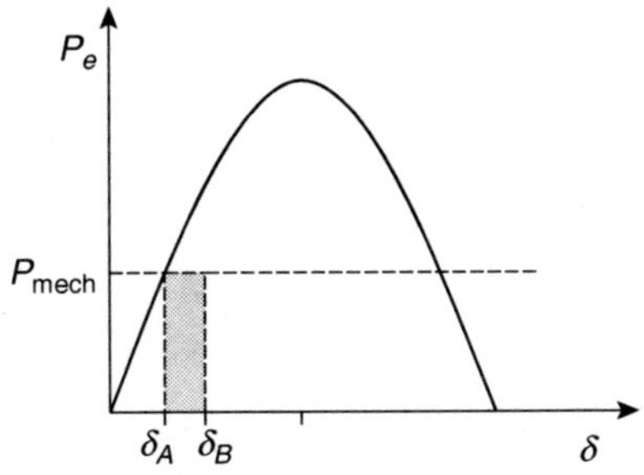

2.8 Acceleration of the rotor of a synchronous machine during a network fault.

surplus input power causes an acceleration of the rotor of the machine, i.e. it gains kinetic energy. An acceleration to above synchronous speed means that the rotor angle δ increases. If the protection system on the grid operates correctly (normally affecting fault clearance in less than 100 ms on a transmission system) the fault will quickly be 'cleared', i.e. the faulted section of the line should be isolated and the current passing into the fault interrupted (see Section 2.6 for a discussion of network protection). Let us suppose in this case that, during the fault, V collapsed to zero (meaning that P_e would also be zero) and that, by the time the fault is cleared, the rotor angle has reached δ_B as shown in Fig. 2.8.

The total kinetic energy gained by the rotor of the generator during the fault is the integral of the difference between the input power P_{mech} and the output power P_e while the fault exists on the system. Because P_e during the fault is simply zero and P_{mech} is constant, the integral of the difference between P_{mech} and P_e can be identified as the shaded area in Fig. 2.8.

If the generator had been connected to the grid via two or more lines and the fault had affected only one (which has now been isolated), the terminal voltage V would now be restored. However, its new, post-fault value would depend on the impedance to the rest of the grid. Although P_e would now be non-zero again, it, too, will depend on the post-fault impedance between the generator and the rest of the grid. (If the grid connection had been entirely lost due to the fault, there would be no sink for the power into the generator, and generator and prime mover protection would be relied on to reduce the input power and safely shut down the generator.)

In the post-fault condition, generally with a lower V and higher net impedance X to the main grid, P_e might be described by the second, lower curve shown in Fig. 2.9. For $\delta = \delta_B$ the electrical output power will be P_{eB}. If P_{mech} remains at its initial value (not an unreasonable assumption in the approximately 100 ms that would typically have elapsed before fault clearance, at least for a transmission fault), there will now be less mechanical input power

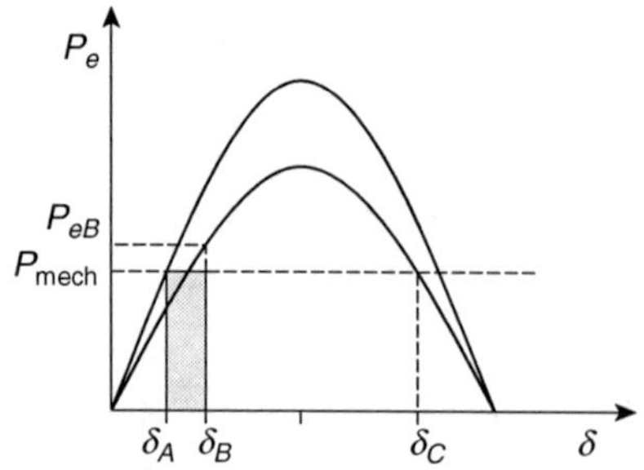

2.9 The second, lower curve shows the electrical power after clearance of the fault.

than electrical power. Some of the rotor's kinetic energy is used to meet the shortfall, i.e. the rotor now slows down and the rate of increase of δ will fall. If all the kinetic energy that had been gained during the fault is expended before δ becomes such that P_{mech} once again exceeds P_e, the rotor's speed will fall below synchronous speed and δ will continue to reduce. It will do so until P_{mech} again exceeds P_e and acceleration – a gaining of kinetic energy – begins again. On the other hand, if the rotor were still rotating faster than synchronous speed at the point at which P_{mech} again exceeds P_e, i.e. not all the gained kinetic energy has been lost, the renewed acceleration will cause δ to increase until the machine 'pole slips' and completely loses synchronism with the system. This is an unstable condition. Generator protection will be relied on to shut down the machine safely, and the generator's infeed of power to the power system will have been lost.

The point at which δ would be such that P_{mech} would exceed P_e is shown by δ_C in Fig. 2.9. However, the loss of kinetic energy during the period in which P_e exceeds P_{mech} is the integral of the difference between them. All of the kinetic energy gained during the fault – area A_1 in Fig. 2.10 – will have been lost if area A_2 before δ_C is reached is greater than A_1. The maximum rotor angle reached will be that when, in Fig. 2.11, area A_3 is equal to A_1. This forms the 'equal area criterion' for determining the peak rotor angle following the fault.

Practical generators will normally be set up with an automatic voltage regulator (AVR) that will modify the excitation voltage E in order to reduce the error between the measured terminal voltage and some chosen reference. Under fault conditions, when V falls to a small value and the error is very large, the excitation will have been increased to its maximum value. This would change the magnitude of the electrical power output, reduce the amount by which the mechanical input exceeds the electrical output, and hence reduce the acceleration and the kinetic energy gained. This means that less would have to be lost for stability to be retained. (An earlier fault clearing time would also reduce the kinetic energy gained, and hence that which needs to

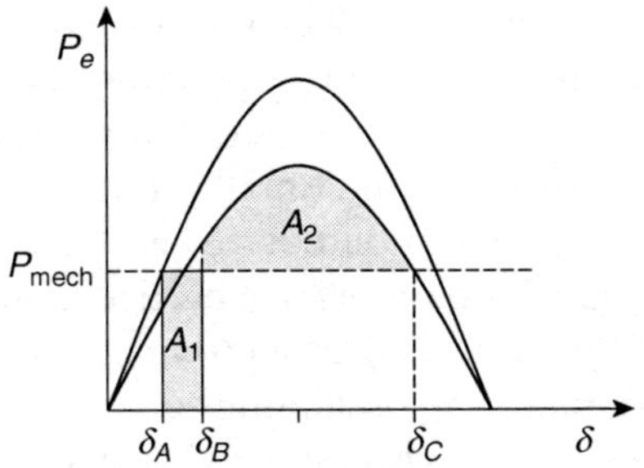

2.10 Rotor angle transient stability is assured if area A_2 > area A_1.

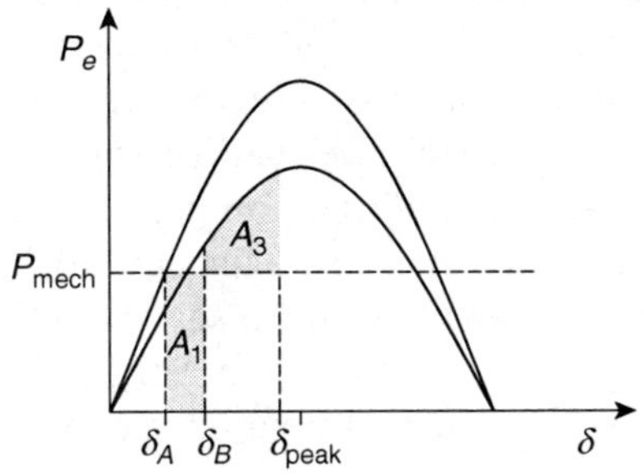

2.11 Use of the equal area criterion to find the peak rotor angle.

be lost.) The same control system (along with power losses within the generator) should also contribute to damping out the oscillations of δ if pole slipping has been avoided, thus allowing it to settle to a new, steady value.

Normally, the first swing of δ would take it to its maximum value and determine whether a new steady state can be reached, hence 'transient stability' is often known as 'first swing stability'. However, under some circumstances, especially if the control system is poorly tuned, subsequent swings of δ might take it to higher and unstable values.

In the above discussion, it has been assumed that 'the grid' can absorb any level of output power from the generator and will provide a strong, constant voltage at the receiving end of the generator connection. However, on practical power systems, other generators (and loads) will show some response to the changes and there will be changes to voltages. The interactions will be highly complex, hence detailed simulation of all large machines on the system and of the changes to voltages and currents around the network will be required to determine if stability would be retained following some particular disturbance. This involves solution of a large set of ordinary differential equations for each of a large number of discrete time steps and some appropriate treatment of non-linearities such as the saturation of transformer magnetic cores or the reaching of limits on controls.

The ability of a generator to survive a particular disturbance will be helped by the excitation system being capable of responding very quickly and achieving a very high maximum voltage. However, these same characteristics will make the AVR very sensitive to quite small variations in terminal voltage and will hence cause variations in output power. Under certain conditions, this might lead to a group of generators in a particular area making a response in one direction at the same time as a group in another area makes an opposite response. This can lead to oscillations of power between the two areas, usually at quite low frequencies (somewhat below the system's nominal frequency). If these oscillations grow in magnitude they can lead to the two areas losing synchronism with each other and becoming electrically isolated, in turn leading to deficit or surplus of power in different electrical islands. This is an example of 'oscillatory' or 'small signal' instability and is usually combatted by the addition of a further control loop on the generator, known as a power system stabiliser (PSS). This is designed to slow down changes in output power. In addition, on parts of the system known to sometimes exhibit power oscillations, specialist monitoring equipment may be installed to allow the system operator to see when the condition has arisen and to take appropriate corrective action. This is generally to reduce the transfer of power across the affected boundary over which the power oscillations are taking place.

2.4 Control of voltages

The magnitudes of voltages on a power system need to be controlled for two main reasons:

1. to ensure that users of electricity can be confident that their equipment works correctly without being over-stressed;
2. to ensure that insulation capabilities of power system equipment are not exceeded.

The above requirements are commonly encoded in statutory voltage limits, typically to ensure that voltages within 5% or 10% of the nominal voltage at particular locations on the system, most notably at electricity users' points of connection. In addition, while maximum voltage limits should be enforced at all system locations for reasons of safety, power system operators tend also to observe minimum voltage limits at locations other than users' points of connection. Although ensuring that such limits can be adhered to in a post-fault condition is often regarded as a good proxy for ensuring some margin from 'voltage instability', it does not actually guarantee it. (Voltage instability is described further below).

The majority of sources of power on an AC power system are 'voltage sources'. That is, they can be thought of as having a controllable, constant voltage with varying output current. This is generally true of synchronous generators, for example. The action of the generator induces a voltage that interacts with the voltage on the machine terminals to produce a net voltage. The magnitude of the voltage is strongly determined by the voltage applied to the rotor of the machine, i.e. the 'field voltage' that is a characteristic of the 'excitation system'. This, in turn, is set by the control system mentioned above – an AVR – that compares the measured terminal voltage with a reference set by the operator of the generator and changes the excitation voltage in order to minimise the difference between the terminal voltage and the reference. The excitation and the conditions on the rest of the system determine the current produced by the generator (see, for example, Kundur (1994) for a fuller description of the generator, excitation system and AVR).

The instantaneous voltage and current determine the instantaneous net injection of power into the system and, under steady state conditions over a cycle of the essentially sinusoidal waveform, the average power injected. However, it should be noted that not only can the magnitude of the injected current vary, but also its phase angle relationship with the voltage.

The relationship between voltage and current at a source of power – a generator – will be determined by the voltage and current relationships for each of the loads on the system and the flow of power on the network. The current drawn by a particular load will be determined by the voltage supplied to it and its impedance. In general, a load has both inductance and capacitance and the final relationship of current to voltage is determined by their relative magnitudes. If the net impedance of the load is inductive, the current will lag the voltage; if it is capacitive, it will lead it. If a particular voltage is to be maintained, the source of power must supply this current.

A convenient way of representing not only steady state magnitudes of voltages and currents but also their phase angles relative to each other is by means of 'phasors', described by complex numbers, the 'apparent power' (measured in volt-amperes (VA)) being the product of the voltage and the current, the real part of it representing the components of voltage and current in phase with each other that will do 'work' being known as the 'active power' (measured in watts, or W), and the component perpendicular to this being known as 'reactive power' (measured in volt-amperes reactive (VAr)). Actually, in order that sign conventions for reactive power are respected, the apparent power is the product of the voltage and the conjugate of the current. An inductive load will absorb reactive power while a capacitive load will generate it.

An applied voltage may be described by

$$v(t) = V_{\max} \cos \omega t$$

and the current through a branch of the circuit to the voltage is applied by

$$i(t) = I_{\max} \cos(\omega t + \varphi)$$

where V_{max} is the maximum value of the sinusoidal voltage waveform with frequency ω (in radians per second), t is time, v and i are the instantaneous values of voltage and current respectively, I_{max} is the maximum value of current and φ is the phase angle difference between the voltage and current. (A positive value indicates that the current leads the voltage meaning that the circuit is capacitive.) Via a series of trigonometric substitutions (see, for example, Glover *et al.*, 2007), the average active power is then

$$\begin{aligned} P &= V_{\max} I_{\max} \frac{\cos \phi}{2} \\ &= VI \cos \phi \end{aligned} \qquad [2.2]$$

where

$$V = \frac{V_{\max}}{\sqrt{2}} \qquad [2.3]$$

is the 'root mean squared' (RMS) voltage and

$$I = \frac{I_{\max}}{\sqrt{2}} \qquad [2.4]$$

is the RMS current.

Similarly, the magnitude of the reactive power into and out of a load is

$$Q = VI \sin \varphi \qquad [2.5]$$

and the active power, P, reactive power, Q, and 'power factor angle', ϕ, are related to each other and the apparent power, S, as shown in Fig. 2.12. In addition, the ratio of P to S is known as the 'power factor'.

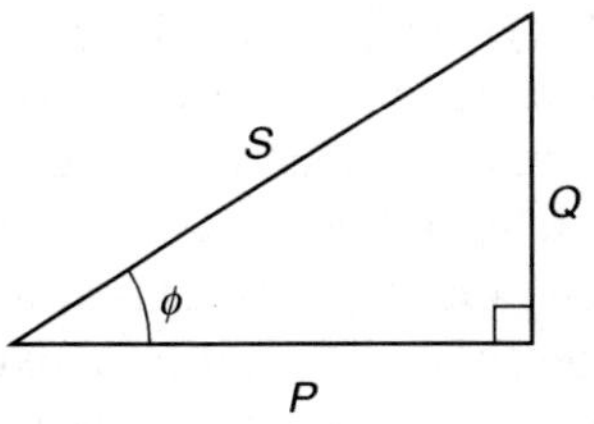

2.12 Relationship of active, reactive and apparent powers.

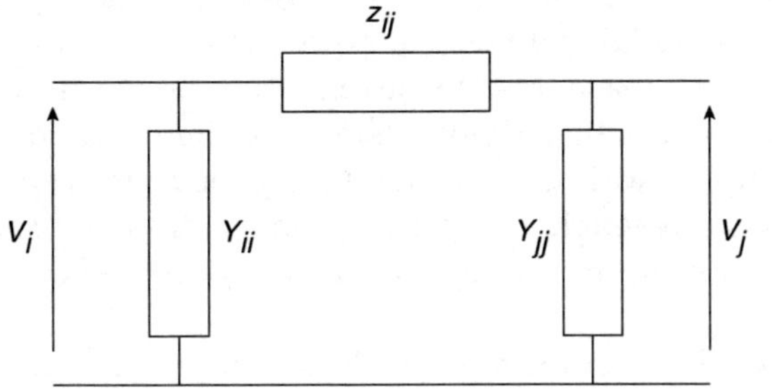

2.13 'pi' representation of the branch of an AC power network.

On practical, large power systems, sources of power are rarely located at the same places as loads; there are some network elements connecting the two. These, too, have characteristics that include inductance and capacitance, the former due to a voltage being induced in the conductor that drives a current in opposition to the changing current through it, the latter due to the charging and discharging of the conductor to a voltage relative to earth. These effects are typically represented by series impedance z_{ij} and shunt susceptances Y_{ii} and Y_{jj} in a 'pi' model of a branch of the network between nodes i and j, the values of impedance and susceptance being determined by the nature of the conductor, the length of the network branch and the nature of the insulation (see Fig. 2.13).

The main difficulty in the control of voltage lies in the interaction of current through a network branch and the voltage at the receiving end. As the power consumed by a load increases, the current that must be carried by the network must increase to serve it. An increased current along a network branch will cause an increase in the voltage drop along it. This is manifested as an increased loss of reactive power on the branch due to its inductive effect. On the other hand, the capacitance of the branch (as shown in Fig. 2.13, this is typically represented in two lumps of susceptance, one at either end) appears as a source of reactive power.

The nature of electrical loads is beyond the scope of this chapter, but it can be noted that the aggregate load seen at, for example, a 'grid supply point' that is the interface between T&D (typically at a voltage of 110 kV or 132 kV), or at a primary substation on a distribution network, is made up of a large number of individual items of connected electrical equipment. In general, these loads will not be purely resistive, and hence the current drawn will be out of phase with the voltage applied, something that, as noted above, reveals itself as a demand for reactive power in addition to active power. Satisfaction of the demand for reactive power means that the magnitude of the current supplied will be greater than it would have been if it had been satisfying only a demand for active power.

In practice, a 'voltage source' such as a generator cannot supply an unlimited current. However, additional reactive power can be generated by the connection of banks of capacitors (the current entering the devices leads the voltage at the terminal). In addition, changes to the tap ratios of transformers that are equipped with suitable monitoring and control systems and tap change mechanisms capable of operating under full load can correct differences between a measured voltage and a voltage reference set by the system operator.

While a first approximation for most loads is to represent them as constant impedances, so that a fall in the supplied voltage is accompanied by a fall in the current drawn, the effect of changes of tap ratios to restore voltages within a distribution system is to make the aggregate load seen at the interface between transmission and distribution seem like a constant power load. If sufficient reactive power is not supplied to it, the voltage at the transmission/distribution interface will fall. Normally, this would lead a great voltage difference between that location and a nearby voltage source, and the drawing of more current from that source. However, it would also lead to less reactive power contribution from the capacitances of overhead lines and cables between source and the transmission/distribution interface and higher series reactive power losses, making the situation worse. It is this type of phenomenon that is known as a voltage instability, and means that sources of reactive power must be adequately distributed around the network and the generation of reactive power suitably shared. This, in turn, entails the careful setting of voltage targets so that reactive power resources are not exhausted under critical conditions.

2.5 Control of currents

The passage of current along a conductor that has some resistance (which is true for all conductors except superconductors, and, even for superconductors, superconductivity only takes place at extremely low temperatures), leads to heating. If that heat is not dumped somewhere, the temperature of the conductor will increase.

The most noticeable effect of an increase in temperature for an overhead line is an expansion of the conductor, i.e. a lengthening. For each span of the line between successive towers supporting the conductor, it leads to a greater sag between the towers and a reduced distance between the conductor and whatever is underneath. If that clearance reduces too much, there is a risk that the insulation afforded by the air between the conductor and whatever is underneath will be broken down and a current will flow down through the ground rather than along the conductor. Such fault currents can be very large indeed[2]; as well as being extremely dangerous for anybody nearby, they can lead to a rise in the potential of 'earth' (which, in turn, could lead to other currents flowing through electrical equipment connected nearby) and to system stability problems such as those described previously.

In a generator, transformer or cable, an excessive increase in temperature could lead to a failure of the insulating material or even to it igniting.

It can thus be seen that it is very important to ensure that currents flowing through any item of equipment are not excessive. However, while it is the flow of current that is critical, because they are accustomed to thinking of loads at particular locations and times of day in power terms, most system operators are accustomed to thinking of the flow of power around the network and will want to compare these power flows with the corresponding limits – derived from the current limits and nominal operating voltages – on each network branch. These limits are expressed in terms of VA, more generally tens or hundreds of MVA for high voltage systems, i.e. the apparent power, and, in view of heating being the key concern, are often described as 'thermal ratings'.

In practice, rises in temperature do not happen instantaneously – a certain amount of heat must be absorbed first. Thus, the maximum current – or MVA – that can be tolerated on a particular branch is conventionally articulated for different durations. For example, a '5 min rating' will be higher than a '20 min rating', which will in turn be higher than a 'continuous rating'. Moreover, the current and, hence, the power flowing through a line that can be tolerated will be dependent on the ability to dump heat. This will be greater when the temperature of the air around an overhead line conductor or the soil around an underground cable is low; thus, different ratings might be expressed for different times of the year, with higher ratings in the winter and lower ratings in the summer.

Of course, temperatures vary through a day, and from day to day. How is it then possible to define a single rating? Normally, it would be set

[2] See Chapter 3 for a discussion of detection and isolation of faults.

conservatively, but there will always be some inherent risk due to the possibility of temperatures in winter being higher than normal (or vegetation underneath a line having grown higher than expected, for example). Exactly what risk will be tolerated by management in a particular utility will depend on their policy and access to information to be able to characterise the risk with confidence.

If power flowing from a source – such as a generator or a neighbouring, exporting part of the system – is so high that the ratings of one or more branches from that source are likely to be exceeded, the power flow should clearly be reduced. That would mean reducing the export, which eventually means reducing the output of one or more generators. If the overall balance of generation and demand is to be maintained, that reduction must be matched by an increase somewhere else. Similarly, if the excessive power flow is serving a particular load, there would be no choice but to reduce that load. (At a transmission level, in order to avoid physically interrupting supplies, the distribution network operator responsible for further flow of power down to the end consumer might be asked either to reconfigure their network in order to reduce the load at one location and switch it to another, provided it is possible, or to reduce voltages on the assumption that the character of the final loads is mostly one of constant impedance so that less current will be drawn. However, this is only allowable if voltages are still kept within tolerable limits, except in particular system-critical situations.)

Reduction of generation normally means using less power from a cheaper source and replacing it with a more expensive one. This is clearly not a desirable outcome, especially if it means reducing output from a low carbon source of power, such as a wind farm, and replacing it with increased output from some fossil-fuelled plant. A network planner will therefore try to design the system using suitable network routes and conductors that typically give particular ratings such that sub-optimal dispatches of power are only rarely needed. In practice, there will be a trade-off between the costs of imposing constraints on the dispatch of power and the cost of the physical network equipment needed to enhance the ratings of critical branches. Normally, the overall least-cost solution would involve at least some restriction of the dispatch of generation for some limited period of time.

The extent of re-dispatch of power could be reduced if branch ratings were determined based on actual conditions rather than some percentile of the range of conditions that arise in a particular season. This would require suitable monitoring of critical branches; for overhead lines that cover many kilometres, if the cost of monitoring is not to be excessive, only the critical spans (on which limits are most likely to be breached) would be monitored.

Various researchers, equipment suppliers and utilities have proposed different monitoring schemes for quantification of 'real-time' or 'dynamic'

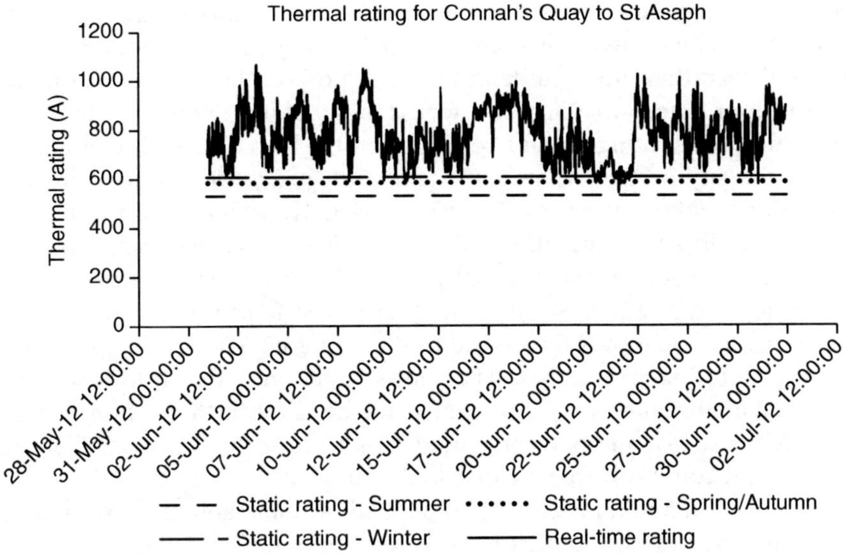

2.14 The effect of real-time ratings in increasing power transfer capacity on an overhead line. (*Source*: Courtesy of Scottish Power Energy Networks.)

ratings (Adapa *et al.*, 2006; Cloet *et al.*, 2010; Michiorri *et al.*, 2009). These might involve any of the following:

- direct measurement of conductor temperatures;
- inference from measurement of conductor strain of what the overhead clearances would be;
- measurement of actual conductor clearances;
- measurement of ambient conditions and online application of rating calculation algorithms.

An example of the effects of one particular scheme operated in North Wales by Scottish Power Energy Networks is shown in Fig. 2.14. The standard, seasonal 'static' ratings are shown as dashed lines for a particular overhead line conductor type along with the dynamically calculated 'real-time' rating for two weeks at the beginning of June in one particular year. The increases in safe current-carrying capability over the summer ratings can be clearly seen.

The fact that higher loadings on network branches can be tolerated for shorter durations than for long periods might be exploited. In Chapter 1, it was noted that one way in which system operators have ensured a relatively reliable supply of power has been to operate the system such that an unplanned outage of any one item of plant, e.g. an overhead line or a generating unit, would not cause any loss of supply to loads or breach of the system limits described in the present chapter. A conservative approach to this would be to dispatch generation before any unplanned fault outage occurred such that, were the fault to occur, no breaches would take place. In respect of branch loadings, this would mean ensuring not only that pre-fault continuous ratings would be respected but also post-fault continuous ratings for any particular 'credible' fault outage. In other words, preventive action would have been taken. The alternative to this would be to take action to re-dispatch generation only after a disturbance has occurred, i.e. to take corrective action. This is made possible by short-term thermal ratings. For example, a power flow up to the 5 min post-fault rating can be tolerated provided the power flow is reduced within that 5 min period to below a longer duration rating. (Normally, flows would be reduced to below the post-fault continuous rating to minimise the need for further action, though eventually the system operator should seek to re-secure the system, i.e. to make sure it could survive a further fault outage.)

Utilisation of 5 min ratings would depend on some automatic responses, which, in turn, would depend on real-time monitoring. Otherwise, if the operator is depending on receiving information and then making a decision, it would be wiser to squeeze the system only to, for example, 20 min ratings.

Some schemes are now being implemented to carry out automatic 'power flow management' (Currie *et al.*, 2010). These are generally most suited to situations in which there is a clear relationship between a particular line being overloaded and a particular generator re-dispatch being sufficient to relieve it. The value of such a scheme is particularly evident when trying to avoid carrying out a preventive limitation of the output of a wind farm, and can be a very cost-effective alternative to reinforcement of the network, especially when a wind farm or a group of wind farms operates at their maximum output only for some relatively small proportion of the time (Ault *et al.*, 2007).

Other means of controlling power flows, at least to some extent, are afforded by phase shifting transformers and some 'flexible AC transmission system' (FACTS) equipment. These have the effect of modifying the angle difference between the voltages at the two ends of a branch, i.e. δ in Equation [2.1], and would hence redirect the sharing of power between parallel network branches. FACTS devices are discussed further in Chapter 7.

2.6 Power system operation and coordination of control

Practical power systems rely on a high degree of automation in their operation. The facilities for frequency response and local voltage regulation, described in outline above, and automatic fault detection and isolation have been applied for decades. However, the choice of settings or control references for individual controls requires coordination. This, in turn, will depend either on conservative settings that apply across a wide range of conditions under which the system might be operated, or a view of the present state of the system at regular intervals.

Protection settings can be reasonably well-determined for a wide range of system conditions. Moreover, as was described above, given the importance of safety, and the need to guarantee operation, that may be a wise course of action anyway.

The need for coordination in real time can be seen in respect of frequency regulation, voltage control and power flow control.

On a meshed network, such as most transmission systems, the exact flow of power on any particular branch of the network depends on the combination of loads and generation at different locations, and the characteristics of the branches. Only the combined effect of a number of generators at different nodes might lead to overloads, and then only for certain levels of demand for electricity, or only under certain outage conditions – knowing how to dispatch one generator depends on knowing how others are dispatched. The effectiveness of a generator action in relieving an overload will depend on the network configuration at the time – if there is a particular planned outage, it may be considerably less effective. On the other hand, a particular action to address an overload on one branch might cause one on another. Often, it is sufficient to know the MW production of different generators to know if there might be overloads; however, particular dispatches of generation might give rise to voltage or stability problems. Resolution of a voltage problem might require starting up a generator. For stable performance, the prime mover of most conventional, thermal generators requires that it operates at some particular level of power. This implies that, in order for its reactive power to be available, the cost of production of at least some active power must be met and would require a review of the active power dispatch of other generators to retain the overall generation/demand balance.

Regarding frequency control, sufficient headroom for low frequency response and reserve must be ensured. However, the operator must also make sure that to use it would not cause power flow problems, i.e. an increase in output from some particular generator in order to correct an under-frequency on the system would not cause an overload of any branch.

It was noted above that the transfer of reactive power entails larger current magnitudes on network branches, which, in turn, would lead to greater network reactive power losses. Thus, reactive power sources remote from reactive power loads are often quite ineffective. However, careful setting of voltage targets – the achievement of which requires generation (or, under low load conditions, absorption) of reactive power, can increase the gain from the capacitances of network branches. Finally, while many different voltage control devices can regulate voltage effectively, they respond to variations at different rates, or might achieve coarse or fine grain control. If a number of devices within an area are not to fight with other, they must be coordinated.

Thus, it can be seen that while respect of thermal, voltage and stability limits for a particular state of the system might each be checked in turn, and that only one of those limitations requires some change to the system state, finally all the limits must be respected.

In Chapter 3 it is noted that power networks, especially at transmission voltages that carry high amounts of power, often have multiple protection devices in order to provide back-up in the event of failure of any of them. However, they should be set in such a way that the most discriminating devices have the chance to act first, in order not to spread the extent of an unplanned outage unnecessarily.

Responses to variations of controlled variables might be made by continuously acting control systems (see Kundur (1994) for descriptions of some examples) or by discrete actions such as opening or closing of one or more circuit breakers, e.g. to isolate a faulted network branch, switch in (or out) a bank of capacitance or, via a pre-programmed sequence and the availability of suitable communication, automatically trip a generator or reconfigure the connectivity of the network. Some degree of coordination can be achieved through the design of the respective controls. For example, large deviations of controlled states away from targets might be responded to more quickly than smaller ones, with fine-tuning taking place only once the system has settled down.

As described above, respect of the system's limits while meeting the demand for power requires appropriate setting of the whole set of control, notably the power outputs and voltage targets of generators. As the system's condition changes, e.g. due to variation in the electrical load, forced outages of generation, or faults on network branches, the system operator must be sure limits continue to be respected. That requires full visibility of all the key system states. This does not actually require measurement of everything but of a sufficient minimum set of measured states – collected via the 'Supervisory Control and Data Acquisition' (SCADA) system that also allows dispatch of certain control settings – that allows others to be derived. In practice, as will be discussed in the next section, the measurements are generally vulnerable to errors. The procedure for derivation of a single, self-consistent set of states – 'state estimation' – can do so in such a way as to place more trust in the less error prone measurements.

As power systems develop with a greater number of 'active' variables that can be chosen, the number of possible combinations of system states also increases. In addition, in order not to limit the utilisation of available renewable generation and avoid the building of unpopular new overhead lines wherever possible, there is pressure to operate power systems closer to their limits, which requires a more precise understanding of what those limits are. This will be aided by more accurate and more prevalent system monitoring, but also requires adequate understanding of the complex interactions of the large number of system states on meshed systems. On portions of power systems that are operated radially, such as most distribution systems, the interactions are less complex but, historically, measurements and communication have been much less available. With such distribution system developments as 'flexible' or 'responsive' load (where consumers of electric power can be influenced to change their demand, something that promises to be extremely useful in responding to variations in the power coming from, in particular, wind farms) and much more generation embedded within the distribution system (such as photovoltaic cells), a judgement must be made as to whether the information gathered by widespread monitoring should be brought back to one place for centralised, coordinated decision making or whether some form of decentralised control is possible via suitably programmed logic devices. If the programs can be written in such a way as to give confidence that correct responses can be ensured for a wide range of conditions, critically those that require guarantees of safety, the reduced need for reliable communications, and the possibility of faster responses than centralised approaches suggest a preference for decentralised solutions. On the other hand, especially within meshed network structures that are typically used when seeking to minimise the number of interruptions to supply load that might be caused by faults, the need for coordination may strictly limit the scope for decentralisation.

2.7 Measurement, monitoring and communications

2.7.1 Overview of measurement and monitoring

The variety of control and operation functions required on power networks outlined above requires adequate measurement, monitoring and communications. Transmission networks, due to the consequences of unexpected failure, have been, and will continue to be, monitored, controlled and protected comprehensively, with an abundance of measurements, often using a degree of redundancy in the measurement equipment, being made available both locally and remotely by (expensive) voltage and current transformers and the associated communications infrastructure and SCADA systems.

In contrast, measurements taken at distribution voltage levels (typically 33 kV and below) are extremely sparse in nature, but this will need to change in the future. Until relatively recently, distribution networks were almost entirely passive in nature. That is, there were no active decisions made by distribution network operators on a minute-by-minute, or even hour-by-hour, basis other than to respond to faults. (Reconfiguration of a network to accommodate maintenance outages is typically planned some days or weeks in advance.) Under normal circumstances, the behaviour of the network downstream of measurement locations could be accurately estimated. Accordingly, other than dedicated, and expensive, measurements being provided for protection systems, which for economic reasons are only implemented only at certain locations – ideally, protection and isolation facilities would be provided at each and every network node, but this is not the case – measurements have normally only been taken at locations such as the outgoing 11 kV feeders from primary substations and used for metering and by engineers at centralised control centres to estimate present and future load levels, to check for overloads, and to monitor the system performance over long time periods. In some cases, other than for metering purposes, almost no measurements have been available from the distribution systems.

More recently, there has been a growing requirement for a greater number of measurements to be made in order for a more comprehensive understanding of system and device behaviour to be made available. To date, this has been driven mainly by two things: growth in small scale generation 'embedded' within the distribution network (sometimes known as 'distributed generation' (DG)); and the need to better utilise existing network assets in the face of capital constraints on reinforcement or replacement and, in urban areas, limits on the available land. In addition, the expected growth in electric vehicles and more electric heating will, if not carefully managed (e.g. by means of 'smart charging' of electric vehicles), increase loading on the network at critical times. Additional measurements are therefore required to provide a more comprehensive picture of the state of the network, enable monitoring of the condition of plant, and facilitate improved protection with higher degrees of discrimination and stability. Voltage and current remain the primary quantities of interest, but increasingly, functions that monitor other parameters, such as temperature, vibration, partial discharge activity and other parameters, are emerging.

Measurement of voltage and current

As mentioned previously, the main quantities of interest are voltage and current, from which a number of other measurements (e.g. real and reactive power, frequency, harmonic distortion, etc.) are derived. Voltage transformers (VT) and current transformers (CT) use long-established principles based on electromagnetic induction.

In the case of VTs, conventional wound transformers (with the entire system phase voltage being dropped across the primary winding) can be

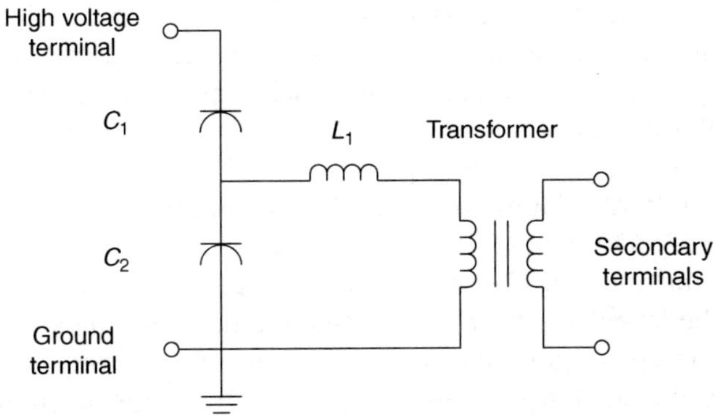

2.15 Schematic of a capacitor voltage transformer (Created by Cedars using OmniGraffle and released into the public domain on Wikipedia).

used up to voltages in the tens of kV range; above this voltage level, use of a wound transformer is generally economically prohibitive, so capacitive dividers are used to reduce the voltage to a suitable value (10 kV+), and then a wound transformer is used to transform this intermediate voltage to the standard input levels of monitoring and protection devices (typically 110 V line-line in a three-phase system). Tuning circuits are required to avoid resonance between inductive and capacitive elements of the transformer and to 'tune' the impedance of the overall circuit to be a minimum (i.e. $1/\omega(C_1 + C_2) = \omega L_1$) at the system frequency in order to minimise measurement errors when current is drawn from the secondary terminals. The layout of a capacitor voltage transformer is shown above in Fig. 2.15, which illustrates the capacitive voltage divider, the tuning inductor and the conventional wound transformer, the secondary terminals of which provide the output voltage as an input to monitoring and protection functions.

Similarly, CTs also employ conventional transformer technology; in the case of a current transformer, the primary 'winding' is the conductor itself, with a secondary winding arranged around a core through which the primary conductor (or winding) passes. There is no voltage drop across the primary, so such CTs behave as a current source. The output from current transformers is usually standardised at 1A or 5A for their rated primary currents, and CTs are available in a wide range of ratios, from 5:1 up to 2000:1 (or even greater).

CTs, like any transformer, are prone to saturation of the magnetic circuit under very high flux levels and will exhibit losses and measurement errors which, if the transformer becomes saturated, can be large and could lead to problems, for example, non-operation or unwanted operation of protection. CTs for different applications (for example, metering as opposed to protection) have hugely different construction: protection CTs are by necessity larger than metering transformers, due to their requirement to provide accurate measurements at currents far in excess of rated currents, for example up to 30 times rated current, which requires a much larger core cross-sectional area to avoid saturation. They are therefore costly items of plant (as are VTs).

There are standards relating to CT and VT performance, and examples of these include IEC 60044 and IEEE C67.13.6–2005.

As stated, conventional CTs and VTs are relatively expensive items of plant. Given the levels of power transferred and the consequences of interruptions, investment in good coverage of a transmission network is justified. However, the challenge of providing cost-effective measurements for the distribution system is one that continues to be faced. Several researchers have developed alternative (often termed 'non-conventional') methods of sensing voltage and current, many of which employ optical, or hybrid optical/electromagnetic, sensing techniques, examples of which are based on Rogowski coils, polarimetric, interferometric, magnetostrictive and piezoelectric mechanisms (Jiao *et al.*, 2006; Niewczas and McDonald, 2007). Such sensors offer the potential to provide a relatively much lower cost per measurement within power systems and may offer a cost-effective solution to the future measurement requirements detailed in the next section. Conventional CTs and VTs typically possess limited bandwidth, and non-conventional CTs and VTs often overcome this limitation. Finally, the IEC produces standards relating to 'electronic' CTs and VTs (i.e. devices producing outputs directly in digital format) and specifies how these devices can integrate with the 'digital substation' (IEC, 2002, 2003). Devices producing outputs in a digital format offer opportunities for remote and/or multiple uses of the data by several functions.

Phasor measurement units

Phasor measurement units (PMU) have been the subject of great interest in recent years. These devices, which normally interface with conventional VTs (and in some cases CTs), use accurate time reference signals provided by the Global Positioning System (GPS) so that accurate time-synchronised measurements of voltages (and currents) from a wide geographic area can be made available to system monitoring, control and protection functions. The applications of PMUs are numerous and are suggested to include (Phadke and Thorp, 2008):

- wide area monitoring systems (WAMS): real-time power flow monitoring, control and detection of instability;
- post-disturbance analysis and diagnosis;

- analysis and tuning of PSS performance;
- protection, including loss of mains, out-of-step, UFLS, etc.

While these devices are costly, the potential that they offer means that they will undoubtedly have a large part to play in providing the data required for advanced functions in the future. There are a large number of PMU devices available, and work continues on standardisation of PMUs, the output data format from PMUs and the development and standardisation of other devices such as phasor data concentrators (PDCs) (Adamiak *et al.*, 2011; IEEE, 2006), to enable large-scale PMU systems to be developed. The number of PMUs deployed has grown significantly, and this growth is expected to continue in future.

The optimal amount, optimal placement and development of functions that can exploit PMUs within power systems remain major research topics, and more information on this can be found in Madani *et al.* (2011).

2.7.2 Future measurement and monitoring requirements

As mentioned in the previous sections, and elsewhere in this book, a far greater and more accurate degree of visibility of system parameters will be required to enable effective monitoring, control and protection of future systems. Whether this visibility, in the form of measured states (and/or estimated – via state estimation), is provided only locally (e.g. to generator control systems) or centrally (e.g. for centralised monitoring and management of voltage, generation dispatch, monitoring and remedying violation of thermal limits of feeders, etc.), is an ongoing debate. Undoubtedly, different variations and combinations of centralised and decentralised systems will emerge as manufacturers develop products in this area.

With the widespread introduction of DG, storage and so-called FACTS devices at distribution levels, the behaviour of the distribution network is no longer passive, and in order to enable safe and reliable utilisation of DG, rapid restoration of any disconnected load following faults, and the accommodation of different kinds of loads such as electric vehicles and heat pumps, a similar level of visibility and control to that historically found on transmission systems, will be required. However, the very large number of circuit km, individual substations and loads on a distribution network mean that 'smart grid' functionality that minimises dependency on human supervision will be required. Much of this, e.g. 'demand side management', will be of great value also to transmission system operators in the accommodation of large volumes of highly variable renewable resources. In any case, a large quantity of measurements will be needed.

In some cases, the measurements may only be required periodically and with low resolution, in others the measurements may require to be continuous and with high resolution. Furthermore, many 'smart grid' functions can share data acquired by the same measurement equipment, but this will require extensive use of communications and use of standard communications that will enable sharing of data across multiple functions and with equipment provided by several manufacturers. Communications will also be required to effect control of equipment from remote locations, and to communicate status changes, indications and alarms, in addition to measurements of analogue quantities.

2.7.3 Communications requirements

Internet based technologies are proposed as being capable of offering cost-effective solutions for 'smart grid' communications, and can provide the low latency and high security (and in some cases, deterministic latencies over packet-switched networks) required by power system monitoring, control and protection applications (Alcatel-Lucent, 2010).

For example, Internet Protocol/MultiProtocol Label Switching (IP/MPLS) provides a connection-orientated deterministic service which improves on the nondeterministic behaviour of traditional Internet Protocol (IP) and Ethernet packet communication. The drawback of IP/MPLS solutions is that they require a dedicated communications medium, and standard enterprise routers are often not capable of implementing IP/MPLS. This is not an obstacle in power system applications where it becomes cost effective to install a private communication infrastructure employing IP/MPLS capable routers, but it does mean that IP/MPLS could not be used presently at the domestic level using existing household routers. However, costs are continually reducing and market players are already promoting offerings that target LV smart grid applications.

Remote household energy monitoring and control is already being employed in smart metering and demand side management schemes. It is possible that the technology used in these schemes could be used for other functions within the context of a smart grid.

There are several different communication protocols and media that are being proposed and demonstrated for transferring information between smart meters and the point of central data processing/collection – Wi-Fi, GPRS, ZigBee, power line carrier and Bluetooth are examples of the candidate technologies being proposed. An excellent overview and comparison of communications technologies for smart grids is available at Aviat Networks. There are also universal metering interfaces (Cambridge Consultants Ltd, 2012) available that are capable of interfacing meters to different communications systems via peripheral devices.

While there are many pilot implementations, there remains much work to be done before the vision of the smart grid can be realised, and standards must continue to be defined that will allow equipment from different manufacturers to inter-operate effectively.

2.8 References

Adamiak, M.G., Kanabar, M., Rodriquez, J. and Zadeh, M.D. (2011). Design and implementation of a synchrophasor data concentrator. *IEEE PESD Conference of Innovative Smart Grid Technologies – Middle East (ISGT Middle East)*, 17–20 December 2011.

Adapa, R., Douglass, D.A. and Reppen, D.N. (2006), 'Applying Dynamic Thermal Ratings in System Operations', Paper C2-306, CIGRE Session, Paris.

Alcatel-Lucent (2010), *Dynamic Communications for Smart Grid: Driving Smarter Energy Management and Usage* [pdf]. Available at: http://enterprise.alcatel-lucent.com/private/images/public/si/pdf_powerUtilities.pdf (Accessed 19 August 2012).

Ault, G.W., Bell, K.R.W. and Galloway, S.J. (2007), 'Calculation of economic transmission connection capacity for wind power generation', *IET Renewable Power Generation*, **1**(1), 61–69.

Aviat Networks (n.d.), *Smart Grid Wireless Technology Comparison Chart.* Available at: http://us.aviatnetworks.com/solutions/smart-grid-solutions/ (Accessed 19 August 2012).

Cambridge Consultants Ltd (2012), *Universal Metering Interface.* Available at: http://umi.cambridgeconsultants.com/ (Accessed 19 August 2012).

CIGRE WG C1.3 (2006), *Electric Power System Planning with the Uncertainty of Wind Generation*, Technical Brochure 293, CIGRE, Paris.

CIGRE WG C1.17 (2010), *Planning to Manage Power Interruption Events*, Technical Brochure 433, CIGRE, Paris.

Cloet, E., Lilien, J.-L. and Ferrieres, P. (2010), 'Experiences of the Belgian and French TSOs using the "Ampacimon" real-time dynamic rating system', Paper C2-106, CIGRE Session, Paris.

Currie, R.A.F., Ault, G.W., Foote, C.E.T., McNeill, N.M. and Gooding, A.K. (2010), Smarter ways to provide grid connections for renewable generators, 2010 IEEE Power and Energy Society General Meeting, Minneapolis, 25–29 July 2010.

Dent, C.J., Bell, K.R.W., Richards, A.W., Zachary, S., Eager, D., Harrison, G.P. and Bialek, J.W. (2010), 'The Role of Risk Modelling in the Great Britain Transmission Planning and Operational Standards', *Proc. 11th International Conference on Probabilistic Methods Applied to Power Systems*, Singapore, 14–17 June.

Ernst, B. (2012), Wind power prediction. In: Ackermann, T., ed., *Wind Power in Power Systems*, Wiley, Chapter 33.

Glover, J.D., Sarma, M.S. and Overbye, T.J. (2007), *Power Systems Analysis and Design*, Nelson.

Gubina, A.F., Keane, A., Meibom, P., O'Sullivan, J., Goulding, O., McCartan, T. and O'Malley, M. (2009), 'New tool for integration of wind power forecasting into power system operation', *IEEE Power Tech,* Bucharest.

IEC (2002), IEC 60044-8. Instrument transformers – Part 8: Electronic current transformers, Geneva: International Electrotechnical Commission.

IEC (2003), TR 61850-1 Communication networks and systems in substations –Part 1:Introduction and overview, Geneva: International Electrotechnical Commission.

IEEE (2006). IEEE Standard C37.118–2005: IEEE Standard for Synchrophasors for Power Systems, Piscataway, IEEE.

IEEE Task Force on Blackout Experience, Mitigation and Role of New Technologies, Report, Blackout Experiences and Lessons (2007), *Best Practices for System Dynamic Performance, and the Role of New Technologies*, Final Report, IEEE.

Jiao, B., Wang, Z., Liu, F. and Bi, W. (2006), 'Interferometric fiber-optic current sensor with phase conjugate reflector', *IEEE International Conference on Information Acquisition*, 20–23 August 2006.

Kundur, P. (1994), *Power System Stability and Control*, McGraw Hill.

Kundur, P., Paserba, J., Ajjarapu, V., Andersson, G., Bose, A., Canizares, C., Hatziargyriou, N., Hill, D., Stankovic, A., Taylor, C., Van Cutsem, T. and Vittal, V. (2004), 'Definition and classification of power system stability IEEE/CIGRE joint task force on stability terms and definitions'. *IEEE Transactions on Power Systems*, **19**(3), 1387-1401.

Madani, V., Parashar, M., Giri, J., Durbha, S., Rahmatian, F., Day, D., Adamiak, M. and Sheble, G. (2011), 'PMU placement considerations – A roadmap for optimal PMU placement', *IEEE Power Systems Conference and Exposition (PSCE)*, Phoenix, 20–23 March 2011.

Michiorri, A., Taylor, P.C. and Jupe, S. C. E. (2009), 'Overhead line real-time rating algorithm: description and validation', *Proc. IMechE Part A: J. Power and Energy*, **24**(3), 293–304.

National Grid (2012), *The Grid Code*. Available at: http://www.nationalgrid.com/uk/Electricity/Codes/gridcode/gridcodedocs/ (Accessed 15 August 2012).

Niewczas, P. and McDonald, J.R. (2007), 'Advanced optical sensors for power and energy systems'. *IEEE Instrumentation and Measurement Magazine*, **10**(1), 18–28.

Phadke, A.G. and Thorp, J.S. (2008), *Synchronized Phasor Measurements and Their Applications*. New York: Springer.

Wood, A.J. and Wollenberg, B.F. (1994), *Power Generation, Operation and Control*, Wiley.

3

Protection of transmission and distribution (T&D) networks

C. BOOTH and K. BELL, University of Strathclyde, UK

DOI: 10.1533/9780857097378.1.75

Abstract: This chapter describes the behaviour of power systems during faults and illustrates the requirements for power system protection. The components of protection systems and the typical schemes used to protect power systems are described. An overview of the future challenges relating to protection is provided, with specific comments relating to the impact of distributed generation on network protection systems.

Key words: power system protection, faults.

3.1 Introduction

All electrical power systems require automatic means of detecting the presence of fault or abnormal conditions and subsequently isolating the faulty equipment. Faults may be short circuits or partial short circuits, open circuits, unbalanced conditions, or any other circumstances deemed to be undesirable, such as operation of an element of the system in islanded mode following disconnection from the main grid (loss of mains (LOM)).

Short circuit faults are the most common and potentially most damaging type of fault, as the resulting excessive current can cause thermal and mechanical damage to the plant carrying it and, if the fault is not removed quickly, the overall stability of the system may be compromised, possibly increasing the risk of partial or complete system collapse. Historically, a large proportion of system blackouts include incorrect (which may be unnecessary) operation of protection as a contributing factor (Atputharajah and Saha, 2009).

The occurrence of faults is inevitable, because the power system suffers from the effects of natural phenomena (e.g. electrical storms), the effects of ageing on insulation, human error, etc. The protection system is thus designed not to prevent faults, but to respond to their occurrence and minimise their effects. The consequences of faults and other undesirable conditions include excessive current flows, voltage and/or frequency depressions or deviations from normal values, all of which can be measured and used by the protection systems to detect the presence of faults on the system.

This chapter describes the behaviour of power systems during faults, how faults may be detected, key elements of protection systems, the performance requirements for protection systems and the components, methods and schemes that are most commonly used to provide protection for power system equipment, along with a brief overview of present and emerging research challenges and solutions in the field, in the context of power systems comprising increasing penetrations of distributed generation (DG) (and storage).

The approach taken is to describe the main concepts and basic operation of protection schemes: detailed theory of operation is not presented. There are several textbooks focussing solely on protection and the reader is advised to consult there for more detailed information regarding the subject (Anderson, 1999; Alstom, 2011).

3.2 Fault detection and isolation

Faults on the power system normally result in a marked change in the measured voltages and currents in the vicinity of the fault. These can be detected by protection systems in order to initiate the actions required to isolate the faulted element(s) of the network. The remainder of this section outlines the behaviour of the power system during faults, and how faults can be detected and isolated.

3.2.1 Short circuits on power systems

Short circuits on power systems are inevitable and system protection must react by detecting the presence of such faults and providing an appropriate reaction, which is typically in the form of tripping a circuit breaker, or multiple circuit breakers, to interrupt all fault current paths from the sources

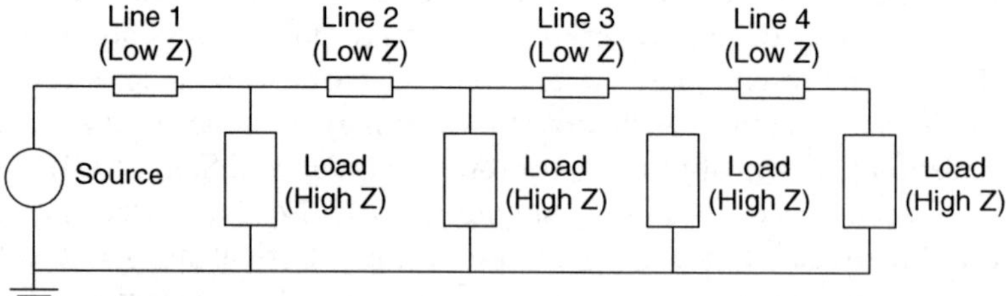

3.1 Simple one-line representation of a section of power system.

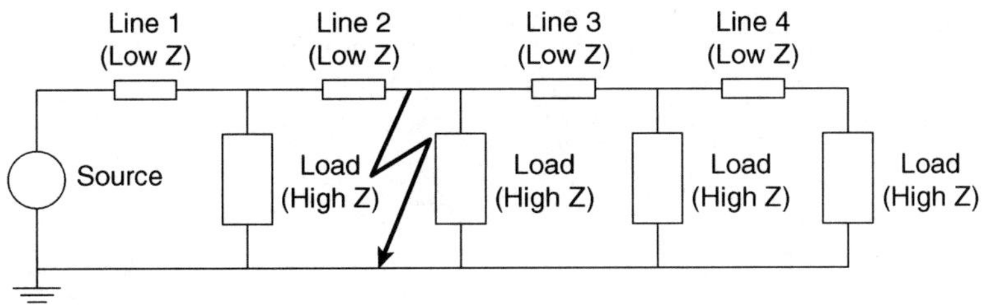

3.2 Simple one-line representation of a faulted power system.

of fault current to the fault location. Short circuits typically involve breakdown of insulation and arcing and must therefore be removed quickly from the system, both to minimise physical damage to plant and surrounding equipment and to reduce system integrity risks due to generators becoming unstable.

Figure 3.1 illustrates a simple one-line single phase representation of a power system:

Consider a situation where the system above is short circuited, in this case via a phase-earth fault (assuming the line at the bottom of the figure is a combined neutral/earth return path) at the location as shown in Fig. 3.2.

For a fault such as this, the current flowing in the circuit would only be restricted by the short circuit capabilities of the source and the impedance to the fault, which in this case is represented by the cumulative impedance of Lines 1 and 2. Assuming the system is solidly earthed and that the impedance of the fault and the return path are both negligible (which would not be the case in a practical situation), then the voltage at the point of the fault would effectively be zero. In a practical situation, the impedance of the fault and the return path would both require to be taken into account, and this normally entails the use of simulation software which can provide a reasonably accurate estimation of the fault currents (and voltage depressions) within a large interconnected power system for faults at various locations.

3.2.2 Detection of faults on the system

The detection of faults on the system is invariably achieved through measurements of voltage, current, or both. These measurements are processed by relays, which are provided with inputs from current transformers (CT) and/or voltage transformers (VT). The relay is responsible for making a decision as to whether a fault is present on the system, and, if so, what subsequent action should be taken (if any is deemed necessary) in response to the detected fault. Modern relays are numerical (i.e. microprocessor based)

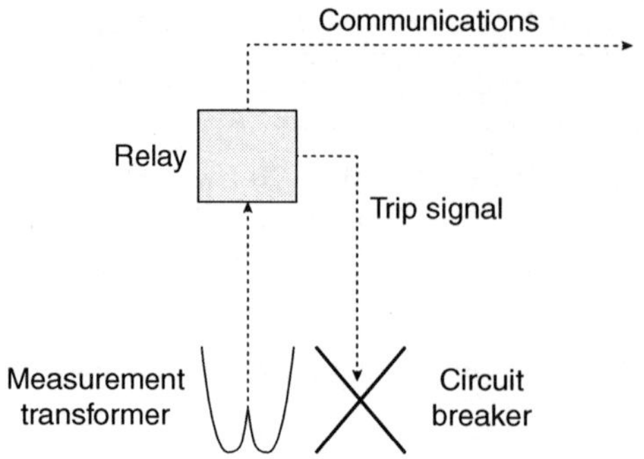

3.3 Main components of a protection system.

and are increasingly replacing their electromechanical and electronic predecessors, although there are still a large number of older relays in service, particularly at the lower voltage levels in the system.

CTs and VTs measure directly from the T&D systems and provide their associated relay(s) with standardised inputs, typically of 110 V line-line voltage for VTs (for nominal input voltages) and 1 A or 5 A for CTs (for nominal current rating of the CT). CTs and VTs are available in a wide range of ratios, for VTs from 11 kV:110 V up to 750 kV (or more):110 V, while CTs are available with transformation ratios up to 2000:1 or more. Standards for the performance of CTs and VTs are available (IEEE, 2005). Conventional CT and VT construction and operation has remained relatively unchanged in recent years and is based on derivations of electro-magnetic power transformers; however, 'non-conventional' CTs and VTs, largely employing optical sensing techniques, have been developed and applied in an increasing number of cases. Finally, the IEC 61850 standard (which is published in several parts) encompasses the provision of measurement data to protection relays in a digital format, and this enables CTs, VTs and relays to communicate via a digital substation/process bus. An introduction to the standard is available at IEC (2003). A simple representation of a protection system is presented in Fig. 3.3, where it is shown how a protection relay continually monitors inputs (obtained directly or via some form of communications system) of current (and/or voltage) from a measurement transformer (or transformers). The protection relay detects the presence of faults and, if deemed necessary, sends signals to trip its associated circuit breaker(s) when the appropriate conditions are detected. In some cases, as also illustrated in Fig. 3.3, communications may be used. Depending on the particular protection system, this may be for the purposes of exchanging measurement data with other protection relays, to provide facilities to remotely trip other circuit breaker(s), or to send information such as indications of operation, alarms and recorded fault data to remote locations.

3.2.3 Isolation of faults from the system

In transmission systems, the degree of interconnection invariably means that, for a fault on a transmission line, circuit breakers at both ends of the faulted line must be opened. If the line is a multi-terminal circuit, then circuit breakers at all line ends must be opened when a fault is detected. Other system components, such as generators, busbars and power transformers, must also be protected, and this is usually provided by protection schemes dedicated to the protection of the components, tripping circuit breakers to provide isolation of the faulted equipment when required.

Traditionally, for a fault on the power distribution system (i.e. the system at 33 kV (sometimes 132 kV) and below in the UK), fault currents flow 'down' to the fault position from the 'upstream' transmission system, to which generators (the sources of fault current) are connected. In such cases, only the immediately 'upstream' circuit breaker from the fault location requires to be tripped in the event of a fault. However, this relatively simple situation has now been complicated by the fact that DG is being increasingly connected to the distribution system, meaning that fault currents may now flow from locations both 'upstream' and 'downstream' of the fault, requiring more complex and costly network protection systems. The DG must, of course, also be protected and its protection systems must coordinate with network protection to ensure that the appropriate generators are quickly disconnected from (or in some cases, remain connected to) the system in the event of a fault on the adjacent network.

The presence of DG or storage can also influence network fault levels and this can, in some cases, adversely affect the operation of network protection, potentially leading to loss of coordination, non-operation of network protection devices for network faults, or spurious operation of protection devices for faults on the system to which they should not normally react. LOMs (or anti-islanding) protection is also a requirement in most utility networks for distributed generators that can operate in parallel with the utility supply system. This also complicates the overall protection function, and is discussed later in Section 3.8.

3.3 Protection system requirements

The protection system must fulfil the requirements of:

- rapidly and automatically disconnecting the faulty item(s) of plant or section of the power network;
- minimising the disconnection of non-faulted equipment, thus ensuring maximum availability and security of supply to consumers, and

minimising the potential for overload on the remainder of the system following a fault event.

The degree to which any protection system satisfies these requirements can be described by four interrelated parameters – discrimination, stability, sensitivity and operating time.

- **Discrimination** is the degree of ability of the protection system to select, on the basis of the power system conditions (e.g. current, voltage, etc.) whether or not to operate for a given measured system state, that is, to send a trip signal to its associated circuit breaker(s). A protection system that is highly discriminative would, for example, be capable of identifying faults that are located on the protected plant item (e.g. a transmission line) as opposed to those not located on the protected plant item (e.g. a fault on the immediately adjacent line to that being protected). A highly discriminative protection system always operates when it should, and never operates when it should not.
- **Stability** is a measure of the ability of the protection system to remain inoperative under certain faulty conditions, because the fault is of such a nature that some other protection system is intended to affect tripping. Additionally, there is also the potential for normal system transients, such as motor starts and transformer magnetising inrush currents, for which the protection system should not operate. Thus, stability in this context is related to discrimination, and a stable protection system would never operate when it should not.
- **Sensitivity** is a measure of the ability of the protection system to identify the presence of a fault or other undesirable condition, even though that condition may be only slightly different from an apparently healthy condition. For example, if a protection scheme is based on the measurement of primary system current, then a sensitive protection scheme would operate for currents slightly higher than the rated system current, whereas a relatively insensitive protection scheme would only operate for currents approaching the maximum possible fault current levels for that particular part of the power system. A sensitive protection system always operates when it should.
- **Operating time** is the total time taken from the onset of the fault to the protection relay sending a trip signal to the circuit breaker(s). It is not possible to state simply that low operating times are good and high operating times are bad, because high and low are relative terms in this context. All operating times must be low enough to ensure the safety of plant, equipment and personnel, but the use of intentional time delays in the operation of the protection system offers useful means of discrimination in certain applications (described later in this section).

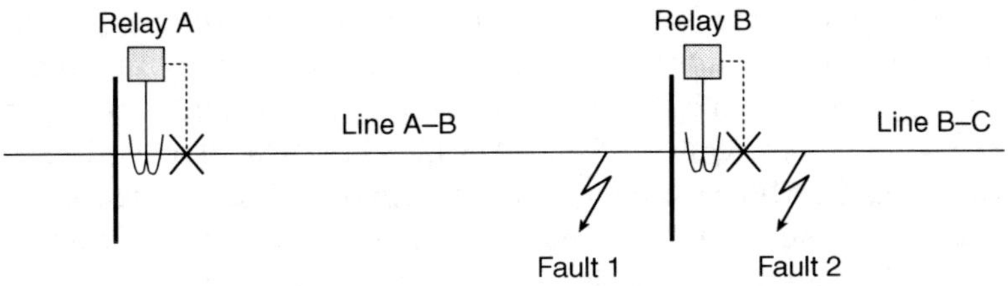

3.4 Protection arrangements for a multi-section radial feeder.

Figure 3.4 can be used to clarify the concepts of discrimination, stability and sensitivity. The figure depicts two serially connected lines, each protected by relatively simple overcurrent protection systems, which would be typical of a distribution system protection arrangement. Also illustrated on the figure are the components of each protection system, which in this case comprises a measurement (current) transformer, protection relay and circuit breaker which is operated by a signal from the relay upon detection of a fault that it should clear.

For the faults shown in the figure, the *sensitivity* of both protection systems must be high enough to detect faults on their protected feeders, but not so sensitive that it will operate incorrectly for temporary overloads or short term transient overcurrents, e.g. due to motors starting or transformer inrush. Furthermore, in the case shown in the figure, the relay at A must be sensitive enough to detect faults on both feeders, as it must operate (albeit after a time delay) if a fault on Line B-C is not cleared due to the failure of the protection system at B.

Ensuring adequate sensitivity for a protection system can be challenging. For example, a resistive fault, which may be encountered when the fault current returns to the source via vegetation that has impinged on the live conductor and caused a connection to earth, may result in fault currents that are not much higher (and indeed in some cases may be lower) than maximum load currents. In such cases, alternative methods of protection, such as detection of imbalance between the three phases, may be required to facilitate detection of these types of faults.

Discrimination and stability can be explained by considering the faults at Locations 1 and 2. Fault 1 should be cleared by the relay at A, while Fault 2 should be cleared by the relay at B to maximise supply availability to consumers and isolate only the faulted element of the system. The relay at A must be able to *discriminate* between the faults at 1 and 2: for Fault 1, the relay at A should operate relatively quickly, while for Fault 2, it should only operate after a time delay, during which, if everything functions correctly, the relay at B will clear the fault and the relay at A will not operate, but will reset after clearance of the fault by the relay at B.

Similarly, for Fault 1, the relay at B must remain *stable*, and never operate. This should not be difficult in this case, under the assumption that fault current flows only from left to right on the system depicted in the figure, that is, the source is 'upstream' of Relay A, which is typical of a distribution system containing no DG. However, if the system were interconnected and/or had DG, then the relay at B may be required to operate in such cases.

Stability and discrimination can also be challenging. For example, if the fault is very close to Location B (e.g. a few metres either side of it), then the questions of how both relays can discriminate between these locations, if they are only measuring current magnitude, and how can Relay B remain stable if the fault is just 'behind' it, must be addressed. Furthermore, other phenomena, such as large motors starting and transformers drawing large inrush currents upon energisation, can also present challenges to the discrimination and stability of protection systems.

The operating time, sensitivity, discrimination and stability performance characteristics of each of the most popular methods of protection are discussed in more detail later in this chapter.

3.3.1 Economic considerations

Protection systems can be relatively simple and cheap, or they can be extremely complex and expensive. The choice of the complexity (and hence cost) of the protection system to be applied to a particular item, or items, of a power system plant depends mainly on two factors:

- the cost of faults;
- the desired level of supply security.

Fault costs are a combination of the potential cost of damage to plant, the cost of lost revenue through supply disconnection, the cost of replacement plant (e.g. generation) and the cost of loss of consumer goodwill. Economic considerations dictate that the higher the fault costs, then the greater the expense invested in providing adequate protection. Generally this means that the higher the mega-volt-amperes (MVA) rating of the plant to be protected, the more complex and costly will be the protection system. This general trend is evident by considering the protection schemes applied at various levels in the power system, as summarised in Section 3.6.1.

3.4 Protection system components and philosophies

A protection system consists of a number of individual components and these components are normally arranged such that the protection scheme behaves in accordance with one of two main operating philosophies. The components and philosophies are described in the remainder of this section.

3.4.1 Overview of protection system components

As illustrated earlier in this chapter in Fig. 3.3, the majority of protection systems contain the following main subsystems:

- equipment for measuring/monitoring the power system conditions (e.g. CTs and VTs);
- equipment for translating the signals derived by the measuring equipment into decisions on the state of the power system and subsequently taking actions if fault or undesirable conditions are deemed to exist (i.e. protection relays);
- circuit breakers and other switchgear, which disconnect and isolate the faulty item(s) of plant in response to signals from the protection relays;
- communications systems for acquiring measurement data, sending tripping signals, and for communicating with remote locations for data/information transfer.

3.4.2 Protection system philosophies

There are two main protection philosophies to which all protection systems adhere.

Unit protection

The philosophy of unit protection defines that the protection system should only detect and react to primary system faults within the zone of protection, while remaining inoperative for external faults. The protection scheme illustrated in Fig. 3.5 represents a simple unit protection scheme. Unit schemes

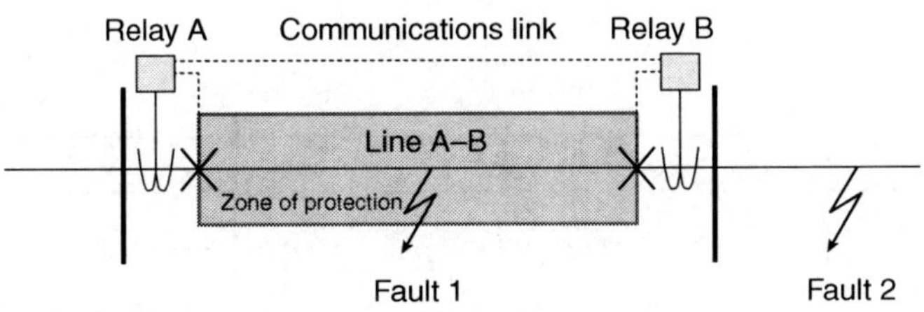

3.5 Unit protection system (current differential protection).

typically involve protection relays that monitor the primary system conditions at each 'end', or boundary, of the protected zone. The relays measure some parameter (normally current – in terms of magnitude, phase, or both) and perform a comparison with the parameter(s) being measured by the other relay(s) within the unit scheme. If some threshold criterion is violated, for example the measured currents are not equal or the vector summation of the measured currents is not equal to zero (neglecting capacitive charging currents associated with the protected line), then the protection relay(s) initiate the process which will lead to isolation of the plant within the protected zone. Because of this requirement for comparison (or some other function) of parameters from each 'end' of the protected zone, almost all unit protection schemes have a requirement for relay-to-relay communications facilities, which may be achieved using a number of methods.

For the system above, the zone of protection is clearly defined. To be exact, the zone of *detection* is between the measurement points, while the zone of *protection* is between the circuit breakers; normally, the CTs and circuit breakers are at almost identical locations. If, for example, each relay in the system above was comparing the magnitude of its measured current with the other relay's measured current (a form of current differential protection), then for Fault 1, the currents would not be equal and the relays would trip.

For the case of Fault 2, while both measured currents would be far in excess of normal load current levels, they would still be equal, and the protection relays should remain stable. Unit protection systems offer high levels of discrimination and stability, ensuring that the protection system only operates for faults within the protected zone, while remaining are inoperative for 'external' faults. The main negative aspect associated with unit schemes is that they do not possess backup protection capabilities, and there is a considerable expense associated with the use of communications. Furthermore, reliance on communications for operation gives rise to concern over the reliability of the communications link, which is sometimes addressed through the use of multiple communications systems, further increasing costs. The

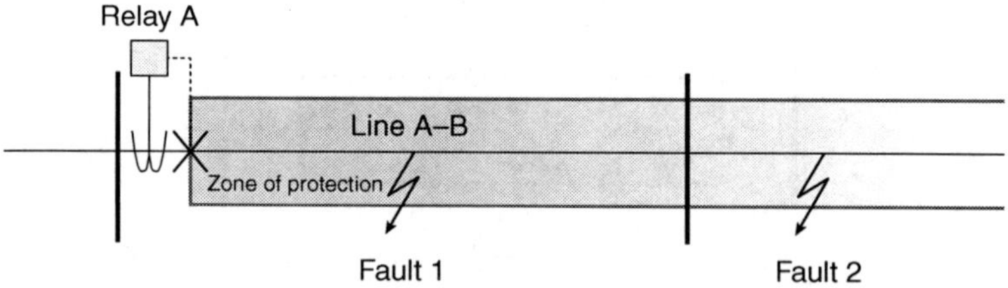

3.6 Non-unit protection system (overcurrent protection).

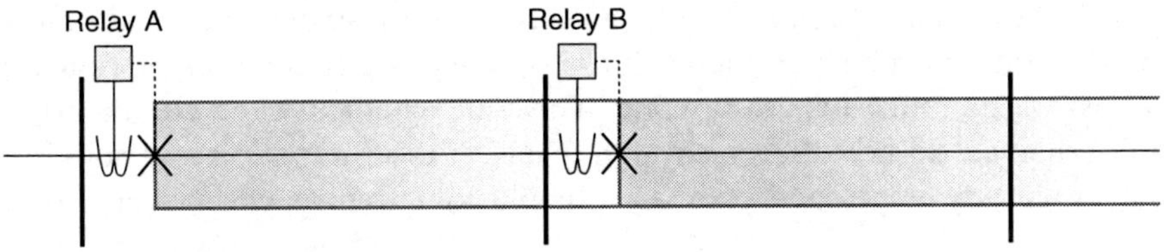

3.7 Overlapping zone of protection in a non-unit arrangement (overcurrent).

lack of backup capabilities is usually addressed through the use of non-unit schemes in addition to the unit scheme.

Non-unit protection

The protection scheme arrangement shown in Fig. 3.6 represents a simple non-unit protection scheme. The main difference between unit and non-unit schemes is that individual non-unit schemes do not independently protect one clearly defined part (or zone) of the system.

In Fig. 3.6, the zone of protection (as illustrated by the depth of the shading within the zone), certainly covers Fault 1, but in this case, the 'zone' gradually 'fades' on the second line and, for Fault 2, it appears that the protection may provide coverage, but in the case above, this is not certain. The 'reach' of non-unit schemes can be varied by altering the settings on the relay (more details are included later in this section of relay settings), but non-unit schemes invariably exhibit characteristics whereby the reaction of the protection system varies as the location of the fault changes. In the example above, if the protection relay were an overcurrent relay, then it would operate quickly for Fault 1, but with an increasing time delay as the fault location moved further along the system to the right, operating with a relatively longer time delay for Fault 2, and ceasing to operate as the fault moved further along Line 2.

Impedance, or distance protection, is also a non-unit scheme, but rather than a continuously decreasing operation time, it exhibits a stepped characteristic, operating with fixed delays as the fault position moves from one (not exactly defined) zone to another in terms of its distance from the relay's measurement point (more on this later in this section).

Adjacent non-unit protection schemes on an interconnected power system have an element of overlap with respect to their respective zones of protection as shown in Fig. 3.7.

This overlap is useful for providing backup protection in the event of failure of one element of the overall protection system. However, the criterion of selectivity, or discrimination, must still be satisfied and the non-unit scheme(s) closest to the fault should always trip before any other non-unit schemes, which must remain stable until the closest schemes have operated, only operating (in backup mode) if the primary protection (i.e. the relay(s) closest to the fault) fails to operate. Non-unit schemes, when compared to unit schemes, do not offer such high levels of discrimination and stability, but, as already mentioned, provide valuable backup protection capabilities.

3.5 Overview of protection techniques

Using measurement inputs from CTs and VTs, protection relays detect the presence of a fault on the system using a number of techniques. From the previous discussion of unit and non-unit schemes, it is clear that both categories of protection are required. Both unit and non-unit are used (in parallel) at transmission levels to provide high levels of discrimination and stability from the unit schemes, with backup being provided by non-unit schemes, although it is important to emphasise that non-unit schemes can also operate extremely quickly, depending on the settings employed. At distribution and consumer voltage levels, non-unit schemes are normally employed.

Each of the three main categories of fault detection and protection that are used to protect networks and components are summarised below, with descriptions of their applications and operating characteristics in terms of operating time, discrimination and stability.

3.5.1 Overcurrent protection

As the name suggests, this method is based on measurement of current magnitudes, and a fault may be deemed to exist when the measured current exceeds a pre-determined threshold level. It is a non-unit scheme as it does not have exact boundaries defining its zone of protection. Overcurrent relays may respond with different time delays, i.e. the delay from initial fault detection until a tripping command is issued to the associated circuit breaker(s), for different levels of measured fault current. This characteristic is normally used to ensure that relatively fast reactions and tripping are achieved for faults closer to the measurement point (indicative of a fault relatively close

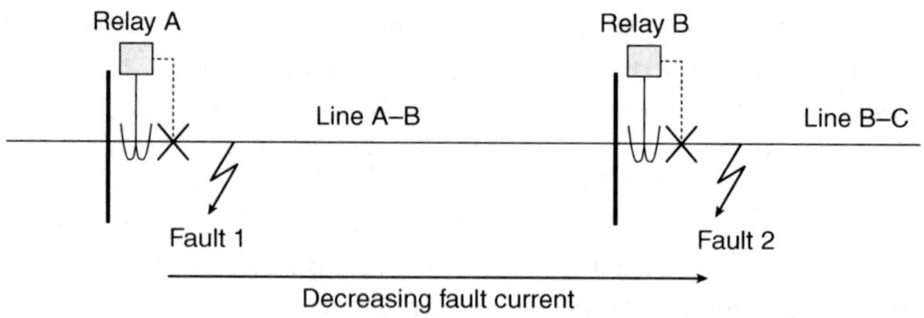

3.8 Overcurrent protection for two serially connected feeders.

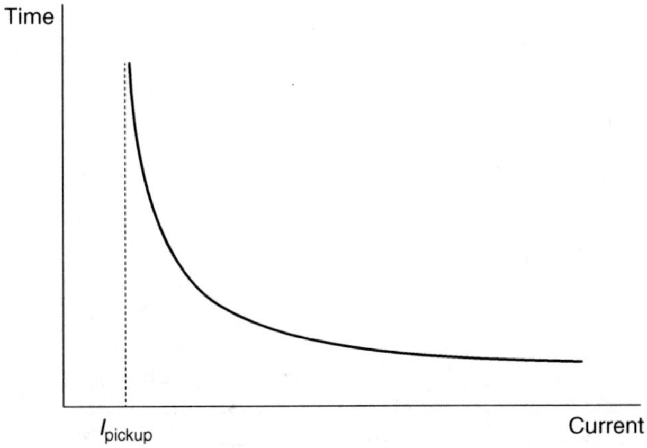

3.9 Simplified time-current characteristic of standard inverse overcurrent protection.

to the relay), with a longer time delay before tripping for lower values of measured current (still above the threshold, so indicative of a fault, but perhaps further away from the relay location).

The slower response time normally allows another relay (or other relays) situated closer to the fault to trip its breaker first, upon which the relay(s) operating with longer time delays will observe the drop in current due to fault clearance and reset. Several relays can therefore be configured to provide primary and backup protection within a network. Consider the section of system illustrated in Fig. 3.8.

If it is assumed that Relays A and B are standard overcurrent relays, then they will have operational characteristics similar to that shown in Fig. 3.9 below. This time-current characteristic is standardised according to, for example (IEEE, 1997), and various shapes of standard characteristics, with accompanying equations, are available. An approximated view of a standard inverse characteristic is shown in Fig. 3.9.

As shown in Fig. 3.9, the relay will begin to operate when the measured current exceeds the threshold (or 'pickup') current, and the time of operation will decrease according to the inverse characteristic as the measured current increases. Accordingly, such relays are extremely useful for distribution system applications, where the fault current magnitude generally decreases as the location of the fault moves further away from the source(s) of fault current, which are typically the transformer(s) that supply a section of the distribution network from the transmission system.

The operation of overcurrent relays can be influenced via two configurable settings. The first is the 'pickup' or 'plug' setting, which modifies the level at which the relay will begin to operate, effectively shifting the

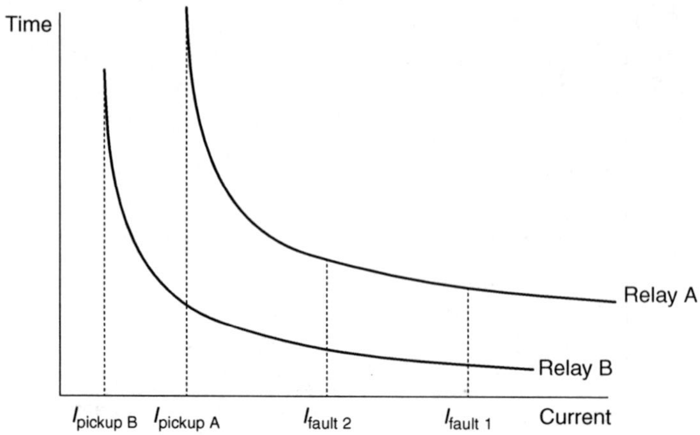

3.10 Time-current characteristics with settings and fault currents for two coordinated overcurrent relays.

characteristic curve in the horizontal plane in Fig. 3.9. The second setting is the 'time' or 'time multiplier' setting, which modifies the time that the relay will take to trip for a given input current that exceeds the pickup setting.

Using these settings, and the fact that fault current decreases within a distribution system as the location of the fault moves further from the source(s) of fault current, a number of overcurrent relays can be configured to protect a series of feeder sections effectively, with the closest upstream relay acting to isolate faults on its protected feeder, while other upstream relay(s) can be configured to operate with a relatively longer time delay in order to provide backup should the relay that is intended to clear the fault fails to operate.

Referring once again to Fig. 3.8, Relays A and B would have settings that would result in the characteristics as shown above in Fig. 3.10, where both relays' characteristics are plotted on the same graph, along with indicative fault current magnitudes (on the x axis) associated with Faults 1 and 2 on Fig. 3.8.

Referring to the Fig. 3.10 and considering the faults at Locations 1 and 2 as shown on Fig. 3.8, it is clear that the current associated with Fault 1 is relatively higher than the current for Fault 2, due to the increased impedance between the source(s) of fault current and the fault location for Fault 2. If the relays' settings are correctly configured, then the characteristic for Relay *A* (the upstream relay) will always be 'above' the characteristic for Relay B (and any other protective devices situated further downstream) when plotted on the same time-current graph, which is usually represented using a log-log scale. The characteristic curves of relays should never intersect; if they do, then proper coordination for all fault situations cannot be guaranteed. In such cases, both Relays A and B may operate, or Relay A may operate before Relay B for a fault downstream from Relay B, which are obviously undesirable situations.

The general guidelines for the setting of overcurrent relays are to ensure that the relays will only begin operating at a current greater than 125–150% of the protected line's rating, with the fastest operating time possible being desired for the furthest downstream relay, taking into account that it may be required to coordinate with other downstream protection devices (for example, protection on the low voltage (LV) side of a distribution transformer, consumer device protection systems, etc.).

The major benefit of overcurrent protection schemes is that relatively fast clearance is provided for faults, but a drawback is that, particularly for multi-section feeders employing several relays, the operating time of the protection increases as the fault location moves closer to the source; this drawback is compounded by the fact that such faults involve relatively higher fault currents. Consequently, faults closer to the source, involving relatively higher fault currents than for faults at downstream locations, will remain on the system for relatively longer durations before being isolated, which is obviously undesirable from the perspectives of the increased risk of permanent physical damage at the point of the fault and the increased duration of undervoltages and possible disruption to consumers' supplies in the system local to the fault location.

To mitigate this problem and enhance the overall operation of the protection scheme, inverse overcurrent protection is often used in conjunction with 'instantaneous' (or 'high set') protection. This function is often embedded within the same protection relay device as the overcurrent function, and is set to operate instantaneously for fault currents that are indicative of faults that are definitely on the protected line (and not on the next line). Such schemes are only applicable to systems where there are relatively large differences in fault level between adjacent relay locations. The pickup setting of instantaneous overcurrent relays (which, for obvious reasons, have no time setting) is typically set to a value of approximately 130% of the fault current at the next downstream relay's location; this setting being selected to ensure that the instantaneous element will never operate for faults on the next line. More information on this is contained in Chapter 9 of Alstom (2011).

Figure 3.11 shows the arrangement of Fig. 3.10, but with an instantaneous element being deployed in addition to the inverse element within Relay A.

In this case, faults with a current higher than the setting (A_{inst}), equating to all faults located to the left of '*Relay* A_{inst}' on Line A–B in Fig. 3.11 (assuming bolted short circuits with zero fault impedance) will lead to instantaneous operation of Relay A, and will not result in delayed operation, as would be

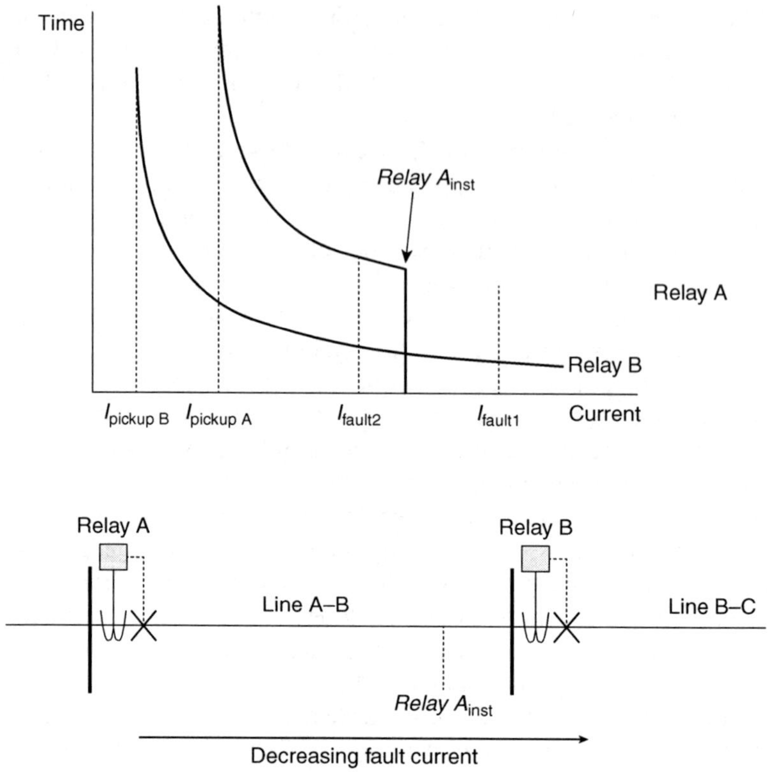

3.11 Illustration of operation of instantaneous overcurrent protection.

the case if the instantaneous element was not used. This addresses the aforementioned problem of long time delays in operation for faults closer to the source, although this method of setting the instantaneous element means that the relay at A will still operate with a time delay for faults towards the end of the protected line (i.e. between the location of *Relay* A_{inst} and the substation at B, with a time delay that is shown on the top half of Fig. 3.11 as the segment of Relay A's time-current characteristic curve between I_{fault2} and the point at *Relay* A_{inst}).

Distribution networks can be complex and the volume of circuits and economic considerations dictate that overcurrent protection relays and circuit breakers cannot be used to protect every section of the network. Figure 3.12 illustrates a typical section of UK distribution network and indicates the protection arrangements used.

In Fig. 3.11, overcurrent relays are used at the head of each feeder where it connects to the main 11 kV busbar at the primary 33/11 kV distribution substation (R-A, R-B and R-C in the figure). These are coordinated with

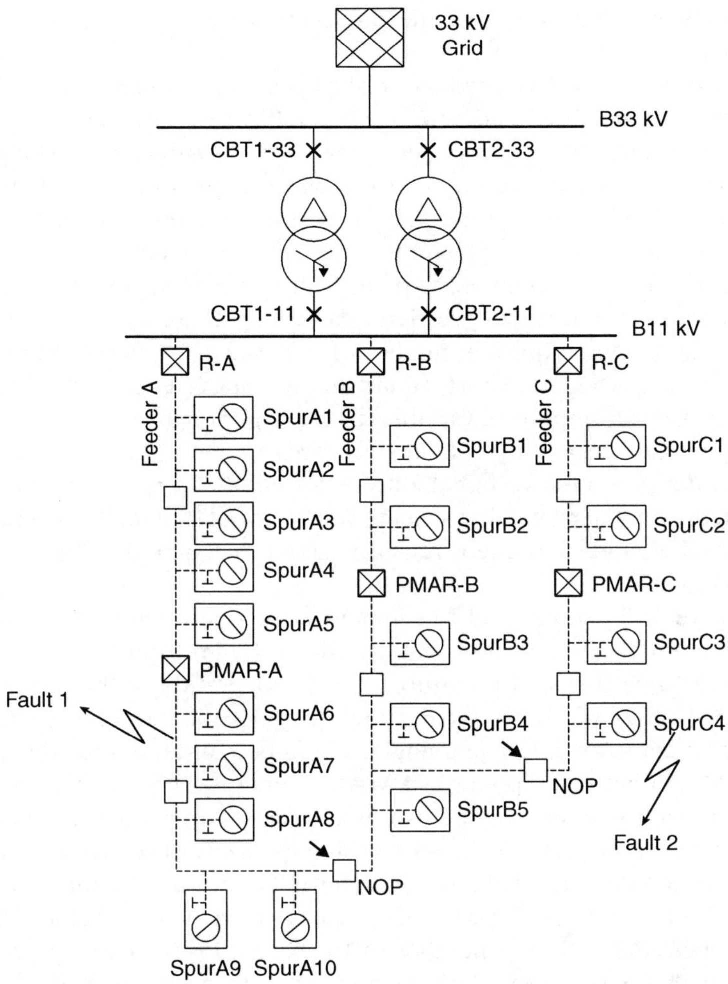

3.12 Typical section of UK distribution network with protection arrangements.

the downstream pole mounted auto recloser (PMAR) devices (PMAR-A, PMAR-B and PMAR-C in the figure) and with the protection on the spurs, which may be either by fuses sized to coordinate with upstream protection, or by section switches, which often have embedded software to detect the flow of fault current and can provide isolation of faulted sections of the network by opening (while an upstream breaker is open) after a predetermined number of auto-reclose attempts have been detected. There are also switches (shown as empty boxes in the figure) which are not capable of interrupting fault current, but that can be used to alter the configuration of the network. The 'NOP' switches in the network indicate 'Normally Open Points'.

Auto-reclose is often used on overhead networks (deployed with both the relays at the heads of feeders and with the PMARs) as many faults are temporary in nature, for example being caused by conductors clashing in high winds, lightning induced flashovers to earth, vegetation or animals causing a connection between phases and/or between phases and earth. Once the flow of fault current is interrupted, the arc or connection that caused the fault will not exist upon subsequent restoration of supply, and the system is returned to normal service. Some faults (e.g. those caused by animals or vegetation) may require a number of reclosures to effectively burn away the debris causing the short circuit, and for this reason, multi-shot reclose schemes, often employing variable (increasingly longer) durations of reclose attempts, are used to affect fault clearance.

If, after a number of failed auto-reclosure attempts, the system remains faulted, then the protection scheme will 'lock out', the fault is deemed to be permanent, and subsequent remedial action is required before the system can be restored.

Figure 3.12 can be used to summarise the operation of an auto-reclose scheme for two scenarios. If Fault 1 (downstream from PMAR-A) is considered to be transient in nature, then the operation of the system is simple. PMAR-A will quickly trip to interrupt the flow of fault current, with overcurrent relay R-A 'beginning' to operate (in order to provide backup to PMAR-A) but not tripping, as PMAR-A operates before this can happen. As the fault is transient, and assuming it clears after the first interruption of fault current, then PMAR-A will simply reclose successfully after a predetermined time delay and the system will be returned to normal service.

If Fault 2 (on Spur C4) is assumed to be a permanent fault, then PMAR-C will open and reclose a number of times. If a fuse is used to protect the spur and it is correctly sized, then it will melt during one of the (failed) reclosure attempts while fault current flows, and the subsequent reclosure will be successful as the fuse at C4 will have isolated the permanent fault on the spur. As already mentioned, many network operators are now using 'smart' section switches instead of fuses, and if one of these was used to protect C4 instead of a fuse, then it would use its internal logic to open after a pre-determined number of failed reclosure attempts (during a period when PMAR-C is opened), isolating the faulted element and allowing the subsequent reclosure to be successful. If for some reason the fuse/section switch failed, then the PMAR-C would simply lock out after its maximum number of reclosure attempts had been exhausted. Gers and Holmes (2004) contains an excellent overview of the protection of distribution networks.

In transmission systems, overcurrent protection relays are normally used as a 'last line of defence' backup, and they are not normally required to coordinate with other relays in the system, being configured with settings that will allow for clearance of faults on the feeder, normally in less than 1 s, at minimum fault level conditions, e.g. a fault at the remote end of the feeder with minimum infeed conditions.

Overcurrent protection is used to provide backup protection for transformers, generators and busbars in transmission systems, and as the main means of protecting these components in distribution systems.

3.5.2 Impedance or distance protection

This method of protection is based on measurement of currents and voltages, in terms of both magnitudes and phases. It relies on the fact that the measured impedance (calculated from the measured voltages and currents) will drop if a fault is on the system, and if the measured impedance is below certain thresholds, it can be deduced that there is a fault close to the relaying point that may require tripping.

As with overcurrent protection, distance protection is classed as a non-unit scheme, as it does not have exact boundaries defining its zone of protection, although communications facilities can be used to enhance the performance of distance schemes, effectively changing the distance scheme from being non-unit to unit in nature, with the zone of protection clearly defined as the element(s) of the primary between the communicating relays. Normally, in modern numerical relays, some form of Fourier transform is used to allow the relative magnitudes and phases of the measured voltages and currents to be calculated and therefore a measure of the complex system impedance from the perspective of the relay's measurement point can be made available.

Using knowledge of the protected lines' impedances, which are used to calculate the settings of the relay, the measured value of impedance can be used to ascertain whether there is a fault on the system, and if so, the approximate location of (or distance to) the fault with respect to the relay's location. For example, if the impedance measured by a distance relay protecting a single transmission line equates to 90% of the (known) line impedance, then it can be deduced that there is a fault on the line, located at approximately 90% of the length of the line from the perspective of the measurement end, assuming a fault with zero fault impedance has occurred.

Measurement errors and variability in fault resistance mean that there will always be a degree of uncertainty associated with the exact location of faults, and for this reason, distance relays are normally set to react instantaneously to faults that have an impedance of less than 80% of the protected line's impedance. The 80% figure, as opposed to 100%, arises from assuming maximum errors of 5% in the current and voltage measurements, a 5% error in the relay's computations, and a 5% error in the impedance data

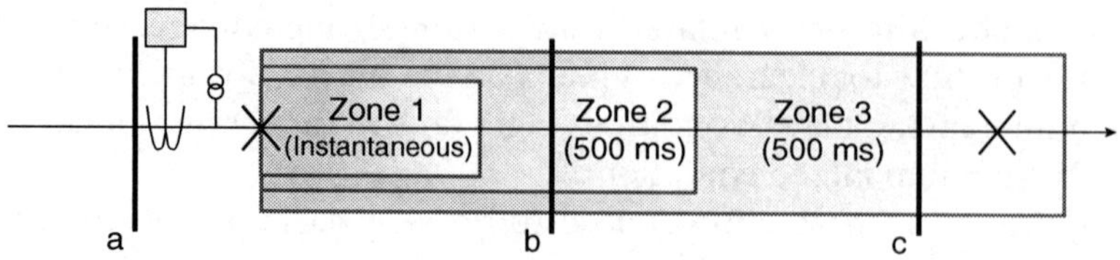

3.13 Zones of protection of distance protection with indicative time delays of operation for faults detected in each zone.

for the line, which result in a cumulative maximum error of 20%, hence the 80% figure used.

Distance relays normally have a number of individual settings that can be used to protect different elements of the system. Typical settings employed are 80% of the main protected line's impedance for Zone 1 (instantaneous operation), 125–150% of the line's impedance for Zone 2 (delayed operation – e.g. 500 ms) and 200–250% of the line's impedance for Zone 3.

Care must be taken with settings to ensure that the 'reach' of the relay does not extend into the distribution system through the distribution transformers in a transmission line protection application. For example, if the lines that are connected in an onward fashion from the main protected line are very short, or if there is a transformer feeding the distribution system connected at the end of the protected line, then the Zones 2 and 3 settings may require to be reduced to prevent the protection system operating incorrectly for faults on the distribution system, or faults located on the third or fourth line 'away' from the relaying location (which should never be reacted to).

Figure 3.13 presents a simple distance protection scheme in terms of its zones of protection and the time delays associated with operation of the relay for faults detected in each zone.

In the above example, if the transmission line being protected has a complex impedance of $8 + j20\ \Omega$ (equivalent to an impedance of 21.5 Ω magnitude with an angle of 68.2° in polar form), then the zone settings, assuming the settings policy is to set the zones to 80%, 125% and 220% reach settings, would be $6.4 + j16\ \Omega$, $10 + j25\ \Omega$ and $17.6 + j44\ \Omega$, respectively.

Figure 3.14 below illustrates the zone boundaries for each of the three zones in the complex impedance plane. Assuming a fault with zero fault impedance occurred at Location B, then the impedance measured at the relay (located at A) would be $8 + j20\ \Omega$ and the relay would therefore trip in Zone 2, with a 500 ms time delay. The reason that the zone boundaries are circular is historical. When electromechanical relays were used, they effectively compared only the magnitudes (and not the relative phases) of the measured currents and voltages, and thus could only determine the

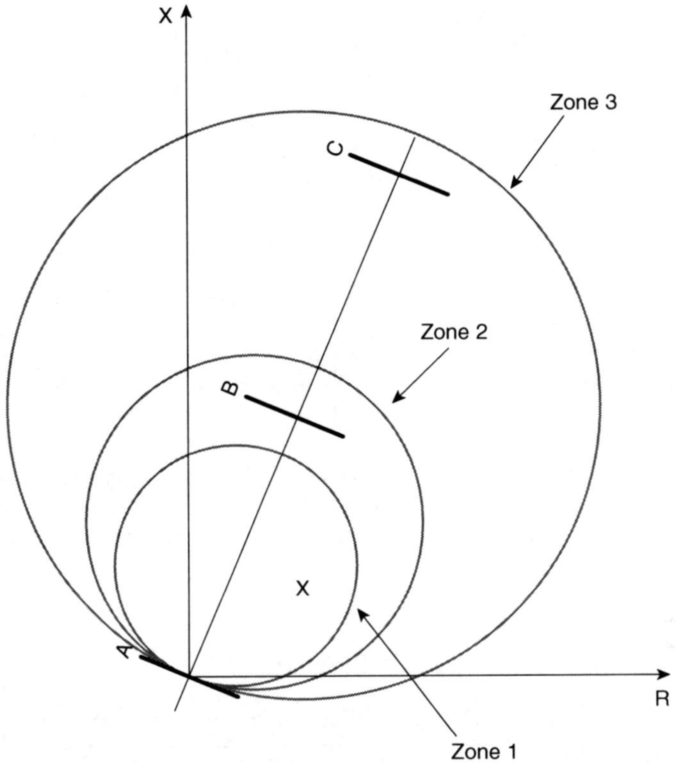

3.14 Illustration of distance protection zone boundaries in the complex impedance plane.

magnitude of the impedance, which equates to a circle when plotted in the impedance plane as shown in Fig. 3.14. This circular characteristic is also a useful feature, as resistive faults can also be detected to a certain degree. For example, if a fault in the middle of Line 1 had a resistive element, then the locus of the measured impedance may lie in the vicinity of the cross indicated in Fig. 3.14, and would still be detected. Modern microprocessor relays can of course determine the real and imaginary components of the measured impedance by computing the magnitudes and relative phases of the measured voltages and currents, so in theory any shape of characteristic is achievable using software. Figure 3.15 illustrates a selection of common impedance characteristics that are available in modern distance protection relays.

To overcome the major shortcoming of distance protection, which is the fact that, for reasons of security, only 80% of the line is protected (with an instantaneous tripping reaction), a number of methods can be used. All

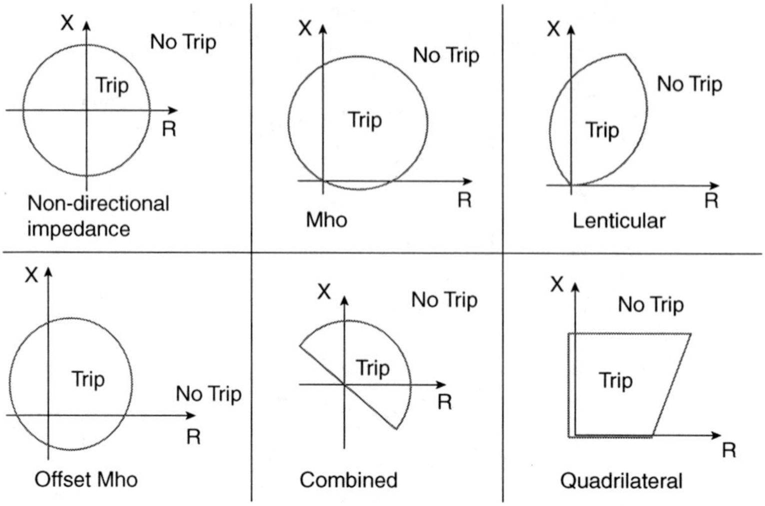

3.15 A selection of commonly used distance protection characteristics.

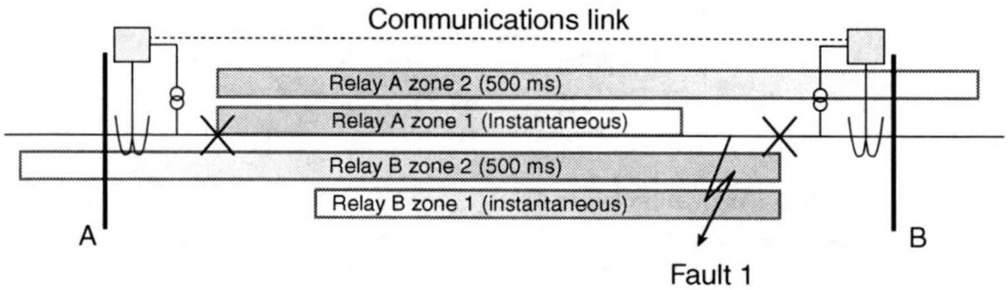

3.16 Distance protection scheme employing communications to enhance performance.

methods require that there are distance protection relays at each end of the protected line, both looking 'into' the line, and communications are used to enhance the performance of the overall scheme. For the purposes of this book, only one method will be presented, although there are several methods in use.

The particular method described here is referred to as 'accelerated' distance protection, or as a 'permissive under-reach transfer tripping scheme'. The basic configuration and operation of the system is illustrated below in Fig. 3.16.

For the configuration above, the fault lies within the protected line, but is detected by Relay A in Zone 2, and by Relay B in Zone 1. The acceleration scheme operates through each relay sending an 'acceleration' signal to the other upon local detection of a fault in Zone 1. Each relay monitors its communications link for receipt of an acceleration signal from the other relay(s). If any relay detects a fault in Zone 2, *and* the acceleration signal is received, then it overrides the time delay normally associated with Zone 2 and trips instantaneously, thereby speeding up the clearance of faults located in the 'last' 20% of the line end that the relay (that receives the acceleration signal) is protecting.

This (Zone2 *and* acceleration signal receipt) logic is used to prevent spurious operation (for example if an acceleration signal was incorrectly sent by one relay) and thereby increases the security of the system.

As already mentioned, there are other methods of using communications to enhance the operation of distance schemes, and more information relating to such schemes can be found in Chapter 12 of Alstom (2011).

Distance protection is used as one of the main means of protecting transmission lines and to provide backup protection for transformers, generators and busbars in transmission systems, and in some cases as a means of providing main and backup protection for distribution systems.

3.5.3 Differential protection

This method of protection relies on the continuous comparison of measurements (normally of current, but in some cases voltage is used) to establish whether there is a fault on the protected equipment. It is a unit scheme, highly discriminative and stable, but does not possess any inherent backup capabilities, although modern multi-functional relays may also include overcurrent protection for backup purposes. Differential protection relies on communications between relays to perform the protection function.

The measurements are taken from the boundaries of the protected zone (i.e. the ends of a protected line, the terminals of a protected busbar, the high

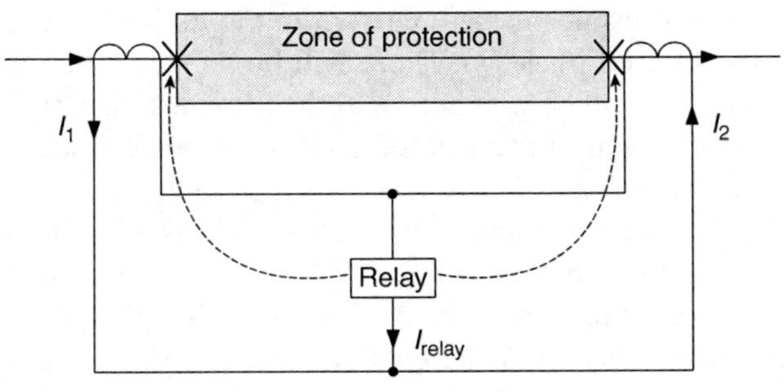

3.17 Single-relay current differential protection scheme.

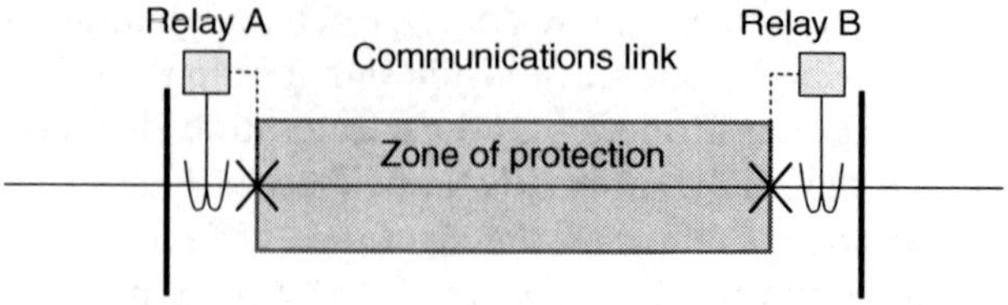

3.18 Dual relay current differential protection scheme.

voltage (HV) and LV terminals of a power transformer, the ends of a protected generator winding, etc.) and compared. If a difference in magnitude and/or phase is detected between the measurements, then it is deemed that there is a fault in the protected equipment and tripping is instructed. In contrast with non-unit schemes, there are no deliberately applied time delays to the operation of differential schemes and such schemes can normally affect tripping very quickly.

Differential protection may be implemented as a single-relay scheme, as shown in Fig. 3.17, which may be applied to the protection of single items of plant (e.g. busbar, transformer, generator) or relatively short lines, and uses direct connections between the relay and the CTs. Multiple relay schemes, as shown in Fig. 3.18, are typically used for the protection of relatively longer lines, where direct connection of CTs would not be feasible.

The principles of operation of both schemes shown in Figs 3.17 and 3.18 are identical; only the implementation differs. In the single-relay scheme, known as a circulating current scheme, it is clear that the current presented to the relay, I_{relay}, is the vector sum of I_1 and I_2, which is zero for no-fault situations and for external faults, but non-zero for internal faults (referring to Fig. 3.17, $I_{relay} = I_1 - I_2$).

For an external fault, the values of $I_1$and I_2 will be much larger than normal, but still equal in magnitude and phase, so the condition $I_{relay} = I_1 - I_2 = 0$ will remain satisfied. For an internal fault, the values of I_1 and I_2 will be different. If there is only one source of fault current (e.g. to the left of the figure), then I_1 will represent the fault current in the faulted phase(s), while I_2 will be 0 on the faulted phase(s). There will be a clear difference in the measured currents, I_{relay} will be non-zero and the relay will trip. If the system is interconnected and fault current is fed from both ends of the protected zone, then the values of I_1 and I_2 will again be different, but I_2 will be non-zero and will be 180° out of phase with I_1. So again there will be a clear difference in the measured currents, I_{relay} will be non-zero and the relay will trip. For the arrangement presented in Fig. 3.18, the principle is the same, but rather than the relays measuring the currents directly, the values are measured locally and communicated, as encoded messages, to the other relay(s) in the scheme for comparison.

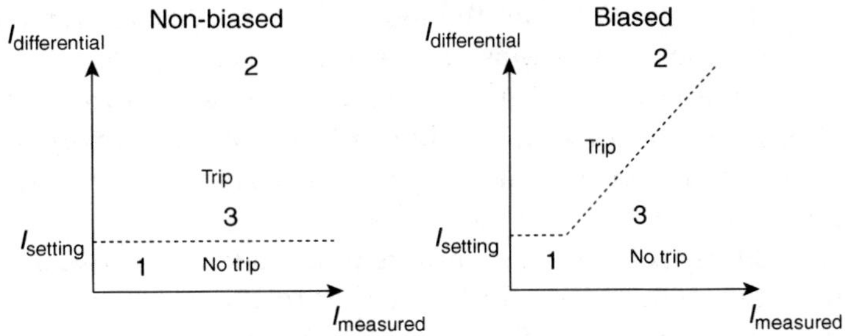

1: No fault (load current) condition
2: Internal fault condition
3: External fault condition

3.19 **Use of biasing to enhance the stability of differential protection.**

Some differential schemes compare only current magnitudes, some compare magnitudes and phases, and others compare only the phases of the measured currents.

The main area of concern associated with differential protection is the potential for incorrect operation of the scheme for an external fault condition, when the currents measured by each CT are much higher than nominal values. The potential for CT errors is greater at such levels of primary current, leading to the potential for a differential current to be incorrectly detected and for the scheme to operate for external faults. The method of mitigating this risk is known as biasing, and this effectively means that the differential current required for operation is increased when the absolute levels of measured current are higher.

Consider Fig. 3.19, which illustrates the concept of biasing. As stated above, the amount of differential current to cause tripping is higher at higher values of measured current and this is illustrated by the sloped boundary characteristic dividing the 'trip' and 'no trip' regions on the right hand side of the figure. Consider the three cases (no fault, internal and external fault), numbered as 1, 2 and 3, respectively on Fig. 3.19, it is clear that the external fault leads to an erroneous measurement of differential current (due to CT errors at such high levels of current) and this would cause tripping in the non-biased case. Introducing the bias eliminates the risk of such tripping. Biasing can be introduced via modern relays' software algorithms. In electromechanical relays, biasing was achieved through passing the measured current through a winding, which acted to restrain the relay from operating as the measured current levels increase (i.e. it was in direct opposition to the coil carrying the differential, or operating, current).

It should also be noted from the figure that there is an element of the characteristic which is not biased (towards the origin of the graph on the right hand side of the figure). This is necessary so that very small values of differential current, e.g. due to line charging currents or small CT errors at rated load currents, will not cause spurious operation of the protection.

The main settings on a modern differential relay are the differential current setting (the value of differential current for the non-biased portion of the characteristic), the value of measured current at which the biased (sloped characteristic) will be introduced (i.e. its starting point on the x axis in the figure) and a value which is used to specify the gradient of the biased characteristic (e.g. steeper slopes may be required if the CT errors are known to be relatively large).

Differential protection is used as the primary means of protection for transformers, generators and busbars in transmission systems (such components would be protected using non-unit schemes at distribution voltages). It is applied in distribution systems when the consequences of faults (if not cleared quickly) are deemed to be grave enough to warrant the expense of differential schemes, or the nature of the system dictates that differential protection is required. Further information relating to different protection can be found in Alstom (2011).

3.6 Typical protection schemes and further considerations

This section outlines the typical protection arrangements as applied in the UK at the various voltage levels.

3.6.1 Summary of typical protection schemes and applications

Consumer level (400 V in UK)

Fuse-based protection of individual items of equipment and circuits is used to protect the LV network to which domestic consumers connect. It should be noted that the fuse is a remarkable device, acting as a combined CT, protection relay and circuit breaker, while also limiting fault current.

Also employed at this level are miniature circuit breakers (MCB) with selectable overcurrent tripping characteristics. Residual current devices, which measure the live and neutral currents and trip when an imbalance is detected (indicative of an earth fault, where the supply current is returning via earth and not through the neutral), are extensively employed to protect consumers' circuits and individual devices. Backup is provided on a system-wide basis by adjacent non-unit protection devices.

Distribution level (11, 33 and 132 kV in UK)

Fuses remain common, particularly at 11 kV, for the protection of spurs connected to main feeders, although fuses are being replaced with section switches and other disconnecting devices, which provide disconnection (but not interruption of fault current) of the faulted section, often as part of a distribution automation scheme. For main 11 kV feeders radiating from 33/11 kV primary substations, overcurrent relays, often in conjunction with auto-reclosers and fuses and/or section switches, as described earlier in Section 3.5.1, are used to provide fault detection and interruption, with the aforementioned fuses and section switches providing isolation of the faulted network section if it is located downstream from these devices.

At 132 kV, which can be classed as either distribution or transmission network, the protection that is akin to that used for the protection of transmission networks is employed. Current differential and/or distance protection (sometimes employing communications) is normally used to provide main protection, with backup being provided by distance and overcurrent protection.

Transmission level (275 and 400 kV in UK)

The consequences of plant damage and loss of availability of part of the network are so high that complex and expensive protection schemes are typically used at transmission voltages. The protection must be fast in order to maintain system stability (this problem increases with system voltage and line MVA capacity) and must possess high levels of discrimination in order to minimise disconnection of non-faulted equipment and the risk of supply disruption. Every part of the transmission network is under the supervision of more than one protection system. Typically, two (or in some cases, three) main protection systems, usually one differential and one distance, are applied, with an additional backup protection system (or systems) also being used for the protection of a single item of transmission system plant. Backup may also be provided by dedicated circuit breaker fail protection, which checks for current flow after the protected circuit breaker has been instructed to trip, and if current still flows then the circuit breaker fail protection will directly trip all other circuit breakers required to provide isolation of the fault, using communications to trip remote circuit breakers.

In summary, for a fault on a transmission line or item of plant, at least two (and sometimes three) main protection systems will detect the fault and 'race' to trip the breakers, normally within 70–80 ms of the fault occurring. If a circuit breaker fails to open when instructed to do so by the main protection(s), then adjacent distance protection, backup overcurrent protection, and circuit breaker fail protection will all act to remedy the situation by tripping other circuit breakers, all normally within 500 ms or less. Using such redundant arrangements, the frequency of occurrence of catastrophic system failures due to non-operation is minimised (but such events, of course, occur on occasion).

3.7 Standard requirements for protection of generators and their interfaces to the utility network

In the UK, there are a number of standards relating to the connection of generators. The main document that contains recommended practice for DG is G59/2 (Energy Networks Association, 2011), which outlines a number of stipulations and recommendations for the operation, control and protection of generators (of different capacities and connected at different voltage levels) when connected to the electrical system of licensed distribution network operators (DNO) in the UK. Grid codes and other similar documents are used by power utility companies throughout the world to stipulate how generators (and consumers and other users of the system) must connect and operate.

From a protection perspective, the G59/2 recommendation concentrates on the protection of the interface between the generators and the DNO's system and recommends a number of protection functions and settings, which are summarised below:

- LOMs (using a variety of optional methods – voltage vector shift, rate of change of frequency, neutral voltage displacement)
- short circuit overcurrent protection and provision of backup protection for network faults
- reverse power protection
- under- and over-voltage and frequency
- phase unbalance protection
- requirements for synchronising.

In addition to the protection requirements summarised above, the document also contains information relating to connection applications and arrangements, earthing, the design of the connection, power quality, stability, operation in islanded modes, control, testing and commissioning.

In the UK, a primary objective is to ensure that DG does not operate in islanded mode, as this could lead to the operation of the islanded system with no intentional earth connection (and therefore no means of detecting earth faults), the possibility of reclosing of two non-synchronised systems, excessive variations in system fault levels, etc.

The G59/2 recommendation therefore specifies that some suitable form of LOM protection is provided, which is typically based on monitoring the system frequency to identify when there is a rate of change in the measured frequency that indicates that the system is no longer grid connected. LOM protection is important as, if the system is allowed to operate (inadvertently) in islanded mode, then the islanded element of the system may be out of synchronism with the main system, and if a reconnection of the (presumed non-energised) island and the main system is made, this could result in damage to the reclosing circuit breaker and/or the generator(s) in the island due to the unsynchronised reclose, presenting an obvious safety hazard. There may also be uncertainty surrounding the identify of all circuit breakers that would be involved in reconnecting the islanded part of the system to the main system, and installing synchronism-checking equipment on all circuit breakers would not be economically viable. Furthermore, an inadvertently islanded system may operate with generators that do not have a deliberate connection to earth, rendering the system unsafe as earth faults may not be detectable. Furthermore, the fault levels in an island may be markedly reduced, meaning that any faults in the island may not be cleared safely by the protection systems.

3.8 Future trends: Impact of distributed generation (DG) and storage on protection

The scope of this book and chapter does not permit a detailed treatment of the potential issues and problems that may be encountered as the penetration of DG and storage increases. However, a summary of the main issues, with references to guide the reader to more information on each of the potential issues, is included.

As the penetration of DG on distribution networks increases, it is important to ensure this will not adversely influence the operation of the network protection. Furthermore, the DG units themselves must also have the capability to 'ride through' network faults in certain circumstances. As mentioned in the previous section, there are several standards and recommendations, for example G59/2 in the UK (Energy Networks Association, 2011) and a variety of grid codes (Eirgrid, 2011; EON Netz, 2006) in Europe and elsewhere relating to protection of DG and the DG-utility interface. While some of these standards are defined and some are in the process of being refined, it is clear from the activities of the research and industrial communities that there are fears over the ability of future networks, incorporating large penetrations of DG (and energy storage), to be protected adequately, with specific concerns being expressed over network protection discrimination, coordination, speed of operation and stability (Brahma and Girgis, 2004; Dysko *et al.*, 2007; Laaksonen, 2010).

Furthermore, the protection of the DG interface is also subject to investigation, with researchers reporting problems relating to the ability of the DG interface to operate correctly under all encountered scenarios, with protection discrimination and selectivity, stability, ride through and reduction of unnecessary disconnection and finally, the ever-present concern over LOMs protection all being cited as potential problems (Jennett, Coffele and Booth, 2012). At the transmission level, increased connection of large wind farms, usually via some form of power converter interface, has been causing some concern over the impact that this may have on the stability of the system; for example, through reduced system inertia as mentioned in Section 3.2, which may require faster protection operation times in order to preserve system stability during and after fault events. Frequency based LOM protection may also suffer as a result of the increase in DG (and due to the increased connection of converter-interfaced wind energy at transmission levels). The reduced inertia of the system could lead to increased magnitudes of frequency perturbations during non-LOM events, leading to unnecessary operation of LOM protection. One solution may be to decrease the sensitivity of the LOM protection, but this could also cause problems in that the protection may not operate during actual LOM events, which is obviously undesirable. Increased penetration of renewables may also lead to system fault levels reducing (or in some cases increasing) from their present levels, perhaps requiring protection settings to be revised or adapted.

Figure 3.20 illustrates a number of the protection-related issues that may be encountered through the increased penetration of DG and, to a certain extent, storage, if the storage unit acts to contribute significantly to fault current.

Referring to the figure below, the fault at the position indicated would, without DG being connected, result in fault current that would be supplied solely from the grid, and this current would be limited in magnitude by the source impedance and the impedance of the path to the fault. The overcurrent protection at B would operate to clear the fault, with the protection at A providing backup if required.

However, the presence of DG at B and E as shown may increase the total fault current, perhaps speeding up the operation of the protection at A to levels where coordination may become an issue. The increased fault current

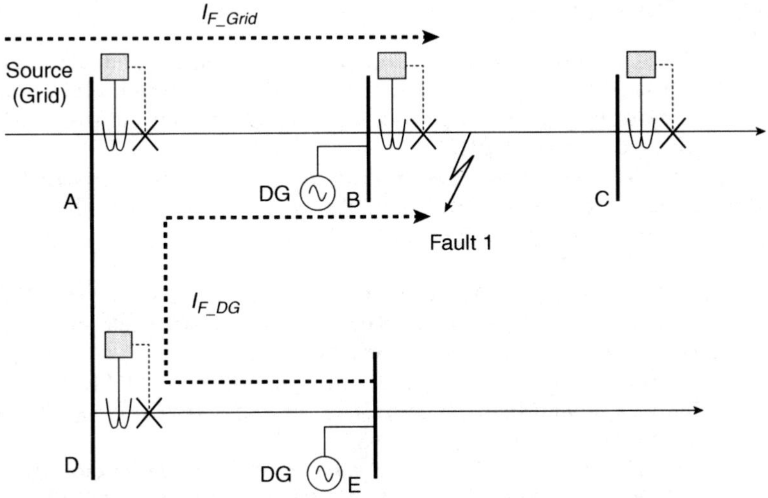

3.20 Potential impact of DG on protection.

measured by Relay A may violate the instantaneous protection setting at A, leading to unnecessary tripping at A for a fault beyond B. Furthermore, if there was significant DG at B, this could reduce the amount of current supplied by the grid, perhaps, in extreme situations, leading to the 'blinding' of the protection at A to Fault 1, possibly compromising backup protection operation.

Finally, the undervoltage that would be apparent at Locations B and E (and all other locations in the vicinity of the fault) during the fault could lead to problems associated with unnecessary 'sympathetic tripping' of the DG units if the fault were not cleared quickly enough by the network protection. Furthermore, as a consequence of the DG being unnecessarily removed, loads on the conductors upstream from the DG could increase to beyond overload levels, leading to yet more tripping and possibly to widespread disruption. These, and other problems, remain the subject of research, and many proposals for solutions to remedy these problems, including the use of directional protection, communicating protection and adaptive solutions, have been made by researchers working in this field (Hussain, Sharkh, Hussain and Abusara, 2010).

The increase in fault levels that accompanies large penetrations of DG has led to the research and development of fault current limiting technology, including the use of superconducting material (which transits to a resistive state above critical current levels), saturated magnetic circuits, extremely fast-acting fuses and switching devices. Some devices are now installed as prototypes at locations throughout the world (Xin *et al.*, 2009).

3.9 References

Alstom (2011), *Network Protection and Automation Guide.* Available (via download request) at: <http://www.alstom.com/grid/products-and-services/Substation-automation-system/protection-relays/Product-configurator-eCORTEC/> (Accessed August 19, 2012).

Anderson, P.M. (1998), *Power System Protection.* Piscataway, Wiley-IEEE Press.

Atputharajah, A. and Saha, T.K. (2009), Power system blackouts – literature review. *International Conference on Industrial and Information Systems (ICIIS)*, Sri Lanka, 28–31 December.

Brahma S.M. and Girgis, A.A. (2004), 'Development of adaptive protection scheme for distribution systems with high penetration of distributed generation,' *IEEE Transactions on Power Delivery*, vol. **19**, pp. 56–63.

Dysko, A., Burt, G.M., Galloway, S., Booth, C. and McDonald, J.R. (2007), UK distribution system protection issues. *IET Journal on Generation, Transmission & Distribution*, vol. **1**, no. 4, pp.679–687.

EirGrid (2011), *Eirgrid Grid Code.* Available at: <http://www.eirgrid.com/media/Grid%20Code%20Version%204.pdf> (Accessed 15 August 2012).

Energy Networks Association (2011), G59/2-1 *Recommendations for the Connection of Generation Plant to the Distribution Systems of Licensed Distribution Network Operators*. London: Energy Networks Association.

EON Netz (2006), *Grid Code, High and extra high voltage*. Bayreuth: EON Netz.

Gers, J. M. and Holmes, E.J. (2004), *Protection of Electricity Distribution Networks.* 2nd ed. London: Institution of Engineering and Technology.

Hussain, B., Sharkh, S.M., Hussain, S. and Abusara, M.A. (2010), Integration of distributed generation into the grid: Protection challenges and solutions. *10th IET International Conference on Developments in Power System Protection (DPSP 2010). Managing the Change.* Manchester, 29 March 2010–1 April 2010.

IEC (2003), TR 61850-1 *Communication Networks and Systems in Substations – Part 1:Introduction and Overview*, Geneva: International Electrotechnical Commission.

IEEE (1997), IEEE Std C37.112–1996: IEEE Standard Inverse-Time Characteristic Equations for Overcurrent Relays, Piscataway, IEEE.

IEEE (2005), IEEE Std C57.13.6–2005: IEEE Standard for High-Accuracy Instrument Transformers, Piscataway: IEEE.

Jennett, K., Coffele, F. and Booth, C. (2012), Comprehensive and quantitative analysis of protection problems associated with increasing penetration of inverter-interfaced DG. *11th International Conference on Developments in Power Systems Protection,* Birmingham, 23–26 April 2012.

Kundur, P., Paserba, J., Ajjarapu, V., Andersson, G., Bose, A., Canizares, C., Hatziargyriou, N., Hill, D., Stankovic, A., Taylor, C., Van Cutsem, T. and Vittal, V. (2004), Definition and classification of power system stability IEEE/CIGRE joint task force on stability terms and definitions. *IEEE Transactions on Power Systems*, vol. **19**, no. 3.

Laaksonen, H.J. (2010), Protection principles for future microgrids. *IEEE Transactions on Power Electronics*, vol. **25**, pp. 2910–2918.

National Grid (2012), *The Grid Code* available at <http://www.nationalgrid.com/uk/Electricity/Codes/gridcode/gridcodedocs/> (Accessed 15 August 2012).

Xin, Y., Hui, H., Gong, W.Z., Ye, F., Wang, J.Z., Tian, B., Ren, A.L. and Zi, M.R. (2009), Superconducting cable and superconducting fault current limiter at Puji Substation. *International Conference on Applied Superconductivity and Electromagnetic Devices*, Chengdu, China, 25–27 September 2009.

4
Integration of distributed energy resources (DER) to the grid

K. KAUHANIEMI, University of Vaasa, Finland

DOI: 10.1533/9780857097378.1.108

Abstract: This chapter discusses the technical issues of successful integration of distributed energy resources (DER) to the grid. The effects of DER on the power system operation and protection are first identified, and then the technical requirements arising from these effects are introduced. Also the relevant grid codes and standards are discussed.

Key words: distributed generation, power systems, grid interconnection.

4.1 Introduction

Interest in small scale renewable energy sources has increased rapidly over the last 15 years. The main reason for this is concern about global climate change and the need to reduce the CO_2 emissions from fossil fuels. From the electricity distribution point of view, the use of local small generators means transition from traditional centralised system to a decentralised system. This provides several technical challenges, which are discussed in this chapter.

Distributed generation (DG), sometimes also referred to as embedded generation (Jenkins *et al.*, 2000), means primarily small size generation connected to the distribution network. Typical for these technologies are modular production units, which are relatively small in size (from 10 kW up to about 10–20 MW) and located near the consumption of energy. Many of the generation technologies are based on primary energy sources that are not available continuously. A good example of this is wind power. In order to have higher security for the power supplied from such sources, it has become evident that some kinds of energy storage facilities are needed. Considering both the DG and energy storage facilities, the term DER is preferred here. Frequently demand response (DR), which covers the means to manage customers' power consumption, is included in DER as well (Rahimi and Ipakchi, 2010).

In this chapter the main focus is on the technical issues of successful integration of DER with the grid. First the different DER technologies and their effect on the grid are introduced. After that, the requirements for grid connection of DER units are discussed. The various grid codes and standards relating to the topic are also covered. Special attention is paid to electricity distribution networks, where the integration of DER causes certain fundamental changes in the system operation and protection.

4.2 DER technologies

There are several DG technologies, which are actually in different development phases. For example, wind power is now in rapid growth stage, while fuel cells are still awaiting a breakthrough. On the other hand, the use of local renewable energy sources, such as waste or wood, has boosted the use of downscaled traditional generating technologies in the form of small combined heat and power (CHP) plants. In the following, some most interesting generation technologies are briefly introduced for further reference. After that, an overview of the energy storage technologies is given.

Wind power technology is developing rapidly, and the size of the units has increased from below 1 kW up to 5 MW. In the power range below 1 MW, a typical design has been the fixed speed stall regulated wind turbine with asynchronous generator. In order to withstand the mechanical stress, most wind turbines above 1 MW are equipped with a variable speed system, incorporating power electronics in combination with pitch control. During past years, this has become the most common concept in new installations (Ackermann, 2005). There are several ways to realise variable speed functionality, but the most common are the doubly fed induction generator (DFIG) and permanent magnet generator with full power converter (FPC). In the DFIG concept, only part of the generator power is fed through the converter, while in systems applying FPC technology all the power fed to the grid goes through a converter. DFIG had previously been a widely used concept, but FPC is now becoming the more common solution (Troedson, 2009). A typical FPC-based wind power plant consists of a permanent magnet synchronous generator (PMSG), AC/DC converter (the generator converter), DC intermediate circuit (usually a capacitor) and the DC/AC converter (the grid converter) as shown in Fig. 4.1.

Photovoltaic (PV) systems consist of solar cells, which are grouped in to larger modules more commonly known as solar panels. The silicon-based solar cells convert sunlight radiation directly into electricity. These cells are connected together in series and parallel in order to give the desired output. A PV panel outputs direct current that is almost directly proportional to the intensity of sunlight (Jayarama, 2010). This current is fed into grid applying first a DC/DC converter in order to achieve desired voltage level and then DC/AC converter to get proper AC output.

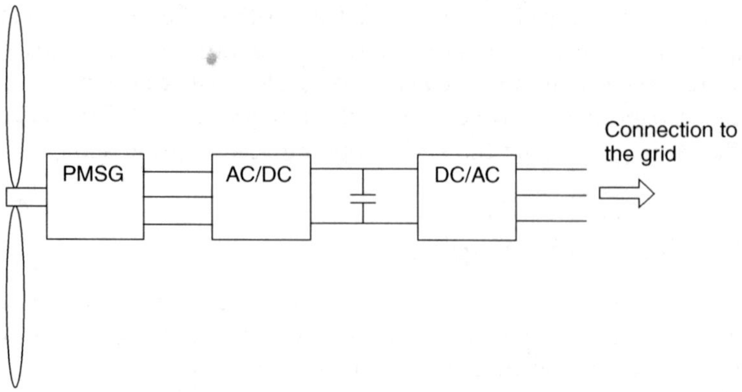

4.1 An example of the FPC concept for wind power plant.

Reciprocating internal combustion engines (ICE) were developed more than 100 years ago, and they are considered to be the earliest DG technology. Nowadays, ICEs are widely applied and are the most inexpensive DG solution (Masters, 2004). They are used on many scales, ranging from small units of 1 kVA to large, several tens of MW, power plants. Reciprocating engines are usually fuelled by diesel or natural gas. Typically, synchronous generators are applied with ICEs, although some examples can be found where induction generators are applied.

The basic CHP plant consists of a system with boiler and steam turbine. In the boiler the incoming water is transformed into dry steam under high pressure. The steam is transmitted to the turbine where it expands and, as a result of this, electricity is produced by the generator. The wet steam leaves the turbine and passes through a heat condenser where it exchanges heat with water in the heating system. There are different types of steam boilers and almost any fuel can be used. The CHP plants are usually driven by the heat demand so that the electricity is a by-product. Thus the controllability of the electric power output is usually limited.

Micro-turbines represent a new modular generation technology that is also capable of operating in CHP mode. For each kilowatt-hour of electricity produced, a micro-turbine will produce 2 kWh of heat. A typical construction of a micro-turbine consists of a turbine mounted on the same shaft as the compressor and a high-speed generator rotor. The shaft spins up to 100 000 rpm. The generated high frequency AC needs to be rectified to DC and then converted back to a three-phase AC with grid frequency.

Fuel cells generate power through the electrochemical reaction between hydrogen and oxygen. The conversion is highly efficient and leaves only water and heat as by-products, which is the main motivation for the increasing interest in this technology. All fuel cells generate a direct current, and the output voltage depends on the cell voltage and the number of cells in series. Furthermore, the voltage varies with the load and also to some extent with time as the fuel cell stack ages. To obtain AC output, a DC-to-AC converter is needed.

The grid connected energy storages are still not very commonly applied mainly due to economic reasons. However, recent surveys about the benefits and costs of energy storages, (Eyer and Corey, 2010; Schoenung, 2011), indicate that situation is about to change. One of the oldest storage technologies is the pumped hydro, which uses large water reservoirs. Flywheels are also applied as energy storages, especially when high power is required for a short duration. Battery technology is developing rapidly, due to mobile applications and, more recently, electric vehicle development. Battery energy storage is not yet widely applied in power grid applications, but it is expected to play a significant role in future smart grids employing large amounts of renewable energy sources (Divya and Østergaard, 2009). Traditionally, battery energy storage has been used in uninterruptible power supply (UPS) applications, but there exists technology that enables large scale (several megawatts) grid connected energy storage. From the grid connection point of view the energy storages are usually based on DC/AC converters at the grid interface. In the pumped hydro applications directly connected rotating machines were earlier common, but the trend is towards variable speed applications due to better controllability (Suul *et al.*, 2008). In such a case, the grid interface is again based on the DC/AC converter.

As can be seen from the above, DER technologies can be categorised from the grid point of view into two groups. In the first, the grid interface is based on a power electronic device: the power converter. In the second, the AC power is generated by a rotating machine. Furthermore, two types of rotating machines are usually applied: synchronous generators and asynchronous generators. Most of the issues relating to grid integration can be studied considering only these three grid interface technologies (power converters, synchronous generators and asynchronous generators), omitting the primary energy source. The given division of grid interface technologies is not unambiguous, but it will be used here for convenience. There are also technologies that do not fit in these categories, such as DFIG or the directly driven permanent magnet generator.

4.3 Effects of DER on the grid

The effects of DER are significant only in distribution networks where traditionally there have rarely been any generation facilities. The distribution networks are usually designed to have unidirectional power flow, i.e. the power flows from a high voltage transmission grid to a medium voltage (MV) network and further down to a low-voltage (LV) network and customers. Another key aspect of traditional distribution systems is that a radial topology is applied. There are only relatively few backup connections, and despite the partial ring topology, the system is always radially operated. This means that there are open switches in certain places of the system, so that the operational topology is always radial. This makes it easier to manage the voltage levels in various parts of the system, and simplifies the protection arrangement.

If there is generation in some part of the distribution system, the direction of the power flow may vary depending on the power balance, i.e. the difference between power produced and power consumed. Thus, the generation placed in the distribution system makes the system configuration always non-radial, practically approaching a meshed system like the high voltage transmission networks.

In addition to the changed configuration, the generation placed into the distribution network may cause several power quality issues:

- too high voltage level,
- large and frequent voltage variations, and
- harmonics.

The first refers to a situation where the power flow in the system is reversed, causing also a reversed voltage drop along feeders. This may raise the voltage at the customer site beyond the allowed limits. This is illustrated in the Fig. 4.2.

Voltage rise is a problem only in cases where the network is relatively weak when compared to the size of the generator. Thus, there is always a certain limit for the size of the generator that can be placed on the network, and that size depends on the characteristics of the network. However, the

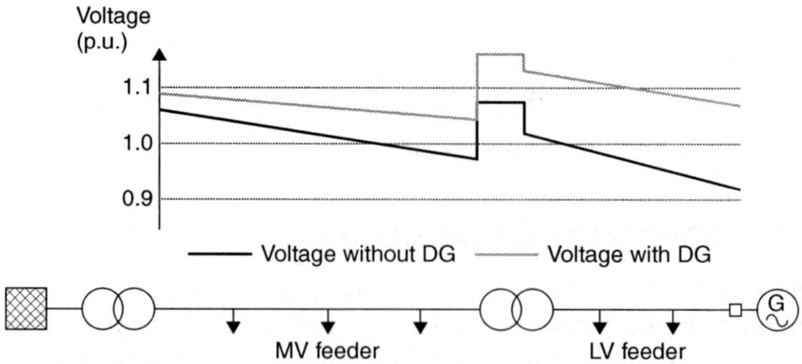

4.2 The effect of DG on the voltage level at the electricity distribution system.

voltage problems can be mitigated if the generator actively participates in the network voltage control. This is achieved simply by producing or consuming suitable amounts of reactive power. Of course, the applied generation technology must be capable and equipped for this, and also the regulatory and energy market environment must enable this kind of activity.

Depending of the type of generator, there may be large uncontrollable variations in the power output. These may be caused, e.g., by the wind speed variations in a wind power plant. Variations in power output cause corresponding variations in the voltage of the network. In a weak network the voltage variation is larger than in a strong network with high short circuit level. Also, the start-up and shut-down of power generation units may cause voltage variations. The rapid voltage variations experienced by the customers are called flicker, for which there are limits defined in power quality standards. The possible flicker emission from the power generating plant must be taken into account even in the design phase.

If the grid interface of the generating unit is based on the power converter the possible harmonics must be taken into account. The output current from a converter always contain harmonics and they should be properly filtered out. Usually passive filters are applied consisting of serial inductances and parallel capacitances. Despite the filtering there may be situations where harmonics in the grid increase due to resonances in certain frequencies.

In addition to voltage waveform distortions, supply outages can be also considered a power quality problem. Electricity supply interruptions have adverse effects on many activities in modern society, and thus the requirements on power system reliability are increasing. The effect of DER in the power system reliability is a complicated issue, as indicated, e.g., in Dugan, (2002). Local generation can act as a backup supply in case of grid failure, which increases supply reliability. On the other hand, it may be argued that the increased system complexity increases the number failures. Ultimately, it is the protection system that has the central role in maintaining the level of supply reliability.

The protection system ensures that the effects of a fault in the system are minimised. Modern microprocessor-based protection relays, nowadays often called intelligent electronic devices (IED), are the brain of the protection system. They are responsible detecting the fault, locating it and disconnecting the faulty part of the system. The connection of DER to the grid may have adverse effects on the correct operation of the protection system if the effects of the DER are not taken into account. Problems may arise, especially in the radially operated distribution networks. These kinds of system enable also quite a straightforward protection concept. Especially in overcurrent protection, it can be assumed that the only source for the fault current is behind the protection equipment. If some DG is added into the network, the situation is different. There are now several sources of fault current in the network, and therefore a more complex protection system design is necessary.

Studies have shown that DG causes several potential problems to the protection of distribution networks. The most commonly mentioned problems are the following (Kauhaniemi and Kumpulainen, 2004):

- false tripping of feeders,
- nuisance tripping of production units,
- blinding of protection,
- increased or decreased fault currents,
- unwanted islanding,
- prohibition of automatic reclosing, and
- unsynchronised reclosing.

The appearance of these problems depends on both the characteristics of the network and DER. In short-circuit faults a DER unit generates fault current that depends strongly both on the grid interface technology (or generator type) and the network configuration. Synchronous generators are able to feed a large sustained fault current, while power converter based systems may be controlled so that their output is limited to only 2–3 times the rated current (Loix *et al.*, 2009).

When a large production unit, or several small ones, are connected to an MV feeder, the fault current seen by the feeder protection relay at the substation may be reduced, which can prevent or delay the operation of the overcurrent relay (blinding of the protection). This is also called protection under-reach.

A typical example of false tripping is shown in Fig. 4.3. The short-circuit fault occurs in the Feeder 2, but the relay at the beginning of Feeder 1 is also tripped because of the overcurrent fed by the generation unit at the feeder. False tripping (sympathetic tripping) is typically caused by synchronous generators, which are capable of feeding sustained short-circuit current. False tripping of healthy feeders can be solved by directional overcurrent relays, but also the protection against the bus faults may have to be changed. There should be a transfer trip from the main infeed relays (overcurrent relays at secondary side of the primary transformer) and from the arc protection relays to the feeder relays for all the feeders having a significant amount of DG connected.

In the case of sudden loss of grid connection, a part of the distribution network may keep on operating as an island with the help of local generators. This kind of situation is not desirable, especially considering the safety of maintenance personnel, so loss-of-main (LOM) or anti-islanding protection is usually considered necessary. LOM protection is usually specifically required in the grid codes. Basic LOM protection is based on detecting

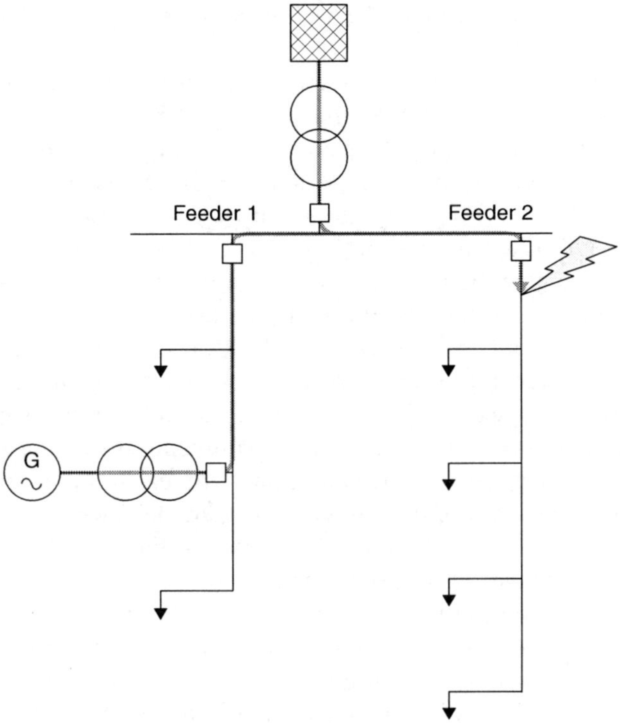

4.3 An example of false tripping.

abnormal system voltage or frequency. Immediately after transition to the island operation, the power production tries to match the power consumption in the islanded part of the system. This usually causes deviations and transients in the voltage and frequency of the islanded part of the system. However, if the power mismatch in the islanded part of the system is close to zero, the passive LOM methods based on voltage and frequency are not able to detect the islanding. This is called the non-detection zone (NDZ). The islanding detection time and NDZ can be decreased with so-called active methods, but they are considered to decrease power quality and system stability (Lee and Park, 2009). An ideal way to arrange the LOM protection is a transfer trip scheme, which requires a telecommunication link between the primary substation and the generating unit. A more detailed description of LOM methods can be found, e.g., in Geidl (2005).

The problems with LOM protection are closely linked with the problems with the auto-reclosing. The auto-reclosing function of protection relays takes care of large amount of temporary faults occurring in overhead line feeders. The extinction of fault arcs is possible when the auto-reclosing function switches off the voltage for few hundreds of milliseconds. The auto-reclosing will fail to remove the fault if there are generators maintaining voltage in the feeder.

From the grid operation point of view, it is essential to maintain system stability. Basically, this means that the amount of power produced must continuously match the power consumed. Fortunately, there is certain flexibility in the system so that in case of some major changes, e.g., due to faults in some parts of the system, it is still possible that the system will reach a new equilibrium after short transients. However, if the change is large enough, system stability may be endangered and a larger blackout is resulted. Recently, when the amount of DER, especially wind power, has increased in some countries, the effect of DER on the system stability has become an issue. Earlier, the DER units were protected against grid faults, so that when the voltage drops below some threshold the unit is tripped off. This is part of the LOM protection, but it also prevents any damage that the abnormal voltage may cause to the generating unit. From the system stability point of view, this kind of behaviour is not beneficial when the amount of DER is significantly high. When a severe three-phase short-circuit fault in the system, a voltage dip is experienced in a relatively wide area, and this may lead to a situation where a large amount of DER is lost due to a single failure. From the system stability point of view, it is necessary that as much generating power stays on the grid during the fault as possible. For this reason, many of the grid companies define so-called fault ride through (FRT) or low voltage ride through (LVRT) requirements. These are further discussed in Section 4.5.

Another problem from the grid operation point of view is the intermittency of some of the DER technologies. For example, the output of a wind power plant depends on the wind speed, which can fall to zero during calm periods. The output is not controllable and does not usually match the consumption pattern. The only thing that can be done is to limit the output power of the DER units if the system capacities are exceeded. This means in practice loss of energy sales for the DER owner. A solution for this is the use of energy storage in order to capture the energy during hours of over-production.

4.4 DER connection to transmission and distribution networks

In order to mitigate the adverse effects of DER to the grid, the grid companies usually define a set of rules and guidelines for making the connection. These are presented as grid codes, which are further discussed in Section 4.5. In this section the focus is on the various technical requirements that need to be taken into account when connecting DER either to the transmission or to the distribution network.

When connecting DER to transmission networks, generally the same requirements apply as for conventional power plants. The generation unit should be capable of participating in the grid control functions by adjusting its real and reactive power according to the system state. However, this might not be possible for certain types of DER, e.g., wind power, and therefore it is often treated separately in the requirements. In practice, this means that the active power produced by a wind power plant may be limited, depending on the system state, to some value below the available power. On the other hand, for the reactive power there are certain minimum limits that depend on the active power exchange (Ackerman, 2005; de Alegría *et al.*, 2007).

Voltage level control is usually required in the transmission systems, where the generators should respond to the voltage variation by adjusting their reactive power output. At the distribution networks, voltage control is traditionally very simple. The voltage level is controlled by the on-load tap-changer (OLTC) at the primary substation, and the voltage drop along the lines is designed to be low enough to achieve a suitable voltage level at the customer site. DER increases the voltage variation if no additional measures are taken. To handle the voltages at the distribution system, an active voltage control system is required, where distribution system operator (DSO) also utilises the capabilities of DER units for voltage control. This requires that the system state is monitored, and also that the control and measurement signals are transmitted between various locations in the system. This kind of advanced voltage control system proposed in Caldon *et al.* (2004) is illustrated in Fig. 4.4. In this system, the master regulator adjusts the voltage setting of the OLTC controller and gives a power factor (cosϕ) references to the DG units based on the measured average network voltage. A prior condition for this kind of ancillary service is that suitable regulatory and commercial arrangements have been developed (Peças Lopes *et al.*, 2007).

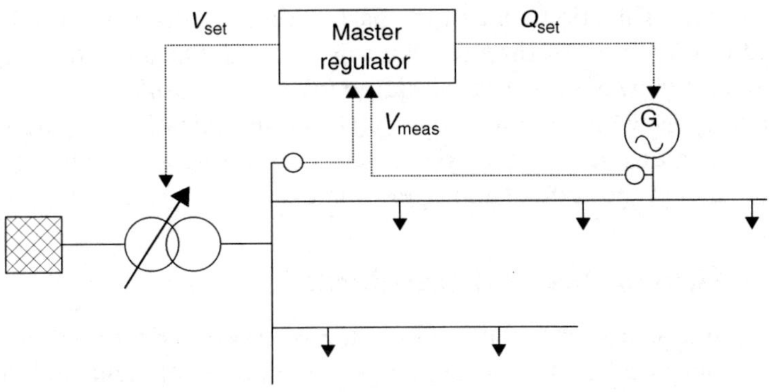

4.4 Coordinated voltage control system for distribution network containing DG.

As presented earlier, the protection problems caused by DER arise mainly in the distribution networks. These problems were identified in Section 4.3, where some suggestions for solutions were also given. Most problems are avoided by careful design of the needed protection system revisions when a DER unit is connected to the network. In many cases, the proper operation of the protection system requires new devices to be installed and application of more advanced protection functions. Ultimately, this affects also the training requirements of the grid company's personnel.

Especially in the case of larger DER units, the grid company may require suitable communication capability between the DER unit and company's central control room. At least the billing measurements are collected by remote read meters, but for various grid control and management functions various status information and on-line measurements of voltage, current and power may be required as well. Furthermore, a possibility to connect and disconnect the generating unit remotely might be required. For ensuring the interoperability and easier integration of the grid company's supervisory control and data acquisition (SCADA) system and the control system at the DER unit, specific telecommunication standards have been developed recently. For wind power plants there is IEC 61400–25, while IEC 61850-7-420 covers various kinds of DER systems.

Summarising the requirements for grid connection of DER, the following three categories of requirements can be identified:

- operational requirements,
- protection requirements, and
- communication requirements.

The operational requirements are related to the normal operation of the DER unit. They define the generator output in various system states. The protection requirements are necessary for maintaining the safe operation of the network protection. One of the basic requirements is that the DER unit should not keep on feeding the fault in the grid. On the other hand, from the point of view of system stability, the DER unit should support the grid if the fault is elsewhere in the network. The communication requirements are necessary to ensure that the system operator has access to all the information needed in the control and operation tasks.

4.5 Grid codes and standards

A grid code is a set of rules, requirements and procedures to enable efficient, economical and coordinated planning, development and operation of the grid. For a transmission grid the codes are set by the transmission system operator (TSO) and for the distribution network they are set by the local DSO. Multinational grid codes are necessary to maintain conformity of practice within a large interconnected power system. In such a case, the grid code is defined in cooperation with several TSOs. Furthermore, certain aspects of transmission grid codes should also be taken into account in the distribution network codes. The practices relating to the distribution network codes made by DSOs differ from country to country. Obviously, it is beneficial to target compliance with the rules at the national, and even at the international, level. For this purpose, the codes should be based on international standards when possible.

In practice, the grid codes cover a wide range of topics including planning, operation, measurements, data exchange, etc. Here the focus is on the grid connection of DER, and for that purpose the grid code includes typically a separate 'connection code'. The connection code establishes all the requirements for connection of a generating unit. It includes the technical, design and operational requirements for the generating units to be connected. Detailed requirements are also defined to many other issues such as the communication and automation system and the planning and agreement procedures.

In the past, small scale production and the characteristics of DER were not taken into account in the grid codes. However, the situation has changed during the last 10–20 years and new codes are being developed especially for the DER and DG. Sometimes specific codes for certain type of generating plants, e.g., wind power, have been developed separately (Altín *et al.*, 2010). A good example of a technology specific connection code is the Danish code 'Wind turbines connected to grids with voltages below 100 kV – Technical regulations for the properties and the control of wind turbines', which is intended especially for wind power plants (Elkraft and Eltra, 2004). The German guidelines are also frequently referred to when discussing the grid connection of wind power. The Danish and German codes have been used as a basis for many codes made later in other countries (de Alegría *et al.*, 2007).

It is also worth noticing that since the related technology has been rapidly developing, as well as the amount of DER is increasing, the codes are also being revised regularly. The stage of development seems to differ from country to country, depending on the amount of DER in the grid. From the manufacturer's point of view, this can be seen as problematic since the technical requirements are different in different market areas, which makes the development of global concepts difficult. In Europe, the joint association of TSOs, European Network of Transmission System Operators for Electricity (ENTSO-E), is developing a harmonised grid connection code, of which the draft version (ENTSO-E, 2011) is available at the time of writing this. It will greatly reduce the variability of the codes among European countries, but also enable the consideration of the local grid characteristics specifying, e.g., different limits for different regions and leaving some issues to be specified by the local TSO.

From the technical point of view, the connection code ensures that the connected generating unit will operate as part of the whole system without degrading system security, reliability and stability. This is also the main concern of the TSO or DSO. On the other hand, the primary target of the DER owner is to get connected with the minimum cost. This is best achieved with easy and standardised solutions, such as the plug-and-play concept proposed in the CERTS MicroGrid Concept (Lasseter *et al.*, 2003). However, these kinds of solutions are not easy to achieve, especially in the distribution networks, where the characteristics of the connection point may vary drastically depending on the location.

The standardisation has an important role for lowering the technical barriers against the DER, since it enables harmonisation of the grid codes among different countries. In the grid codes it is very convenient to only refer to the standards to be applied. One of the earliest standards for DG interconnection was the IEEE standard 1547 (IEEE, 2003). It is applicable for generating units up to 10 MVA that are connected to distribution networks. In Europe the standard EN 50438 (2007) for generators connected to LV networks has been introduced for units rated up to 16 A per phase. For a three-phase system, this means rated power up to 11 kVA. European Committee For Electrotechnical Standardization (CENELEC) is also preparing European standards for units above 16 A per phase, and also requirements for generators connected to MV distribution networks. In the Europe there are certain differences in the distribution systems in different countries, and thus it is necessary to specify some national deviations. Typically, these are introduced in the appendices of the associated standard.

In addition to the general grid connection standards, there are specific standards for different types of generators. A large number of standards have been developed, for example, for wind power. These standards specify many aspects relating not only to the grid connection but also to the whole generating plant. In addition to this, various measurement methods and test procedures have been defined so that conformance with the requirements given can be validated.

Due to the large variety of the connection codes and their constant evolution, only some key requirements given in the codes will be introduced in the following. The focus is on the requirements relating to the operation and protection of the DER. From the grid operation point of view the requirements for a generating unit are typically as follows:

- It should be possible to limit the active power output.
- The up or down control of active power should be possible with certain ramping rates or within some short time frame.
- The active power control based on the system frequency must be possible.

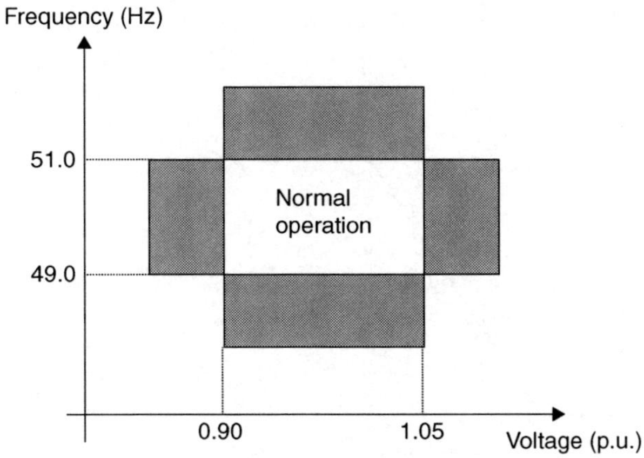

4.5 Principle of defining normal operating ranges in grid codes.

- The reactive power must be either controllable to zero (at the interconnection point) or it should be controllable based on the system voltage.

The codes also define the operating ranges of generators in terms of system frequency and voltage. Figure 4.5 shows the principle of defining the normal operation area. Beyond this area, there are ranges (grey boxes in the figure) where the generators must stay connected at least for a specified time (e.g. 30 min). A more complex example of the operating ranges may be found in, e.g., the Nordic grid code (Nordel, 2007).

The above-mentioned conditions are applicable in normal operating state. Considering the non-normal operating states, disturbances or faults, there are requirements relating to fault tolerance and specifications for the protection. Since a short-circuit fault causes in the system a voltage dip, which can be seen to some extent throughout the system, it is vital that the generator tolerates short voltage dips. This is usually defined by the FRT requirements, which are given as voltage-time curves. The principal format of these curves is illustrated in Fig. 4.6. The FRT curves essentially define the conditions where the generator should stay connected during the voltage dip. The voltage dip starts at t_0 and the generator must stay connected as long as the voltage stays above the curve. The time from t_0 to t_1 is usually 150 ms, which is a typical operating time of protection relays and during that time the generator must tolerate a minimum voltage V_{min}, which may vary from zero to 0.25 p.u. depending on the specific code (Altín *et al.*, 2010). In the curve the slope between t_1 and t_2 ranges typically from 1 to 1.5 s from the start of the dip. It should also be noticed that many codes employ some more complicated form of the curve, as can be seen in Altín *et al.* (2010). Furthermore, the grid code may contain some requirements relating to the output of the

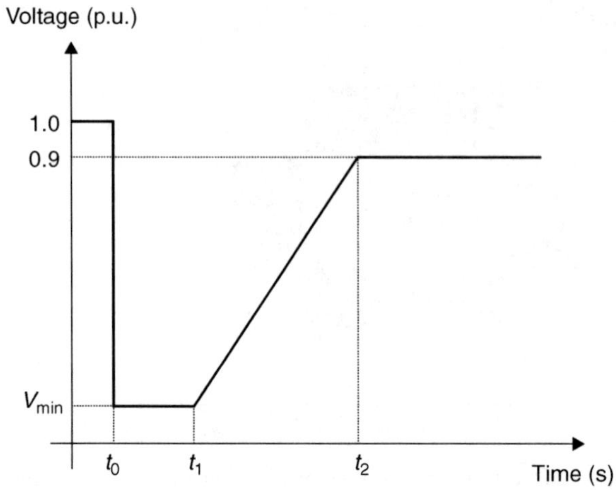

4.6 A voltage-time curve defining FRT requirements.

generator during the voltage dip. For example, it may be required that the generator should produce a certain amount of reactive power during the voltage dip (Tsili *et al.*, 2008). It is also worth noticing that FRT and LOM requirements are interrelated, since they both define the behaviour of the DG unit under abnormal grid states. While the LOM was the main concern earlier, the FRT requirements were first adopted in areas with a high amount of wind power (European Wind Energy Association (EWEA), 2005) and are currently widely applied (Tsili *et al.*, 2008). The FRT requirements are in many areas only applicable for wind parks connected to high voltage (HV) network (Tsili *et al.*, 2008) and not for small units like, e.g., domestic PV installations, for which only the LOM requirements are applied.

For many types of DER the control of the active power output is primarily dictated by the availability of power from the primary energy source. However, from the power system point of view it is necessary to continuously maintain the balance between the produced and consumed electric power. Any imbalance is indicated by the system frequency deviation from the nominal value, and thus the power control of generators can be based on frequency measurements. In an interconnected power system there must be always enough generating units participating to the primary frequency control, which is based on a simple frequency droop. This means that when the frequency deviates from the nominal frequency f_0 the output power of the generator is changed in proportion to the deviation (Fig. 4.7).

If the DER is based on a primary energy source that cannot be controlled (such as wind or solar) the realisation of frequency droop control is complicated. In principle this means that a certain margin must always be left between the available power and the actual output. This in turn means lost

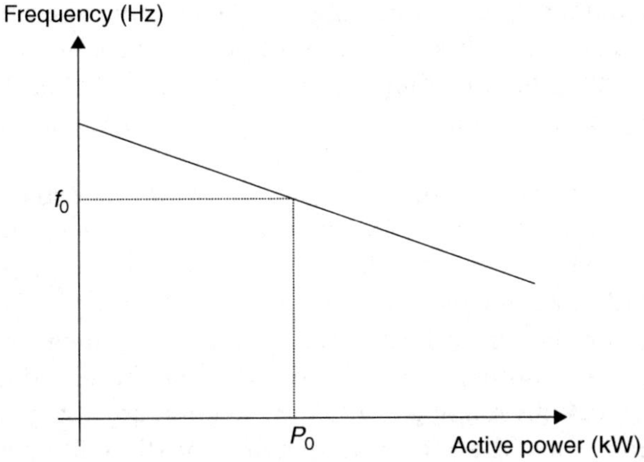

4.7 Frequency droop control of generator active power.

revenues for the DER owner if there is no monetary compensation for the participation to the primary frequency control.

The reactive power should preferably be produced near the location of the consumption and therefore it is usually desirable to have the reactive power output of a generator either at zero or equal to the local reactive power consumption. From the system control point of view the target is to keep the voltage within given limits in various parts of the system and the voltage level can be controlled by the reactive power. If the voltage tends to be too high it is desirable that the generator consumes reactive power. On the other hand, if the voltage is below the target value, the generator should support the voltage by injecting reactive power to the network. Since the reactive power capabilities are not dependent on the primary energy source any type of DER unit can be equipped with suitable voltage dependent reactive power control if required. The grid codes define control bands for the reactive power output that are dependent on the active power output level.

As said before, the protection of DER units is considered problematic, especially at the distribution network level. The associated standards and codes usually give only the voltage and frequency ranges beyond which the generator must be disconnected. In addition to this, a specific LOM protection may be required: the generator should be disconnected from the local network when the connection to the main grid is lost. Thus the required protection functions are usually the following:

- under- and over-voltage,
- under- and over-frequency, and
- LOMs.

In the codes and standards, the setting ranges and maximum operating times for these functions are given. In terms of LOM protection, at least the minimum time for detecting the islanding is specified and sometimes also a specific type of the protection function is recommended or required (Stråth, 2005).

The requirements included in the codes and standards depend on the size of the DER unit and also the voltage level. For units connected to the LV network, the requirements are somewhat different from those at higher voltage levels. At the LV level, the main concern is to maintain acceptable power quality and to ensure the safety. Therefore, e.g., the requirements regarding harmonics are usually considered more important than power controllability. Also, the specifications relating to LOM protection are always included in the requirements, while the FRT requirements are still seldom extended to LV level.

4.6 Challenges and future trends

The increasing use of DER creates different kind of challenges for the industry, depending on the phase of development. The phase of development here means the relative amount of DER in the grid. For certain areas, this amount can be stated by so-called penetration degree (Gomes *et al.*, 2009), which denotes the ratio of local generation to the total amount of consumption in the area. In the first phase of development, the first DER units are introduced and the system operator has to adopt suitable codes and guidelines. At this phase the main concern is making the interconnection as simple as possible without endangering the safe and secure operation of the grid. In the second phase the penetration degree increases, so that the DER must be taken into account in the system operation. This necessitates the creation of rules and practices for various management functions considering the power and energy management as well as for the voltage level control. In the third phase the DER units are considered as an integrated part of the system. Very often the term 'active distribution networks' is mentioned in this context (Chowdhury *et al.*, 2009). This means that there are large numbers of active elements in the system taking care of the abovementioned management tasks. In addition to the DER units, the active customer participation has a key role as well. Ultimately the future smart grid is achieved where all management and control actions are based on the use of modern information and telecommunication technology. In practice, this means that in parallel with the primary system there is a telecommunication network that enables the seamless communication with all the intelligent nodes in the system.

One of the most interesting technologies in the future employing DER is the microgrid concept first introduced in Lasseter (2001). Since then, microgrids have been researched and piloted in many international projects. Recently, the research focus in power industry has transferred more towards smart grids, while microgrids are considered as one of the building blocks of the smart grids (Giga, 2010; Colson and Nehrir, 2011). Microgrids provide a way to accomplish the self-healing feature of the smart grids. When there is a fault in the main grid the microgrid can continue the operation as an island. This is possible if there is enough generation capacity in the islanded microgrid. Furthermore, in order to ensure stable operation of the microgrid, there must be enough fast responding generating capacity, which is most conveniently achieved by suitable energy storage systems. In addition to technical difficulties, also the legislation, lack of standards, and high costs are seen as barriers to microgrid development (Chowdhury *et al.*, 2009). Despite this, the latest news indicates first signs of a trend from pilot and research installations towards commercial projects (Pike Research, 2011). Still, it seems apparent that microgrids will stay for quite a long time as a special solution, suitable for remote locations or small communities having vast amount of local renewable energy sources.

To achieve the future scenario illustrated above, it must be ensured that no extra barriers exist against the use of DER, considering among others the interconnection rules and applied technology. The basis of rules and technical solutions should preferably be based on international standards, which enable globally compatible system solutions and equipment, and reduce the need for costly solutions tailored for local requirements. An ultimate solution would be a plug-and-play grid interconnection, which might be one of the techniques employed in the future smart grid. In practice, this would require the definition and development of an interface that exchanges not only electric energy but also management, control and energy market data.

4.7 Conclusion

In this chapter the grid interconnection issues of DER were introduced from the technical point of view. A major concern in the grid interconnection is to enable the operation of a DER unit without endangering any of the operational or safety requirements put on the transmission or distribution network. From the network operation point of view, both the normal and faulted operating states must be taken into account. In normal state the main target is to maintain system stability and, for this purpose, the DER units need to be equipped with real and reactive power control facilities. As a result of a fault situation, stable operation may also be lost if a large amount of DER is disconnected from the grid during the fault. In order to avoid this, DER units need to meet specific FRT requirements. Considering the safe and secure operation of the system, it is also essential that the protection system at the DER interconnection point is compatible with the network protection. It should be also noted that in many cases the distribution network protection system must also be redesigned to maintain proper operation after the connection of DER. This means usually slightly more complex protection systems than have been traditionally used in distribution networks. However, this can also be seen as a first step towards the future smart grids, where DER takes an active role in the grid management functions.

4.8 Sources of further information and advice

One of the first books written about DG was the 'Embedded Generation' by Jenkins *et al.* (2000). A comprehensive review of wind power technology can be found in Ackermann (2005). Since the majority of DER are power converter based grid inverter a good source of further information is the recent white book about grid converters by DERlab (Strauss, 2009). Future microgrids and active distribution networks are discussed in the book by Chowdhury *et al.* (2009).

Among the grid codes published by the system operator the most well-known are the Danish (Elkraft and Eltra, 2004) and German rules. The latest version of German rules is published by BDEW and also an English translation is now available (BDEW, 2008). The first standard relating to the DER interconnection was the IEEE standard 1574 (IEEE, 2003). There is also an EN standard 50438 for DER units connected to LV network (EN 50438, 2007).

Further sources of information are the web-sites maintained by relevant organisations or research projects etc. A selection of recommended sites is given below:

- DERlab – European DER Laboratories e.V.: http://www.der-lab.net/
- US DOE Energy Efficiency and Renewable Energy (EERE): http://www.eere.energy.gov/
- Smart Grids: European Technology Platform: http://www.smartgrids.eu/

It is also worth noticing that up to date information about grid codes and guidelines is available from the www-sites of relevant TSOs and DSOs and latest standards are available on the www-sites of the standardisation organisations IEC, CENELEC and IEEE.

4.9 References

Ackermann T (ed.) (2005), *Wind Power in Power Systems*, Hoboken, John Wiley and Sons Ltd.

Altín M, Göksu O, Teodorescu R, Rodriguez P, Jensen B-B and Helle L (2010), 'Overview of recent grid codes for wind power integration', *12th International*

Conference on Optimization of Electrical and Electronic Equipment (OPTIM), 1152–1160, 20–22 May 2010, Brasov, Romania.

BDEW (2008), Technical guideline: generating plants connected to medium-voltage network – guideline for generating plants' connection to and parallel operation with medium-voltage network, June 2008 issue, Available from: http://www.bdew.de/ bdew.nsf/id/DE_7B6ERD_NetzCodes_und_Richtlinien/$file/BDEW_RL_EA-am-MS-Netz_engl.pdf (Accessed 28 February 2011).

Caldon R, Turri R, Prandoni V and Spelta S (2004), 'Control issues in MV distribution systems with large-scale integration of distributed generation', *Proceedings of Bulk Power System Dynamics and Control – VI*, 583–589, 22–27 August 2004, Cortina d'Ampezzo, Italy.

Chowdhury S, Chowdhury S P and Crossley P (2009), *Microgrids and Active Distribution Networks*, The Institution of Engineering and Technology, Stevenage, Herts, United Kingdom.

Colson C M and Nehrir M H (2011), 'Algorithms for distributed decision-making for multi-agent microgrid power management', *IEEE Power and Energy Society General Meeting*, 1–8, 24–29 July 2011, San Diego, CA, USA.

de Alegría I M, Andreua J, Martín J L, Ibañez P, Villate J L and Camblong H (2007), 'Connection requirements for wind farms: A survey on technical requirements and regulation', *Renewable and Sustainable Energy Reviews*, **11**, 1858–1872.

Divya K C and Østergaard J (2009), 'Battery energy storage technology for power systems – An overview', *Electric Power Systems Research*, **79**, 511–520.

Dugan R C (2002), 'Distributed resources and reliability of distribution systems', *IEEE Power Engineering Society Summer Meeting*, 106–108, July 2002, Chicago, IL, USA.

Elkraft and Eltra (2004), Wind turbines connected to grids with voltages below 100 kV – Technical regulations for the properties and the control of wind turbines, Regulation TF 3.2.5, Available from: http://www.wt-certification.dk/ Common/Regulation%20for%20Windturbines%20TF%203.2.6.pdf (Accessed 28 February 2011).

EN 50438 (2007), Requirements for the connection of microgenerators in parallel with public low-voltage networks, CENELEC, BS EN 50438:2007 published by BSI, London, UK, ISBN 978 0 580 54535 1.

ENTSO-E (2011), *ENTSO-E Draft Requirements for Grid Connection Applicable to all Generators*, 27 October 2011, Available from: https://www.entsoe.eu/fileadmin/user_upload/_library/news/ Network_Code_on_Connection_Requirementsapplicable_to_all_Generators_-_working_draft.pdf (Accessed 5 January 2012).

EWEA (2005), *Large Scale Integration of Wind Energy in the European Power Supply: analysis, issues and recommendations*, European Wind Energy Association, Available from: http://www.ewea.org/fileadmin/ewea_documents/documents/publications/grid/051215_Grid_report.pdf (Accessed 4 January 2012).

Eyer J and Corey G (2010), Energy storage for the electricity grid: Benefits and market potential assessment guide, SAND2010-0815, Sandia National Laboratories, Albuquerque, New Mexico, USA and Livermore, California, USA, Available from http://www.sandia.gov/ess/publications/SAND2010-0815.pdf (Accessed 9 July 2013).

Giga O M (2010), *Microgrids: Building blocks of the Smart Grid*, Available from: http://gigaom.com/cleantech/microgrids-building-blocks-of-the-smart-grid/ (Accessed 5 January 2011).

Gomes P, Martins A C B, Zani C R and Sardinha S L A (2009), 'Connection requirements and Grid Codes for distributed generation', *CIGRE/IEEE PES Joint Symposium on Integration of Wide-Scale Renewable Resources Into the Power Delivery System*, 1–12, 29–31 July 2009, Calgary, Canada.

IEEE (2003), IEEE standard for interconnecting distributed resources with electric power systems, IEEE Std 1547–2003, The Institute of Electrical and Electronics Engineers, Inc., New York, USA, ISBN 0-7381-3720-0.

Jayarama R P (2010), *Science Technology of Photovoltaics*, Hyderabad, India, Global Media.

Jenkins N, Allan R, Crossley P, Kirschen D and Strbac G (2000), *Embedded Generation*, London, The Institution of Electrical Engineers.

Kauhaniemi K and Kumpulainen L (2004), 'Impact of distributed generation on the protection of distribution networks', *Eighth IEE International Conference on Developments in Power System Protection*, Amsterdam, Netherlands, 315–318, 5–8 April 2004.

Lasseter B (2001), 'Microgrids (distributed power generation)', *IEEE Power Engineering Society Winter Meeting*, 146–149, 28 January–01 February 2001, Columbus, OH, USA.

Lasseter R, Akhil A, Marnay C, Stevens J, Dagle J, Guttromson R, Meliopoulous A S, Yinger R and Eto J (2003), 'Integration of distributed energy resources: The CERTS MicroGrid concept', Available from: http://certs.lbl.gov/pdf/50829.pdf (Accessed 28 February 2011).

Lee S-H and Park J-W (2009), 'New Islanding detection method for inverter-based distributed generation considering its switching frequency', *IEEE Industry Applications Society Annual Meeting*, 1–8, 4–8 October 2009, Houston, TX, USA.

Loix T, Wijnhoven T and Deconinck G (2009), 'Protection of microgrids with a high penetration of inverter-coupled energy sources', *CIGRE/IEEE PES Joint Symposium on Integration of Wide-Scale Renewable Resources into the Power Delivery System*, 29–31 July 2009, Calgary, Canada.

Masters G M (2004), *Renewable and Efficient Electric Power Systems*, Hoboken, Wiley.

Nordel (2007), *Nordic Grid Code 2007*, Available from: https://www.entsoe.eu/fileadmin/user_upload/_library/publications/nordic/planning/070115_entsoe_nordic_NordicGridCode.pdf (Accessed 28 February 2011).

Peças Lopes J A, Hatziargyriou N, Mutale J, Djapic P and Jenkins N (2007), 'Integrating distributed generation into electric power systems: A review of drivers, challenges and opportunities', *Electric Power Systems Research*, **77**, 1189–1203.

Pike Research (2011), *Microgrid Deployment Tracker 4Q11*, Available from: http://www.pikeresearch.com/research/microgrid-deployment-tracker-4q11 (Accessed 5 January 2012).

Rahimi F and Ipakchi A (2010), 'Demand response as a market resource under the smart grid paradigm', *IEEE Transactions on Smart Grid*, **1**, 82–88.

Schoenung S (2011), *Energy Storage Systems Cost Update*, SAND2011-2730, Sandia National Laboratories, Sandia National Laboratories Albuquerque, New Mexico, USA and Livermore, California, USA. Available from http://prod.sandia.gov/techlib/access-control.cgi/2011/112730.pdf (Accessed 9 July 2013).

Stråth N (2005), *Islanding Detection in Power Systems*, Licentiate thesis, Lund University.

Strauss P (ed.) (2009), *International White Book on the Grid Integration of Static Converters*, European Distributed Energy Resources Laboratories (DERlab) e.V., Available from: http://www.derlab.eu/media/pdf/docs/ DERlab_D2.8_whitebook_static_converters_rev0.pdf (Accessed 28 February 2011).

Suul J A, Uhlen K and Undeland T (2008), 'Variable speed pumped hydropower for integration of wind energy in isolated grids – case description and control strategies', *Proceedings of Nordic Workshop on Power and Industrial Electronics, NORPIE 2008*, Espoo, Finland, 9–11 June 2008, Espoo, Finland.

Troedson A (2009), *PM Generator and Full Power Converter – The New Drive Train Standard*, Next Generation Utilities Summit, White Paper, Available from: http://www.ngusummitapac.com/media/whitepapers/The_Switch_NGUAPAC.pdf (Accessed 6 January 2012).

Tsili M, Patsiouras C and Papathanassiou S (2008), 'Grid code requirements for large wind farms: A review of technical regulations and available wind turbine technologies', *Proceedings of EWEC'08*, Brussels, 31 March–3 April 2008, Brussels, Belgium.

Part II

Advances in materials and technologies for electricity transmission and distribution

5

Development of advanced materials for transmission and distribution (T&D) networks equipment

J-L. BESSÈDE, Schneider Electric, France

DOI: 10.1533/9780857097378.2.133

Abstract: The evolution of electricity T&D networks over the last few decades has greatly advanced research in the field of material engineering. This chapter reviews the use of state-of-the-art materials for high voltage (HV) applications, focusing on circuit breakers. The first section will discuss the switchgear principle and the roles materials can play in this process, and will review the ways that the interaction arc and materials used can influence the endurance of HV circuit breaker chambers. Examples of arcing contact materials will also be discussed, and composite and thermoplastic insulation, and ester oil, are presented. The final section of this chapter describes some of the challenges and future trends predicted in the development of advanced materials, as the result of research into the T&D network.

Key words: high voltage (HV) switchgear, circuit breaker, arcing contact, contact material, carbon contact, composite insulators, insulation, ester oil, eco-design.

5.1 Introduction

SF_6 HV circuit breakers have undergone major developments since they were first patented 40 years ago (Dufournet *et al.*, 2003). However, despite the extensive optimization of materials, (Aeschbach *et al.*, 2002), a great deal of research continues to be undertaken to find new technologies and materials for arcing contacts, which must be inexpensive and have good switching performances.

Various principles can be applied to electrical interrupting apparatus (Browne, 1984). For circuit breakers to cope with currents under regular network service conditions, as well as to ensure the required interrupting performances under fault conditions, optimized application of materials with highest possible life endurance and designs requiring reduced drive energies, are necessary.

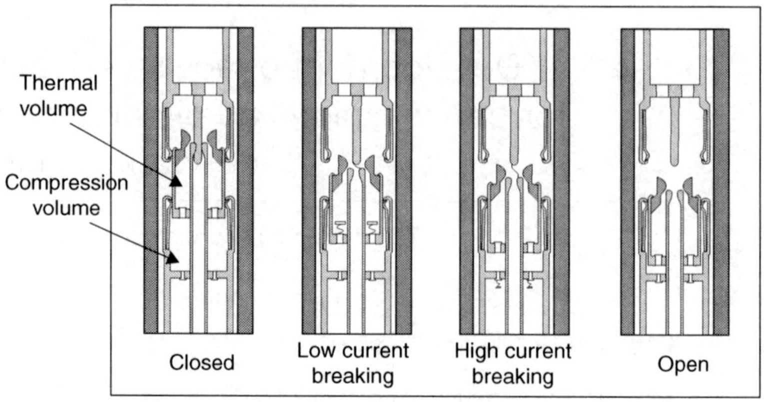

5.1 Self-blast interrupting principle (ThinkT&D).

Figure 5.1 shows the switching operation during the high current phase. In blast devices, the current is interrupted by separating two contacts. The current is then carried through the arc, before it is interrupted by a blast of gas, which causes the arc to cool. In SF_6 'puffer' devices, the pressure rise is obtained by compressing the gas with a piston and cylinder. For high current interruption, arc energy is used to produce the overpressure necessary to generate an efficient blast that cools the arc, leading to an interruption in current. This high overpressure is produced in the expansion volume; it closes the valve that isolates this volume from the compression volume mainly used for low current interruption. The overpressure necessary for breaking the current is obtained through the optimal use of the thermal effect, and of the nozzle-clogging effect produced when the cross-section of the arc significantly reduces the exhaust of gas in the nozzle.

To optimize the region where the hot gas puffs take place, and also to predict successful or unsuccessful switching, sophisticated numerical programs are now used (Robin-Jouan and Kairouani, 2002).

5.2 Switchgear materials: properties, types and performances

5.2.1 Generalities

In the arcing chamber of a circuit breaker, the nozzle and arcing contacts are the main parts. The materials used are selected specifically for their designed use, which must cope with the stresses imposed during the switching operations: the materials must withstand the high temperature (that may exceed some thousands of K for high currents), the HV or electrical strength, ultraviolet (UV) radiation induced by the de-excitation of gas molecules, and must withstand other arc constituents such as metal vapour and material eroded from the nozzle and ablation of the chamber walls, as well as metallic vapours and droplets.

The main properties needed in the arcing contacts in this very aggressive environment are: sufficient electrical conductivity, good mechanical properties, and a low friction coefficient, as well as good thermal resistance during the arc switching period.

The microstructure of the contact materials influences the performance of the arc extinction, because the presence of the arc induces, on the one hand, the extraction of metallic particles from the contacts and their mixing with the insulating gas and, on the other hand, the ablation of the nozzle due to the interaction with the UV radiation (Ferry *et al.*, 1996).

5.2.2 Contact materials and parts

Electrical contacts allow current to pass from one conductor to another, and therefore the contact materials used have to be soft, show high-conductivity, and be oxidation-resistant. Electrical contacts and contact materials are available in many different forms, and have wide electrical applications. They often have a second phase in order to provide anti-welding and/or arc resistance.

Current, conductivity, resistivity, tensile strength, flexural strength, outer diameter (OD), length, width, and thickness are important parameters to consider when selecting electrical contacts and contact materials. Good contact materials include aluminium, copper, gold, nickel, silver, and tin (Table 5.1).

Table 5.1 Specific characteristic of some conducting materials

	Density	Specific capacity (300 K)	Electrical resistivity (300 K)	Thermal conductivity (300 K)
	(g/cm^3)	($J.mol^{-1}.K^{-1}$)	($n\Omega.m$)	($W.m^{-1}.K^{-1}$)
Al	27	24.20	28.20	237
Cu	8.94	24.44	16.78	401
Ag	10.49	25.35	15.87	429
Ni	8.91	26.07	69.30	90.9
Sn	5.77–7.37	27.11	115	66.8

5.2.3 Insulation

The materials used as electric insulators in electric switchgear have to be able to block the flow of electric current across it. The electric insulator separates electrical conductors from one another and redirects electric currents. This non-conductive material provides the equipment with electrical, mechanical, and heat-dissipation functions.

Insulators can be solid, fluid, or gaseous. Most commonly used insulating materials are glasses, porcelain, polymeric materials, SF_6, and fluorocarbons (such as C_2F_6, C_4F_8). Critical factors in selecting electrical insulation materials are their resistivity, dielectric strength (breakdown voltage), and dielectric constant (relative permittivity).

Insulators also have to be able to transfer the heat that is generated in the electrical devices.

5.3 Development and impact of advanced switchgear materials

5.3.1 Arcing materials

In the arcing chamber, the interaction between the different materials and the arc will have an impact on the erosion rate of the contacts and nozzles. This will in turn have impact on the mechanical and electrical endurance of the HV circuit breaker, and will also introduce surface pollution through contact erosion (in droplets, metallic vapours, and other components). The combined effects of these can have a negative influence on the general behaviour of the arcing chamber (Aeschbach *et al.*, 2002).

Taking into account all the material requirements mentioned above, the range of arcing materials from which to choose is very poor. Copper/tungsten (W-Cu) materials are quite common in the HV field. These materials have a composite structure consisting of a tungsten skeleton filled with copper. Figure 5.2 illustrates the homogeneous distribution of the tungsten particles in the matrix. The porous surface of the material is filled in with copper to increase the conductivity of the material.

Carbon materials for arcing contacts have also been used for over 20 years. However, in order to improve the performances of the arcing contacts, new technologies such as copper/carbon composites (C-Cu) have to be developed. For such materials the microstructure of the materials is different from that of the copper/tungsten material.

Figure 5.3 illustrates a composite made of carbon fibres interlaced in three directions; copper is used to fill in the pores and thus increases the electrical conductivity of the material.

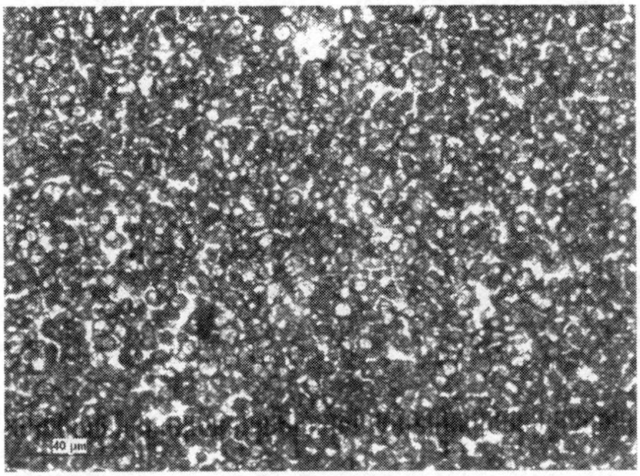

5.2 Microstructure of W-Cu material.

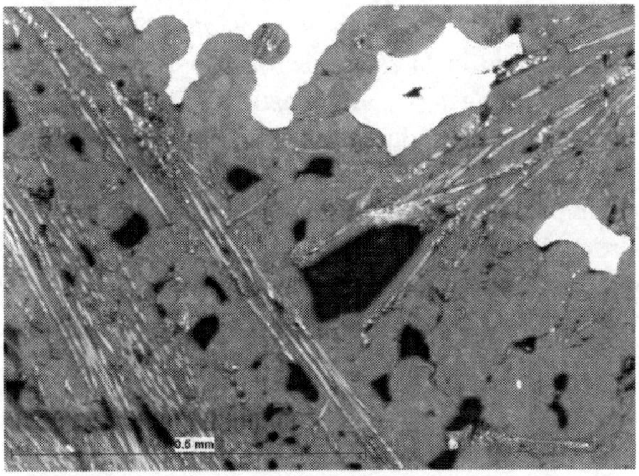

5.3 Microstructure of composite C-Cu. The white zone represents the copper.

5.3.2 Composite insulators

For HV circuit breakers, composite insulators are used increasingly in place of ceramic insulators. Indeed, composite insulators are light, resilient, do not explode under impact, have good seismic behaviour, and withstand pollution well. However, the ageing of such insulators is not well known, and their air-tightness is difficult to manage. The design of insulators, as well as choice of contact material, must be considered, because when the arc is

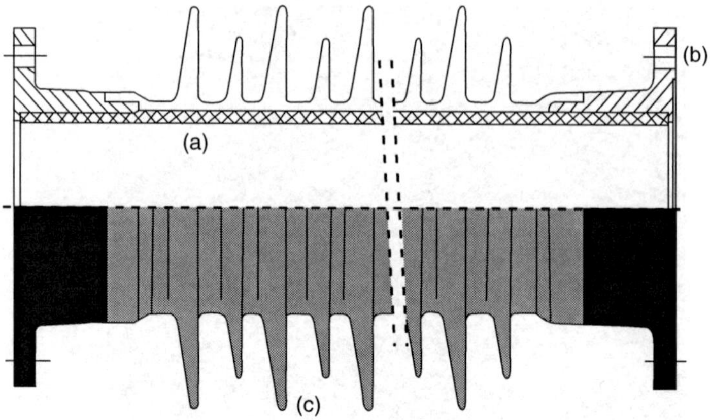

5.4 (a–c) Composite insulator structure. See text for explanation.

formed, metal droplets of contact and nozzle materials will be in contact with decomposed and polluted SF_6 (Domejean *et al.*, 1997).

Composite insulators used in HV circuit breakers generally comprise a composite tube (a), metal flanges (b), and elastomeric silicon sheds (c) – see Fig. 5.4. The composite is made of glass fibres and epoxy resin. The sheds are made out of silicone rubber.

During their lifetime of about 30 years, insulators must withstand the temperature requirements and heavy atmospheric conditions, and support mechanical stresses. Therefore, the qualification procedures of such insulators are very demanding, and are described in the IEC 61462 International Standard.

5.3.3 Thermoplastic insulators

Growing eco-design initiatives are aimed at taking account of environmental constraints at the design stage of the products (Froelich, 2000). Among the different aspects of eco-design, improving the recyclability of the HV apparatus is a major concern for the manufacturers, and is especially important in the case of polymeric materials.

In gas insulated substations (GIS), thermosets – epoxy resins are used to manufacture support insulators. To improve their recyclability, it is best to replace the thermoset with a thermoplastic material (Huet *et al.*, 2005).

Considering the main physical, mechanical, and electrical characteristics of the material, polyethyleneterephtalate (PETP) is considered a good candidate for replacing the current solution based on alumina filled epoxy resins. Under operating conditions, it has been shown that the main electrical properties of the material are not affected by ageing in the presence of SF_6. Crystalline PETP presents good electrical and thermal capability (Shugg, 1995) and is commercially available in thick plate form. It is also recyclable. Thus, the drawbacks of epoxy, such as poor recyclability, moulding cost, and shrinkage, can be avoided by using PETP.

For some applications, technical thermoplastics such as polyoxymethylene (POM) or polyetherimide (PEI) are also used in HV switchgear. However, these materials are expensive, and must be manufactured with an injection process.

Polyethylene (PE) is largely used in HV and medium voltage (MV) polymeric insulated cables (in the form of cross-linked material cross link polyethylene, XLPE). However, for application as a support insulator in HV switchgear, PE lacks the necessary mechanical characteristics.

Polyimide (PI), such as Kapton, is also used to a large extent, but this product in only available in the form of film.

5.3.4 Ester oils

Oil circuit breakers have significant use in our power systems. Thanks to their lubricating power for moving parts, as well as to their resistance to fire in some cases, they can be used as an arc extinguishing medium.

Immersing the interrupting contacts in oil will not prevent arc formation, but the heat of the arc immediately evaporates the surrounding oil, which dissociates in carbon and gaseous hydrogen. Oil has the following advantages as an arc extinguishing medium: formed hydrogen helps the arc extinguishing process, and also provides insulation for the live exposed contacts from the earthed portion of the container.

Ester oil has a high insulating capability and has replaced mineral oil in electric power apparatus because of its environmental advantages, as well as its high dielectric strength. Esters are also non-toxic to aquatic life, readily biodegradable, and have a lower volatility and higher flash points than mineral oil. Commonly used oils are rapeseed ester oil, lenne, and soya. The specifications for ester oils with electrical use are defined in IEC 61009.

5.4 Challenges and future trends

In the last 10 to 20 years, electricity markets and the electricity supply industry have made remarkable advances globally. The increase in energy consumption per capita, the growing world population, and the ageing of the networks have had the effect of increasing congestion and decreasing the power quality. In addition, energy deregulation, including unbundling of national markets, greenhouse gas reduction schemes, and the increase in use of renewable energy, has also changed the approach that electricity networks must now take. This renewal is a driving force behind finding advanced solutions and technologies in the field of electricity networks, encouraging energy efficiency as well as cost-efficient solutions and environmentally friendly infrastructures. The sections below review briefly three advances in the area of switchgear materials for T&D networks.

5.4.1 Simulations

The interruption of high short-circuit currents depends on the intensity of the gas blast (characterized by the volume in pressure Vt) at the instant of current passage through zero (Dufournet, 2002).

5.4.2 Use of nano-materials

Nano-technologies are gaining increasing importance in various economic sectors, including the field of energy production. Research into this area focuses on structures from hundreds of nanometres down to a few nanometres in size, with many potential applications, such as new conducting and insulating materials and coatings, and high-performance dielectrics.

Being able to create structures and materials at the atomic level has enhanced the development of materials that have less weight and improved stability and functionality, which in turn increases the efficiency of energy production. Recent developments include improvements in fire retardation and increased resistance to surface arcing, biodegradability, and resistance to corrosion, as well as improved mechanical and dielectric performance. There are, in fact, a number of areas where nano-technology finds useful application. Advanced performance nano-composites, using nanofillers instead of microfillers, demonstrate a number of attractive qualities, including improved resistance to surface discharge and thus a greater resistance to erosion. Nanostructures, for example inexpensive nano-diamond, could be used as particulates in dielectric fluids to increase thermal conductivity, which in turn increases the power densities in power transformers. Their use has also been explored in oil recovery. However, there are some limitations and there is still no set of parameters for the practical application of nano-materials. Nanostructures lack specification, purity, quality, and reproducibility, and it is currently only possible to produce them on a relatively small scale. Although nano-composites have great potential, we need to fully understand their properties and long-term ageing behaviour before their use in critical applications.

5.4.3 Eco-friendly design

Eco-friendly design aims at reducing the environmental impact of products and systems throughout a product's life cycle (from raw material extraction to the final disposal), in order to provide solutions consistent with the needs of customers involved in the operation and development of power systems and ensuring the security of supply (Directive 2005/32/EC, 2005).

Future practice, and growing environmental initiatives, will lead to the development of products and services which have environmental advantages, such as wider recyclability, lower energy consumption, and a longer life time (Froelich, 2000).

5.5 References

Aeschbach, H., O. Visata, D. Dufournet, J-L. Bessede and J. Blatter (2002), *Arcing-Contact/Insulating-Wall Interactions in High Voltage Circuit Breakers*, Proceeding of 21st International Conference on Electrical Contacts, Zurich, Switzerland, September 2002, pp. 511–517.

Browne, T.E. (1984), *Circuit Interruption: Theory and Techniques*, New York, Marcell Dekker.

Domejean, E., P. Chevrier, C. Fievet and P. Petit (1997), Arc-wall interaction modeling in a low-voltage circuit breaker, *Journal of Physics D: Applied Physics*, vol. **30**, 2132–2142.

Dufournet, D., P. Kirchesch and C. Lindner (2003), *Switching Technologies for HV Switchgear*, 4th International Conference on Power Transmission & Distribution Technology, Changsha, Hunan Province, China, 14–16 October 2003.

Dufournet, D., J.M. Wilieme and G. Montillet (2002), Design and implementation of a SF 6 interrupting chamber applied to low range generator circuit breakers suitable for interruption of current having a non-zero passage, *IEEE Transactions on Power Delivery*, vol. **17**, no. 4, 963–968.

Ferry, L., G. Vigier and J-L. Bessede (1996), Effect of ultraviolet radiation on polytetraflouroethylene: morphology influence, *Polymers for Advanced Technologies*, vol. **7**, 493–500.

Froelich, D. (2000), *Towards Eco-design of Products*, MEIE 2000 2nd European Conference on Industrial Electrical Equipment and Environment, Paris, June 2000, pp. 107–109.

Huet, I., H. Aeschbach, C. Tschannen, K. Pohlink and J.L. Bessede (2005), *Application of the Concepts of Eco-Design to a Gas Insulated Substation 72.5 kV*, IEEE General Meeting, San Francisco, USA, June 2005.

Robin-Jouan, Ph. and N. Kairouani (2002), *Numerical and Experimental Analysis of the Propagation of Hot Arc Plasma in High Voltage Circuit-breakers*, Proceedings Congress Gas Discharges, Liverpool, UK, September 2002, pp. 103–105.

Shugg, W.T. (1995), *Handbook of Electrical and Electronic Insulating Materials*, 2nd Ed., Willey-IEEE Press.

6
High Voltage Direct Current (HVDC) electric power transmission systems

D. VAN HERTEM, University of Leuven, Belgium and M. DELIMAR, University of Zagreb, Croatia

DOI: 10.1533/9780857097378.2.143

Abstract: HVDC is a technology which has been used for long-distance bulk power transfer, connecting different synchronous zones or submarine connections. Lately, HVDC has experienced a revival. In Europe, this revival is being driven by the strong increase in generation from renewable energy sources, the liberalization of the energy system, and the difficulty in investing in new overhead transmission. In countries such as China, India and Brazil, the development is driven by a fast growth in energy consumption and the need for large amounts of energy, often from remote locations. HVDC transmission uses power electronic converters to interface between the alternating current (AC) grid and the direct current (DC) grid. Two types of HVDC exist: the traditional line commutated converter (LCC), which uses thyristors, and the voltage source converter (VSC), which uses insulated gate bipolar transistors (IGBT). LCC HVDC has the highest available ratings, but VSC offers better controllability and easier integration of cross-linked polyethylene (XLPE) cables for bidirectional power flow. HVDC is seen as the transmission technology of the future, specifically as the connection technology for renewable energy sources but also for a new backbone system. This new backbone system is expected to be in the form of a DC grid, with considerable parts of it being built using cable connections.

Key words: high voltage direct current (HVDC), electric power transmission, grid development.

6.1 Introduction

HVDC electric power transmission systems use DC for bulk transmission of electrical power, as opposed to the common AC systems.

Most of the early power systems were based on DC, but due to the invention of AC motors and transformers in mid 1880s, the famous 'War of Currents' broke out, involving not only Edison and Tesla, but a large number companies in Europe and North America whose investments in one type of system led them to hope that the use of the other type would decline. In the early 1890s it was clear that AC had won and that power systems would continue developing mostly on AC technology, which indeed was the case. Nevertheless, DC technology never died out, and recently DC systems have been growing in importance in modern power systems.

HVDC allows power transmission between unsynchronized AC systems, and can increase system stability by controlling power flow and help to prevent failures typical for AC grids. For long-distance transmission, HVDC systems may be less expensive as they suffer lower losses. For (underwater) power cables, HVDC avoids the problems caused by the cable's capacitance.

The first transmission of electrical energy over a large distance used DC. The Miesbach-Munich Power Transmission was developed in 1882, connecting a steam engine situated near Miesbach and the glass palace of Munich, where world's first international electricity exhibition was held. 2.5 kW of power was successfully transmitted with a 2 kV DC line over a distance of 57 km (Arrillaga, 1998).

However, AC became the main technology to transmit power because of the ease of changing voltage levels, the possibility to form a rotating field, and the ease of interrupting AC currents (compared to DC).

In the 1950s, DC technology made a comeback. Early commercial installations on HVDC lines include a 100 km, 200 kV, 30 MW line between Moscow and Kashira, and a 98 km line between Gotland and mainland Sweden, respectively built in 1951 and 1954 (Hingorani, 1996).

The longest HVDC links in the world are currently the Xiangjiaba-Shanghai 2071 km, 6400 MW link connecting the Xiangjiaba Dam to Shanghai (China) which has been in operation since 2010, and the Rio Madeira link connecting the Amazonas to the São Paulo area, where the length of the DC line is over 2500 km. This installation was expected to be completed in 2012.[1]

A large number of HVDC lines are present in the European power system and many more are expected in the near future (see Fig. 6.1).

6.1.1 Traditional uses of HVDC

Existing high voltage transmission systems consist largely of AC lines, with a few DC connections. There are three important cases in which it is preferable to use HVDC over AC:

[1] Information about reference projects can be found on the websites of the main manufacturers: ABB (http;//www.abb.com/hvdc), Alstom Grid (http://www.alstom.com/grid/products-and-services/engineered-energy-solutions/hvdc-transmission-systems/) and Siemens (http://www.energy.siemens.com/us/en/power-transmission/hvdc/index.htm)

6.1 List of installed HVDC systems in northern Europe (full lines) and proposed HVDC lines (dashed lines). (*Source*: Adapted from J. Messerly, 'HVDC Europe', Wikimedia Commons. Released under Creative Commons Attribution – ShareAlike.)

1. The first application is the connection of asynchronous networks. It is clear that systems at a different frequency cannot be connected using an AC connection. Even systems operating at the same frequency, but not operating synchronously, cannot be connected using AC technology. HVDC by contrast separates the two systems by a DC link, making the connection feasible. In some cases, a back-to-back HVDC connection is used, with the two converter stations directly connected to each other.
2. A second application is the use of long-distance cable. As the length of an AC cable increases, the charging current increases. At a certain point this charging current 'fills the cable' with reactive current. As a consequence, no more 'useful' power can flow through the line. Therefore, long cable connections need reactive compensation, spread over the length of the cable. This compensation usually takes the form of shunt reactors. For submarine cables, installing this compensation is not trivial.
3. The third application is the transmission of bulk power over long distances. AC is significantly cheaper over short distances, while HVDC is more economical over longer distances. The 'break-even' point is about 400–800 km for land connections, and 40–80 km for submarine cables. Also, as stated before, there are no length limitations for DC connections (Section 6.2).

6.1.2 Revival of HVDC technology

Although HVDC has been in use for well over 50 years, its application has been limited. Recently, different evolutions have driven the development of HVDC technology.

In the developed world, and especially in Europe, the targets given by energy policy are aimed at a more sustainable energy system that is competitive and cost-effective and, at the same time, offers sufficient security of supply. This has resulted in a considerable increase in renewable energy production, specifically wind and solar, which are variable in nature. The increased penetration of undispatchable distributed generation resources (e.g. wind power and heat driven combined heat and power) gives rise to a high amount of uncertainties in the international grid. Variable energy generation results in variable energy flows throughout the meshed power system. Furthermore, the large and concentrated renewable sources are often placed at locations far from the load centres (e.g. offshore wind farms), and thus require additional transmission lines. The liberalization of the electricity sector has resulted in unbundled systems in which more market players are active. This causes decoupling of generation and load planning, and increasing transmission distances as well as reduced (free) control options for the system operator. Furthermore, market operations, with hourly commitments of generation and load based on price incentives, also increase the variable energy flows.

Although these developments have increased the need for electricity transmission as the system is working closer to its limits, investments in new transmission lines have been lacking, mainly because of opposition to new overhead lines, and regulatory issues in obtaining permits.

HVDC is seen as a potential solution, because HVDC cable solutions are possible whereas AC solutions are not. An example is the INELFE project between France and Spain: after a permission struggle of more than 20 years, it was decided to develop the line using VSC HVDC. This new circuit will be roughly 60 km in length, using underground cables (Labra Francos, 2012).

Another example is the recent development of offshore wind. As the locations are increasingly farther from shore, for these systems, HVDC forms the most techno-economical solution, and several such projects are currently ongoing (e.g. Borwin 1 and 2, Dolwin 1 and 2), with more being planned.

In countries in which the development of the economy is causing a fast growth in the energy consumption, such as China, India and Brazil, HVDC is seen as a viable solution for the transmission of bulk electricity over large distances. The fast evolution in these countries is driving the development of HVDC towards larger systems, both electrically (power and voltage) and with respect to line lengths.

6.2 AC or DC?

When installing a new transmission line, the grid owner or the investor has to decide which technology is going to be used: AC or DC. AC systems have the advantage of being a well-known technology, offering a cheap and reliable solution for high power transmission using classic overhead lines. AC systems are capable of carrying large quantities of electric power over large distances when ultrahigh voltage (UHV), 1000 kV AC or even higher, is used. AC systems using UHV voltages were developed in the 1970s, and are now planned and installed in China and Japan. However, the higher the voltage is, the larger are the towers, thus requiring more space and leading to higher visibility. Given the existing objections against overhead transmission lines, it is very difficult to get approval (and permits) for such investments (Van Hertem, 2010).

The technical reasons why UHV AC is experiencing a reduced acceptance as a suitable transmission system investment technology compared to HVDC are:

- DC lines have lower losses (no skin effect, no proximity effect);
- DC systems have no practical length limitations. When AC cable connections are used, a line of some tens of km is already considered long (at transmission voltages);
- over longer distances, DC systems are cheaper to build and operate;
- DC systems do not emit varying electromagnetic fields;
- HVDC offers an inherent active power control, making it more flexible (controllable) in use. HVDC can be used to alleviate overloads elsewhere in the system;
- AC cables experience a continuous high charging current which limits their length;
- long AC cables at very high voltages are difficult to construct and are expensive;
- offshore resources, as well as connections outside the main continent, are virtually inaccessible when using AC.

A simple economic comparison shows the benefits of DC cables over AC (Fig. 6.2). The power that can be transmitted through a DC connection (three DC circuits, six wires) is 1.5 (overhead lines) to three (cables) times that of an AC link (two AC circuits, six wires) when the same insulation level and equal losses are considered. When comparing an equal power transfer, DC link losses are 0.3 (cable) to 0.7 (overhead line) times that of a comparable AC line. However, power electronic converters have significantly higher losses than the corresponding AC substation.

HVDC systems also have a number of disadvantages:

- power electronic converters are (very) expensive;
- power electronic converters have considerable losses (0.8–1% per converter);

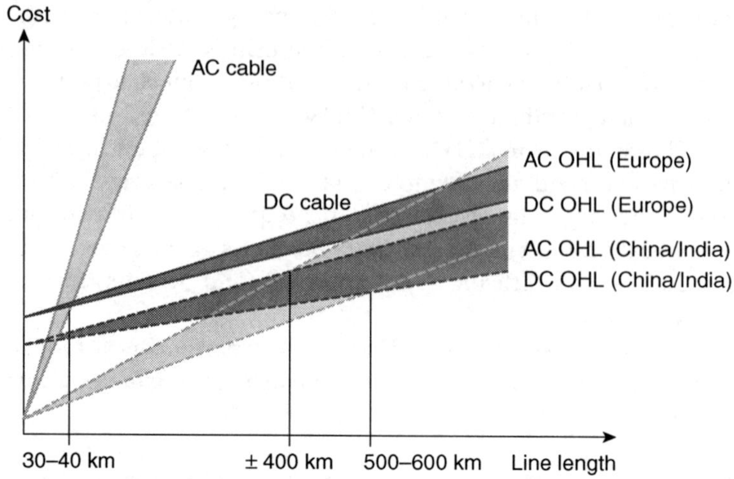

6.2 Comparing AC and DC technology costs, both for overhead and cable systems. The difference between costs for Europe and China/India are based on the additional requirements that are posed to overhead lines, and which in Europe often lead to additional costs or detours. OHL, Overhead line.

- depending on the scheme (see Sections 6.4.1 and 6.4.2), harmonics are generated which need to be filtered, significant reactive power compensation might be needed and commutation failures might occur;
- the ability to control the HVDC link makes system operations more complex, especially if multiple devices are present and when multiple zones are influenced;
- creating multi-terminal configurations or even grids is not as simple as with AC grids.

6.3 HVDC configurations

A HVDC connection can be built in different configurations, depending on the earthing, converter connections, asymmetric or symmetric connections, etc., (Fig. 6.3).

An asymmetric monopolar link, or simply monopole, has one conductor at a high DC voltage and uses the ground or the sea as the return path. In many cases, a separate return wire is used, also called metallic return, which avoids currents through the ground under normal operation. Since the corona effect is more significant with positive than negative voltages, the monopolar link is normally operated at negative polarity in case overhead lines are used (and when the flow is predominantly in a single direction).

Although systems with a single conductor and earth or sea return are a technically easy and economically beneficial solution, they are less used because of the possible effects of corrosion on metallic pipes in the ground, and the possible negative environmental effects on living creatures. However, several monopolar configurations are currently in operation, amongst others the Baltic cable (connecting Germany and Sweden power systems) and the GRITA connection between Italy and Greece.

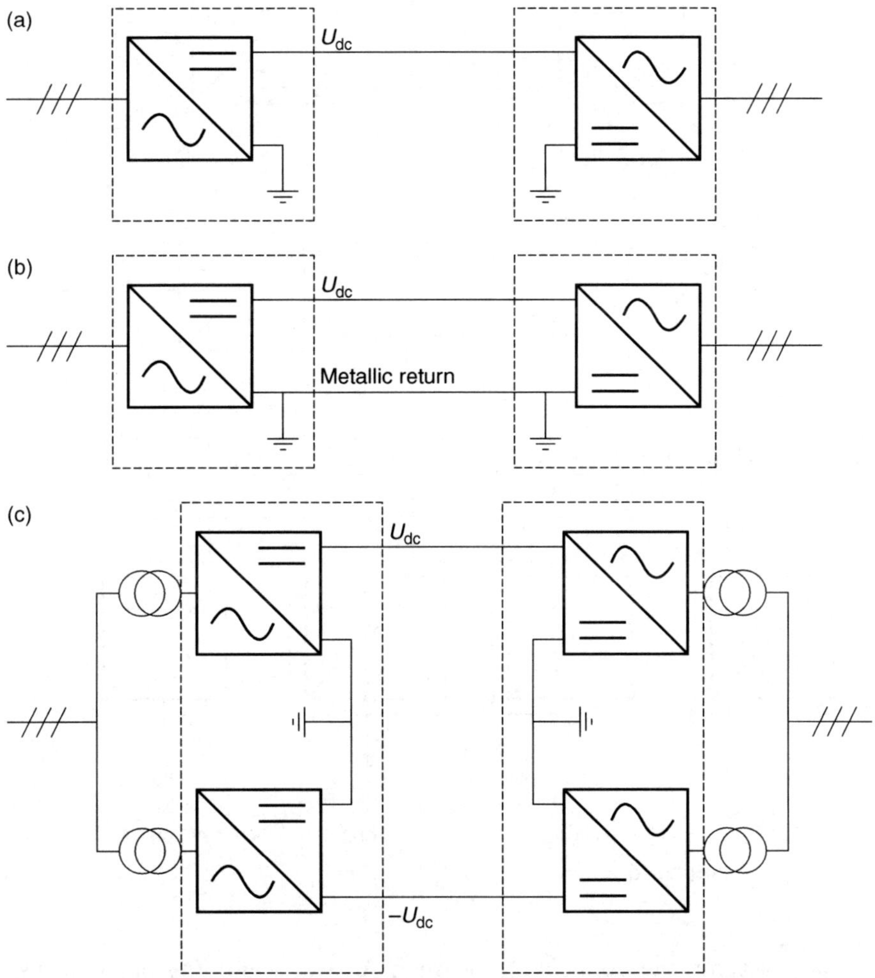

6.3 Possible configurations of HVDC systems (a) monopolar, (b) monopolar with metallic return, (c) bipolar, (d) bipolar with metallic return, (e) symmetric monopolar, (f) multi-terminal.

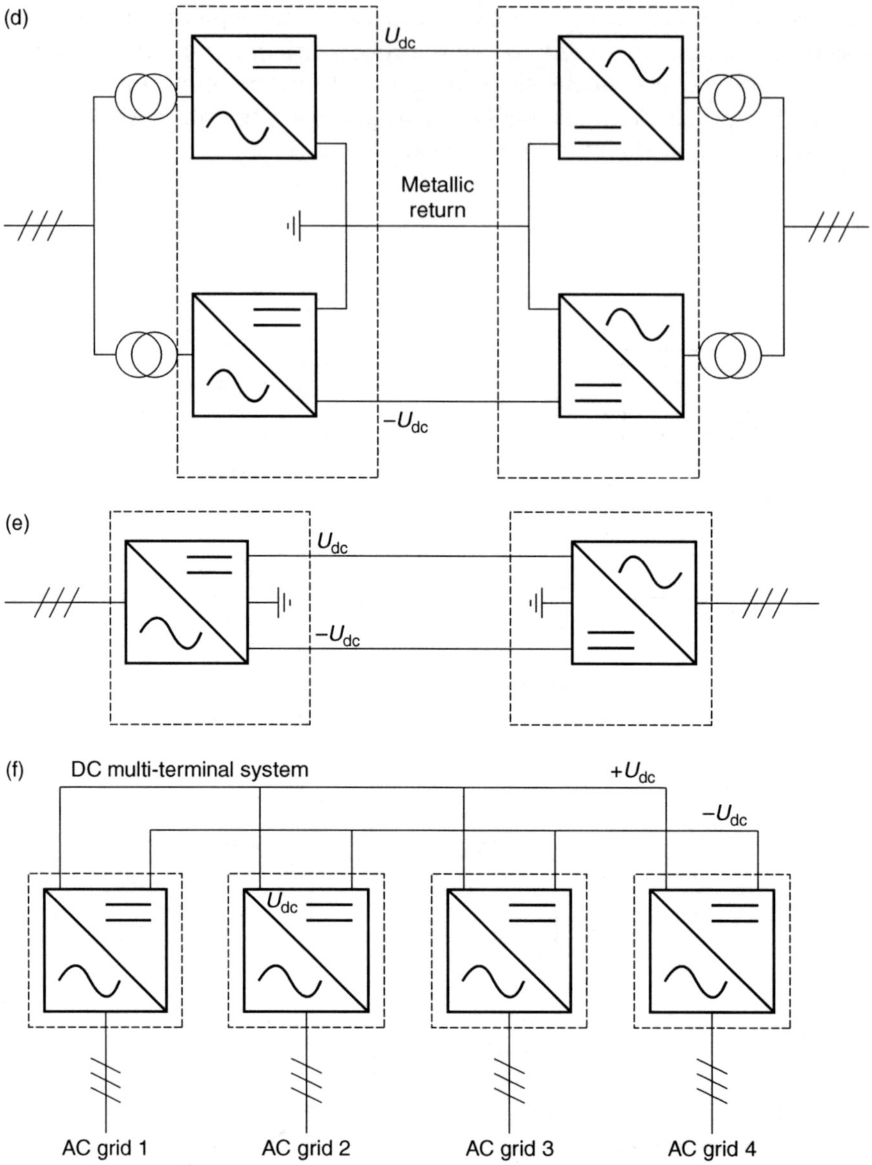

6.3 Continued

The Basslink, the connection between Australia and Tasmania (in use since 2005), is also of the monopolar type with metallic return. Also the connection between Sweden and Poland (SwePol) is of this type.

With the advent of VSC HVDC, symmetrical monopoles were introduced. This configuration uses a monopole (a single converter per terminal) with a midpoint earth connection. As such, the voltage on the transmission lines is equal but opposite (as in a bipolar configuration). In case of an outage on a single line, the earth cannot be used as a return wire.

The bipolar link is the most commonly used topology for modern HVDC connections of high power ratings. Each terminal has two converters of equal rating. Between the converters, the midpoint can be grounded on one or both sides. In case both sides are grounded, there is no current flowing through the ground during normal operation and the two converters are operated symmetrically. If one line is out due to a fault, the other circuit can remain in operation and the system works at half capacity. Here the ground serves as an emergency conductor.

The advantage of systems with a metallic return wire is that the metallic return only has to be insulated for a very low voltage (few kV, R·I). This significantly reduces the need for insulation material around that wire, and hence the cost. However, this needs to be weighed against the following consideration: for the same power transfer, the maximum voltage rating (and with it the insulation) of a single conductor 'under voltage' has to be double that of the two conductors in a bipolar set-up. The metallic return is more commonly used as a backup connection so that during maintenance/fault of one module/cable of a bipolar link, 50% of the power can still be delivered. It is also used when a staged development occurs (first monopolar and later bipolar configuration).

In a back-to-back connection, the two converters are located in close proximity to each other. The typical application is the separation of two independent power systems. These systems need not be synchronized and can operate on a different nominal frequency. The DC voltage in the intermediate circuit can be selected freely at HVDC back-to-back stations because of the short conductor length. Because of the short distance, the line resistance is low, and the DC voltage is kept relatively low. This allows for a small valve hall and avoids parallel switching of valves. For these reasons, HVDC back-to-back stations use valves with the highest available current rating.

Multi-terminal connections have multiple (parallel) connections in which, generally, power can be injected or withdrawn from the DC system. Multi-terminal HVDC will be addressed in greater detail in Section 6.6.

6.4 HVDC equipment and components

An HVDC system converts the three-phase AC into DC using power electronic converters called rectifier. A DC circuit then transports the energy towards another power electronic converter which in turn converts the DC

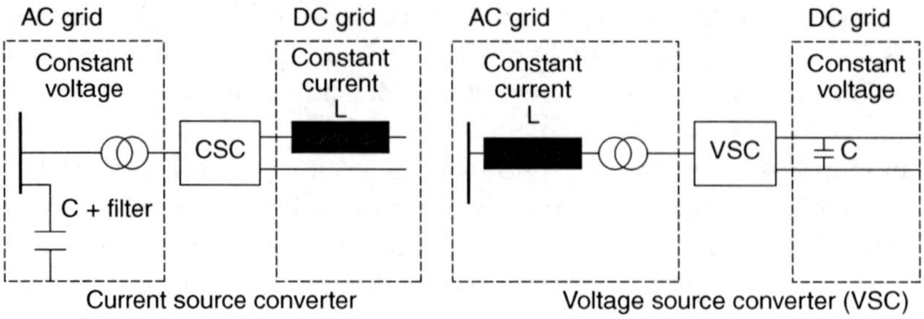

6.4 CSC (LCC) and VSC for HVDC systems.

power into AC. The latter device is called the inverter. This section describes the different components, technologies used and their limitations.

6.4.1 Two types of converter technologies

Converters are the most important component of the DC system as they transform AC into DC and vise versa. Power electronic AC–DC converters (for transmission systems) can be divided into two main categories (Fig. 6.4): Current Source Converter (CSC) and VSC. The difference is related to the choice of power electronic components (valves or switches) and their operation. This choice in technology also has an influence on the DC bus voltage and current and the equipment that is needed.

For HVDC applications, the CSC converter is more widespread. It utilizes the thyristor as a switching device. Installations using this topology are usually referred to as LCC HVDC, or as classical HVDC (Fig. 6.5).

Basic VSC converter technology has been available for quite a long time, as it is also used in variable speed drives. However, only after the development of self-commutating power electronic components, such as the gate turn–off thyristor (GTO) and the IGBT, for sufficiently high power ratings, together with the increased computational power of digital signal processors (DSP), it was possible to use this technology in power transmission. The first VSC HVDC system was built at the end of the 1990s (Arrilaga, 2011).

As both technologies have a different set-up and use different equipment, they will be treated separately.

6.4.2 LCC HVDC

Converters

LCCs use thyristor valves to perform the conversion from AC to DC and vice versa. Thyristors can be switched on using an impulse, but need current

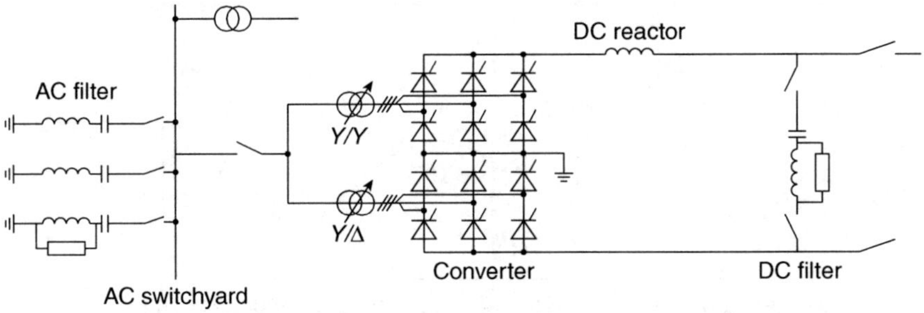

6.5 LCC HVDC system configuration.

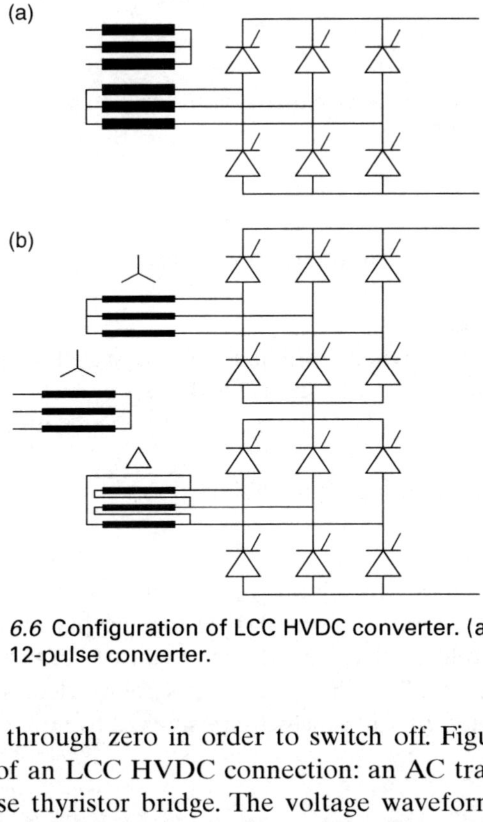

6.6 Configuration of LCC HVDC converter. (a) 6-pulse converter, (b) 12-pulse converter.

to pass through zero in order to switch off. Figure 6.6a depicts the basic circuit of an LCC HVDC connection: an AC transformer is connected to a 6-pulse thyristor bridge. The voltage waveforms for the 6-pulse bridge are given in Fig. 6.7. In order to reduce harmonics in the AC system, additional harmonic filters are needed. A significant reduction of the harmonics can also be obtained by utilizing converters with a higher order of pulses: 12-, 18- or even 24-pulse converters. Most LCC HVDC configurations use 12-pulse converters (Fig. 6.6b), which cancel out the 5th, 7th, 17th, 19th, … (6*k +/– 1) harmonics.

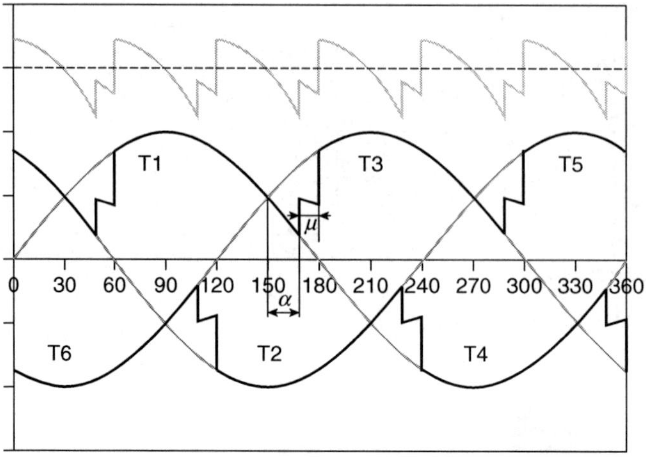

6.7 Voltage waveform of a 6-pulse thyristor bridge with firing angle α, and commutation angle μ.

Thyristors

The thyristor valves have to perform the following tasks:

- connect AC and DC sides;
- conduct the rated current when switched on (no parallel operation within one converter arm); up to 4.5 kA is available;
- block a high voltage when not conducting (rated voltages up to 10 kV are available);
- control the DC voltage via the ignition angle;
- fault tolerant and robust.

Since the DC voltage in HVDC installations is higher than the maximum blocking voltage of a single thyristor, several thyristors are placed in series. Similarly, the current capabilities of a single thyristor can ask for the connection of several thyristors in parallel, which however is not needed. Normally the thyristors are installed in modules, which are stacked to make maintenance easier.

Transformers

Theoretically, transformers are not always needed in an HVDC installation. However, they are (virtually) always used in order to optimize the voltage of the DC transmission grid independently of the AC system. Furthermore they allow the use of tap-changers, which reduce the losses through optimal voltage control, using rectifiers with a higher pulse number, and they limit the short circuit current.

Depending on the situation, different transformers and rectifiers can be used to obtain a techno-economically optimal solution. Because of their size, the choice of a transformer often depends on the available transport means, the end location, the construction site, or a combination of the above.

Filters and reactive compensation

An LCC HVDC link requires several filters and reactive compensation devices. As the current is always lagging the voltage due to the firing angle delay of the converters, the LCC HVDC link always consumes reactive power. If there are no nearby generators that can compensate for this power, additional capacitors, a Static Var Compensator (SVC) or STATCOM (Static Compensator) has to be installed. Furthermore, the switching of thyristors causes high harmonic distortions on the power system voltage. These harmonics have to be mitigated by additional filters. These installations add a considerable cost and considerable volume to the substation.

On the DC side, another filter is placed to smooth the current. This filter is formed by a large DC reactor.

6.4.3 VSC HVDC

The development of new self-commutated semiconductor switches for high power applications, especially the IGBT, has led to the emergence of VSC-based applications and has provided a new HVDC technology (Fig. 6.8).

Both VSC and LCC HVDC enable full active power flow control. An additional advantage of the VSC HVDC technology is its ability to control the reactive power at both end terminals, independently of the active power flow (within the operating limits). Furthermore, the control of a VSC HVDC link can be far more dynamic than that of an LCC HVDC link. A result of this improved control is the possibility to connect to weak power systems, and even to provide a rotating field for offshore wind farms. This is not possible when using LCC HVDC as the thyristors are not able to commutate unless a strong AC grid is present.

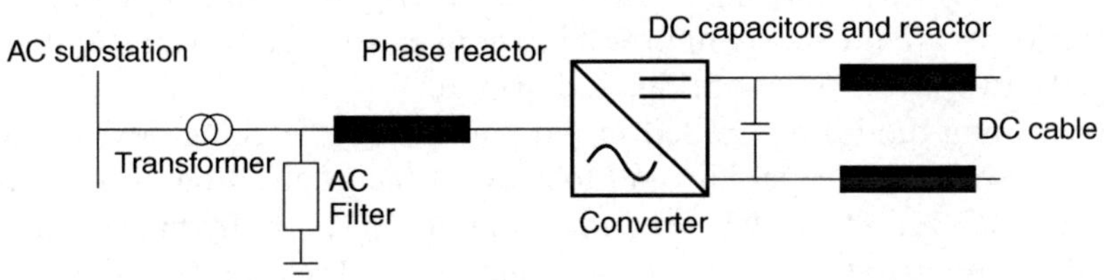

6.8 VSC HVDC configuration.

Currently, the maximum installed power rating is 400 MW using ± 200 kV. This limit is still rising as more installations are being built. At the moment, VSC HVDC installations of up to 1.2 GW are available and planned. However, the VSC HVDC installations come at a higher cost and incorporate higher losses. The losses in the newest VSC HVDC systems are around 1% per converter at full load, while LCC systems have about 0.8% losses per converter.

The first VSC HVDC installation started its operation in 1997. It has a rating of 3 MW, with a ±10 kV DC voltage, and was a prototype. Currently there are about ten active installations worldwide, with several new projects ongoing or planned.

IGBT valves

Although theoretically other semiconductor valves, such as MOSFETs or GTOs, could be used, VSC HVDC always uses voltage-driven IGBT technology. Currently, IGBTs are available up to 6 kV, although usually valves with a lower rating are used. Depending on the current rating, a number of IGBTs are placed in parallel (usually 2, 4 or 6). According to the desired DC voltage, several IGBTs are installed in series. For a 150 kV converter station, this implies about 300 series components. A few extra, redundant components are placed in the series chain in order to reduce the number of shutdowns due to IGBT failure. In the Cross Sound VSC HVDC (New York, USA) installation, an average IGBT failure rate of 0.25%/year was measured during seven years of operation (Dodds, 2010).

In order to improve the electromagnetic compatibility, the valves are shielded by a steel and aluminium enclosure. To carry away the losses, the IGBTs need to be water cooled.

VSC converter

The first VSC HVDC converters used PWM modulation, very similar to traditional variable speed drives. The switching frequency of these converters can be up to 2 kHz. This high frequency reduces the generated harmonics significantly, so that transformers with tertiary windings (such as with the 12-pulse configuration) are not needed. However, as each switching operation causes losses, the high switching frequency increases total losses. Switching losses are the most important losses in existing converter stations. In Fig. 6.9 a two-level 3-phase bridge is depicted.

Free-wheeling diodes are added in parallel to the switching devices to ensure reverse current capability and to prevent the application of reverse voltage. The IGBTs are switched at a fixed multiple (typically higher than 20) of the fundamental grid frequency. For a two-level bridge, the AC

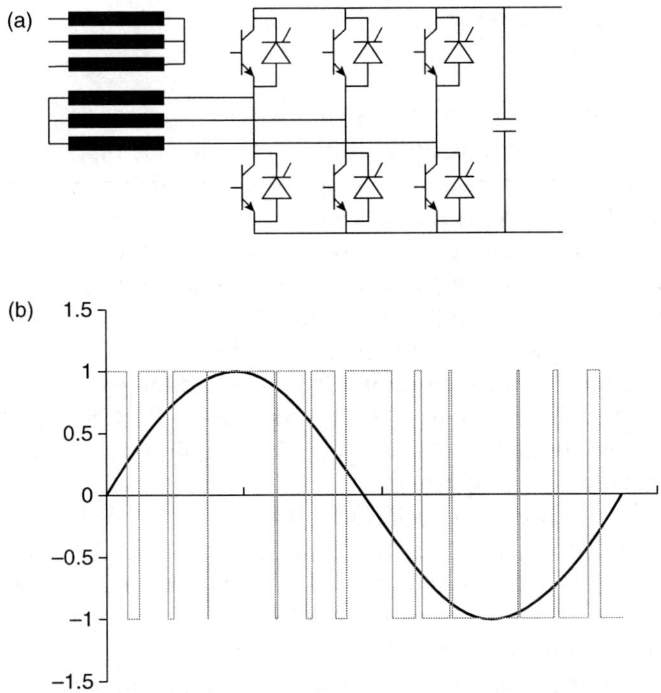

6.9 (a) 6-pulse bridge for the VSC HVDC converter and the PWM signal. (b) An AC voltage waveform built out of positive and negative DC pulses for a two-level bridge.

voltage waveform is built out of positive and negative DC pulses as shown in Fig. 6.9b. In order to lower the harmonic distortion of the output voltages, higher level converter bridges are used. A three-level neutral point clamp was used for second-generation VSC HVDC configuration such as the Murray Link in Australia and the Cross Sound Cable (USA).

The current trend is to move towards multi-level converters, which consist of a number of series-connected cells that individually switch their components to obtain the three-phase waveform. The multi-level converters reduce not only the harmonic content of the AC signal, but also have significantly lower losses as they reduce the number of switching operations per switch. However, with multiple levels of switches that are operated differently, the converter complexity increases rapidly with the number of voltage levels. These higher level topologies have the following advantages:

- additional degrees of freedom (amplitude);
- fewer switching operations;

- lower voltage per switch;
- lower power losses.

The three manufacturers of VSC HVDC (ABB, Alstom Grid and Siemens) have recently moved towards the multi-level converter topology, each different on a power electronic level, but operating in a similar manner towards the AC and DC grid side. These different implementations of VSC converters are described in more detail in respectively Jacobson (2010), Alstom Grid (2011) and Dorn (2008), and a review comparing them is provided in Glasdam (2012) and Ahmed (2011). In this chapter, only a short description of the different topologies and their schematics is given.

The traditional 2- (or 3)-level converter switches the entire (or half) the DC voltage at a high frequency. These high voltage steps at high frequencies are the cause of higher losses, higher filter requirements and additional requirements for the transformers.

The Modular Multilevel Converter (MMC or M2C), that was proposed by Lesnicar and Marquardt (Lesnicar, 2003) and first introduced for HVDC by Siemens, is now gaining popularity due to its advantages over conventional topologies. The basic component of an MMC is called a submodule. The number of submodules can be increased or decreased as the number of levels increases or decreases to get the desired output voltage.

The MMC technology (Fig. 6.10a) avoids the high switching voltages by individually switching the submodules and, as such, creating smaller voltage steps. These separate voltages are not only smaller, the required frequency is also considerably lower (100–150 Hz compared to 1–2 kHz), both having a positive effect on the losses and the equipment requirements. The cascaded two-level topology developed by ABB uses a similar concept to the MMC topology. One important difference is that it uses press-pack IGBTs in the valves as in the traditional two-level converter.

The multi-level topology as commercialized by Alstom is called 'MaxSine'. Alstom has also proposed a second-generation converter (Fig. 6.10b), which is a hybrid converter. It tries to take advantage of the MMC (lower losses and lower equipment requirements), as well as from the traditional two-level converter (fewer power electronic components). This is done by using a two-level converter with a low switching frequency as the main switching component and adding the remainder of the sinus waveform with a multi-level converter.

Although the different multi-level designs show differences regarding the components used and the internal controls (switching of the components), the system characteristics are similar. The converter topologies are, however, continuously evolving. One potential evolution is the move towards full-bridge converters instead of half-bridge converters (Fig. 6.10a). The full-bridge converter is able to limit currents during faults, but it is more expensive and experiences higher losses.

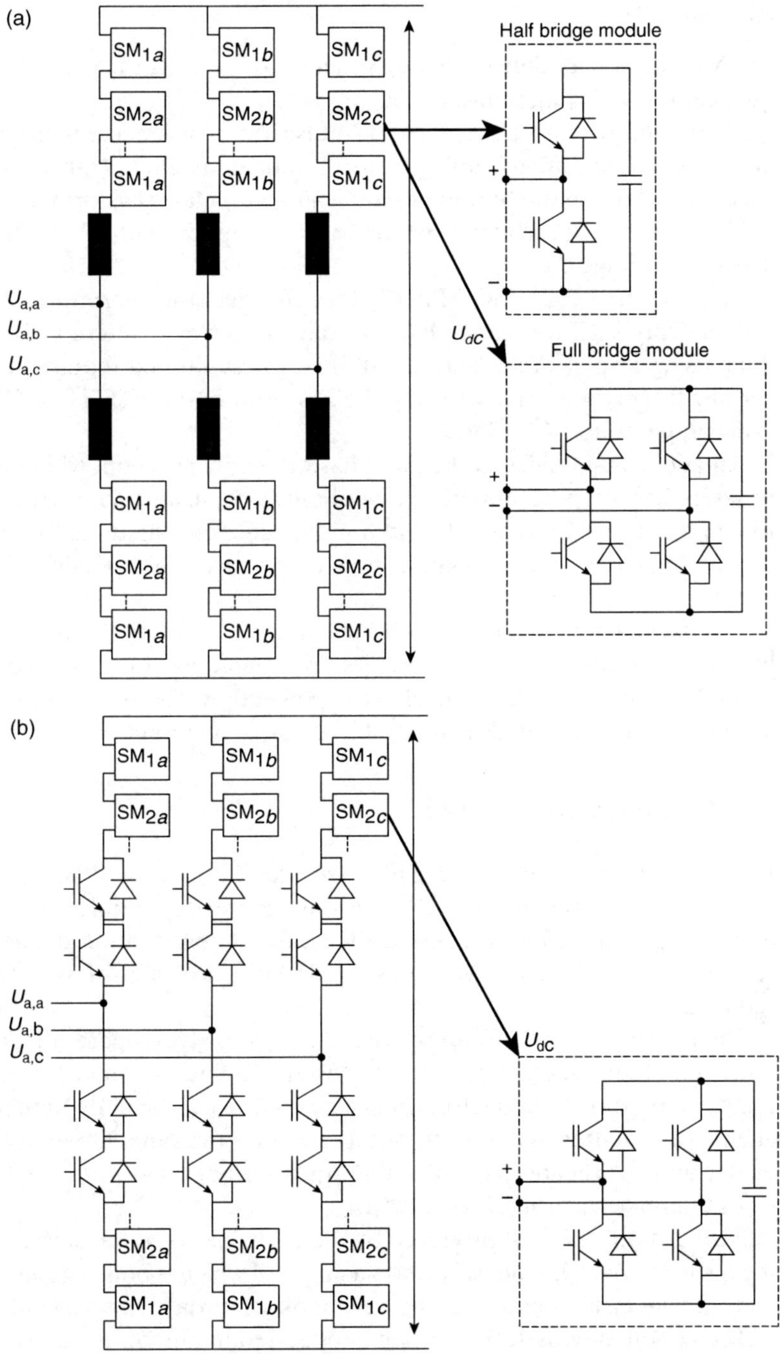

6.10 Concept schematic of (a) the VSC multi-level converter, which can make use of the standard half-bridge or full-bridge modules (b) a hybrid multi-level converter.

AC equipment

The VSC converter delivers a much cleaner AC waveform, which has consequences on the equipment needed.

Firstly, there is no need for a 12-pulse transformer configuration, and hence no special three-winding transformers are needed. Transformers also do not need to handle harmonics and can be regular transformers.

However, the transformer needs to be equipped with a tap-changer to assist the voltage control.

Contrary to LCC, VSC HVDC does not generate harmonics of lower orders. Only high frequency PWM signals have to be filtered. This can be done using a single, relatively small filter. These limited requirements concerning filters significantly reduce the footprint of the VSC HVDC installation compared to LCC HVDC.

An additional device is the AC phase reactor. This coil, which is placed in series, has a typical impedance of about 0.15 pu, and is mostly cooled by forced air. The AC reactor has two main functions: firstly it forms a filter for the high frequency switching signal, and secondly it facilitates voltage control.

The AC reactors block the PWM signal from entering the power system, limit the rise rate of the short circuit current and provide a constant fundamental frequency for the control of the power flow. The control of the power exchange at the terminals is treated in detail in Section 6.5.

6.4.4 Lines and cables

Both DC overhead lines and cables are used in practice. When comparing AC and DC overhead lines, DC lines are more susceptible to flash-overs due to the dust on insulators and the constant polarization of the air surrounding the conductors. When using HVDC, the Right-of-Way (RoW) is reduced.

The permissible load of an overhead line, at a given operating voltage, is limited by thermal expansion (and consequently its sag) and the annealing temperature of the conductor. Cables by contrast, are mainly limited by the ageing of insulator material. In order to assure the same life expectancy of cables and overhead lines, the cable must be operated at lower temperatures than bare overhead conductors.

Furthermore, the heat produced by the cable has to be transferred through the insulator (which is nearly always a material which is both a good electrical and thermal insulator), and in the case of direct burial, the surrounding soil.

The conducting material can be copper or aluminium for cables as well as for overhead lines. Copper has a higher conductivity, resulting in thinner cables, while aluminium has a lower density and cost per kilogram. This lower density is especially important for overhead lines as it reduces the weight.

Currently the relative pricing of both materials makes aluminium the preferred choice, especially for overhead lines where almost no copper is found anymore. Copper is still used for cable connections with very high power ratings to avoid the cables becoming too thick.

Since the heat requirements of a cable system are more stringent than those of an overhead line, a cable must have a higher conductor section for an equal power. As a consequence, the cable resistance is typically lower than that of overhead lines of equal rating.

Because of the narrower spacing possible with cables, the inductance is generally lower. However, a cable has a much higher capacitance as the distance between the conductor and the ground/shielding is much smaller. As a consequence, the charging current of cables is significantly higher, even up to a degree in which 400 kV systems need reactive compensation every 20–40 km. Of course, inductance, capacitance and charging current are less important when considering DC connections. As such, one of the main applications of HVDC is long interconnections where no overhead connection is possible, such as long-distance underwater cables.

There are three main types of cable used for HVDC connections: mass impregnated, self-contained fluid filled (SCFF) and extruded (Peschke, 1999).

Mass impregnated (MI) cables

MI cables are the most commonly used and reliable cable solution for HVDC systems. The insulation material is paper, impregnated with a high viscosity compound. These cables are available for up to 500 kV DC, with a maximum cable section of 2500 mm^2. This large cross-section makes the cable very heavy (30–60 kg/m) and difficult to install. The maximum length of a single section segment of land cable is not limited by production limitation or the reactive properties of the cable, but by the maximum weight and diameter of the cable drum. This means that the thickest cables are limited to about 1 km. The different pieces are connected using joints. The jointing of the different pieces forms an important part of the costs and the duration of the laying process and the reliability of the cable. Submarine cables can be much longer (up to 100 km) as they can be placed on special cable boats.

Self-contained fluid filled (SCFF)

SCFF cables also use paper as the insulating material, but instead of high viscosity oil, low viscosity oil is used. They are used for very high voltages (up to 600 kV qualified, up to 500 kV used) and for short connections, in which there are no hydraulic limitations in order to cool the cable during thermal transients. The SCFF cable is available in sizes up to 3000 mm^2 and typically weighs 40–80 kg/m for the largest sections. The diameter of a single conductor can be up to 160 mm.

Cross-linked polyethylene (XLPE)

A new development for HVDC applications, mainly for VSC HVDC, is the use of extruded cables (XLPE). Compared to oil-filled cables, these cables are currently only available for relatively low voltages (200 kV installed, 320 kV available), and can only be used when the power flow can be reversed without reversing the voltage polarity on the cable. This is important because the extruded insulation (and PE material in general) can be subjected to an uneven distribution of charges, which can migrate inside the insulation due to the electric field. Therefore, these charges can accumulate in localized areas inside the insulation (space charges) which, especially with rapid reversal of the voltage polarity, can give rise to high localized stress. This in turn leads to accelerated ageing of the insulation material. The main problems of the XLPE cable for DC are given by Terashima (1998):

- the DC breakdown strength at high temperatures is low and largely depends on the insulation thickness;
- the DC breakdown strength drops sharply when the polarity reverses, or when an opposite polarity lightning is superposed on the DC pre-stress.

These properties make the use of extruded cables inappropriate in systems in which the voltage polarity changes regularly. Since power reversal with LCC HVDC is done by changing the voltage polarity, XLPE cables are less suited for this type of HVDC. With VSC HVDC on the other hand, the voltage keeps the same polarity, independent of the power flow direction. This makes extruded cables better suited for VSC HVDC applications. Extruded cables for HVDC are still under active development.

The use of XLPE cables also has the following advantages:

- lighter;
- smaller bending ratio;
- no environmental risks (e.g. oil spill);
- easier and faster installation and maintenance;
- cheaper.

Currently, all except one VSC HVDC schemes are developed using XLPE cables. The exception is an overhead line in the Caprivi system between Zambia and Namibia. The main reasons for using cable connections are:

- low RoW and visibility resulting in faster permitting;
- lower costs for VSC HVDC cables (XLPE) compared to oil-filled cables, making the price difference smaller; as overhead lines are more prone to lightning strikes, the VSC installation needs extra protection, which is also quite costly.

Unlike LCC HVDC, a VSC-based system has no inherent capability to clear DC line faults. A line-to-line or line-to-ground fault causes the VSC DC capacitor to be discharged. The fault current is maintained through the free-wheeling diodes until the AC protection clears the DC line. Using cables reduces the number of such interruptions.

No need for extra protection against lightning strikes which could damage the power electronic devices; and, especially, the reduced environmental and social effects of these cables, which is one of the major selling points.

The XLPE cables can be put in close proximity and installation can be quite straightforward by 'ploughing' the cable in the ground.

6.4.5 Ratings

In Fig. 6.11, the current ratings for HVDC systems are given, and both LCC and VSC systems, as well as overhead lines such as XLPE and MI cable types, are depicted. It is important to note that in general the voltage and current dependency need to be dealt with separately. The voltage limit is mostly related to the cable technology as the power electronic converters are expected to be able to handle larger voltages without innovative changes in converter design. The current limit, however, is limited by the maximum current through the power electronic components.

Although overhead lines for LCC HVDC are commonly used, for VSC HVDC systems only one such system exists: the Caprivi link between Zambia and Namibia. The figure shows that the voltage limit for cables is the limiting factor for the development of VSC HVDC of high power ratings. Higher voltage ratings (up to 500 kV) for DC XLPE cables are under development. The numbers provided are indicative, as they are subject to the installation and the environmental conditions. Specifically for cables, the ratings are dependent on a number of factors, such as the burial (direct burial, sea burial, tunnel, etc.) and the material used (copper or aluminium). Oil-filled cables are not suited for long-distance transmission as they require regular oil refilling (Van Hertem, 2010).

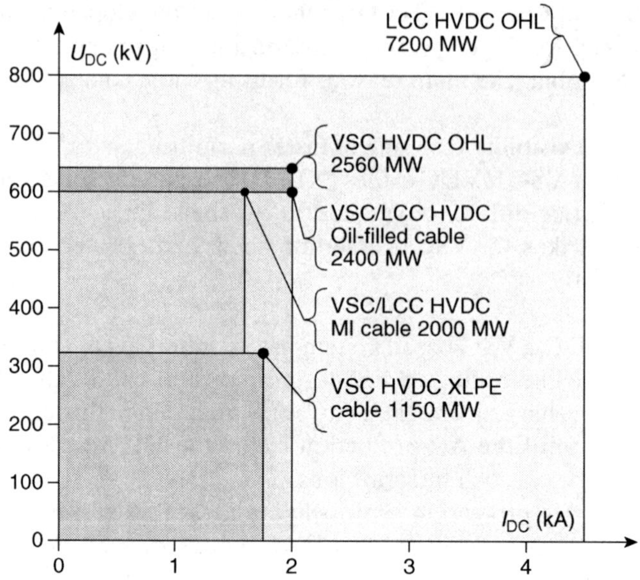

6.11 Available ratings for HVDC equipment.

6.5 Operation of HVDC

HVDC connections offer significant control freedom which enable the system operator to optimize his controls to the actual grid situation. In this section the operation of the converters, the link as a whole and the contribution to dynamic stability are addressed.

6.5.1 Control of the HVDC system

An HVDC scheme can be represented by a simple DC network. The LCC and VSC converters operate differently. Where the LCC operates with a constant current, the VSC operates with a constant voltage (in normal steady-state operation). In Fig. 6.12, the simplified scheme for the LCC and the VSC connection is shown.

For both systems, Ohm's law is valid: the current through the DC line is equal to the difference of the voltage at the rectifier (U_{rec}) and inverter (U_{inv}) side, divided by the resistance of the line.

$$I_{dc} = \frac{U_{rec} - U_{inv}}{R_{dc}} \tag{6.1}$$

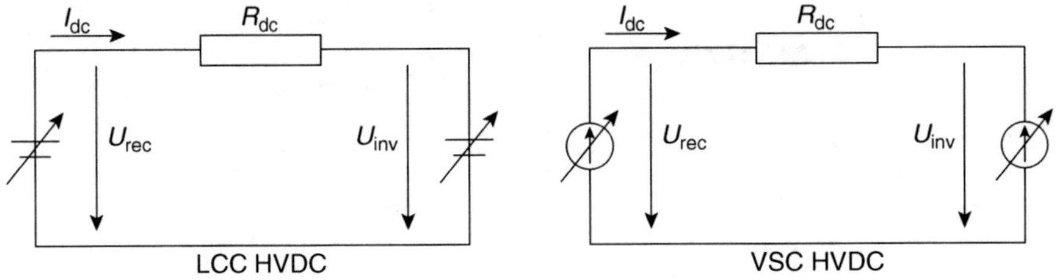

6.12 The simplified DC system representation for the LCC and VSC link, respectively.

The power through the DC system can be expressed in a very similar manner:

$$P_{loss} = \mathbf{P}_{rec} - P_{inv} \qquad [6.2]$$

or

$$(U_{rec} - U_{inv}) \cdot I_{dc} = U_{rec} \cdot I_{dc} - U_{inv} \cdot I_{dc} \qquad [6.3]$$

The LCC system operates by altering the firing angle of the thyristors and thereby changing the voltage at both sides of the DC link. A feedback-loop keeps the current constant and by altering the firing angle, the power flowing through the DC link is altered. An important consequence of this operation is the power reversal. In order to achieve this, the polarity of the voltage needs to be changed. This brings additional stress on cable systems, which require special power reversal schemes. Also the use of XLPE cables is not possible for this reason.

The VSC HVDC system operation is the dual form of the LCC system, as the voltage at the nodes is kept constant through altering the current in the inner control loop. This is possible because the VSC converter uses IGBT switches that can shape the waveforms as they can both open and close the valve.

Also the behaviour of the DC link towards the AC side is different for the LCC and VSC converter. The current towards the LCC converter is always lagging the voltage because of the thyristor bridge configuration. The converter always consumes reactive power as a function of the switching angle. The reactive power needs to be compensated elsewhere in the grid. Typically capacitor banks, SVCs or STATCOMs are used.

The behaviour of the VSC converter towards the AC system can best be compared with a synchronous machine. The AC voltage at the converter

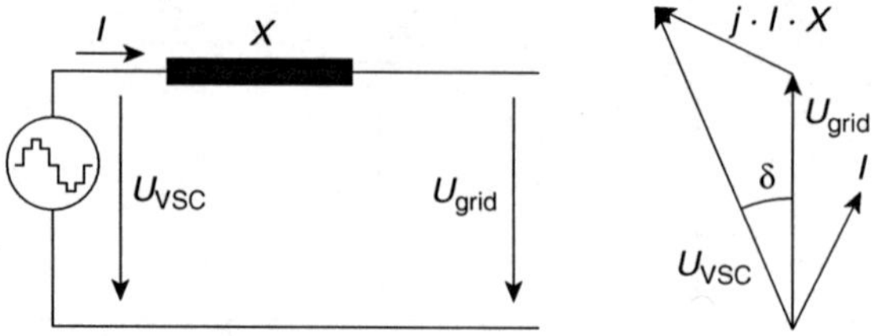

6.13 Equivalent scheme and phasor diagram of the VSC terminal. The VSC terminal acts in a similar manner as a synchronous machine.

side can be fully controlled in both amplitude and phase angle. The AC waveforms at the converter side are shaped independently to obtain the desired operation. The simplified equivalent circuit of Fig. 6.13 can be used. The basic equation for the transport of electrical power over an inductive transmission line is:

$$P = \frac{U_{grid} \cdot U_{vsc}}{X} \sin(\delta) \quad [6.4a]$$

$$Q = \frac{U_{grid}^2}{X} - \frac{U_{grid} U_{vsc}}{X} \cos(\delta) \quad [6.4b]$$

In Equation [6.4], U_{grid} is the system voltage RMS value, and U_{vsc} is the voltage RMS value of the sine wave generated by the converter. The impedance X is that of the AC circuit between converter and grid, i.e. AC phase reactor and transformer. δ is the angle between the voltages U_{grid} and U_{vsc}.

Equation [6.4] shows that both active and reactive power can be controlled by the VSC link as both δ and U_{vsc} can be altered through the controls of the converters. Clearly, the actual converter control schemes are more complex than this, involving the control of the switches, the current through the converter, the power or DC voltage controller, additional grid related converters and the necessary limiters and special control schemes for non-normal behaviour.

The active power exchanged between DC link and AC transmission system can be altered by introducing a phase angle between the AC voltage and the VSC voltage. Changing the RMS value of the VSC voltage has only a minor influence on the active power flow. The control of the AC system voltage is therefore independent of the active power flow control. As shown by Equation [6.4b], the reactive power exchange is more directly linked to the voltage amplitudes and less to the angle between them (when considering the voltage angles to remain relatively low, and the voltage amplitudes close to 1 pu).

Equation [6.4] can be rewritten to describe a circle in the PQ-plane (Fig. 6.14). The VSC HVDC system can deliver the desired active and reactive power whenever they are within this circle. Clearly the VSC HVDC terminal can operate in all four quadrants. In Quadrant 1, it delivers active and reactive power (behaving as a capacitor) to the system. In Quadrant 2 it absorbs active, and delivers reactive power. In Quadrants 3 and 4, the terminal absorbs reactive power (behaving as an inductance), while it withdraws, respectively delivers, active power from/to the grid. The active power exchanges at both terminals are linked: $P_{\text{Terminal1}} - P_{\text{Terminal2}} = P_{\text{Loss}}$. However, the reactive power outputs at both sides of the DC link are independent.

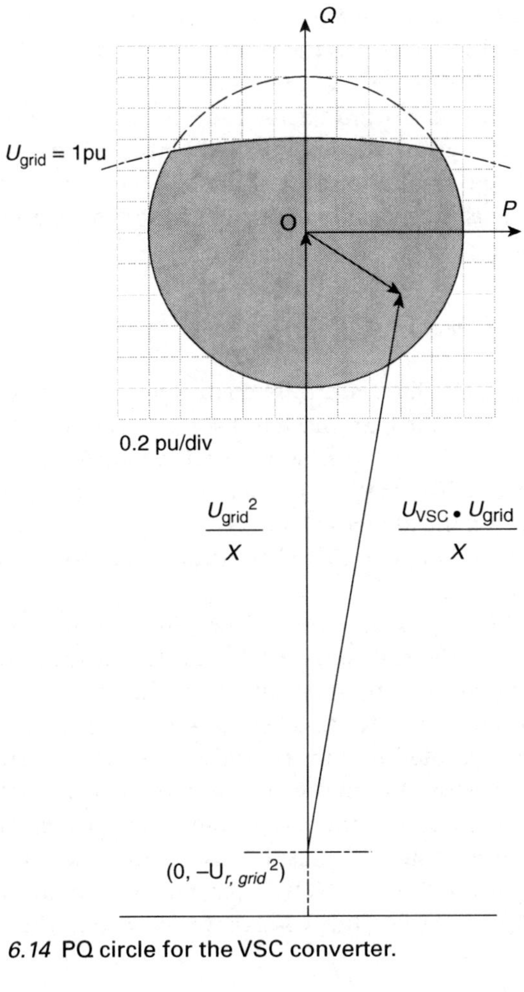

6.14 PQ circle for the VSC converter.

The control of the VSC link is limited by the following physical properties of the converter:

- The maximum current through the converter. With the impedance between converter and grid side being constant, this results in a maximum voltage $\Delta U = U_{vsc} - U_{grid}$ across this component. In the PQ diagram of Fig. 6.14 the maximum current is represented by a circle around O. As the grid voltage increases, the maximum power flow increases, for an equal voltage drop across the converter impedance.
- The maximum voltage across the IGBTs, and consequently the maximum AC voltage at converter side U_{VSC}.
- The maximum DC voltage and current the DC cable can withstand.

Next to the design properties, also the voltage at the grid side U_{grid} has an influence on the power control range. In normal grid conditions, the voltage can vary between 0.9 and 1.1 pu across the semiconductors.
The advanced controllability of the VSC HVDC link makes it more suitable for smaller systems, for instance wind farms. The VSC converter also has a smaller impact on the AC grid (when controlled correctly). An important secondary advantage is the small footprint of the VSC converter compared to the LCC converter. This is especially important for offshore applications where space is expensive.

6.5.2 Operation of the HVDC link

Contrary to AC lines, HVDC links are fully controllable. This has a consequence that the power flow through the line (and in case of VSC HVDC also the voltage at the terminals) is now a variable which can be set by the line owner or operator. The HVDC line as such allows the operator to adjust the system, based on the objective of the operator. Possible objectives are minimization of losses, keeping a fixed transfer, minimizing the consequences of contingencies, etc.

As the HVDC link influences the power flows through the system fundamentally, the control of the link must be integrated in the grid scheduling. In case multiple grid operators are involved (directly or indirectly), this becomes an important parameter to take into account. This is especially so when the link is not operated by the transmission system operator, but rather by a merchant investor. As the flow through the HVDC link can be changed significantly, the potential influence of these changes on the operational security in the power system(s) must be taken into account.

The mutual influence of multiple HVDC links in close proximity further complicates the control of HVDC lines. Furthermore again, the situation becomes more complex when multiple operators are present, each setting the control of their link.

Nevertheless, the additional degrees of freedom give the grid operators the opportunity to adjust the system to the current situation (e.g. high wind or solar infeed from a certain region for a number of hours), while maintaining operational security. Greater flexibility comes at the cost of higher complexity.

6.5.3 Contribution to system stability

The HVDC connection is a highly dynamic component which interacts with the AC systems. Without going into detail on the dynamics of HVDC systems, it is important to recognize that the interaction of HVDC with the remainder of the system needs to be carefully investigated. Through proper control, an HVDC link is often able to improve the overall dynamic behaviour of the power system. As examples, the voltage control of the VSC HVDC link can be used to damp oscillations at the AC side, and the HVDC link can provide frequency support to other asynchronous zones through adequate control of the power flow through the link.

6.6 HVDC grids

The development of VSC HVDC was a turning point in the transmission of electrical energy. It allows the connection of remote energy sources (such as offshore wind parks) and it facilitates long cable connections to locations which were previously not reachable by traditional AC connections or LCC HVDC. The use of VSC HVDC to achieve these connections is now in the stage of deployment.

The next step in the development is the move towards DC grids. DC grids are DC systems with multiple terminals connected to a single DC circuit. The DC grid in its simplest form is a multi-terminal system, but meshed DC grids are also possible. The advantage of DC grids over point-to-point DC lines is the reduced number of converters needed, which results in lower investment costs. For many, the VSC HVDC grid is seen as the key driver for the future development of a sustainable energy supply (see for instance, Desertec (2007) and Woyte (2008)).

Regardless of the potential, there are still a number of questions remaining with respect to VSC HVDC based grids. For example, it is not clear what the optimal configuration or layout of the grid connection of offshore wind farms is, and how the actual controls of the VSC converters in the DC grid will be implemented (Jovcic, 2011). Some specific concerns are related to protection, reliability and control of these fundamentally new transmission systems.

6.7 Future trends

The development of HVDC is happening in different domains. A first evolution is the increase in rated power of the equipment. The current limitations are related to the converter voltage for systems with overhead lines and the cable voltage for cable systems. XLPE cables, but also other types of cables, are available for increasingly higher ratings. The improvements which are needed are a higher voltage rating for cables, and this is through better insulation materials. As the cable for transmission systems require a significant amount of insulation, the thickness of that material (and in turn the admissible voltage gradient (kV/cm)) is of utmost importance to allow the cable system to be manageable during installation and production. A further innovation which would benefit the cable technology is the development of an insulation material which is good insulator against electricity, but a good conductor for heat.

A second evolution is related to the HVDC converter stations. Especially VSC HVDC converters are still being developed, with the aim of lower losses and cheaper designs.

A third development is the use of new semiconductor materials for the valves. As well as the improvement of existing valves, new types of power electronic valves are also being developed. Wide-bandgap semiconductor devices are constantly under development, and have the potential to replace the existing power electronics in transmission systems. They have several advantages over traditional power electronic components: they can withstand higher reverse voltages for the same layer thickness, they have lower conduction losses, the junction temperature can be higher, the devices are less influenced by the temperature and they can be switched faster. Main materials used for these developments are silicon carbide (SiC) and gallium nitride (GaN), for both of which small scale tests have been developed, but no commercial application at transmission voltages exists.

The control and utilization of VSC HVDC in a more dynamic manner is also of interest to the research community. Especially with regard to the control of systems with low inertia, e.g. offshore wind farms, an optimal control of the VSC terminal can result in a higher yield and a more reliable operation.

A final research item for the HVDC community is the development of the HVDC grid. A significant number of researchers, transmission system operators and other stakeholders are considering offshore grids, and even a pan-European Supergrid (also called the pan-European Electricity

Highways System 2050) based on HVDC technology. If successful, this DC grid could become the backbone of Europe's electric power transmission (Van Hertem, 2010; Cigré B4-52, 2012).

6.8 Conclusion

This chapter provides an introduction to HVDC transmission systems. First, the drivers for HVDC development are given. This is often due to the policy that aims at sustainable and competitive power systems that offer sufficient security of supply. At the same time, traditional investments have been difficult. In developing areas, the driver is often the fast increase in energy consumption which demands new transmission lines to be built. Because of the large quantities and the long distances, HVDC is the preferred technology.

The two main technologies for HVDC have been described. The traditional technology is LCC HVDC which is still the most commonly used. It also presently has the highest available power ratings. VSC HVDC on the other hand is more dynamic and allows the use of XLPE cables. This makes this option more suitable for such applications as offshore wind farms. The equipment that is needed for HVDC systems is covered in detail.

The HVDC system offers additional control to the grid operators, allowing it to optimize the grid operation, yet this comes at a cost of a more complex operation of the system.

HVDC is not a new technology; already in the nineteenth century, early electric power systems were based on DC technology. Over time, AC technology became dominant; yet in the 1950s, HVDC technology started to play an important role in connecting power systems that could not be connected by AC lines. Still it remained a 'special' solution. In recent years, the importance of HVDC is again rapidly growing. The development of VSC HVDC has opened the possibility to develop an HVDC grid. This HVDC grid can develop into a future offshore grid connecting major renewable energy sources, or even a new backbone of the transmission system.

6.9 References

ABB, 'It's time to connect – Technical description of HVDC Light technology,' ABB, Tech. Rep., 2006, (Last checked on 07/11/2008). [Online]. Available: http://www.abb.com/hvdc

Ahmed, N., Haider, A., Van Hertem, D., Zhang, L., Nee, H.-P., 'Prospects and challenges of future HVDC SuperGrids with modular multilevel converters,' Proceedings of the 2011-14th European Conference on Power Electronics and Applications (EPE 2011) EPE 2011.

Andersen, B. R., 'VSC Transmission,' Cigré Workgroup B4: HVDC and power electronics, Tech. Rep. 269, Apr. 2005, working group B4.37.

Alstom Grid, HVDC-VSC: transmission technology of the future, Thinkgrid magazine #8, 2011.

Arrillaga, J. (1998), High Voltage Direct Current Transmission. Institution of Engineering and Technology (IET). p. 1. ISBN 978-0-85296-941-0.

Arrillaga, J., Liu, Y.H., Watson, N.R., *Flexible Power Transmission: The HVDC Option*, John Wiley & Sons, ISBN 978-0-47005-688-2.

Asplund G., (convenor Cigré B4-52), 'HVDC grid feasibility study', 2012, Cigré Working group B4-52 (In preparation).

Bahrman, M. P., Johnson, B. K., 'A brief look at the history of HVDC to help understand its bright future.' IEEE Power and Energy Magazine, vol. **5**, no. 2, pp. 32–44, Mar./Apr. 2007.

Bahrman, M. P., Johnson, B. K., 'The ABCs of HVDC transmission technologies,' IEEE Power and Energy Magazine, vol. **5**, no. 2, pp. 32–44, Mar./Apr. 2007.

Beerten J., 'Modelling and Control of HVDC Grids, PhD dissertation,' 2013, KU Leuven.

Desertec Foundation, 'Clean power from deserts—the DESERTEC concept for energy, water and climate security, whitebook.' Trans-Mediterranean Renewable Energy Cooperation (TREC), 2007

Dodds, S., Railing, B., Akman, K., Jacobson, B., Worzyk, T., Nilsson, B., 'HVDC VSC (HVDC Light) transmission – Operating experiences', Cigré sessions 2010, paper B4-203, Paris.

Dorn, J., Huang, H., Retzmann, D., 'A new Multilevel VSC Topology for HVDC Applications,' CIGRE Paris Session 2008.

Henry S., (convenor Cigré JWG C4/B4/C1), 'Influence of Embedded HVDC Transmission on System Security and AC Network Performance' Working group Cigré JWG C4/B4/C1 (In preparation).

Hingorani, N.G., 'High-voltage DC transmission: a power electronics workhorse,' Spectrum, IEEE, vol. **33**, no.4, pp.63–72, Apr 1996. doi: 10.1109/6.486634. URL: http://ieeexplore.ieee.org/stamp/stamp.jsp?tp=&arnumber=486634&isnumber=10407

INELFE – Europe's first integrated onshore HVDC interconnection P. Labra Francos, S. Sanz Verdugo, H. Fernández Álvarez, S. Guyomarch, J. Loncle, IEEE PES general meeting 2012, San Diego.

Jacobson B., P. Karlsson, G. Asplund, L. Harnefors, T. Jonsson: 'VSC-HVDC Transmission with Cascaded Two-Level Converters', Cigré sessions 2010, paper B4-110, Paris

Jakob Glasdam, Jesper Hjerrild, Łukasz Hubert Kocewiak, Claus Leth Bak, 'Review on Multi-Level Voltage Source Converter Based HVDC Technologies for Grid Connection of Large Offshore Wind Farms', IEEE PES Powercon 2012.

Jovcic, D., Van Hertem, D., Linden, K., Taisne, J.-P., Grieshaber, W., 'Feasibility of DC Transmission Networks,' in Proceedings of the IEEE PES Innovative Smart Grid Technologies Europe 2011, Manchester, United Kingdom, December 2011.

Lesnicar, A., Marquardt, R., 'An innovative modular multilevel converter topology suitable for a wide power range,' Power Tech Conference Proceedings, 2003 IEEE Bologna, vol. **3**, 2003, p. 6.

Peschke, E., von Olshausen, R., 'Cable systems for high and extra-high voltage' 1999, Pirelli, ISBN: 3-89578-118-5.

Terashima, K., Sukuki, H., Hara, M., Watanabe, K., 'Research and development of +/-250 kv DC XLPE cables,' IEEE Transactions on Power Delivery, vol. **13**, no. 1, pp. 7–16, Jan. 1998.

Van Hertem D., Ghandhari M., Delimar M., 'Technical limitations towards a SuperGrid – A European prospective,' Energy Conference and Exhibition (EnergyCon), 2010 IEEE International vol: **1**, IEEE ENERGYCON edition, Bahrain, December 18–23, 2010.

Woyte, A., De Decker, J., Vu Van, T., 'A North Sea electricity grid [R]evolution – electricity output of interconnected offshore wind power: a vision of offshore wind power integration,' Greenpeace – 3E, 2008

7
Modern flexible AC transmission system (FACTS) devices

K. WANG and M. L. CROW, Missouri University of Science and Technology, USA

DOI: 10.1533/9780857097378.2.174

Abstract: This chapter describes the primary modern FACTS devices currently deployed worldwide. Modern FACTS devices are based on the concept of the voltage source converter (VSC) to inject a controllable AC waveform at the fundamental frequency into the power system to alter the power flow at a bus or across a transmission line. This chapter reviews the three primary modern FACTS devices: the static synchronous compensator (STATCOM), the static synchronous series compensator (SSSC), and the unified power flow controller (UPFC). The state-space nonlinear model is provided for each device as well as a typical control method. The chapter concludes with a brief overview of future trends in FACTS devices including the interline power flow controller and VSC-based high voltage DC systems.

Key words: flexible AC transmission systems, power electronics, power system control, voltage source converters.

7.1 Introduction

In bulk power transmission systems, the use of power electronics-based devices can potentially overcome limitations of the present mechanically controlled transmission system. These flexible networks help delay or minimize the need to build more transmission lines and enable neighboring utilities and regions to economically and reliably exchange power.

In the decentralized control of transmission systems, FACTS devices offer increased flexibility. As the vertically integrated utility structure is phased out, centralized control of the bulk power system will no longer be the only option. Transmission providers will be forced to seek means of local control to address a number of potential problems, such as uneven power flow through the system (loop flows), transient and dynamic instability, subsynchronous oscillations, and dynamic overvoltages and undervoltages.

The use of FACTS devices in a power system can potentially overcome the limitations of the present mechanically controlled transmission systems.

By facilitating bulk power transfers, these interconnected networks help minimize the need to enlarge power plants and enable neighboring utilities and regions to exchange power. The stature of FACTS devices within the bulk power system will continually increase as the industry moves toward a more competitive posture in which power is bought and sold as a commodity. As power wheeling becomes increasingly prevalent, power electronic devices will be utilized more frequently to ensure system reliability and stability, and to increase maximum power transmission along various transmission corridors.

The rapid development of the high power electronics industry has made FACTS devices viable and attractive for utility applications. FACTS devices have been shown to be effective in controlling power flow and damping power system oscillations (Hingorani and Gyugyi, 1999; Zhang *et al.*, 2010).

Several installations of FACTS devices are currently in service in Japan, Brazil, the United States, and other locations worldwide.

This chapter is organized as follows. First, a brief overview of the VSC is presented. The VSC is the primary structural component of the modern FACTS device. Both conduction mode and pulse width modulation switching strategies are briefly presented. The following sections focus on the more prevalent FACTS devices: the STATCOM, SSSC, and UPFC. The final section introduces several hybrid applications.

7.2 The voltage source converter

The primary building block of the modern FACTS device is the VSC. A VSC consists of a voltage source (battery or capacitor) to provide a near constant DC voltage. Power injection or absorption is controlled by the direction of the DC current into or out of the converter. Most FACTS controllers utilize a three-phase, full-wave VSC to interface between the DC and AC sides of the converter. A typical three-phase converter is shown in Fig. 7.1. Although the transistor switches may be triggered independently, the typical ideal switching pattern is such that $a_1 = \overline{a}_2 = a$, $b_1 = \overline{b}_2 = b$, and $c_1 = \overline{c}_2 = c$. The alternating voltage waveforms can be synthesized by the choice of switching patterns. For example, if $a = 1$ (on), $b = 1$ (on), and $c = 0$ (off), then the converter has the topology shown in Fig. 7.2. Note that in this topology, $V_{ab} = 0$, $V_{bc} = V_{DC}$, and $V_{ca} = -V_{DC}$ (highlighted). The various switching combinations are given in Table 7.1. The resulting line-line voltage waveform is shown in Fig. 7.3. This is often called '6-pulse' operation since there are six different voltage levels possible in the line-line voltage.

If the converter is connected to the AC system through a delta-connected or ungrounded-wye transformer, there are no zero sequence components in the voltage and subsequently all triplen ($3n$) harmonics are suppressed, leaving only harmonics of $6n \pm 1$ with amplitudes of 1/5, 1/7, etc. Therefore,

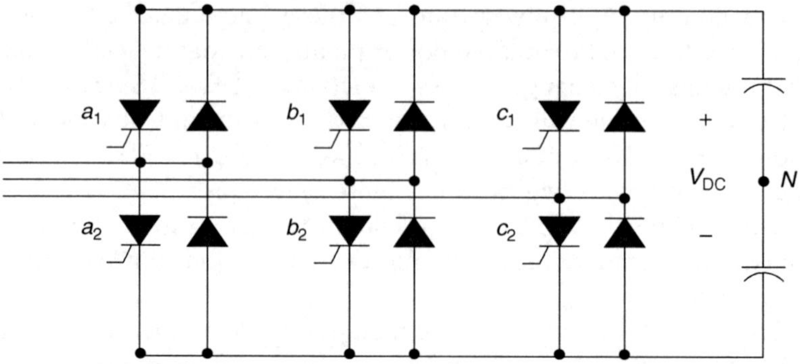

7.1 Three-phase, full-wave VSC.

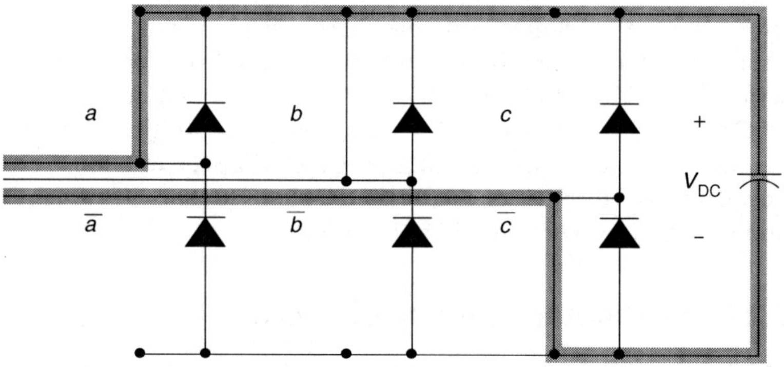

7.2 Three-phase, full-wave VSC with $a = 1$ (on), $b = 1$ (on), and $c = 0$ (off) with V_{ca} highlighted.

Table 7.1 Switching states

Switch			Voltage		
a	*b*	*c*	V_{ab}	V_{bc}	V_{ca}
0	0	0	–	–	–
0	0	1	0	$-V_{DC}$	V_{DC}
0	1	0	$-V_{DC}$	V_{DC}	0
0	1	1	$-V_{DC}$	0	V_{DC}
1	0	0	V_{DC}	0	$-V_{DC}$
1	0	1	V_{DC}	$-V_{DC}$	0
1	1	0	0	V_{DC}	$-V_{DC}$
1	1	1	–	–	–

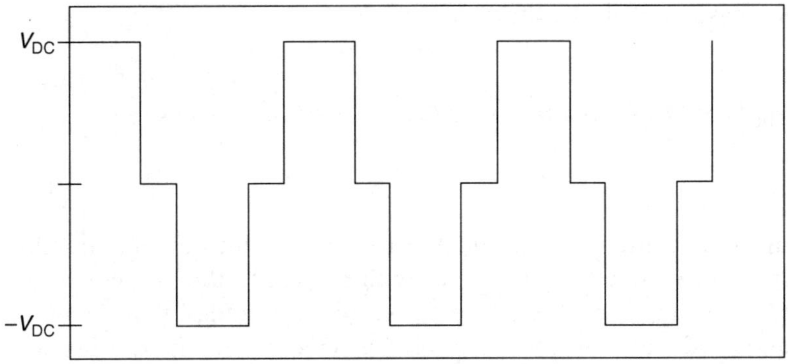

7.3 Line-to-line voltage of the 6-pulse converter.

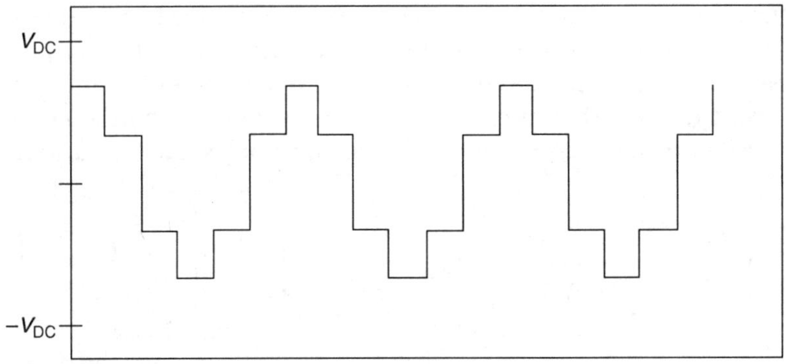

7.4 Phase to neutral voltage of the 6-pulse converter.

the phase to neutral voltage at the transformer secondary will be lacking the triplen harmonics (shown in Fig. 7.4). The corresponding line-to-line voltage is given by:

$$V_{ab} = \frac{2\sqrt{3}}{\pi} V_{\mathrm{DC}} \left[\cos \omega t - \frac{1}{5}\cos 5\omega t + \frac{1}{7}\cos 7\omega t - \frac{1}{11}\cos 11\omega t + \frac{1}{13}\cos 13\omega t - \cdots \right] \quad [7.1]$$

The rms value of the fundamental is given by:

$$V_1 = \frac{\sqrt{6}}{\pi} V_{\mathrm{DC}} = 0.78 V_{\mathrm{DC}} \quad [7.2]$$

where the phase to neutral voltage is given by:

$$V_{ab} = \frac{2}{\pi} V_{DC} \left[\cos \omega t + \frac{1}{5} \cos 5\omega t - \frac{1}{7} \cos 7\omega t - \frac{1}{11} \cos 11\omega t + \frac{1}{13} \cos 13\omega t - \cdots \right] \quad [7.3]$$

The harmonic content can be further reduced by introducing an additional converter with a 30° phase shift with respect to the first converter. If the output voltages of these two converters are added together then a subsequent 12-pulse staircase waveform can be obtained with harmonics on the order of $12n \pm 1$ with amplitudes of 1/11, 1/13, etc. Such an arrangement is shown in Fig. 7.5, in which two 6-pulse converters are connected in parallel on the same DC bus. One has an ungrounded-wye secondary; the other has a delta-connected secondary with $\sqrt{3}$ times the turns of the wye-connected secondary. The output line-to-line voltage is shown in Fig. 7.6. As the harmonics are shifted to higher frequencies, the passive components required to filter the harmonics become smaller and less expensive. Therefore, it is desirable to shift the harmonics as high as possible in the frequency spectrum. Harmonics can be further reduced

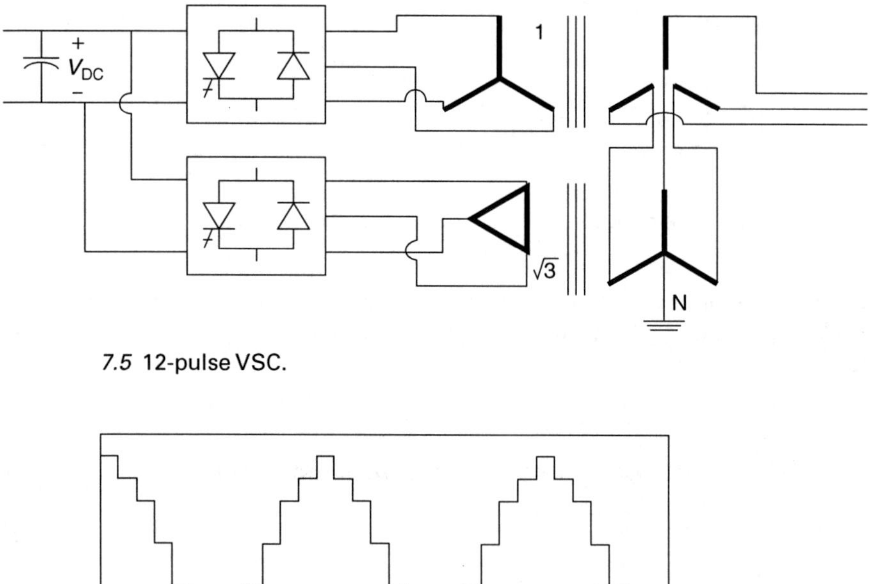

7.5 12-pulse VSC.

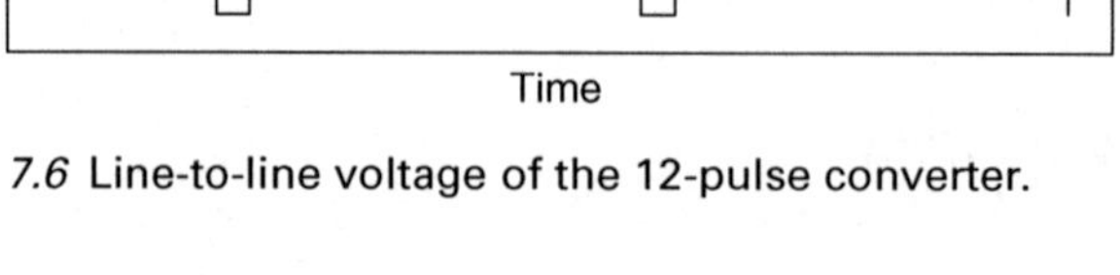

7.6 Line-to-line voltage of the 12-pulse converter.

and higher voltages are achieved by combining 12-pulse converters to achieve a 24-pulse converter. Two 12-pulse converters, each phase-shifted by 15° can produce a 24-pulse waveform with harmonics shifted to $24n \pm 1$ with amplitudes of 1/23, 1/25, etc. Similarly, a 48-pulse converter can be constructed using smaller converters and various phase-shifted transformers. Although the harmonics are significantly increased as the number of pulses increases, the magnetic configuration of the transformers becomes increasingly complex.

Another method of decreasing harmonic content is to use pulse width modulated (PWM) switching of the converters. This is only possible if the power electronics devices from which the converter is constructed can switch sufficiently quickly. This has still not been achieved for extremely high voltage converters, but devices at the distribution level are viable. In a PWM-switched 6-pulse converter, the output voltage can be controlled by varying the width of the voltage pulses. The resulting harmonic content is then determined by the frequency of the modulation frequency, moving the harmonics into the kHz range, making filter construction considerably less expensive. The drawback to the PWM switching is that if the devices are switched while they are conducting (known as 'hard switching') considerable active power losses may result, thereby reducing the efficiency of the converter.

The most common PWM method is the 'sine-triangle' method, in which a triangle waveform is superimposed on the reference three-phase sinusoidal waveforms as shown in Fig. 7.7. The switches are then activated whenever the carrier triangle waveform crosses the reference sinusoidal waveform. The resulting waveforms are still three-level waveforms (consisting of voltage levels of V_{DC}, 0, and $-V_{DC}$) but rapidly switched between the levels as shown in Fig. 7.8. When the high-frequency waveform is filtered using an appropriate low-pass filter, the resulting waveform shown in Fig. 7.9 is sinusoidal. Further refinement can also be achieved by using multilevel PWM-switched converters (Corzine, 2002).

There are several compelling reasons to consider a multilevel converter topology for FACTS. These well-known reasons include lower harmonic injection into the power system, decreased stress on the electronic components due to decreased voltages, and lower switching losses. Various multilevel converters also readily lend themselves to a variety of PWM strategies to improve efficiency and control.

7.2.1 The cascaded converter

A cascaded multilevel FACTS converter is shown in Fig. 7.10. This converter uses several full bridges in series to synthesize staircase waveforms. Because

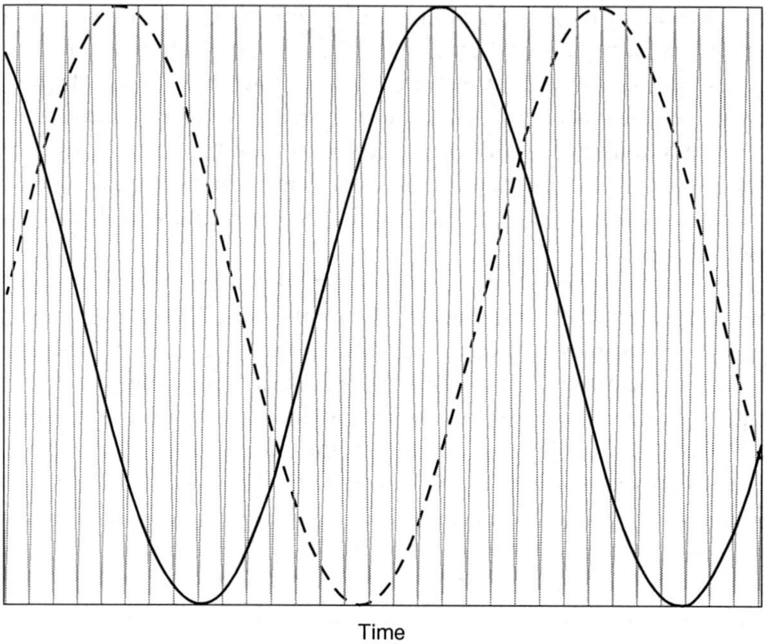

7.7 Reference and triangular carrier waveforms.

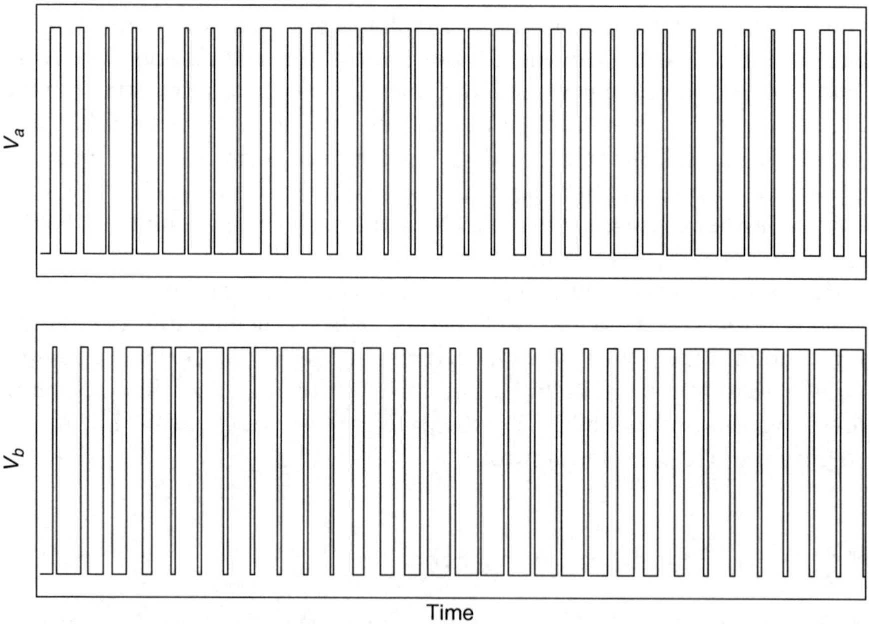

7.8 The PWM-switched waveforms of V_a and V_b.

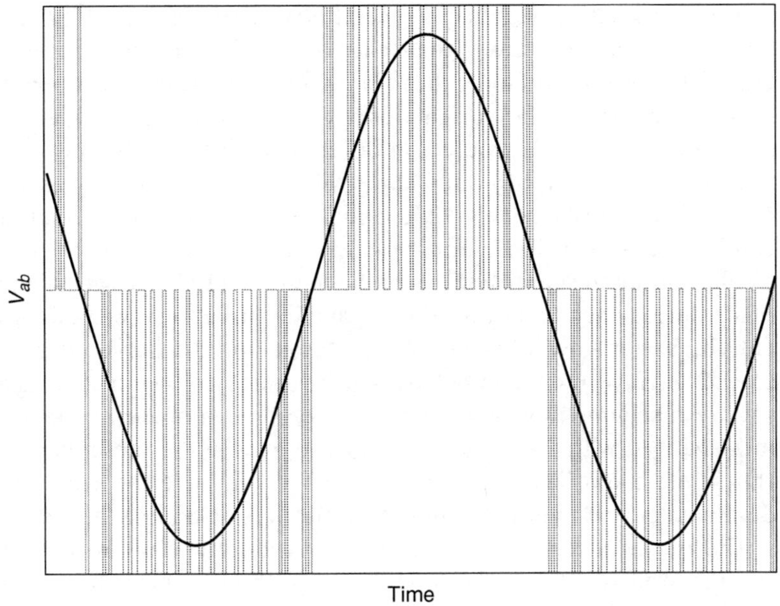

7.9 PWM line-to-line voltages of the 6-pulse converter, pre-filtering (square traces) and post-filtering (smooth (sinusoidal) trace).

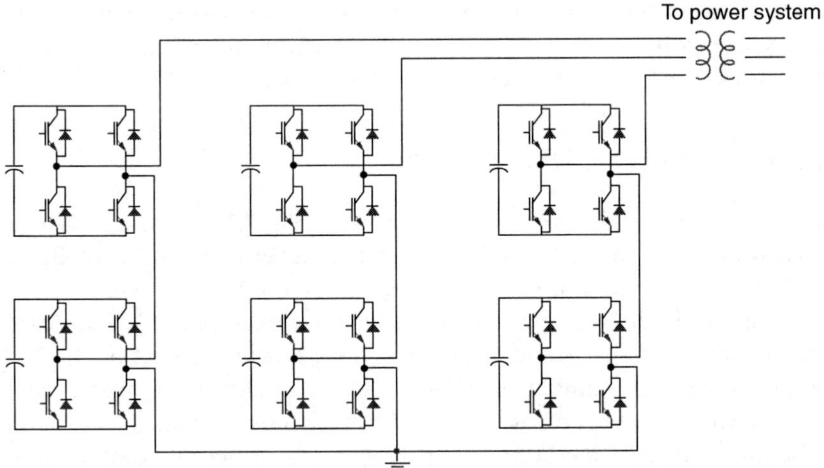

7.10 Cascaded multilevel converter.

every full bridge can have three output voltages with different switching combinations, the number of output voltage levels is $2N + 1$ where N is the number of full bridges in every phase. The converter legs are identical and therefore modular.

The cascaded converter has several attractive features (Rodriguez *et al.*, 2002). One such feature is that it can be easily controlled to mitigate the charge balancing problems that plague many multilevel converter topologies. A charge imbalance originates from the uneven charging and discharging of the multiple DC sources at different voltage levels, causing a voltage imbalance. A DC voltage imbalance will degrade the quality of the voltage output; in severe cases, this could lead to the complete collapse of the power conversion system. The output voltage is a staircase waveform synthesized by alternating the contributions of the multiple H-bridges at each half cycle. Since during each half cycle, the conduction time for each battery is different, the batteries will have different charge and discharge intervals. In the cascaded converter, the imbalance can be mitigated by introducing a rotated switching scheme in which the conduction period of each voltage source is interchanged with the adjacent voltage sources in the subsequent half cycle (Du *et al.*, 2006). Therefore each capacitor will have the same charge and discharge period over the complete cycle, thus balancing the voltage.

The cascaded converter also has the attractive feature that it can be reconfigured to exploit the modularity of the converter topology. A five-level cascaded converter with unequal voltage levels can be converted into a nine-level converter (Liu *et al.*, 2000). For example, if the batteries of the five-level converter are rearranged such that the DC voltages V_1 and V_2 are unequal, then a nine-level staircase output voltage waveform can be synthesized. By careful selection of the DC voltages, the nine-level converter can provide superior harmonic performance. The corresponding nine-level output waveform for $V_1 = 0.6$ V and $V_2 = 0.4$ V is shown in Fig. 7.11.

7.2.2 Diode-clamped converter

The diode-clamped multilevel converter shown in Fig. 7.12 uses a series string of capacitors to divide the DC side voltage into several levels. Normally an N-level diode-clamped multilevel inverter has $2(N - 1)$ main switches and $2(N - 1)$ main diodes per phase. The switches of each phase leg are connected via power diodes to the different voltage level points set by the DC capacitors. When operating, two adjacent switches (for a three-level converter) in each phase leg are 'on' to provide a respective voltage level, therefore the line voltage waveforms are synthesized by different combinations of switches. Diode-clamped converters are used less frequently in industrial applications than cascaded converters due to the potential for charge imbalance of the capacitors. However, recent advances in charge balancing methods have made the diode-clamped converter more attractive (Lin, 2000; Peng, 2000). The two most common approaches for charge balancing are to introduce an external balancing circuit or to use space vector modulation (Botao *et al.*, 2002; Marchesoni and Tenca, 2002).

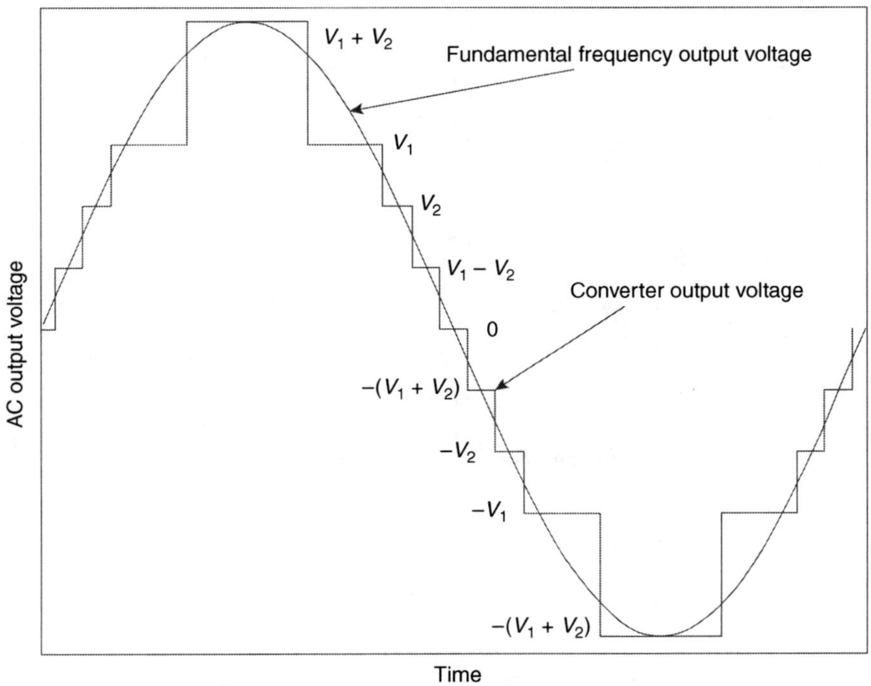

7.11 Nine-level waveform constructed from five-level cascaded converter.

7.3 The static synchronous compensator (STATCOM)

The STATCOM is a shunt-connected FACTS device that is used primarily for reactive power control. The STATCOM has two possible steady-state operating modes: inductive (lagging) and capacitive (leading). STATCOMs have been widely accepted to improve power system operation. The STATCOM consists of a single VSC and its associated shunt-connected transformer. The STATCOM capability is similar to that of the rotating synchronous condenser or static VAR compensator (SVC) and is typically used for providing reactive power compensation for voltage support (IEEE Power Engineering Society FACTS Application Task Force, 1996). The STATCOM achieves this objective by drawing (or injecting) a controlled reactive current from the line. In contrast with a conventional static VAR generator, the STATCOM also has the ability to exchange active power with the line through the charge and discharge of the DC link capacitor. However, unless an external energy storage system is available (such as a battery), the active power must be actively controlled to a value that is zero on average, and departs from zero only to compensate for the losses in the system (Schauder and Mehta, 1993).

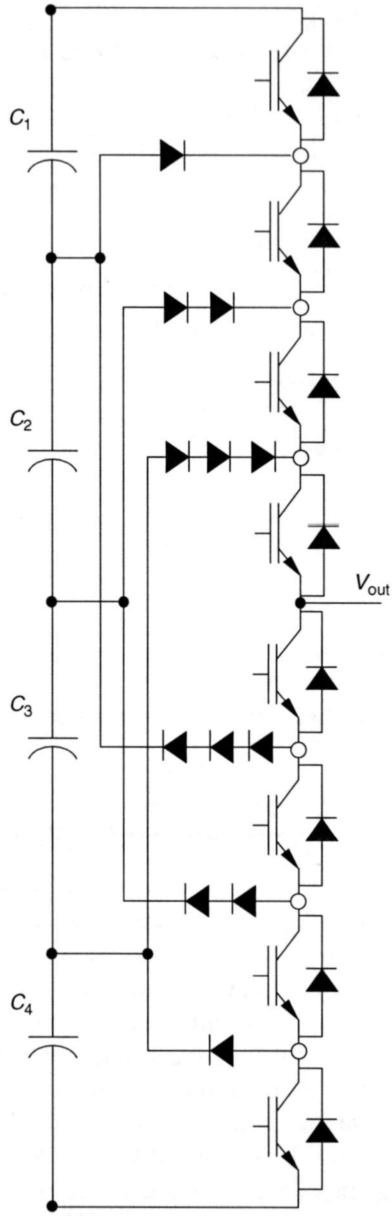

7.12 Diode-clamped converter (single phase).

By injecting a current of variable magnitude in quadrature with the line voltage, the STATCOM can inject reactive power into the power system. The STATCOM does not employ capacitor or reactor banks to produce reactive power as does the SVC, but instead uses a capacitor to maintain a

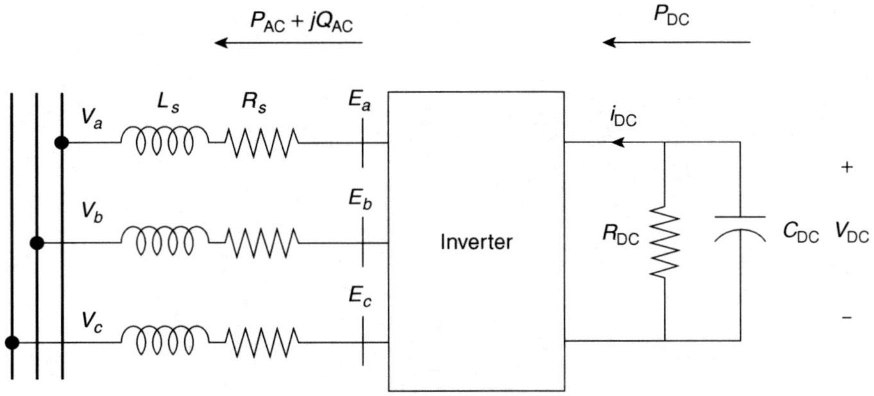

7.13 Equivalent circuit of the STATCOM.

constant DC voltage for the inverter operation. An equivalent circuit for the STATCOM is shown in Fig. 7.13.

The loop equations for the fundamental frequency circuit can be written in vector form as:

$$\frac{\mathrm{d}}{\mathrm{d}t}i_{abc} = -\frac{R_s}{L_s}i_{abc} + \frac{1}{L_s}(E_{abc} - V_{abc}) \qquad [7.4]$$

where R_s and L_s represent the STATCOM transformer losses, E_{abc} are the inverter AC-side phase voltages, V_{abc} are the system-side phase voltages, and i_{abc} are the phase currents. The fundamental frequency STATCOM voltage is given by:

$$E_a = kV_{\mathrm{DC}}\cos(\omega t + \alpha) \qquad [7.5]$$

where V_{DC} is the voltage across the DC capacitor, k is the modulation gain, and α is the injected voltage phase angle. This model can be simplified by converting it to a synchronously rotating reference frame. The reference frame coordinate system is defined in which the d-axis is always coincident with the instantaneous system voltage vector and the q-axis is in quadrature with it.

The first stage of dynamic model development is the transformation that converts the three-phase vector from the *abc* frame into the two-axis orthogonal stationary frame.

$$\begin{bmatrix} V_\alpha \\ V_\beta \\ 0 \end{bmatrix} = C_1 \begin{bmatrix} V_a \\ V_b \\ V_c \end{bmatrix}, \begin{bmatrix} I_\alpha \\ I_\beta \\ 0 \end{bmatrix} = C_1 \begin{bmatrix} I_a \\ I_b \\ I_c \end{bmatrix} \qquad [7.6]$$

where

$$C_1 = \begin{bmatrix} 2/3 & -1/3 & -1/3 \\ 0 & \sqrt{3}/3 & -\sqrt{3}/3 \\ \sqrt{2}/3 & \sqrt{2}/3 & \sqrt{2}/3 \end{bmatrix}$$

and

$$C_1^{-1} = \frac{3}{2} C_1^T$$

If the three-phase system is balanced, then the zero sequence components are zero and the instantaneous power can be expressed in terms of $\alpha\beta$ quantities as follows:

$$P = \frac{3}{2}\left(V_\alpha I_\alpha + V_\beta I_\beta\right) \qquad [7.7]$$

$$Q = \frac{3}{2}\left(V_\alpha I_\beta - V_\beta I_\alpha\right) \qquad [7.8]$$

The second stage is the transformation which converts the $\alpha\beta$ vector from the two-axis orthogonal

$$\begin{bmatrix} V_d \\ V_q \end{bmatrix} = C_2 \begin{bmatrix} V_\alpha \\ V_\beta \end{bmatrix}, \begin{bmatrix} I_d \\ I_q \end{bmatrix} = C_2 \begin{bmatrix} I_\alpha \\ I_\beta \end{bmatrix} \qquad [7.9]$$

where

$$C_2 = \begin{bmatrix} \cos\theta \sin\theta \\ \sin\theta \cos\theta \end{bmatrix}, \quad \theta = \omega t$$

and

$$C_2^{-1} = C_2^T$$

Note that the matrix C_2 is time-varying since the dq-axes are not stationary and rotate with synchronous frequency ω_s. Similarly, the instantaneous power in terms of dq quantities becomes:

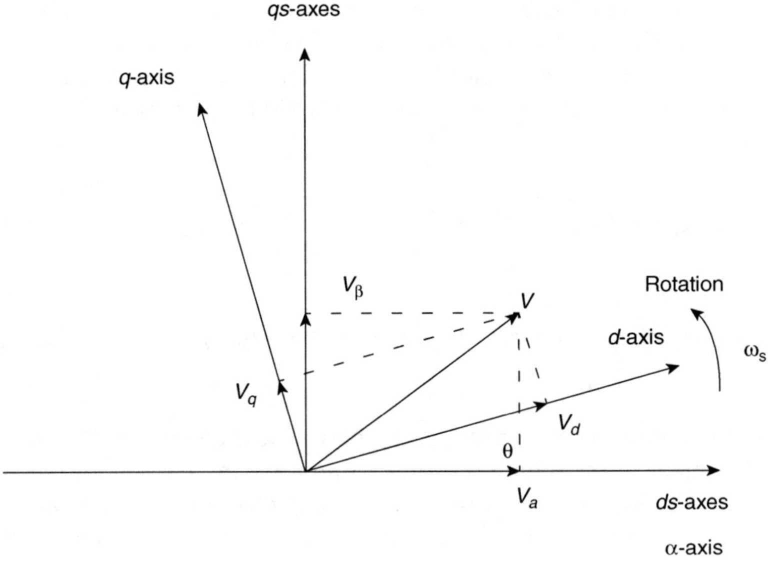

7.14 Synchronously rotating *dq* reference frame.

$$P = \frac{3}{2}\left(V_d I_d + V_q I_q\right) \qquad [7.10]$$

$$Q = \frac{3}{2}\left(V_d I_d - V_q I_q\right) \qquad [7.11]$$

Figure 7.14 shows the phase vectors under *abc*, $\alpha\beta$, and *dq* reference frames. The axes of the *abc* and $\alpha\beta$ frames are stationary but the phase vectors are varying. The axes of the *dq* frame are rotating but the phase vectors are constant.

The nonlinear STATCOM state equations for the equivalent circuit model shown in Fig. 7.13 in the *dq* reference frame are given by:

$$\frac{1}{\omega_s}\frac{d}{dt}i_d = -\frac{R_s}{L_s}i_d + \frac{\omega}{\omega_s}i_q + \frac{k}{L_s}V_{\mathrm{DC}}\cos(\alpha + \theta_i) - \frac{V_i}{L_s}\cos\theta_i \qquad [7.12]$$

$$\frac{1}{\omega_s}\frac{d}{dt}i_q = -\frac{R_s}{L_s}i_q + \frac{\omega}{\omega_s}i_d + \frac{k}{L_s}V_{\mathrm{DC}}\sin(\alpha + \theta_i) - \frac{V_i}{L_s}\sin\theta \qquad [7.13]$$

$$\frac{C_{\mathrm{DC}}}{\omega_s}\frac{d}{dt}V_{\mathrm{DC}} = -k\cos(\alpha + \theta_i)i_d - k\sin(\alpha + \theta_i)i_q - \frac{V_{\mathrm{DC}}}{R_{\mathrm{DC}}} \qquad [7.14]$$

where i_d and i_q are the injected dq STATCOM currents, V_{DC} is the voltage across the DC capacitor, R_{DC} represents the switching losses, R_s and L_s are the coupling transformer resistance and inductance, respectively, and the STATCOM rms bus voltage is $V_i\angle\theta_i$. The STATCOM power balance equations at Bus i are:

$$0 = V_i\left(i_d \cos\theta_i + i_q \sin\theta_i\right) - V_i \sum_{j=1} V_j Y_{ij} \cos\left(\theta_i - \theta_j - \phi_{ij}\right) \qquad [7.15]$$

$$0 = V_i\left(i_d \sin\theta_i - i_q \cos\theta_i\right) - V_i \sum_{j=1} V_j Y_{ij} \sin\left(\theta_i - \theta_j - \phi_{ij}\right) \qquad [7.16]$$

where the summation terms represent the power flow equations, $Y_{ij}\angle\phi_{ij}$ is the (i,j)-th element of the admittance matrix and n is the number of buses in the system. The first set of terms indicate the active and reactive powers injected by the STATCOM (respectively), whereas the summation of power terms on the right are the power flow equations of the power system. Note that Equations [7.15] and [7.16] represent the only coupling of the STATCOM states with the power system.

The control objectives for the STATCOM are to provide independent reactive power support and to maintain constant DC capacitor voltage. This is best accomplished by regulating the PWM switching commands to alter the modulation index and phase angle in Equation [7.5]. Since its inception, a variety of control approaches have been proposed for STATCOM dynamic control (Schauder and Mehta, 1993; Lehn and Iravani, 1998; Rao *et al.*, 2000; Sahoo *et al.*, 2002; Wang *et al.*, 2002; Liu *et al.*, 2003; Dong *et al.*, 2004; Soto and Pena, 2004; El-Moursi and Sharaf, 2005; Cheng *et al.*, 2006; Jain *et al.*, 2006; Lu and Ooi, 2007; Mohagheghi *et al.*, 2007; Saeedifard *et al.*, 2007; Song *et al.*, 2007, 2009; Sternberger and Jovcic, 2009; Hatano and Ise, 2010; Liu and Hsu, 2010; Spitsa *et al.*, 2010; Wang and Crow, 2011). The majority of approaches involve traditional linear control techniques, in which the nonlinear equations of the VSC average value model are linearized at a specific equilibrium (Schauder and Mehta, 1993; Lehn and Iravani, 1998; Rao *et al.*, 2000; Liu *et al.*, 2003; Dong *et al.*, 2004; El-Moursi and Sharaf, 2005; Cheng *et al.*, 2006; Saeedifard *et al.*, 2007; Sternberger and Jovcic, 2009; Hatano and Ise, 2010).

A proportional-integral (PI) STATCOM controller structure was initially proposed in Schauder and Mehta (1993). Since that time, numerous PI controllers have been reported to exhibit satisfactory performance when the parameters are fine-tuned, whether in the applications of two-level converter (Lehn and Iravani, 1998; Rao *et al.*, 2000; Dong *et al.*, 2004) or the multilevel converter (El-Moursi and Sharaf, 2005; Cheng *et al.*, 2006;

Saeedifard *et al.*, 2007; Hatano and Ise, 2010). The drawback of PI controllers is that their performance might degrade with a change of operating conditions, especially a large disturbance such as sudden load change or nearby short-circuit fault.

To compensate for a change of operating conditions, a variety of controls have been reported that provide satisfactory performance over a wide range of operating conditions (Sahoo *et al.*, 2002; Liu *et al.*, 2003; Soto and Pena, 2004; Lu and Ooi, 2007; Song *et al.*, 2007, 2009; Spitsa *et al.*, 2010; Wang and Crow, 2011). One approach is to adaptively change the gains of the PI controller in response to changes in operating condition. A number of intelligent techniques have been proposed to adapt the PI controller gains of STATCOMs such as artificial immunity (Wang *et al.*, 2002), neural networks (Mohagheghi *et al.*, 2007), and particle swarm optimization (Liu and Hsu, 2010). Adaptive control and linear robust controls have been reported in the literature (Jain *et al.*, 2006; Spitsa *et al.*, 2010; Wang and Crow, 2011). In Jain *et al.* (2006), a gradient-based estimation of the load conductance is proposed to account for the load variations. The concept of calculating a complete set of the admissible feedback gains is proposed in Spitsa *et al.* (2010).

One easily implemented control is shown in Fig. 7.15. In this control, the input signals V_{DC} and Q are compared against reference values and used to compute the error signals in i_d and i_q. A PI-based control is then used to produce the control signals k and α. In reality, both the firing angle α and modulation gain k are impacted by changes in i_d and i_q. However, since α is more strongly correlated to changes in i_q and k is more strongly correlated to changes in i_d, the cross-coupling terms can be neglected ($K_d = K_q = 0$), leading to a decoupled control approach.

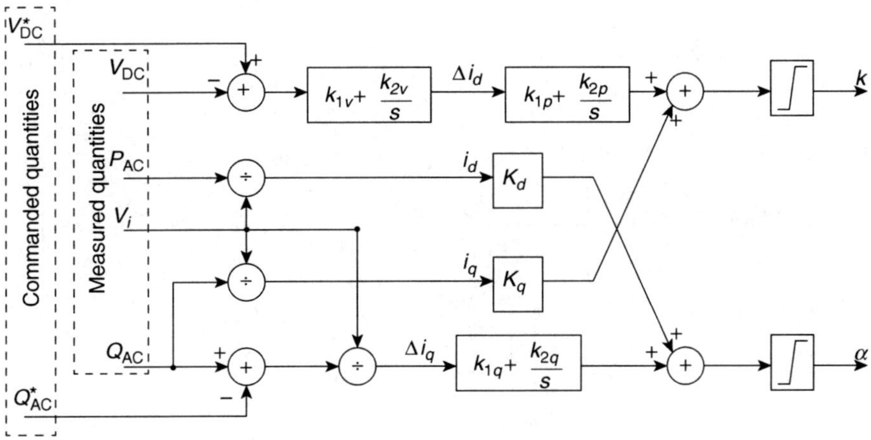

7.15 STATCOM PI control.

7.4 The static synchronous series compensator (SSSC)

The SSSC is a VSC-based serial FACTS device that can provide capacitive or inductive compensation independent of the line current (Larsen *et al.*, 1994; Gyugyi *et al.*, 1997). The SSSC typically has the same power electronics topology as the STATCOM. However, it is incorporated into the AC power system through a series coupling transformer, as opposed to the shunt transformer found in the STATCOM. The series transformer is used to inject an independently controlled voltage in quadrature with the line current for the purpose of increasing or decreasing the overall reactive voltage drop across the line and thereby controlling the transmitted power. In essence, the SSSC may be considered to be a controllable effective line impedance (Gyugyi *et al.*, 1997). Since the SSSC has a VSC topology, the DC capacitor is used to maintain the DC voltage, giving the SSSC the ability to increase or decrease the transmitted power across the line by a fixed fraction of the maximum power, independent of the phase angle. As a result of the SSSC's ability for reactive power generation or absorption, it makes the surrounding power system impervious to classical subsynchronous resonances.

Figure 7.16 shows an SSSC located at the midpoint of a transmission line. The transmission line is modeled with lumped impedances and connects the sending end bus with voltage $V_S\angle\theta_S$ and the receiving bus with voltage $V_R\angle\theta_R$. The voltages $V_1\angle\theta_1$ and $V_2\angle\theta_2$ are the midpoint voltages on each side of the SSSC, and $V_{inj}\angle\theta_{inj}$ represents the voltage injected by the SSSC controller.

When PWM switching is used to govern the switching of VSC, then:

$$V_{\text{inj}} = kV_{\text{DC}} = k_{\text{tr}}m_a V_{\text{DC}} \tag{7.17}$$

$$\theta_{\text{inj}} = \theta_1 + \alpha \tag{7.18}$$

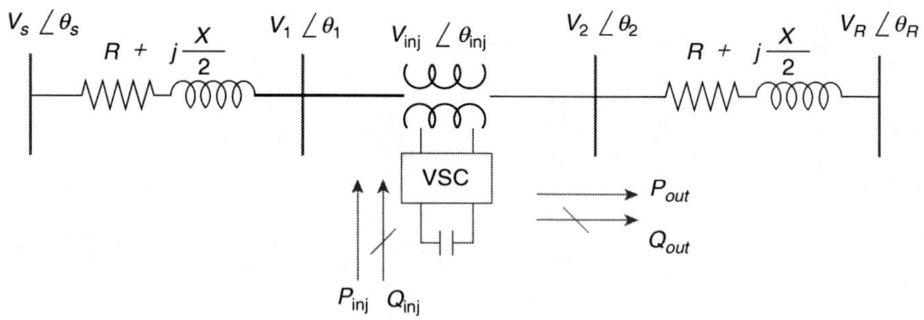

7.16 Two-bus system with midpoint SSSC.

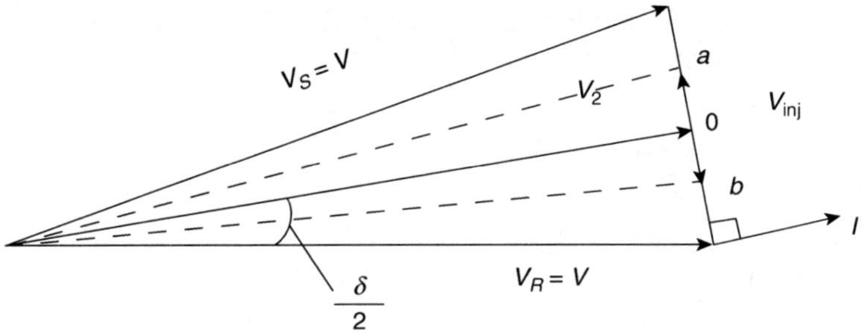

7.17 SSSC phasor diagram.

where k and α are the PWM modulation gain and phase shift. The modulation gain k is proportional to the modulation index m_a and k_{tr} which are determined by the modulation method and the serial coupling transformer winding ratio. The range of m_a and α are constrained by the steady-state reactive power capabilities of the controller. Figure 7.17 gives the related phasor diagram of the SSSC, where it has been assumed that $V_S = V_R = V$ and $\delta = \theta_S - \theta_R$. The phase shift α is referenced to θ_I since it is not practical to synchronize the injected voltage to the system reference. This diagram can be used as a basis from which to derive a control approach. If resistive losses are neglected, then the injected power P_{inj} and Q_{inj}, and the output powers P_{out} and Q_{out} can be determined.

The injected powers are given by:

$$S_{\text{inj}} = V_{\text{inj}}\angle\theta_{\text{inj}}\left(\frac{V\angle\delta - V\angle 0}{jX}\right)^{*} = P_{\text{inj}} + jQ_{\text{inj}} \tag{7.19}$$

$$\begin{aligned} P_{\text{inj}} &= \frac{V_{\text{inj}}V}{X}\left(\sin\theta_{\text{inj}} - \sin\left(\theta_{\text{inj}} - \delta\right)\right) \\ &= \frac{V\left(E_q - E_q\cos\delta + E_d\sin\delta\right)}{X} \end{aligned} \tag{7.20}$$

$$\begin{aligned} Q_{\text{inj}} &= \frac{V_{\text{inj}}}{X}\left(V\cos\left(\theta_{\text{inj}} - \delta\right) + V_{\text{inj}} - V\cos\theta_{\text{inj}}\right) \\ &= \frac{VE_d\cos\delta + VE_q\sin\delta - VE_d + V_{\text{inj}}^2}{X} \end{aligned} \tag{7.21}$$

where

$E_q = V_{\text{inj}}\sin\theta_{\text{inj}}$, $E_d = V_{\text{inj}}\sin\theta_{\text{inj}}$, and $V_{\text{inj}}^2 = E_d^2 + E_q^2$.

Since $P_{\text{inj}} = 0$, Equation [7.20] indicates that $\theta_{\text{inj}} = \pm(\pi/2)+(\delta/2)$. Similarly, the output powers are given by:

$$\begin{aligned} P_{\text{out}} &= \frac{V^2 \sin\delta + VV_{\text{inj}} \sin\theta_{\text{inj}}}{X} \\ &= \frac{V^2 \sin\delta + VE_q}{X} \end{aligned} \qquad [7.22]$$

$$\begin{aligned} Q_{\text{out}} &= \frac{V_{\text{inj}}^2 + 2VV_{\text{inj}} \cos(\delta - \theta_{\text{inj}})}{2X} \\ &= \frac{2VE_q \sin\delta + 2VE_d \cos\delta + V_{\text{inj}}^2}{2X} \end{aligned} \qquad [7.23]$$

Note that the compensated SSSC voltage phasor V_2 will remain on the line *ab* in Fig. 7.17 since the injected voltage V_{inj} must be perpendicular to the current I at all times. The phasor diagram and the relationships described above provide a framework in which to develop a systematic control scheme for the SSSC.

The SSSC model is similar to that of the STATCOM when transformed into the dq frame. The nonlinear SSSC state equations for the equivalent circuit model shown in Fig. 7.16 in the dq reference frame are given by:

$$\frac{1}{\omega_s}\frac{\mathrm{d}}{\mathrm{d}t} i_d = -\frac{R_s}{L_s} i_d + \frac{\omega}{\omega_s} i_q + \frac{k}{L_s} V_{\text{DC}} \cos(\alpha + \theta_i) - \frac{1}{L_s}(V_2 \cos\theta_2 - V_1 \cos\theta_1) \qquad [7.24]$$

$$\frac{1}{\omega_s}\frac{\mathrm{d}}{\mathrm{d}t} i_q = -\frac{R_s}{L_s} i_q - \frac{\omega}{\omega_s} i_d + \frac{k}{L_s} V_{\text{DC}} \sin(\alpha + \theta_i) - \frac{1}{L_s}(V_2 \sin\theta_2 - V_1 \sin\theta_1) \qquad [7.25]$$

$$\frac{C_{\text{DC}}}{\omega_s}\frac{\mathrm{d}}{\mathrm{d}t} V_{\text{DC}} = -k\cos(\alpha + \theta_1) i_d - k \sin(\alpha + \theta_1) i_q - \frac{V_{\text{DC}}}{R_{\text{DC}}} \qquad [7.26]$$

where $V_1\angle\theta_1$ and $V_2\angle\theta_2$ are the terminal voltages of the SSSC, i_d and i_q are the injected dq SSSC currents, V_{DC} is the voltage across the DC capacitor, R_{DC} represents the switching losses, and R_s and L_s are the coupling transformer resistance and inductance, respectively. The power balance equations at the sending end of the SSSC (Bus 1) are given by:

$$0 = V_1(i_d \cos\theta_1 + i_q \sin\theta_1) - V_1 \sum_{j=1} V_j Y_{1j} \cos(\theta_1 - \theta_j - \phi_{1j}) \qquad [7.27]$$

$$0 = V_1\left(i_d \sin\theta_1 - i_q \cos\theta_1\right) - V_1 \sum_{j=1} V_j Y_{1j} \sin\left(\theta_1 - \theta_j - \phi_{ij}\right) \qquad [7.28]$$

and at the receiving end (Bus 2)

$$0 = V_2\left(i_d \cos\theta_2 + i_q \sin\theta_2\right) - V_2 \sum_{j=1} V_j Y_{2j} \cos\left(\theta_2 - \theta_j - \phi_{2j}\right) \qquad [7.29]$$

$$0 = V_2\left(i_d \sin\theta_2 - i_q \cos\theta_2\right) - V_2 \sum_{j=1} V_j Y_{2j} \sin\left(\theta_2 - \theta_j - \phi_{ij}\right) \qquad [7.30]$$

where the summation terms represent the power flow equations, $Y_{ij}\angle\phi_{ij}$ is the (i,j)-th element of the admittance matrix, and n is the number of buses in the system.

The power flow control capability of an SSSC is constrained by its pure reactive power compensation capability during steady-state operation. The traditional approach to PWM-based SSSC control is to use the modulation index m_a to adjust the compensated apparent impedance, while using the phase shift to charge or discharge the DC capacitor (Rigby, 1998). Since the control effect of m_a and α interact with each other, it is desirable to introduce two new constrained decoupled control variables ΔE_d and ΔE_q to obtain the control target, where

$$\Delta P_{\text{inj}} = \frac{V}{X}\left(\Delta E_q(1 - \cos\delta) + \Delta E_d \sin\delta\right)$$

Under normal operations, the phase angle between adjacent buses is relatively small. Therefore, since δ is small, then $1 - \cos\delta \approx 0$ and

$$\Delta P_{\text{inj}} \approx \frac{V}{X}\Delta E_d \sin\delta \qquad [7.31]$$

Limited by the lack of an active power source on the DC side from which to absorb or inject energy, it is therefore important to control the injected active power close to zero to maintain near constant DC voltage across the VSC capacitor. A nearly constant DC voltage is important, since it directly affects the control speed and effectiveness of the SSSC. From Equation [7.31], ΔE_d is the main factor affecting the injected active power, thus ΔE_d can be used to adjust the DC capacitor voltage around its reference value. By combining the controls for DC voltage and transmission line active power flow, a constrained decoupled PI control algorithm for the SSSC is given by:

$$\Delta E_q = k_{1p}\Delta P_{\text{out}} + k_{2p}\int \Delta P_{\text{out}}\,\mathrm{d}t \qquad [7.32]$$

$$\Delta E_d = k_{1q}\Delta V_{\text{DC}} + k_{2q}\int \Delta V_{\text{DC}}\,\mathrm{d}t \qquad [7.33]$$

In implementation, these quantities are combined with the initial operating point and converted into a modulation index m_a=k/k_{tr} and phase shift α such that:

$$m_a = \frac{\sqrt{E_d^2 + E_q^2}}{k_{\text{tr}} V_{\text{DC}}} \qquad [7.34]$$

$$\alpha = \begin{cases} \sin^{-1}\left(\dfrac{E_q}{\sqrt{E_d^2 + E_q^2}}\right) - \theta_1 & E_d > 0 \\ \pi - \sin^{-1}\left(\dfrac{E_q}{\sqrt{E_d^2 + E_q^2}}\right) - \theta_1 & E_d < 0; E_q > 0 \\ -\pi - \sin^{-1}\left(\dfrac{E_q}{\sqrt{E_d^2 + E_q^2}}\right) - \theta_1 & E_d < 0; E_q < 0 \end{cases} \qquad [7.35]$$

where $E_q = E_{q0} + \Delta E_q, E_d = E_{d0} + \Delta E_d, E_{q0} = k_{\text{tr}} m_{a0} V_{\text{DC}} \sin(\alpha + \theta_1)$, and $E_{d0} = k_{tr} m_{a0} V_{\text{DC}} \cos(\alpha + \theta_1)$. This control is summarized in Fig. 7.18.

7.5 The unified power flow controller (UPFC)

The UPFC is the most versatile FACTS device. It consists of a combination of a shunt and series branches (STATCOM and SSSC) connected through the DC capacitor. The series-connected inverter injects a voltage with controllable magnitude and phase angle in series with the transmission line, thereby providing real and reactive power to the transmission line. The shunt-connected inverter provides the real power drawn by the series branch and the losses, and can independently provide reactive compensation to the system. Since the UPFC is a combination of both the STATCOM and the SSSC, it can be used for either, or both, control functions of these devices. The UPFC can increase or decrease the amount of active power

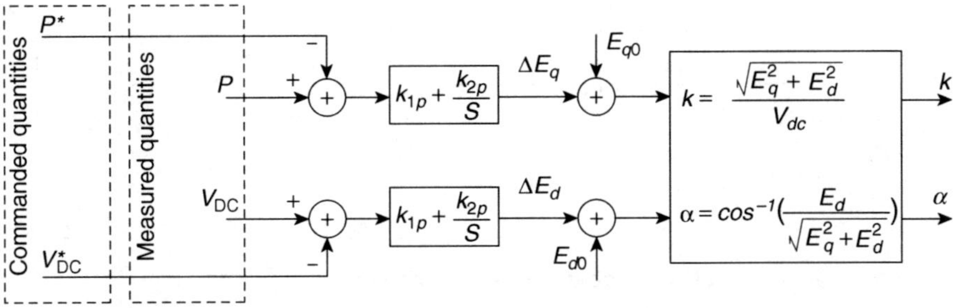

7.18 SSSC PI control.

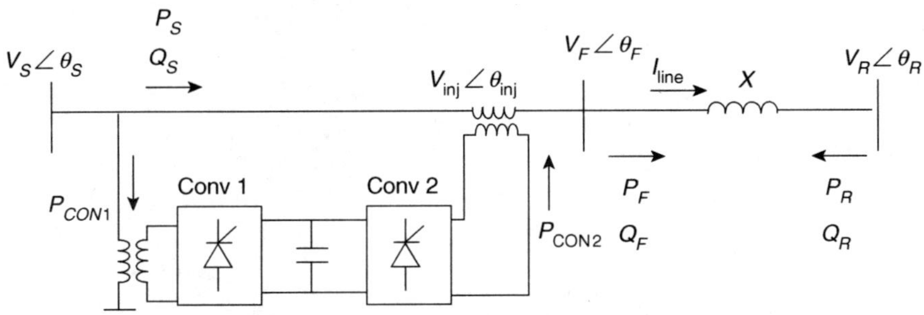

7.19 UPFC system.

transmitted over the line, increase the magnitude of the voltages at either end (i.e. voltage support), and/or reduce or increase the reactance of the line (i.e. line compensation).

To understand the basic capabilities of the UPFC, consider the two-bus AC transmission system shown in Fig. 7.19, where V_S is the sending end voltage, V_R is the receiving end voltage, and X is the line impedance of the adjacent transmission line. The line resistance is assumed to be negligible for simplicity. By convention, the UPFC is installed on the sending end bus.

Converter 2 (the SSSC) provides active and reactive power control by injecting a series voltage source V_{inj} with controllable magnitude and phase. The voltage V_F is the voltage at a fictional bus that sits on the line side of the UPFC series converter. The UPFC itself cannot generate or absorb active power. Therefore, the active power exchanged with the transmission system by Converter 2 must be compensated by Converter 1 (the STATCOM) through the common DC link. If the UPFC losses are neglected, then $P_{\text{conv2}} = P_{\text{conv1}}$. This relationship describes the power balance constraint of the UPFC. Because of the variability of the magnitude and angle, the UPFC can be considered to be a variable series voltage source with arbitrary magnitude and phase angle $\bar{V}_{\text{inj}} = V_{\text{inj}}\angle(\delta+\rho)$, where $0 \le V_{\text{inj}} \le V_{\text{inj}}^{\max}$ and $0 \le \rho \le \pi$ as shown in Fig. 7.20.

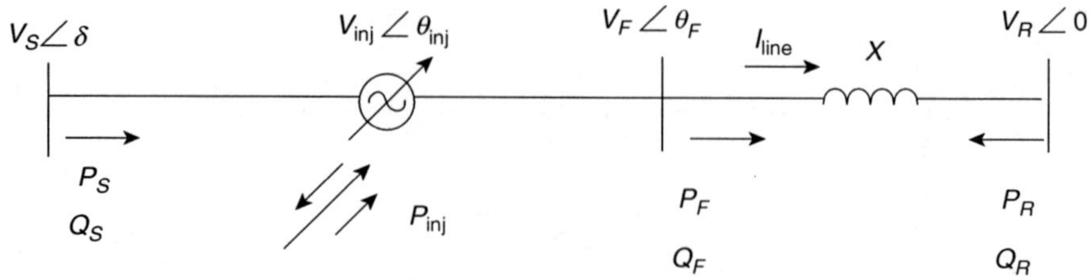

7.20 UPFC as a variable series voltage source.

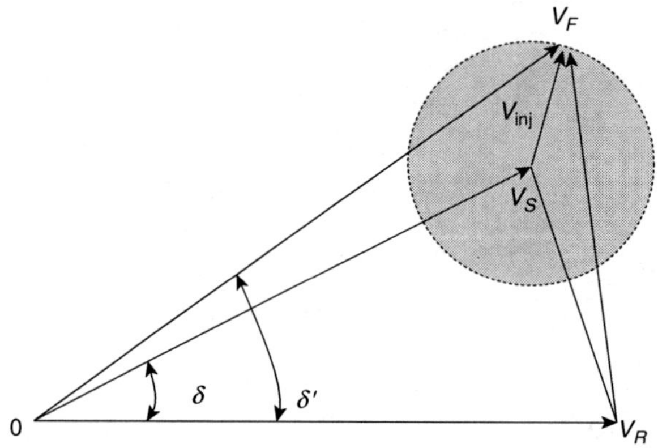

7.21 UPFC phasor diagram.

Figure 7.21 shows the corresponding phase diagram of the UPFC two-bus system. The voltage V_F can be freely controlled within the dashed circle centered at V_S with a radius of length V_{inj}^{max}. The area of the triangle (V_SOV_R) represents the uncompensated receiving end active power P_{r0}, the area of triangle (V_FOV_R) represents the compensated active power P_r, and the area of triangle $(V_SV_RV_F)$ represents the injected power P_{conv2}.

The active power P_{inj} $(= P_{conv2})$ is the active power injected into the system by UPFC Converter 2. Therefore, the total active power absorbed or supplied by the UPFC must be zero. Hence, $P_{inj} = P_{conv2} = P_{conv1}$.

The receiving end complex power is

$$\begin{aligned}
S_R &= \bar{V}_R\left(\frac{\bar{V}_S + \bar{V}_{inj} - \bar{V}_R}{jX}\right)^* \\
&= \bar{V}_R\left(\frac{\bar{V}_S - \bar{V}_R}{jX}\right) + \bar{V}_R\left(\frac{\bar{V}_{inj}}{jX}\right)^* \\
&= S_R^0 + \bar{V}_R\left(\frac{\bar{V}_{inj}}{jX}\right)^* \\
&= P_R + jQ_R
\end{aligned} \tag{7.36}$$

where $S_R^0 = P_R^0 + jQ_R^0$ is the uncompensated complex power at the receiving end, $P_R^0 = (V_s V_R / X)\sin\delta$ is the uncompensated active power at the receiving end, $Q_R^0 = (V_S V_R \cos\delta - V_R^2)/X$ is the uncompensated reactive power at the receiving end, $P_R = P_R^0 + (V_R V_{\text{inj}}/X)\sin(\delta+\rho)$ is the active power at the receiving end, and $Q_R = Q_R^0 + (V_R V_{\text{inj}}/X)\cos(\delta+\rho)$ is the reactive power at the receiving end.

The receiving end active power P_R and reactive power Q_R form a circle in the P–Q plane with a radius $V_R V_{\text{inj}}/X$ and center given by $S_R^0 = P_R^0 + jQ_R^0$, which is the uncompensated complex power at the receiving end. Similarly, the sending end complex power is

$$
\begin{aligned}
S_S &= \bar{V}_S \left(\frac{\bar{V}_S + \bar{V}_{\text{inj}} + \bar{V}_R}{jX} \right)^* \\
&= \bar{V}_S \left(\frac{\bar{V}_S - \bar{V}_R}{jX} \right)^* + \bar{V}_S \left(\frac{\bar{V}_{\text{inj}}}{jX} \right)^* \\
&= S_S^0 + \bar{V}_S \left(\frac{\bar{V}_{\text{inj}}}{jX} \right)^* \\
&= P_S + jQ_S
\end{aligned}
\qquad [7.37]
$$

where $S_S^0 = P_S^0 + jQ_S^0$ is the uncompensated complex power at sending end, $P_S^0 = (V_S V_R / X)\sin\delta$ is the uncompensated active power at the sending end, $Q_S^0 = (V_S^2 - V_S V_R \cos\delta)/X$ is the uncompensated reactive power at the sending end, $P_S = P_S^0 + (V_S V_{\text{inj}}/X)\sin(\delta+\rho)$ is the active power at the sending end, and $Q_S = Q_S^0 + (V_S V_{\text{inj}}/X)\sin(\delta+\rho)$ is the reactive power at the sending end.

The sending end active power P_S and reactive power Q_S form a circle on the P–Q plane with a radius $V_S V_{\text{inj}}/X$ and center given by $S_S^0 = P_S^0 + jQ_S^0$, which is the uncompensated complex power at the sending end.

The injected power is given by:

$$
\begin{aligned}
S_{\text{inj}} &= \bar{V}_{\text{inj}} \left(\frac{\bar{V}_S + \bar{V}_{\text{inj}} - \bar{V}_R}{jX} \right)^* = P_{\text{inj}} + jQ_{\text{inj}} \\
P_{\text{inj}} &= \frac{V_R V_{\text{inj}}}{X} \sin(\delta+\rho) - \frac{V_S V_{\text{inj}}}{X} \sin\rho \\
P_{\text{inj}} &= \frac{V_{\text{inj}}^2}{X} + \frac{V_S V_{\text{inj}}}{X} \cos\rho - \frac{V_R V_{\text{inj}}}{X} \cos(\delta+\rho)
\end{aligned}
\qquad [7.38]
$$

The operating characteristic of the injected power is a circle on P–Q plane with a radius of $\left(|\bar{V}_S - \bar{V}_R||\bar{V}_{\text{inj}}|\right)/X$ and center given by $\left(0, V_{\text{inj}}^2/X\right)$.

Lastly, the fictional bus complex power is given by:

$$
\begin{aligned}
S_F &= \bar{V}_F \bar{I}_{\text{line}}^{*} = S_S + S_{\text{inj}} \\
P_F &= \frac{V_S V_R}{X}\sin\delta + \frac{V_R V_{\text{inj}}}{X}\sin(\delta+\rho) \\
Q_F &= \frac{V_S^2 - V_S V_R\cos\delta}{X} + 2\frac{V_S V_{\text{inj}}\cos\rho}{X} + \frac{V_{\text{inj}}^2}{X} - \frac{V_R V_{\text{inj}}\cos(\delta+\rho)}{X}
\end{aligned}
\qquad [7.39]
$$

which has an elliptical operating characteristic on the P–Q plane centered at $\left(P_S^0, Q_S^0 + \left(V_{\text{inj}}^2/X\right)\right)$.

The UPFC model is a combination of the synchronous static compensator (STATCOM) and SSSC models:

$$\frac{1}{\omega_s}\frac{\mathrm{d}}{\mathrm{d}t} i_{d1} = -\frac{R_{s1}}{L_{s1}} i_{d1} + \frac{\omega}{\omega_s} i_{q1} + \frac{k_1}{L_{s1}} V_{\text{DC}}\cos(\alpha_1+\theta_1) - \frac{V_1}{L_{s1}}\cos\theta_1 \qquad [7.40]$$

$$\frac{1}{\omega_s}\frac{\mathrm{d}}{\mathrm{d}t} i_{q1} = -\frac{R_{s1}}{L_{s1}} i_{q1} - \frac{\omega}{\omega_s} i_{d1} + \frac{k_1}{L_{s1}} V_{\text{DC}}\sin(\alpha_1+\theta_1) - \frac{V_1}{L_{s1}}\sin\theta_1 \qquad [7.41]$$

$$\frac{1}{\omega_s}\frac{\mathrm{d}}{\mathrm{d}t} i_{d2} = -\frac{R_{s2}}{L_{s2}} i_{d2} + \frac{\omega}{\omega_s} i_{q2} + \frac{k_2}{L_{s2}} V_{\text{DC}}\cos(\alpha_2+\theta_1) + \frac{1}{L_{s2}}(V_1\cos\theta_1 - V_2\cos\theta_2) \qquad [7.42]$$

$$\frac{1}{\omega_s}\frac{\mathrm{d}}{\mathrm{d}t} i_{q2} = -\frac{R_{s2}}{L_{s2}} i_{q2} - \frac{\omega}{\omega_s} i_{d2} + \frac{k_2}{L_{s2}} V_{dc}\sin(\alpha_2+\theta_1) + \frac{1}{L_{s2}}(V_1\sin\theta_1 - V_2\sin\theta_2) \qquad [7.43]$$

$$
\begin{aligned}
\frac{C_{\text{DC}}}{\omega_s}\frac{\mathrm{d}}{\mathrm{d}t} V_{\text{DC}} = &-k_1\cos(\alpha_1+\theta_1) i_{d1} - k_1\sin(\alpha_1+\theta_1) i_{q1} - k_2\cos(\alpha_2+\theta_1) i_{d2} \\
&- k_2\sin(\alpha_2+\theta_1) i_{q2} - \frac{V_{\text{DC}}}{R_{\text{DC}}}
\end{aligned}
\qquad [7.44]
$$

The currents i_{d1} and i_{q1} are the dq components of the shunt current. The currents i_{d2} and i_{q2} are the components of the series current. The voltages $V_1\angle\theta_1$ and $V_2\angle\theta_2$ are the shunt and series voltage magnitudes and angles, respectively. The UPFC is controlled by varying the phase angles (α_1, α_2)

and magnitudes (k_1, k_2) of the converter shunt and series output voltages, respectively, and L_{s1}, L_{s2} are the shunt and series transformer inductances; R_{s1}, R_{s2} are the shunt and series transformer resistances; R_{DC} is the resistance representing converter losses; and ω, ω_s are the bus and synchronous frequency (in radians).

Many different UPFC control methods have been proposed that are based on Equations [7.40]–[7.44]. Just as the effectiveness of the UPFC depends on the gating control (the ability of the VSC to synthesize a reference waveform), the effectiveness also depends on the accuracy of the reference waveform magnitude and phase. This accuracy is the *internal* control of the UPFC and the ability of the controller to accurately convert system level setpoints into the shunt and series injected voltages. Most internal control methods for the UPFC are based on linear control techniques (Gyugyi *et al.*, 1995; Liu *et al.*, 2007). These controls arise from small perturbation linearization about the equilibrium of the nonlinear equations of the VSC average value model. Therefore, the feedback gains of linear control strategies may have to change with operating conditions; otherwise the nonlinearities of the VSC may cause the performance of the controller to degrade outside the linearization region.

A considerable amount of effort has been devoted to compensate for a change of operating conditions so that the controller can provide satisfactory

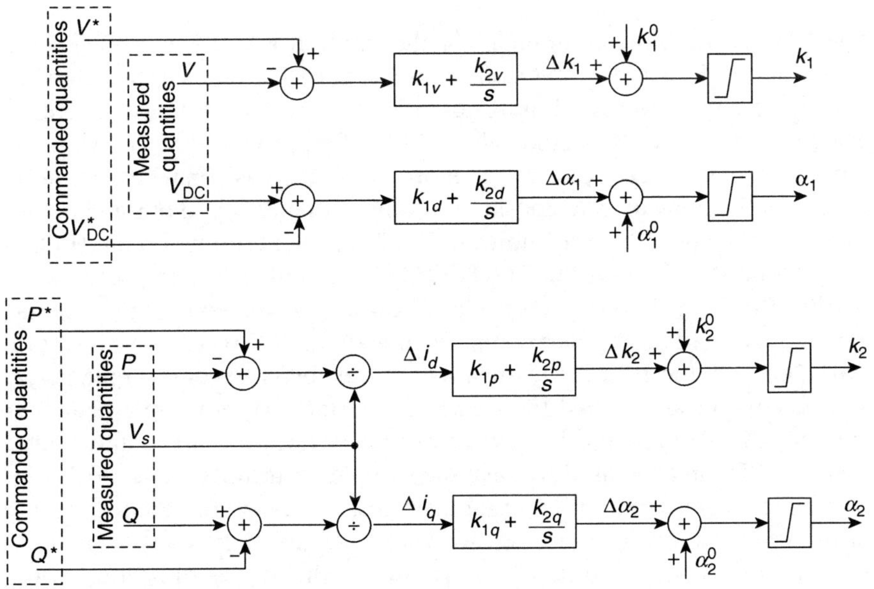

7.22 UPFC PI control.

performance over a wide range of operating conditions (Dash *et al.*, 2000, 2004; Al-Awami *et al.*, 2007; Ray and Venayagamoorthy, 2008). Many intelligent techniques have been reported to adaptively change the PI controller gains of UPFC, such as fuzzy logic (Dash *et al.*, 2004), neural networks (Dash *et al.*, 2000; Ray and Venayagamoorthy, 2008), and particle swarm optimization (Al-Awami *et al.*, 2007).

One of the simplest to implement is the decoupled PI control shown in Fig. 17.22. This control adjusts the switching signal reference to track desired active and reactive power outputs, the sending end bus voltage magnitude, and the DC link voltage. This control works well for slowly changing reference values.

7.6 Hybrid flexible AC transmission system (FACTS) technologies

The voltage source converter provides considerable flexibility in application and topology. By connecting converters and transmission lines in different ways, a number of hybrid topologies can be realized. Two recently proposed hybrid controllers are the interline power flow controller and the multiterminal HVDC system.

7.6.1 The interline power flow controller

The VSC is the basic component of the modern FACTS device, and the UPFC shows how individual shunt and series converters can be interconnected via a common DC link to provide increased flexibility and control. The interline power flow controller (IPFC), first proposed in Gyugyi *et al.* (1999), is constructed from two or more SSSC models connected on adjacent lines that may or may not share a common sending end bus and do not share a receiving end bus as shown in Fig. 7.23. The literature on the IPFC is far more limited than for the STATCOM, SSSC, and UPFC. Reported work on the IPFC considers aspects such as power flow and constraints (Zhang, 2003; Zhang *et al.*, 2006), and comparison with other FACTS devices (Arabi *et al.*, 2002). In the IPFC, only one line has the ability to control both active and reactive power (called the *master* converter). All other lines can control only the active or the reactive power (*slave* lines). It can be anticipated that the IPFC may be used to solve the complex transmission network congestion management problems that transmission companies are now facing in the transmission open-access environment. This is the case of the convertible static compensator (CSC) installed at the Marcy Substation of the New York Power Authority (NYPA) as party of project that will increase

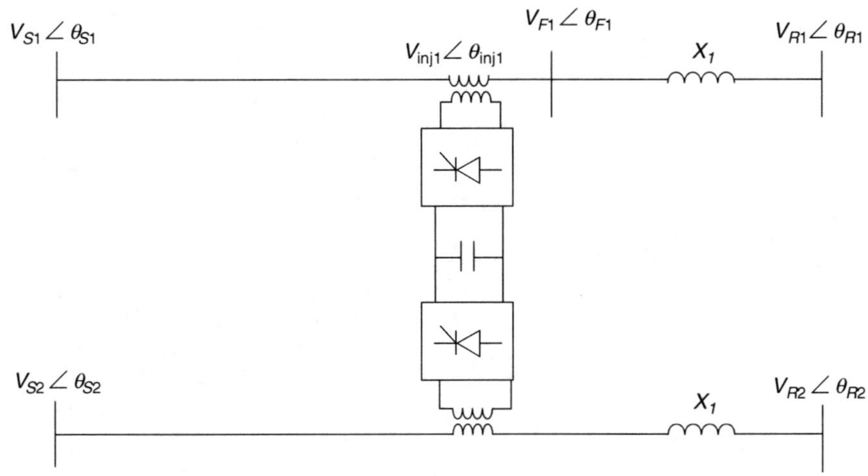

7.23 The IPFC.

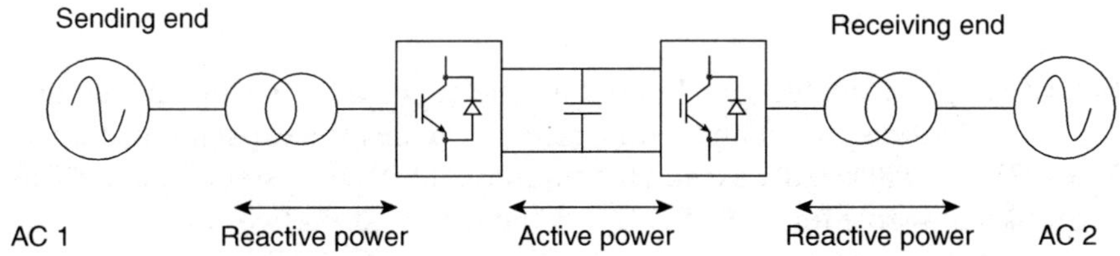

7.24 HVDC system based on VSC technology.

power transfer capability and maximize the use of the existing transmission network (Arabi *et al.*, 2002).

7.6.2 Multiterminal HVDC

Another application of VSC-based devices is HVDC transmission. The development of VSC-based HVDC systems (shown in Fig. 7.24) opens new opportunities. In contrast to the traditional thyristor-based HVDC system, the VSC–HVDC system has the following features: (1) it is very easy to make multiterminal connections; (2) it has the ability to independently control active and reactive flows at its terminals; (3) it has the option to control its terminal bus voltages instead of reactive powers; (4) the costs for filtering of harmonics may be significantly reduced if suitable PWM techniques are used; and (5) construction and commissioning of a VSC–HVDC system takes less time than that for a traditional thyristor-based HVDC system (Zhang, 2004). Presently the capacity of the VSC–HVDC is limited to 300–500 MW,

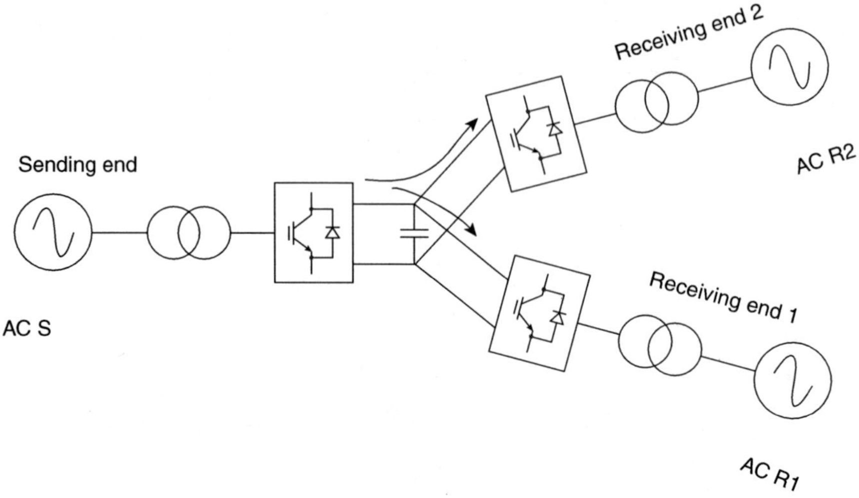

7.25 Three-terminal M-VSC–HVDC.

but with larger insulated-gate bipolar transistor (IGBT) switches and emerging silicon carbide-based high temperature switches, the capacity of the new VSC–HVDC will rapidly exceed the traditional HVDC system capacity. This suggests a new role for VSC–HVDC in the future AC network.

One of the most promising applications of VSC–HVDC models is the development of multiterminal (M-VSC–HVDC) installations (Zhang, 2004). A multiterminal installation is when two or more converters are directly connected with a common DC link and co-located in a substation. Figure 7.25 shows a three-terminal M-VSC–HVDC that has one sending end terminal and two receiving ends. Note that in M-VSC–HVDC 'sending' and 'receiving' end notation is merely for reference since both active and reactive power can flow in either direction.

7.7 Conclusion

This chapter described the family of modern FACTS devices, which are based on the concept of the VSC. The basic description of the VSC is given along with a brief introduction to various switching schemes and multilevel topologies. The three primary FACTS devices (the STATCOM, SSSC, and UPFC) were described in depth and the fundamental frequency state-space models were presented along with simple control strategies. Furthermore, two hybrid devices, the IPFC, and the multiterminal HVDC line, were introduced and briefly described. The VSC-based modern FACTS device holds considerable potential for improved steady-state and dynamic control of increasingly complex power transmission networks.

7.8 References

Al-Awami, A., Abdel-Magid, Y. and Abido, M. A. (2007). A particle swarm-based approach of power system stability enhancement with unified power flow controller. *International Journal of Electrical Power and Energy Systems*, **29**(3), pp. 251–259.

Arabi, S., Hamadanizadeh, H. and Fardanesh, B. B. (2002). Convertible static compensator performance studies on the NY state transmission system. *IEEE Transactions on Power Systems*, **17**(3), pp. 701–706.

Botao, M., Congwei, L., Wang, Z. and Fahai, L. (2002). New SVPWM control scheme for three-phase diode clamping multilevel inverter with balanced DC voltages. Annual conference of the IEEE Industrial Electronics Society, November, pp. 903–907.

Cheng, Y., Qian, C. and Crow, M. L. (2006). A comparison of diode-clamped and cascaded multilevel converters for a STATCOM with energy storage. *IEEE Transactions on Industrial Electronics*, **53**(5), pp. 1512–1521.

Corzine, K. (2002). Multilevel converters. In: T. Skvarenina, ed. *The Power Electronics Handbook*. Boca Raton: CRC Press.

Dash, P. K., Mishra, S. and Panda, G. (2000). A radial basis function neural network controller for UPFC. *IEEE Transactions on Power System*, **15**(4), pp. 1293–1299.

Dash, P. K., Morris, S. and Mishra, S. (2004). Design of a nonlinear variable-gain fuzzy controller for FACTS devices. *IEEE Transactions on Control Systems Technology*, **12**(3), pp. 428–438.

Dong, L., Crow, M. L., Yang, Z. and Atcitty, S. (2004). A reconfigurable FACTS system for university laboratories. *IEEE Transactions on Power Systems*, **19**(1), pp. 120–128.

Du, Z., Tolbert, L. M. and Chiasson, J. N. (2006). Active harmonic elimination for multilevel converters. *IEEE Transactions on Power Electronics*, **21**(2), pp. 459–469.

El-Moursi, M. S. and Sharaf, A. M. (2005). Novel controllers for the 48-pulse VSC STATCOM and SSSC for voltage regulation and reactive power compensation. *IEEE Transactions on Power Systems*, **20**(4), pp. 1985–1997.

Gyugyi, L., Schauder, C. and Sen, K. K. (1997). Static synchronous series compensator: a solid state approach to the series compensation of transmission lines. *IEEE Transactions on Power Systems*, **12**(1), pp. 406–417.

Gyugyi, L., Schauder, C., Williams, S. and Rietman, T. (1995). The unified power flow controller: a new approach to power transmission control. *IEEE Transactions on Power Delivery*, **10**(2), pp. 1085–1097.

Gyugyi, L., Sen, K. K. and Schauder, C. D. (1999). The interline power flow controller concept: a new approach to power flow management in transmission systems. *IEEE Transactions on Power Delivery*, **14**(3), pp. 1115–1123.

Hatano, N. and Ise, T. (2010). Control scheme of cascaded H-bridge STATCOM using zero-sequence voltage and negative-sequence current. *IEEE Transactions on Power Delivery*, **25**(2), pp. 543–550.

Hingorani, N. G. and Gyugyi, L. (1999). *Understanding FACTS: Concepts and Technology of Flexible AC Transmission Systems*. Piscataway, NJ: IEEE Press.

IEEE Power Engineering Society FACTS Application Task Force (1996). *FACTS Applications*. Piscataway, NJ: IEEE Press.

Jain, A., Joshi, K., Behal, A. and Mohan, N. (2006). Voltage regulation with STATCOMs: modeling, control and results. *IEEE Transactions on Power Delivery*, **21**, pp. 726–735.

Larsen, E., Clark, K., Miske, S. and Urbanek, J. (1994). Characteristics and rating considerations of thyristor controlled series compensation. *IEEE Transactions on Power Delivery*, **9**(2), pp. 992–1000.

Lehn, P. W. and Iravani, M. R. (1998). Experimental evaluation of STATCOM closed-loop dynamics. *IEEE Transactions on Power Delivery*, **13**(4), pp. 1378–1384.

Lin, B.-R. (2000). Analysis and implementation of a three-level PWM rectifier/inverter. *IEEE Transactions on Aerospace and Electronic Systems*, **36**(3), pp. 948–956.

Liu, C. H. and Hsu, Y. Y. (2010). Design of a self-tuning PI controller for a STATCOM using particle swarm optimization. *IEEE Transactions on Industrial Electronics*, **57**, pp. 702–715.

Liu, F., Mei, S., Lu, Q., Ni, Y., Wu, F.F. and Yokoyama, A. (2003). The nonlinear internal control of STATCOM: theory and application. *Electrical Power & Energy Systems*, **25**, pp. 421–430.

Liu, L., Zhu, P., Kang, Y. and Chen, J. (2007). Power-flow control performance analysis of a unified power flow controller in a novel control scheme. *IEEE Transactions on Power Delivery*, **22**(3), pp. 1613–1619.

Liu, Y., Zhao, Z., Qian, C. and Zhou, X. (2000). A novel three-phase multilevel voltage source converter. Proceedings of the 2000 Power Electronics and Motion Control Conference, Beijing, China.

Lu, B. and Ooi, B. T. (2007). Nonlinear control of voltage-source converter systems. *IEEE Transactions on Power Electronics*, **22**(4), pp. 1186–1195.

Marchesoni, M. and Tenca, P. (2002). Diode-clamped multilevel converters: a practical way to balance DC-link voltages. *IEEE Transactions on Industrial Electronics*, **49**(4), 752–765.

Mohagheghi, S., del Valle, Y., Venayagamoorthy, G. K. and Harley, R. G. (2007). A Proportional-Integrator type adaptive critic design-based neurocontroller for a static compensator in a multimachine power system. *IEEE Transactions on Industrial Electronics*, **54**, pp. 86–96.

Peng, F. (2000). A generalized multilevel inverter topology with self-voltage balancing. IEEE 2000 Industry Applications Conference, October 8–12, Rome, Italy.

Rao, P., Crow, M. L. and Yang, Z. (2000). STATCOM control for power system voltage control applications. *IEEE Transactions on Power Delivery*, **15**(4), pp. 1311–1317.

Ray, S. and Venayagamoorthy, G. K. (2008). Wide-area signal-based optimal neurocontroller for a UPFC. *IEEE Transactions on Power Delivery*, **23**(3), pp. 1597–1605.

Rigby, B. (1998). An improved control scheme for a series-capacitive reactance compensator based on a voltage-source inverter. *IEEE Transactions on Industry Applications*, **34**(2), pp. 355–363.

Rodriguez, J., Lai, J. S. and Peng, F. Z. (2002). Multilevel inverters: A survey of topologies, controls, and applications. *IEEE Transactions on Industrial Electronics*, **49**(4), pp. 724–738.

Saeedifard, M., Nikkhajoei, H. and Iravani, R. (2007). Space vector modulated STATCOM based on a three-level neutral point clamped converter. *IEEE Transactions on Power Delivery*, **22**(2), pp. 1029–1039.

Sahoo, N. C., Panigraph, B. K., Dash, P. K. and Pan, G. (2002). Application of a multivariable feedback linearization scheme for STATCOM control. *Electric Power Systems Research*, **62**, pp. 81–91.

Schauder, C. and Mehta, H. (1993). Vector analysis and control of advanced static VAR compensators. *IEE Proceedings, Generation Transmission and Distribution*, **140**(4), pp. 299–306.

Song, E., Lynch, A. F. and Dinavahi, V. (2009). Experimental validation of nonlinear control for a voltage source converter. *IEEE Transactions on Control Systems Technology*, **17**, pp. 1135–1144.

Song, Q., Liu, W. and Yuan, Z. (2007). Multilevel optimal modulation and dynamic control strategies for STATCOMs using cascaded multilevel inverters. *IEEE Transactions on Power Delivery*, **22**(3), pp. 1937–1946.

Soto, D. and Pena, R. (2004). Nonlinear control strategies for cascaded multilevel STATCOMs. *IEEE Transactions on Power Delivery*, **19**(4), pp. 1919–1927.

Spitsa, V., Alexandrovitz, A. and Zeheb, E. (2010). Design of a robust state feedback controller for a STATCOM using a zero set concept. *IEEE Transactions on Power Delivery*, **25**(1), pp. 456–467.

Sternberger, R. and Jovcic, D. (2009). Analytical modeling of a square-wave-controlled cascaded multilevel STATCOM. *IEEE Transactions on Power Delivery*, **24**(4), pp. 2261–2269.

Wang, H. F., Li, H. and Chen, H. (2002). Application of cell immune response modelling to power system voltage control by STATCOM. *IEE Proceedings, Generation Transmission and Distribution*, **149**(1), pp. 102–107.

Wang, K. and Crow, M. L. (2011). Power system voltage regulation via STATCOM internal nonlinear control. *IEEE Transactions on Power Systems*, **26**(3), pp. 1252–1262.

Zhang, X. P. (2003). Modelling of the interline power flow controller and the generalised unified power flow controller in Newton power flow generation, transmission and distribution. *Proc. Inst. Elect.*, **150**(3), pp. 268–274.

Zhang, X. P. (2004). Multiterminal voltage-sourced converter-based HVDC models for power flow analysis. *IEEE Transactions on Power Systems*, **19**(4), pp. 1877–1884.

Zhang, X.-P., Rehtanz, C. and Pal, B. (2010). *Flexible AC Transmission Systems: Modeling and Control*. Berlin: Springer-Verlag.

Zhang, Y., Zhang, Y. and Chen, C. (2006). A novel power injection model of IPFC for power flow analysis inclusive of practical constraints. *IEEE Transactions on Power Systems*, **21**(4), pp. 1550–1556.

8

Nanodielectrics and their role in power transmission applications

G. C. STEVENS, GnoSys Global Ltd, UK and University of Surrey, UK and A.S. VAUGHAN, University of Southampton, UK

DOI: 10.1533/9780857097378.2.206

Abstract: This chapter first describe the nature of nanodielectrics, before moving on to consider the range of properties that these material systems are capable of exhibiting and what potential benefits these confer in high voltage electrical equipment used in power transmission applications. A key feature is that if significant property enhancements are to be achieved, then optimal control of the nano-particulate dispersion, the particle-polymer matrix interface and processing are essential. Methods for preparing and characterizing nanodielectrics are also described and the potential benefits of their application in enhancing the performance of high voltage equipment is discussed.

Key words: nanodielectric, nanomaterial, structure, properties, design, processing, composite, permittivity, voltage endurance, electrical treeing, thermal endurance, partial discharge resistance, corona resistance, mechanical properties, thermal conductivity, space charge, electrical breakdown, structure-property relationships, design rules, HVAC, HVDC, water absorption.

8.1 Introduction

Dielectric materials constitute critical elements within electrical and electronic devices that range in size from microscopic field effect transistors in integrated circuits to multi-tonne supergrid transformers. As the technology has evolved, the demands placed upon the dielectric within such devices have increased. For example, the earliest power cables were installed in the second half of the nineteenth century to transmit electrical energy at just a few thousand volts, and relied upon natural materials for their insulation.[1] Modern cable systems using synthetic polymer materials are designed to transmit AC power at voltages up to 500 kV, which requires the use of highly refined superclean polymers that are expected to operate reliably for decades under electric fields in excess of 10 kV mm^{-1}. However, major challenges still exist, for example in connection with the DC power transmission at comparable voltages, that require a radical new approach to the development of materials across a wide range of power transmission applications requiring novel combinations of physical properties, and it is this need that has driven recent interest in so-called nanodielectrics.

Nanocomposites may involve the inclusion of a filler that is nanoscopic, within a host matrix; the term nanodielectric commonly refers to a nanocomposite where the dielectric properties are of primary interest. In the case of technologically relevant systems, these materials generally involve a nanoscopic inorganic filler with dimensions in the size range 1–1000 nm dispersed within a polymeric matrix. More generally, it can relate to one or more structural elements in the material that are nanoscopic, including organic and inorganic polymer, oligomer, dendrimer and other structures in addition to inorganic fillers. These elements may be randomly volume dispersed, or they may be structurally ordered via processes of aggregation and self-assembly, or they may form pseudo-lattices – providing a range of structural ordering types.

Much of the current interest in nanocomposites was initiated by work at Toyota in the late 1980s,[2] in connection with the dispersion of a layered clay material into a polymeric host, in such a way as to break up the primary clay aggregates into their constituent aluminium silicate layers. The result was a nanostructured composite material with improved macroscopic physical properties, particularly mechanical and permeation properties. Since then, interest in composites based upon nanofillers within a host matrix has grown exponentially such that, while the ISI Web of Knowledge database identifies just two papers containing the term 'nanocomposite' in 1986, 3021 papers are found for 2006 and 4952 papers for 2009. The reason for this is that the addition of a nanofiller to a base resin is able to improve not just one material characteristic but, rather, can be used to generate material systems that exhibit combinations of properties that are not attainable by more conventional means. Also, these property enhancements are generally seen at very low filler loading levels (1–10%), which raises the possibility, ultimately, of being able to design unique materials through the use of combinations of nanofiller additives. Examples where dielectric properties directly impact technological advances range from all-organic flexible display devices, which require thin film transistors incorporating flexible, high dielectric constant, low leakage current gate dielectrics, to high electrical power density generators, where the dielectric must exhibit a plethora of characteristics, including high electrical breakdown strength, high thermal conductivity, high resistance to partial discharge (PD) activity, low dielectric loss, good thermal stability and good mechanical properties, etc.

This chapter will first describe the nature of nanodielectrics, before moving on to consider the range of properties that these material systems are capable of exhibiting and what potential benefits these confer in high voltage electrical equipment. A key feature is that if significant property enhancements are to be achieved – or even deterioration in performance is to be avoided – then optimal processing is essential. But what is optimal processing? To answer such a question requires one of the most fundamental challenges of materials science to be addressed: an understanding of the connectivity between chemistry, processing, material structure and material properties. The next section therefore considers methods for preparing and characterizing nanodielectrics since, without mastery of these topics, the technological exploitation of this class of materials is fraught with difficulty. The potential pitfalls of such a strategy are finally outlined, in order to highlight the challenges that are still faced.

8.2 Nanodielectric materials

This section discusses the essential structure of nanodielectrics, the importance of the interphase region between the nanoparticle and the matrix, and how this relates to the enhancement of physical properties, the control of space charge and the production of high voltage performance improvements.

8.2.1 Structural elements

As with all material types, the macroscopic behaviour of a nanodielectric is determined by its structure, which, in turn, is dictated by its composition and processing history. In the case of clay-based nanocomposites, for example, the filler is composed of thin sheets that are nanoscopic in one dimension and microscopic in two; for systems containing carbon nanotubes (CNT), the filler is nanoscopic in two dimensions and microscopic in one; particles of metal oxides, such as alumina or titania, are nanoscopic in all three dimensions. Thus one can immediately consider nanocomposites in terms of their nanoscopic dimensionality, and this has a direct impact on properties and potential application areas. Sheet-like systems, for example, have been considered for inclusion into polymers as a means of improving gas barrier properties for packaging applications, while the addition of CNTs produces a material with an electrical percolation threshold as low as 0.0025%[3] as a consequence of the high aspect ratio of the conducting phase. Similar strategies can be envisaged as a means of increasing the thermal conductivity of an insulating polymer, provided attention is paid to the problem of phonon scattering at interfaces.

However, in reality, it is insufficient to consider these materials as simple two-phase systems since, as pointed out by Lewis,[4,5] a key feature that defines any nanocomposite is the interface between the nanofiller and the matrix. Consequently, it is necessary to include at least one interphase region[6] where the structure, chemistry and properties are typical of neither of the bulk phases that were combined to generate the composite. While this is true of the

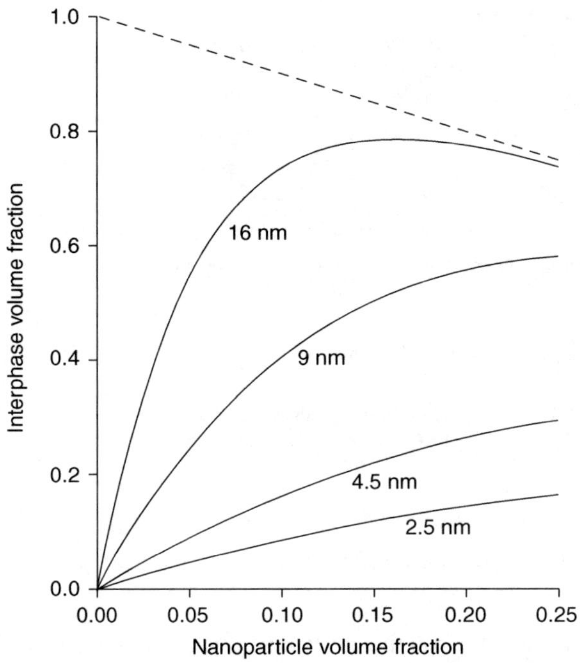

8.1 Interphase volume fraction as a function of nanoparticle volume fraction for different diameter spherical nanoparticles.

interface between any two chemically distinct materials, it is the enormous specific surface area of nanoparticles that means that, even at low volume percentages of filler, the fraction of the material that must be attributed to the interphase region can be extremely large in a nanocomposite. This point is illustrated in Fig. 8.1, which shows plots of interphase volume fraction as a function of particle volume fraction for a random array of different sized particles from 2.5 to 16 nm diameter. For 16 nm diameter nanoparticles, if the interphase surrounding each particle is just 10 nm thick, then it would constitute close to 50% of the total material volume for a composite containing 5% by volume of the nanofiller. Similar conclusions have been drawn previously for the less physically realistic cases of regular arrays of spheres.[5,7]

8.2.2 Permittivity

The response of a dielectric to an AC electric field can be characterized in terms of a complex permittivity, where the real part, ε', is related to the energy stored in the dielectric and the imaginary part, ε'', is related to various processes that dissipate energy, such as coupling between polar moieties within the system and the applied field. As such, the approach can provide much useful information.[8,9]

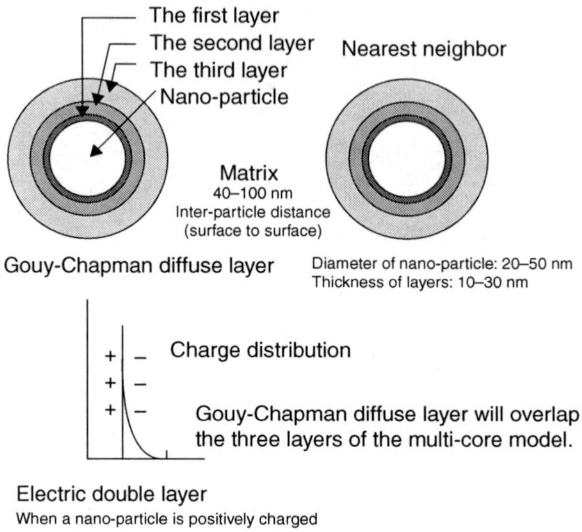

8.2 Core-shell layer interfacial model for nanodielectrics.

The potential significance of the interphase was recognized early in the history of nanodielectrics and models, such as that proposed by Tanaka,[10] have been developed. As illustrated in Fig. 8.2, this envisages a nanoparticle embedded within matrix to be surrounded by three structurally distinct layers, superimposed upon which is a diffuse, Gouy-Chapman charge layer. The two layers immediately adjacent to the nanoparticle are proposed to consist of immobilized species that are either covalently bonded to, or interacting strongly with, their inner neighbour. Between these largely immobilized shells and the unperturbed matrix is a region characterized by increased free volume and enhanced chain mobility. All of these various features have a marked effect on the dielectric behaviour of what is, in effect, a three-phase system.

Many analyses of the behaviour of composite systems are based upon the concept of rules of mixing. That is, that for a composite of two materials, A and B, the property of interest will fall some way between the value exhibited by A and that exhibited by B. In the case of ε' for a composite system, a number of options have been proposed[11,12] but these all indicate a monotonic variation in the real part of the permittivity with composition. In the case of nanocomposites, however, experimentally observed behaviour differs significantly from this, when the percentage of nanofiller is low (below ~5%). For example, the real part of the permittivity value was reported as a function of composition for polypropylene/iso-octyl polyhedral oligomeric silsesquioxane (POSS) systems; the value of ε' fell from ~2.33 for the unfilled system to ~2.19 at 5% filler before increasing again to close to 2.33 at 10%.[13] Srisuwan *et al.*[14] examined the dielectric behaviour of nanocomposites based upon polyimide and silica hybrid, reporting increasing the fraction of nano-silica from 0.06% to 5.58% caused a reduction in ε' from 3.27 to 2.26. Recent studies by Andritsch *et al.*[15] have mapped out the dielectric response of a range of different functionalized nanofillers within an epoxy matrix, which show in detail the form of the local minimum in ε' for each composite type. Many other examples of similar behaviour can be found in the literature;[16–20] although it is generally assumed that this effect is related in some way to interfacial effects in nanocomposites, we are not aware of any quantitative theory that is able to explain the behaviour described above.

8.2.3 Molecular relaxations

In addition to providing a measure of the frequency dependence of the permittivity, dielectric spectroscopy is a powerful technique for the study of molecular dynamics in polar materials and consequently has been used to support the concept that the presence of interfaces does significantly affect molecular mobility. Nelson *et al.*[21] contrasted the dielectric response of epoxy/titania composites containing 10% of nanoscopic and microscale filler and showed that these two systems exhibit markedly different forms of behaviour. Specifically, no significant dispersion peaks were evident in the mid-frequency range in the microcomposite data, whereas broad peaks were seen in the case of the nanocomposite, which shifted to higher frequencies as the temperature increased. These data were interpreted as showing that the polymer in the nanocomposite has enhanced mobility above the glass transition. Interestingly, this paper also considers the effect of degree of cure on measurements of ε', showing that the value of this parameter increases with crosslink density. It has been shown in the case of polyethylene/silica systems that varying the surface chemistry to the silica, and hence the nature of the interface, can have a marked effect on the dielectric spectrum of the material.[22]

However, all nanocomposites, irrespective of the nanosilica surface chemistry, were found to exhibit evidence of reduced chain mobility, compared with micro-filled equivalents. Sengwa *et al.*[23] examined the dielectric properties of montmorillonite (MMT) clay filled poly(vinyl alcohol)/poly(ethylene oxide) blend nanocomposites, similarly concluding that the presence of well-dispersed MMT clay within the polymers matrix resulted in inhibited polymer chain dynamics. Conversely, a study of PMMA-based system containing nanosilica suggested that the polymethylmethacrylate (PMMA) and its nanocomposites exhibited comparable segmental dynamics; however, the nanocomposites were found to display significantly accelerated physical ageing.[24] Numerous other studies relating to the intrinsic dielectric response of nanocomposites have been described in the literature,[24–29] but a complete discussion of the multitude of effects that have been reported goes beyond the scope of this review. Nevertheless, one additional extrinsic process is worthy of brief mention, and likely to be of great importance in the practical utilisation of nanodielectrics, which is related to the exposure of materials composed of relatively polar nanoparticles within a relatively non-polar matrix related to ambient water. The related topic of water treeing provides graphic evidence of the way in which the presence of local hydrophilic inclusions can lead to dielectric failure.[30] In epoxy-based nanocomposites containing absorbed water, the low frequency dielectric response has been interpreted as indicating percolation conductivity and a 'water shell' has been proposed to account for this.[31] Zhang and Stevens[32] contrasted the effect of water absorption on the dielectric response of alumina nanocomposites based upon two polymers: an epoxy resin, which absorbs water readily, and polyethylene, which does not. The presence of the alumina in polyethylene was found to give rise to a low frequency dielectric loss peak, which increases in frequency as the nanoalumina content increases, as shown in Fig. 8.3. Fréchette *et al.*[33] have shown that the presence of water resulted in a significant increase in both the real and imaginary parts of the permittivity, the dielectric loss of epoxy-based composites, containing both micro quartz and a layered aluminosilicate nanoclay.

8.2.4 Short term dielectric breakdown

A key characteristic of any material for use in dielectric applications is its ability to withstand the applied electric field. In addressing this, it is convenient to divide breakdown processes into short term failure modes, which provide some measure of a material's breakdown strength, and longer term mechanisms, which result in progressive degradation prior to failure.

The effect on the AC breakdown strength of adding MMT to polyethylene-based systems has been reported in a number of papers. Green *et al.*,[34] for example, showed that, provided the presence of the MMT does not serve adversely to affect the crystallization of the polymer, a significant increase in breakdown strength can be achieved. This result is shown in Fig. 8.4. Similar improvements in performance have been reported for low density polyethylene (LDPE)/MMT systems.[35] Conversely, Stefanescu *et al.*[36] report that, for multi-functional PMMA/ceramic composites containing mixtures of MMT and calcium copper titanate (CCTO), the breakdown decreases linearly with filler loading, due, they suggest, to the resultant increase in the density of ionic carriers and dipolar species within the system.

Similar reduction in breakdown strength in epoxy-based systems containing MMT or barium titanate has also been reported.[37] Roy *et al.*[22] examined the DC breakdown strength of crosslinked polyethylene (XLPE)/silica systems and demonstrated that while the addition of nanosilica can lead to an increase in breakdown strength, tailoring the nanoparticle surface chemistry can yield further benefits. In contrast, including micrometric silica results in

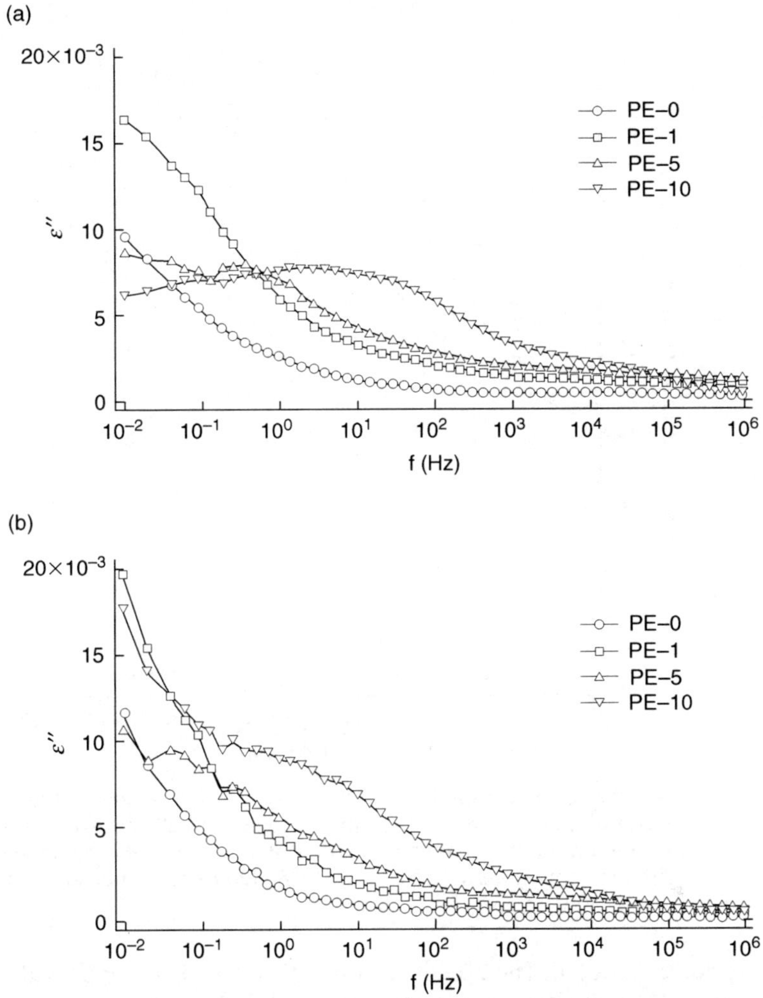

8.3 Effect of water absorption on broad band dielectric response of nanoalumina filled polyethylene as a function of filler content from 0% to 10% b.w.: (a) 10°C and (b) 70°C.

a marked decrease in performance. Similar results obtained by Smith *et al.*[38] for voltage endurance behaviour are shown in Fig. 8.5.

Where the breakdown strength of systems containing poorly dispersed nanofillers has been determined, similar reductions in performance have been reported.[39,40] Numerous studies have considered the topic of breakdown in nanodielectrics and, taking these as a whole, it is not easy to deduce any generally applicable trends, since the response seen is likely to be highly dependent upon the extent of the nanoparticle dispersion, which is rarely

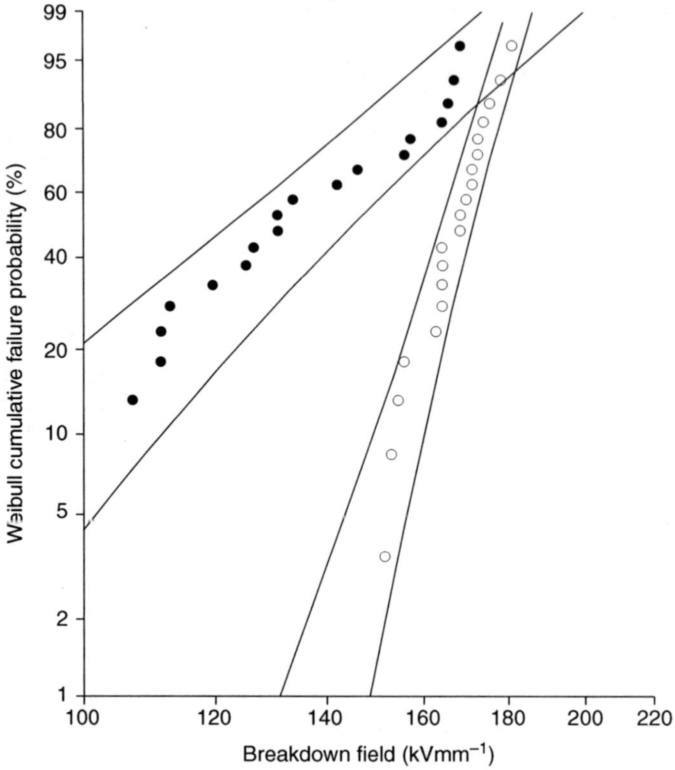

8.4 Effect of montmorillonite (MMT) nanofillers on the short term electric strength of polyethylene. The figure shows Weibull probability plots comparing the breakdown behaviour of an unfilled (●) polyethylene blend with the same material containing MMT (20 parts of masterbatch – ○).

considered in detail. The recent review article by Li *et al.*[41] provides a good summary of both the short and long-term breakdown behaviour of nanodielectrics. Finally, a topic that is beginning to attract increasing attention is the use of nanoparticles as reinforcement for conventional highly filled microcomposites. Imai *et al.*,[42] for example, report that the dispersion of a layered silicate into an epoxy-based microcomposite results in a 7% increase in breakdown strength.

8.2.5 Electrical treeing

The long-term electrical breakdown of polymeric dielectrics occurs through the evolution of tree-like fractal objects, termed trees, which grow through the progressive erosion of the dielectric as a result of PD activity within the

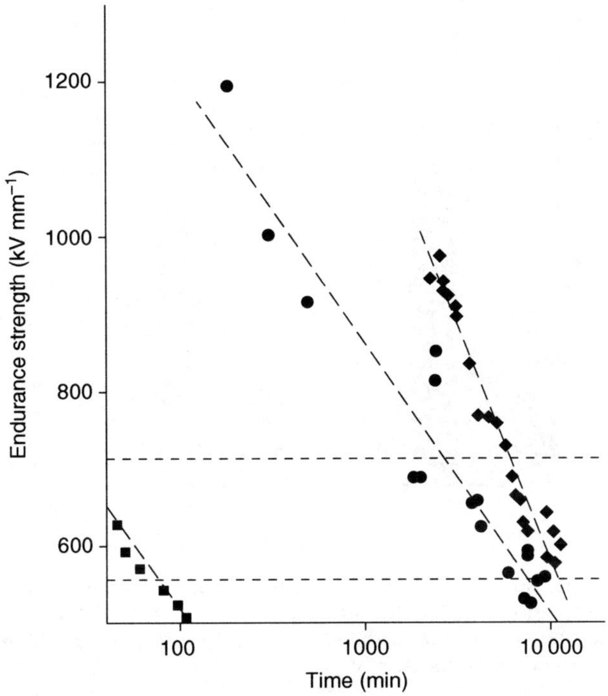

8.5 Voltage endurance results presented as breakdown field as a function of logarithmic time for crosslinked polyethylene unfilled (■), with untreated nanosilica (●) and vinylsilane treated nanosilica (◆) showing two orders of magnitude improvement in time to failure. (*Source*: Adapted from Smith *et al.*[38])

extended, gas-filled tubules that make up the structure. As such, the process of dielectric breakdown by electrical treeing can be divided into two distinct regimes: initiation and subsequent growth. In a study of LDPE containing various nanoparticles of different dimensionalities (spherical, fibrous and layered),[43] both time for tree initiation, as judged from the onset of PD activity, and time to breakdown were examined. Above 15 kV, all the systems studied were found to exhibit some evidence of increased initiation times, together with shorter times to failure, compared with unfilled LDPE. However, these effects were not ascribed to the direct role of the nanoparticles but rather, it was suggested, that these changes in treeing characteristics are caused by changes in polymer morphology resulting from polymer/nanofiller interactions. No detailed morphological study is presented to support this, suggesting that this assertion should be treated with caution. Below 15 kV the tree initiation time and time to breakdown increased in the nanocomposites containing the fibrous and layered nanoparticles, but decreased in the system containing particulate silica.

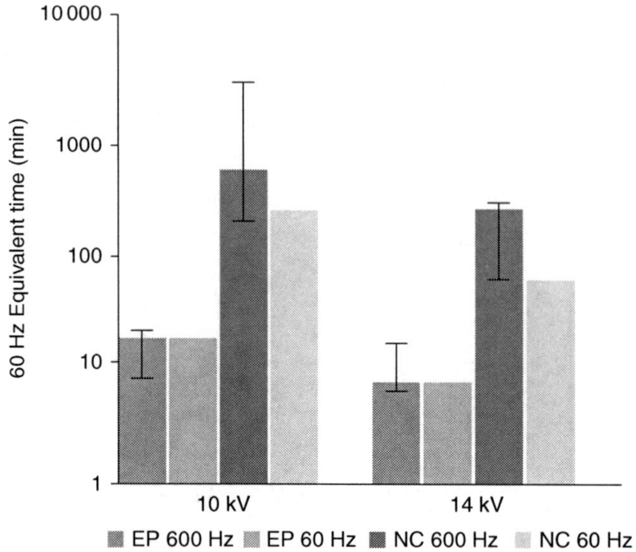

8.6 Comparison of electrical tree initiation in an epoxy-based nanocomposites (NC) containing 5% of the layered silicate bentonite with the behaviour of an unfilled epoxy (EP) reference at 600 and 60 Hz and 10 and 14 kV, respectively.

As shown in Fig. 8.6 Raetzke *et al.*[44] compared electrical tree initiation in epoxy-based nanocomposites containing 5% of the layered silicate bentonite with the behaviour of an unfilled epoxy reference, reporting that the introduction of the nanofiller increased the tree initiation by about one order of magnitude. Although this paper provides some evidence to suggest that the inclusion of nanoparticles can serve to reduce tree growth rate, scatter in the data makes any conclusive analysis impossible. Where increased time to breakdown is reported, this is commonly explained in terms of the nanofiller particles acting as barriers to tree growth.[45] Since much of the epoxy resin that is used technologically contains a large proportion of micrometric filler, the treeing behaviour of nano-reinforced microcomposites has also been studied. In such a system, the inclusion of 5–10% of a nanoalumina filler has been reported to increase tree initiation time dramatically and, thereafter, to have a marked effect on the structure of the trees that form.[46] The effect of adding a layered silicate to a conventional epoxy/silica microcomposite has also been reported;[47] while the addition of micrometric silica reduced the tree initiation time, the inclusion of 9.8 parts of nanofiller per hundred parts resin resulted in a system that exhibited significantly improved performance, compared with the unfilled resin. This study also considered the response of the various materials to PD erosion, and demonstrated a correlation between tree initiation time and PD resistance, leading the authors to conclude that tree formation processes involve degradation of the dielectric by PDs.

8.2.6 Surface erosion and partial discharge resistance

The aspects of nanodielectrics considered thus far have concerned bulk properties. However, the response of nanocomposites to surface discharges and corona has also been considered and significant improvements have been reported for many systems. Exposure of the surfaces of a polymer-based material to energetic events will tend to degrade the organic components of the system and, therefore, it is not surprising that adding an inorganic filler will improve performance. What is worth noting is the fact that the inclusion of very small quantities of nanofiller can have a dramatic effect.

For example, Maity *et al.*[48] studied the effect on surface erosion of adding either nanoalumina or nanotitania to an epoxy resin matrix, showing that even though the 0.5 volume per cent of filler was present, it served significantly to enhance performance. These authors suggest that this effect is related to a number of factors: the nanoparticles themselves are resistant to electrical erosion and, as the surrounding epoxy is removed, their surface concentration increases; epoxy molecules tightly bound to the nanoparticles are also resistant to erosion; the epoxy matrix between neighbouring particles is eroded to give a rough surface topography, and this also inhibits further matrix degradation.

As in previous sections, the formation of a stable surface layer that is resistant to further erosion appears to be associated with the interface between the polymer and the nanofiller. Silane coupling agents and pre-treatment of the nanofiller to remove bound water are reported to improve performance, suggesting that good interfacial adhesion and/or dispersion are important factors.[49–51] Similar improvements in performance have been reported in many other studies of both nanocomposites and materials containing mixtures of nanoscopic and micrometric fillers.[52–55] An example is shown in Fig. 8.7 which illustrates the effect of nanoparticle surface treatment and the influence of micro-nano mixtures on erosion resistance to continuous surface discharge on epoxy resin materials.[56]

8.2.7 Space charge

The increasing interest in the transmission of electrical power over long distances and the desire to establish pan-continental power grids, has increased interest in the high voltage DC transmission of power. However, when a dielectric is exposed to a high DC field for a prolonged period, charges may be injected into the material from the electrodes or from ionization, and charge separation may occur within the bulk. The consequence of the resulting

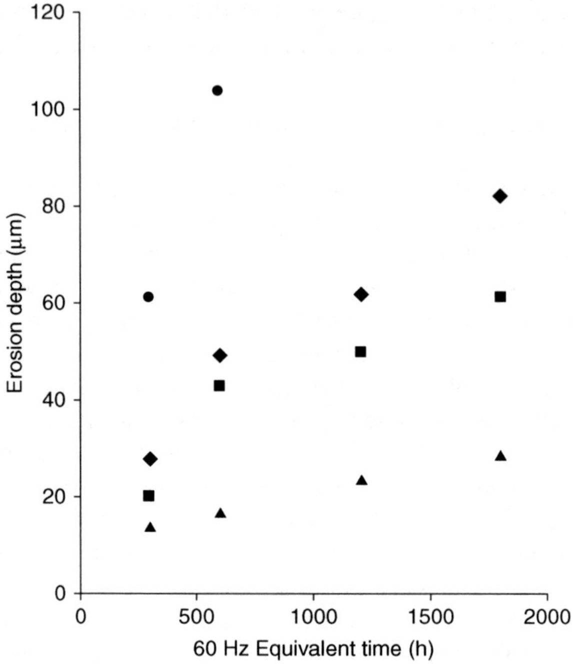

8.7 The effect of nanofilling, organic modification and mixed micro/nano filler addition to anhydride cured epoxy resin, showing progressive improvement in surface discharge induced erosion as a function of exposure time. (*Source*: Adapted from Reference 56.)● – base epoxy resin material acid anhydride cured. ◆ – organically modified nanofiller. ■ – solubilized nanocomposite preparation. ▲ – micro and nano mixed filled material.

space charge is that the local field is intensified at some point within the system leading to accelerated ageing and the initiation of degradation. Consequently, the modification of dielectrics in order to suppress space charge accumulation has attracted considerable attention. Ishimoto *et al.*[57] compared the space charge behaviour of LDPE with LDPE containing micrometric and nanoscopic magnesium oxide fillers. In the unfilled sample, significant quantities of positive space charge were observed throughout the sample, immediately after application of a field of 150 kV mm^{-1}. In the conventional microcomposite, the application of the same field resulted in the formation of negative charge, which was seen to move progressively toward the anode; in contrast, little space charge was observed in the equivalent nanocomposite. Other papers where beneficial effects of adding nanoparticles have been reported are included in the references.[58–61] An example is shown in Fig. 8.8, taken from the work of Chen and Stevens,[58] which shows that the presence of 10% nanotitanium in polyimide reduces the internal space charge in corona charged films.

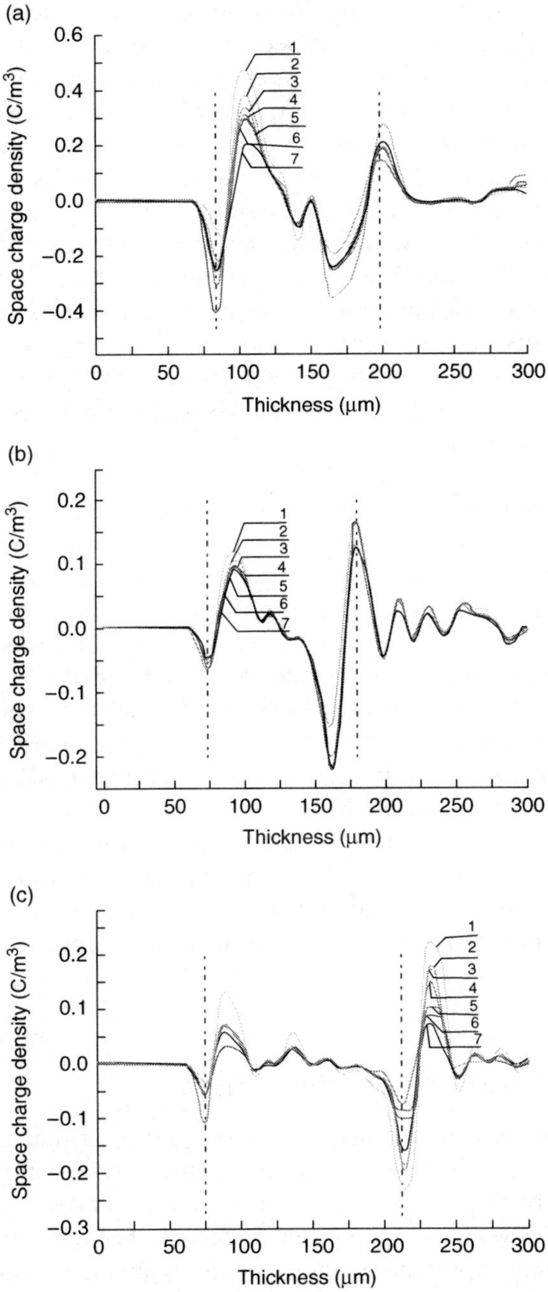

8.8 The effect of adding nanoparticles on the space charge distribution decay in polyethylene following removal of surface corona charging and short circuiting the specimen; (a) pure PI film, (b) PI/TiO2 – 5 wt% nanocomposite film, and (c) PI/TiO2 – 10 wt% nanocomposite film. Note: time after short circuiting (1): 10 s; (2): 30 s; (3): 50 s; (4): 70 s; (5): 90 s; (6): 210 s; and (7): 1790 s, for the same ageing time (10 h).

The effect on space charge accumulation of adding nanoparticles to linear low density polyethylene (LLDPE) has been investigated using nanosilicas with different surface chemistries.[62] In this case, the space charge profile of the unfilled LLDPE exhibited significant homocharge concentrations adjacent to each electrode, whereas in the nanocomposite containing untreated nanosilica, damaging heterocharge dominated; similar results have been reported for a polypropylene nanoclay system.[63] However, both of the samples containing surface-treated nanosilica were found to contain greatly reduced levels of homocharge. While it is suggested that these changes may occur as a direct result of changes in nanofiller surface chemistry directly influencing the density of charge carrier traps or their energy distribution, the authors also point out that indirect effects associated with nanoparticle-induced changes in the morphology of the matrix polymer may also contribute to the effects they report.

Fabiani *et al.*[64] examined the effect of nanofiller aspect ratio and absorbed water of the dielectric response of ethylene/vinyl acetate (EVA)-based nanocomposites. The reported results indicate that increased space charge occurs in samples that have not been appropriately dried in the case of systems containing high aspect ratio layered silicates. In the case of EVA/bohemite nanocomposites, no comparable behaviour was seen. As a result, these workers suggested that the establishment of a percolating nanofiller network can have a marked effect on the dielectric response of a nanodielectric; the higher the aspect ratio of the nanoparticles, the larger the percolation probability. Evidently numerous factors can influence the observed behaviour and, consequently, not all have concluded that adding nanoparticles to a polymer results in a material with improved space charge characteristics; Smaoui *et al.*[65] examined epoxy resin/zinc oxide nanocomposites, concluding that the magnitude of the space charge density in the epoxy nanocomposite samples has increased compared with the unfilled polymer.

8.3 Development of nanodielectrics

A feature that is evident in the account of properties is that there are numerous examples in the literature which demonstrate contradictory trends. Undoubtedly, the reason for this is that differences in material processing methodology have generated materials that are structurally different, even when superficially they appear equivalent. This a key challenge in the future development of nanodielectrics, for their reliable technological use relates to (a) the development of reliable processing strategies, and (b) means by which the resulting materials can be adequately characterized.

8.3.1 Processing of nanodielectrics

Many different strategies have been developed to generate systems composed of nanoparticles dispersed within a polymeric matrix, the most direct of which involves combining the chosen filler material in the form of nanoparticles and mixing it with the required matrix polymer in liquid form in some continuous process, such as extrusion, or via batch mixing. When adopting such a strategy, a key issue concerns the interactions that occur between the polymer and the nanofiller, plus any added solvent. As with any mixing process, if the mixed system is to be thermodynamically stable, then the Gibbs function of the mixed system must be less than that of the two separate components. So, for example, polar water and non-polar oil are immiscible, unless a surfactant is added. This analogy has much in common with many polymer nanocomposites, where the desire is to introduce polar nanoparticles into a non-polar polymer.

Epoxy resins are technologically important materials and, to aid dispersion of materials such as nanoalumina, silane compatibilizers are often used to increase interactions between the two components, and thereby ease dispersion, which is often accomplished through combinations of high shear mixing and sonication, with or without added solvents, which are used to adjust the viscosity of the system.[66–68] An example of these effects is shown in Fig. 8.9, which shows transmission electron microscopy (TEM) images of a well-dispersed surface-treated nanofiller in resin (a sol–gel sourced nanosilica system) with a non-surface-treated nanoalumina in epoxy, which produces dendritic aggregation at a microcluster level.[32]

An area where management of the interactions between the polymer and the nanofiller is particularly important is in connection with nanocomposites based upon polar clay-like systems. Since clays initially exist in the form of aggregates of sheets, processing of the nanocomposite must break up these primary objects and disperse the constituent layers (exfoliate) into the host matrix. In the case of polar materials such as polyamides, the existence of intrinsic clay/matrix interactions assists in dispersion but, for non-polar polymers such polyethylene, it is necessary to modify the filler and/or the host matrix. Clays are generally modified by replacing the sodium cations that reside within the interlayer galleries with appropriate compatibilizer molecules, which resemble surfactants. That is, they are composed of a polar head group, which interact with the clay, plus an organophilic tail which interacts with the polymer. The structure of the molecule must be carefully designed if optimal behaviour is to be achieved.[69–71] Additionally, the introduction of some polar character into the polymer can also be beneficial, such as the grafting of maleic anhydride into the polar backbone[72–74] or through copolymerization with a suitable polar monomer.[75–77] Although such chemical modification eases processing and dispersion of the clay

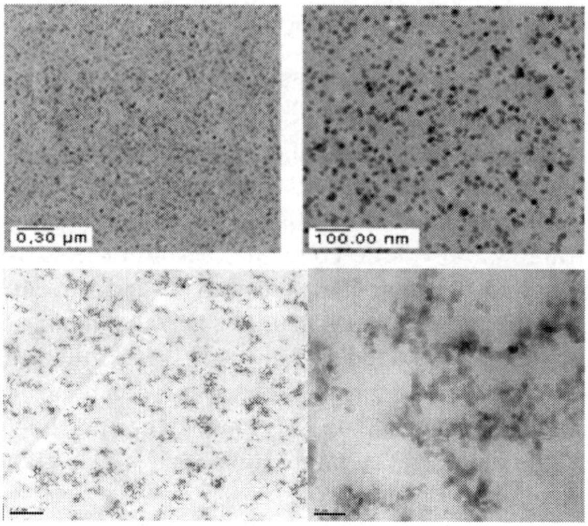

8.9 Comparison of well distributed nanosilica in epoxy resin – TEM images of a cured resin sample with 5% b.v. SiO_2 nanoparticles produced in a sol–gel dispersion route (top micrograph) with low level clustering and dendritic aggregation (bottom micrograph) – TEM – 5% b.w. nanoalumina in an epoxy resin.

platelets, if the resultant polymer/nanofiller interactions are insufficient re-aggregation can occur.[78]

One possible mechanism by which nanofillers become dispersed within a polymer matrix involves the initial diffusion of polymer molecules into the interlayer galleries (intercalation), such that interlayer adhesion is decreased and adjacent layers are more easily separated. A problem with this concerns the sheer size of the polymer molecules and, therefore, an alternative strategy is to diffuse monomers into interlayer galleries and polymerize them *in situ*. Although an appealing approach, it is subject to a number of constraints, related to the compatibility of polymerization catalysts and aluminium silicate systems, which is often low, and the fact that the filler and polymerization chemistries are often incompatible. Nevertheless, such problems can be overcome and numerous systems have been produced successfully,[79–81] although post-synthesis re-aggregation processes can still occur if the resulting system is not appropriately stabilized and if the kinetics allow.

An extension of the idea of *in situ* production of polymer molecules described above is the formation of nanoparticles within a polymeric host using sol–gel chemistry. In this, appropriate metal alkoxide monomers are reacted together to form nanostructured polymer/ceramic composites; the *in situ* production of nanosilica from tetra-alkoxysilane in the presence of polar polymers that readily hydrogen bond with the evolving silica phase has been studied extensively.[82–84] For non-polar polymers, compatibility is again an issue, and common strategies to overcome this involve the use of amphiphilic solvents, the introduction of appropriate chemical functionality into the evolving silica phase through the use of silane coupling agent, or chemical modification of the polymer.[85–87] Alternative nanofillers that have been generated using similar chemical routes include titania,[88] alumina[89] and zirconia/silica.[90] Finally, it is possible simultaneously to polymerize the polymer and use sol–gel chemistry to form the nanofiller, provided the two polymerization reactions are non-interacting. This is particularly important from the technological perspective in connection with the production of nanostructured epoxy resins.[91–93]

8.3.2 Characterization of nanodielectrics

A central theme of the preceding section has been strategies to produce uniform dispersions of nanoparticles with the host matrix. The discussion of breakdown behaviour, for example, has clearly shown that where this is not achieved, material properties are degraded not enhanced by the inclusion of nanoparticles. Many efforts to understand the dielectric response have focused on the role of the interphase, which is also inextricably linked with the issue of particle/matrix interactions and therefore dispersion. If the undoubted potential of nanodielectrics is to be realized technologically, it is of vital importance that we understand the nature of nanodielectric interphase regions and can determine and control the dispersion state of the nanofiller. Further, without the ability to characterize the critical features that make up a nanodielectric a gamut of questions, ranging from the esoteric physics of these materials to practical issues of quality control during manufacture, will remain unanswerable. Many different techniques have been employed in the study of nanocomposites, each of which provides a complementary view of the system; for convenience, these are discussed below under the headings of spectroscopy, microscopy and scattering.

Spectroscopy. Infra-red (IR) and Raman spectroscopy both provide information about the vibrational modes of molecules, which are determined by the constituent functional groups and their local environment. Consequently, these techniques have been used to study many nanocomposite systems and have demonstrated that spectral alterations occur in response to polymer/nanofiller interactions. In polyamide/organoclay systems, reductions in the intensity of the IR organoclay hydroxyl bands have been observed on introducing the clay into the polymer,[94] while additional IR spectral features in the region 960–1140 cm^{-1} associated with MMT/polymer or MMT/compatibilizer interactions have been reported for polyethylene/MMT nanocomposites.[95] It has been suggested that Raman spectroscopy can be used to provide information pertaining to disorder and particle size.[96] In the case of nanocomposites containing CNTs, shifts in the position of the Raman G and D bands of sp^2 carbon have been associated with polymer/CNT interactions.[97] Jeon *et al.*[98] have also suggested that in their high density polyethylene (HDPE)/CNT systems, the Raman breathing mode close to 266 cm^{-1} is related to individual CNTs, while another at 232 cm^{-1} originates from aggregates.

Nuclear magnetic resonance (NMR) spectroscopy probes unpaired nuclear spins and is a sensitive probe of the chemical environment of certain target nuclei. The Fe^{3+} ions that are present in MMT as impurities affect the proton relaxation time in the polymer and, consequently, the better the MMT dispersion, the more the proton relaxation time is affected, providing a measure of nanofiller dispersion.[99] Arantes *et al.*[100] used ^{13}C NMR spectroscopy to study styrene-butadiene rubber (SBR)/titania nanocomposites, reporting significant shifts in peak positions, which are associated with polymer/nanoparticle interactions. Electron spin resonance (ESR) spectroscopy (the electronic analogue of NMR) has been used to probe both interfacial interactions and consequent changes in molecular dynamics. ESR studies of the photochromic behaviour of nanometric polyoxometalates (POM) in polyvinyl alcohol (PVA) have revealed evidence of charge transfer between the PVA and the POM during UV excitation.[101] Miwa *et al.*[102] studied spin-labelled nanocomposites of poly(methyl acrylate) (PMA) containing synthetic fluoromica, and showed that in well-dispersed systems, polymer/nanofiller reduces the local chain mobility, resulting in a rigid interface region that extends ~10 nm into the polymer. Finally, X-ray photoelectron spectroscopy (XPS) studies of PMMA/kaolinite systems have revealed[103] an increase of 0.6 eV in the oxygen 1s energy as a result of hydrogen bonding between the polymer and the nanofiller, while changes in the silicon 2*p* peak were associated with exfoliation.

Microscopy. Since the dimensions of nanoparticles are considerably less than the wavelength of visible light, electron microscopy is most commonly used to probe the structure of nanocomposites in real space. Of these, the most widely used approach is TEM as shown in Fig. 8.9, since the difference in electron density between nanofillers and the surrounding matrix polymer provides a readily accessible contrast mechanism. Many issues associated with nanofiller dispersion have been examined, such as the effect of vinyl acetate comonomer content on nanofiller aggregation in EVA/zinc aluminium layered double hydroxide (LHD) systems,[104] and the distribution of zinc oxide nanoparticles within PMMA.[105] However, while bright field TEM is well suited to imaging the location of the nanoparticles, it cannot provide information pertaining to the effect of the nanoparticles on the morphology of a semicrystalline matrix or partition of the nanofiller between the different phases of an incompatible blend.[106]

Alternative techniques can be used to obtain supplementary information: high resolution TEM (HRTEM) has directly revealed disordered surface layers 1–5 nm in thickness on TiO_2 nanoparticles after grafting a compatibilizer onto the surface;[107] selected area electron diffraction (SAED) has been used to show that, in systems based upon polystyrene and graphite nanosheets,[108] processing affects the crystallinity of the nanofiller; energy filtering the transmitted beam has been shown to enhance image quality in studies of a vinyl ester resin containing POSS nanoparticles.[109]

While scanning electron microscopy (SEM) cannot produce images with as high a resolution as TEM, the relatively straightforward sample preparation and the ability to survey large areas can more than offset this. However, in most cases, imaging in SEM relies upon surface topography and, therefore, some sample preparation is required (a) to reveal a representative internal surface, and (b) to ensure that the included surface features provide meaningful structural information. The simplest sample preparation approach that meets both of these requirements is cryo-fracture; that is, fracture below the glass transition temperature of the material. An example is shown in Fig. 8.10. This has been used with success in studying many different nanocomposite systems.[110–112]

An alternative approach is to cut open a sample using a microtome and then to etch[113] the exposed surface to induce structural detail – an example of this is shown in Fig. 8.11 for polyethylene containing montmorillonite (MMT).[115] Wang *et al.*[114] used cryo-fracture in concert with solvent etching (exposure to boiling toluene) to examine rather complex rubber toughened poly(trimethylene terephthalate) organoclay ternary nanocomposites; in this case, the objective of the approach was specifically to extract the rubber phase to re-examine both the phase structure of the matrix and partition of the organoclay between the two polymers, while Green and Vaughan[115] used a true permanganic etchant to study simultaneously the morphology of an HDPE/LDPE blend and the dispersion of MMT within the system; a typical result is shown in the SEM image of Fig. 8.11.

Scattering. Scattering techniques complement the real space techniques described above in that, while they intrinsically provide statistically representative data, it is not possible to infer a unique real space structure from a diffraction pattern because in recording these intensity data, phase information is lost. Consequently, for systems that possess low degrees of order, idealized equations such as the Bragg equation should be used with extreme caution. For example, X-ray scattering is routinely used in the study of clay-based systems to gauge the aggregation state of the nanoclay. While shift

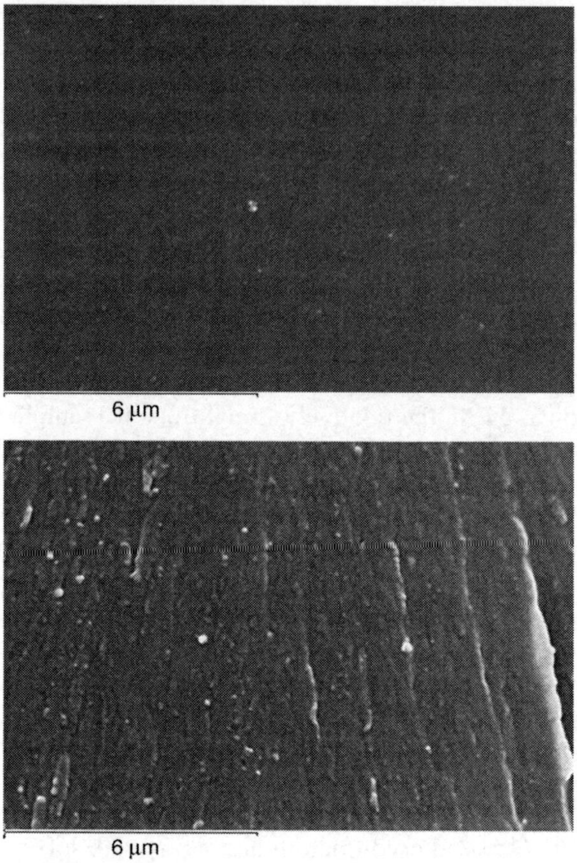

8.10 Comparison of the surface texture of fracture surfaces in an unfilled epoxy (upper) and a system containing nanosilica (lower). In addition to revealing larger nanoparticles directly, the fracture process also generates a different surface texture in the latter case.

in peak position can reasonably be related to the intercalation of monomer, polymer or compatibilizer molecules,[116–118] it is questionable to conclude that 'in the absence of diffraction peaks (indicate) the silicate layer are completely exfoliated'.[119]

Gârea *et al.*[120] used wide angle X-ray scattering (WAXS) and TEM to study the structure of polybenzoxazine/MMT nanocomposites; although the WAXS data do not include any sharp peaks, the TEM clearly reveals extensive MMT aggregates. Small angle X-ray scattering (SAXS) concerns radiation scattered by a sample at lower scattering vectors and, consequently, provides data about larger structural dimensions. Guinier analysis (plotting ln $I(q)q^2$ vs q^2) of SAXS data obtained from poly(butyl methacrylate)/

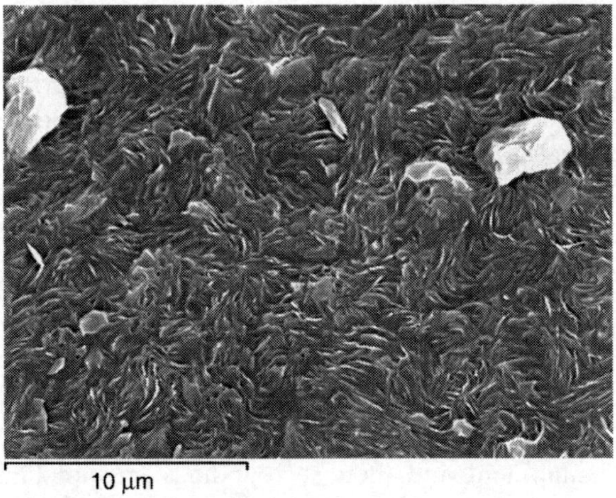

8.11 A permanganically etched polyethylene blend containing MMT. Large MMT tactoids can be clearly seen together with the lamellar texture of the polymer matrix. In this case, crystallization of the polyethylene has been markedly perturbed by the presence of the nanoclay, smaller aggregates of which cannot clearly be distinguished from the surrounding polymer lamellae.

MMT nanocomposite latexes has been interpreted to give the thickness of the constituent MMT platelets and the radius of gyration of the structure.[121] In this case of polyester/exfoliated graphite nanocomposites, plots of log $I(q)$ against log q were interpreted as showing that the graphite layers existed with a fractal structure of dimensionality 2.0–2.3.[122] While X-ray scattering is an electronic process, neutron scattering is a nuclear process and, consequently, involves very different contrast mechanisms. Jouault *et al.*[123] used small angle neutron scattering (SANS) to probe the structure of polystyrene–silica nanocomposites. A log–log plot of corrected scattering intensity against scattering vector gave a q^{-4} dependence at intermediate q values, indicating a sharp well-defined interface between the nanosilica and the matrix, while analysis of the data obtained at smaller scattering vectors suggested that the nanosilica exists as fractal objects of dimensionality 2.5. Similar experiments on polyethylene oxide (PEO)/laponite hydrogels gave a fractal dimensionality of 2.6–2.8.[124]

8.4 Impact of advanced dielectric materials

In this section we discuss the primary drivers for, and likely impact and benefits of, nanodielectric materials in electricity transmission and distribution systems.

8.4.1 Primary transmission and distribution drivers

The key engineering drivers for high voltage equipment performance arising from enhancement of physical properties and performance based on nanodielectric materials fall into the following categories:

- Increased electric strength and voltage endurance enable higher voltage stress, leading either to potentially more compact insulation systems and/or longer life insulation systems. The latter are subject to maintaining quality of materials and insulation system manufacture.
- Increased thermal stability and thermal conductivity materials enable higher power densities to be achieved through higher current and thermal ratings. Provided the heat transfer constraints of individual high voltage (HV) equipment and their surroundings are not limited, the improved thermal performance of insulation materials may have a significant impact on equipment design and operation. This could facilitate enhanced network flexibility and flexible alternating current transmission systems (FACTS) capability, and control over equipment footprint arising from more compact insulation systems for the same power density or increased power density for the same footprint. Greater emergency overload power ratings would also be possible. Higher thermal stability would support higher thermal classification, such as moving from class F to H, with consequential increases in both continuous and peak operating temperatures.
- Increased PD and surface corona resistance may also be possible, and this would support more defect and fault tolerant insulation systems, leading to longer life insulation systems. It would also enable higher surface electrical stresses to be achieved, supporting smaller equipment footprints.
- In some cases the potential ability to control internal space charge and conduction would lead to more enhanced and tolerant high voltage direct current (HVDC) insulation systems and improved stress grading reducing the risk of bulk discharge and surface corona.
- Enhanced static and dynamic mechanical properties may support improved static and impact/impulse mechanical strength leading to higher component stressing or reduced component size particularly in air and gas insulated switchgear, in bushings and in cable joints and terminations.
- Enhanced multi-stress performance – linking materials design to HV equipment design and improved asset operation capability under overload conditions.
- For certain materials types it may also be possible to obtain enhanced fire performance with higher ignition temperatures and lower heat

release rates providing lower fire risk materials and lower potential for collateral fire damage.

8.4.2 Examples of potential impact

The primary benefits for high voltage equipment are in practice derived from enhancements in physical properties and performance. The following examples, taken from a recent consideration of application benefits by the authors in discussion with equipment manufacturers and utilities, are intended to give a flavour of what might be achieved:

- An increase by a factor of 10–100 in voltage endurance and a factor of 3–10 in electric strength could allow an increase in voltage stress by a factor of between 3 and 10 if the threshold field for rapid electrical breakdown of close to 20 kV/mm is avoided.

  In switchgear this could potentially reduce the equipment insulation thickness by a factor of 1.5–3 or footprint by a factor of 30–50% in the absence of constraints to maintain conductor size to contain power losses.
- An increase in composite insulation thermal conductivity from 0.3 to between 0.5 and 1 W/m/K could increase the power density of equipment components by a factor of 1.5–2.0. For the same power rating this could reduce the size of the equipment by up to 25% or more if no serious external heat loss constraints exist.
- An increase of a factor of 3–5 may be achieved in PD and surface corona resistance in a variety of epoxy composite materials used in barriers and bushings. This could extend the lifetime of these components by potentially a factor of 2 or more. For the same lifetime the surface electric stress may be increased by a factor of 2 or so. Such an increase may enable shorter bushing and insulator lengths to be achieved consistent with maintaining flashover constraints.
- In the field of electricity generation, work on turbo-generators has been conducted by various manufacturers targeting the redesign of the mica tape-based groundwall insulation system which is impregnated with epoxy resin to increase the thermal conductivity and reduce the insulation tape thickness while increasing the voltage endurance. Such multifunctional layered insulation systems have some common ground with insulation systems in bushings and barrier systems in transformers.

  In the groundwall case an increase in thermal conductivity from close to 0.3 W/m/K to a higher performance range of 0.5–1 W/m/K, and an increase in thermal classification from class F to class H, and also an increase in field strength from 2.5 to 4 kV/mm or more provides significant opportunity to increase stator current ratings and to reduce the footprint of generator stators, not withstanding rotor constraints.

In all cases the ability to engineer materials using known materials design rules related to actual applications will enable new equipment insulation design options to be explored to optimize the operational profile of the equipment. The latter include multi-functional enhancements with reduced trade-offs centering on enhanced power ratings and/or reduced HV equipment footprint, and higher electrical stresses vs lifetime and insulation thickness.

8.4.3 Potential operational benefits

A power utility could benefit from these enhancements in several ways related to both the specified capacity and the operational flexibility of high voltage equipment. For example:

- Increased continuous, switching and emergency current ratings.
- Higher power density equipment or smaller footprint assets to assist network reinforcement and to aid replacement and retrofitting.
- Longer insulation lifetime and insulation system more tolerant to current and voltage overloads.
- Enhanced flexibility in network operation particularly under emergency overload conditions.
- Greater resistance to power electronic system harmonics particularly in systems containing HVDC technologies.
- Lower capital costs for civil works and lower whole life investment and operational costs – some linked to increased reliability.
- Greater flexibility in network design, planning and consents.
- Higher retained asset value and operational efficiency.

8.5 Challenges and future trends

Over the last decade or so, the concept of a nanometric dielectric has grown from a topic of pure research interest to one that is nearing practical deployment because of the significant opportunities that they undoubtedly could provide. However, three significant questions need immediately to be addressed:

1. Do we understand enough about the mechanisms by which pristine nanodielectrics function to be able to use them reliably in a manufacturing environment?

2. Do we have enough quantitative structure–process–property relationship information to create materials design rules for the design of next generation nanodielectric insulation systems?
3. Do we understand enough about the long-term behaviour of these material systems to be confident that any benefits they bring when pristine will be retained throughout the service life of the device within which they are deployed?

This review has sought to show that nanodielectrics can exhibit enhanced properties, but that even when being examined in laboratory environments, contradictory data has been reported. The reasons for this are two-fold. First, optimal processing routes have not yet been defined for most systems, such that different workers employ different sample preparation procedures to manufacture materials, which they then describe in identical ways. If, for example, we consider say, a 5 wt% of nanosilica is in an epoxy resin, what factors could influence the properties of such a material?:

- The mean size of the nanosilica.
- The size range of nanosilica particles.
- The aspect ratio of the nanosilica.
- The crystallinity and stoichiometry of the nanosilica.
- The dispersion state of the nanosilica.
- The surface chemistry of the nanosilica and its influence on the interphase region.
- The influence of the nanosilica surface chemistry on the epoxy resin curing reaction and final network chemistry and structure.

The above list is not meant to be comprehensive and, equally, many of the factors listed will, probably, have minimal impact. But, which are important and which can be ignored? Without answers to such questions, it is difficult to use nanodielectrics reliably in a manufacturing environment and impossible to decide on the impact of, say, a material supplier subtly changing the nature of the nanoparticles they supply in the years to come. Thus the first major challenges that we would highlight concerns the need to develop appropriate methodologies for the reliable and reproducible processing of nanodielectrics and appropriate characterization tools to enable raw materials and finished compounds to be adequately characterized. We do not believe that either exists at present.

Many of the most fundamental aspects of nanodielectrics are related to the enormous specific surface area of nanometric objects and the effect that this will have on the adjacent polymer. In the short term, this is believed to be strongly related to changes in macroscopic properties; it is easy to visualize how the existence of interfacial charges could lead to enhanced dissipation of space charge – a most desirable characteristic in, for example, a high voltage DC plant. But are these interfaces stable? The discussion of processing makes clear that considerable efforts need to be expended to disperse nanoparticles effectively into polymers because, fundamentally, many nanoparticles are thermodynamically incompatible with many polymers. In which case, how will these interfaces age and what effect will penetrant molecules such as water have over extended periods? In the case of nanoclays, for example, significant quantities of amphiphilic compatibilizers are required to promote exfoliation; this strategy seems entirely counter to the desire to use 'superclean' materials in many high voltage applications.

Assuming that appropriate characterization protocols can be developed to allow the reliable processing of nanodielectrics that are stable over decades, then the potential that this would offer for the active design of nanodielectrics is enormous. For example, thus far, few attempts have been made to produce and study multi-component dielectrics – the addition of a single type of nanofiller to a single polymer is generally complicated enough. But, surely, the next step is to design material systems through the addition of Filler A to improve thermal conductivity, in concert with some of Filler B, say, to give improved resistance to surface discharge erosion. The concept of engineering the surface chemistry of nanoparticles also seems ripe for exploitation; thus far, the surface of nanoparticles has only been modified, rather crudely in most studies, as a processing aid. But why not tag the nanoparticles, prior to their introduction into the host matrix, in order to produce a nanostructured, functional material?

8.6 Conclusion

The term nanometric dielectric, or nanodielectric, was first coined as recently as 1994 and, therefore, the field of research is a relatively young one. It is also significant in that, despite the inherent conservatism of the industries that manufacture and operate high voltage equipment, the concept has grown rapidly to the point where practical deployment is being actively pursued. In this review we have sought to illustrate the current state of understanding of these complex material systems, which offer many possibilities. However, while we accept that the concept of nanodielectrics does have enormous potential, we equally believe that a number of critical questions still remain to be answered if reliable engineering application of these materials is to be realized.

8.7 Sources of further information and advice

Numerous review articles exist that describe specific aspects within the broad general area nanocomposites, such as material subclasses (such as layered silicates[2]) or preparation routes.[125–129] Interest in the topic of nanodielectrics has only really taken off since the early 2000s and, therefore, relatively few reviews exist that address this specific topic in a manner that is comprehensive, critical and up-to-date. The reviews by Tanaka *et al.*,[130] Tanaka[131] and Cao *et al.*[132] provide an interesting historical perspective while the composite article by Frechette *et al.*[133] provides an intriguing melange of perspectives provided by researchers active in nanodielectrics in 2010. A rather more extensive work on nanodielectrics has been published as a result of the activities of CIGRE Working Group D1.24;[134] this article runs to some 115 pages and considers many different aspects of these materials, while the recent review by Li *et al.* focuses exclusively on dielectric breakdown.[135] However, probably the most coherent, contemporary reference work currently available relating to dielectrics is '*Dielectric Polymer Nanocomposites*', edited by J.K.Nelson, that was first published in 2010.[136]

8.8 References

1. R. M. Black, *The History of Electric Wires and Cables*, Peter Perigrinus, London, 1983, 48–67.
2. S. S. Ray and M. Okamoto, Polymer/layered silicate nanocomposites: a review from preparation to processing, *Prog. Polym. Sci.*Vol. **28**, 1539–1641, 2003.
3. J. K. W. Sandler, J. E. Kirk, I. A. Kinloch, M. S. P. Shaffer and A. H. Windle, Ultra-low electrical percolation threshold in carbon-nanotube-epoxy composites, *Polymer* Vol. **44**, 5893–5899, 2003.
4. T. J. Lewis, Nanometric dielectrics, *IEEE Trans. DEI* Vol. **1**, No. 5, 812–825, 1994.
5. T. J. Lewis, Interfaces: nanometric dielectrics, *J. Phys. D: Appl. Phys*. Vol. **38**, 202–212, 2005.
6. T. Tanaka, M. Kozako, N. Fuse and Y. Ohki, Proposal of a multi-core model for polymer nanocomposite dielectrics, *IEEE Trans. DEI* Vol. **12**, No. 4, 669–681, August 2005.
7. S. Raetzke and J. Kindersberger, Role of interphase on the resistance to high-voltage arcing, on tracking and erosion of silicone/SiO2 nanocomposites, *IEEE Trans. DEI* Vol. **17**, No. 2, April 2010.
8. N.G. McCrum, B. Read and G. Williams, *An elastic and dielectric effects in polymeric solids,* Wiley, New York, 1967.
9. F. Kremer, *Broadband Dielectric Spectroscopy*, Berlin: Springer, 2003.
10. T. Tanaka, M. Kozako, N. Fuse and Y. Ohki, Proposal of a multi-core model for polymer nanocomposite dielectrics, *IEEE Trans. DEI* Vol. **12**, No. 4, 669–681, August 2005.
11. K. Lichtenecker and K. Rother, Die Herleitung des logarithmischen Mischungsgesetzes als allegemeinen Prinzipien der staionaren Stromung, *Phys. Zeit.* Vol. **32**, 255–260, 1931.
12. S.-Hui X., Bao-Ku Zhu, J.-B. Li, Xiu-Z. Wei and Z.- K. Xu, Preparation and properties of polyimide/aluminum nitride composites, *Polym. Test.* Vol. **23,** 797–801, 2004.

13. M. Takala, M. Karttunen, P. Salovaara, S. Kortet, K. Kannus1 and T. Kalliohaka, Dielectric properties of nanostructured polypropylene-polyhedral oligomeric silsesquioxane compounds, *IEEE Trans. DEI* Vol. **15**, No. 1, 40–51, February 2008.
14. S. Srisuwan, S. Thongyai and P. Praserthdam, Synthesis and characterization of low-dielectric photosensitive polyimide/silica hybrid materials, *J. Appl. Polym. Sci.* Vol. **117**, 2422–2427, 2010.
15. T. Andritsch, R. Kochetov, P. H. F. Morshuis and J. J. Smit, Short term DC breakdown and complex permittivity of Al2O3- and MgO-epoxy nanocomposites, 2010 *Ann. Rep. CEIDP, IEEE*, Piscataway NJ, 530–533, INSPEC Accession Number: 11851326.
16. N. Tagami, M. Okada, N. Hirai, Y. Ohki, T. Tanaka, T. Imai, M. Harada and M. Ochi, Dielectric properties of epoxy/clay nanocomposites-effects of curing agent and clay dispersion method, *IEEE Trans. DEI* Vol. **15**, No. 1, 24–32, February 2008.
17. S. Singha and M. Joy Thomas, Dielectric properties of epoxy nanocomposites, *IEEE Trans. DEI* Vol. **15**, No. 1, 12–23, February 2008.
18. S. Singha and M. Joy Thomas, Permittivity and tan delta characteristics of epoxy nanocomposites in the frequency range of 1 MHz–1 GHz, *IEEE Trans. DEI* Vol. **15**, No. 1, 2–11, February 2008.
19. M. Roy, J. K. Nelson, R. K. MacCrone, L. S. Schadler, C. W. Reed, R. Keefe and W. Zenger, Polymer nanocomposite dielectrics – the role of the interface, *IEEE Trans. DEI* Vol. **12**, No. 4, 629–643, August 2005.
20. S. Hui, T. K. Chaki and S. Chattopadhyay, Dielectric properties of EVA/LDPE TPE system: effect of nanosilica and controlled irradiation, *Polym. Eng. Sci.* Vol. **50**, 730–738, 2010.
21. J. K. Nelson and Y. Hu, Nanocomposites dielectrics – properties and implications. *J. Phys. D:Appl. Phys*. Vol. **38**, 213–222, 2005.
22. M. Roy, J. K. Nelson, R. K. MacCrone and L. S Schadler, Candidate mechanisms controlling the electrical characteristics of silica/XLPE nanodielectrics. *J. Mater. Sci.* Vol. **42**, No. 11, 3789–3799, 2007.
23. R. J. Sengwa, S. Choudhary and S. Sankhla, Dielectric properties of montmorillonite clay filled poly(vinyl alcohol)/poly(ethylene oxide) blend nanocomposites, *Compos. Sci. Technol.* Vol. **70**, 1621–1627, 2010.
24. V. M. Boucher, D. Cangialosi, A. Alegría, J. Colmenero, J. González-Irun and L. M. Liz-Marzan, Accelerated physical aging in PMMA/silica nanocomposites, *Soft Matter* Vol. **6**, 3306–3317, 2010.
25. A. C. Comer, A. L. Heilman and D. S. Kalika, Dynamic relaxation characteristics of polymer nanocomposites based on poly(ether imide) and poly(methyl methacrylate), *Polymer* Vol. **51,** 5245–5254, 2010.
26. B. Hallouet, P. Desclaux, B. Wetzel, A. K. Schlarb and R. Pelster, Analysing dielectric interphases in composites containing nano- and micro-particles, *J. Phys. D: Appl. Phys*. Vol. **42,** 1–10, 2009.
27. P. Klonos, A. Panagopoulou, L. Bokobza, A. Kyritsis, V. Peoglos and P. Pissis, Comparative studies on effects of silica and titania nanoparticles on crystallization and complex segmental dynamics in poly(dimethylsiloxane), *Polymer* Vol. **51**, 5490–5499, 2010.
28. A. C. Comer, D. S. Kalika, V. A. Kusuma and B. D. Freeman, Glass-transition and gas-transport characteristics of polymer nanocomposites based on crosslinked poly(ethylene oxide), *J. Appl. Polym. Sci.* Vol. **117**, 2395–2405, 2010.

29. D. Prevosto, M. Lucchesi, M. Bertoldo, E. Passaglia, F. Ciardelli and P. Rolla, Interfacial effects on the dynamics of ethylene–propylene copolymer nanocomposite with inorganic clays, *J. Non-Cryst. Solids* Vol. **356**, 568–573, 2010.
30. J. -P. Crine, Electrical, chemical and mechanical processes in water treeing, *IEEE Trans. DEI* Vol. **5**, No. 5, 681–694, 1998.
31. C. Zou, J. C. Fothergill and S. W. Rowe The effect of water absorption on the dielectric properties of epoxy nanocomposites, *IEEE Trans. DEI* Vol. **15**, No. 1, 106–117, 2008.
32. C. Zhang and G. C. Stevens, The dielectric response of polar and non-polar nanodielectrics, *IEEE Trans. DEI* Vol. **15**, No. 2, 606–617, 2008.
33. M. F. Fréchette, E. David, H. D. Martinez and S. Savoie, Post-heat treatment effect on the dielectric response of epoxy samples, *Annual Rept. IEEE-CEIDP* 705–708, 2009, INSPEC Accession Number 11059984.
34. C. D. Green, A. S. Vaughan, G. R. Mitchell and T. Liu, Structure property relationships in polyethylene/montmorillonite nanodielectrics, *IEEE Trans. DEI* Vol. **15**, No. 1, 134–143, 2008.
35. F. Guastavino, A. Dardano, S. Squarcia, P. Tiemblo, J. Guzman, E. Benito and N. Garcia, Breakdown and electrical aging tests on different nanocomposite materials, *Proc. Conf. IEEE-ISEI* 529–533, May 31–June 3, 2009, Montreal, QC.
36. E. A. Stefanescu, X. Tan, Z. Lin, N. Bowler and M. R. Kessler, Multifunctional PMMA-ceramic composites as structural dielectrics, *Polymer* Vol. **51**, 5823–5832, 2010.
37. V. Tomer, G. Polizos, E. Manias and C. A. Randall, Epoxy-based nanocomposites for electrical energy storage. I: effects of montmorillonite and barium titanate nanofillers, *J. Appl. Phys*. Vol. **108**, 074116, 2010.
38. R. C. Smith, C. Liang, M. Landry, J. K. Nelson and L. S. Schadler, The mechanisms leading to the useful electrical properties of polymer nanodielectrics, *IEEE Trans. DEI* Vol **15**, No. 1, 187–196, 2008.
39. A. S. Vaughan, S. G. Swingler and Y. Zhang, Polyethylene nanodielecterics: the influence of nanoclays on structure formation and dielectric breakdown, *Trans. IEE Jap*. Vol. **126**, 1057–1063, 2006.
40. J. -W. Wang, Q. -D. Shen, C. -Z. Yang and Q. -M. Zhang, High dielectric constant composite of P(VDF-TrFE) with grafted copper phthalocyanine oligomer, *Macromolecules* Vol. **37**, 2294–2298, 2004.
41. S. Li, G. Yin, G. Chen, J. Li, S. Bai, L. Zhong, Y. Zhang and Q. Lei, Short-term breakdown and long-term failure in nanodielectrics: a review, *IEEE Trans. DEI* Vol. **17**, No. 5, 1523–1535, 2010.
42. T. Imai, F. Sawa, T. Nakano, T. Ozaki, T. Shimizu, M. Kozako and T. Tanaka, Effects of nano- and micro-filler mixture on electrical insulation properties of epoxy based composites, *IEEE Trans. DEI* Vol. **13**, No. 1, 319–326, February 2006.
43. P. Tiemblo, M. Hoyos, J. M. Gómez-Elvira, J. Guzmán, N. García, A. Dardano and F. Guastavino, The development of electrical treeing in LDPE and its nanocomposites with spherical silica and fibrous and laminar silicates, *J. Phys. D: Appl. Phys*. Vol. **41**, 125208, 8, 2008.
44. S. Raetzke, Y. Ohki, T. Imai, T. Tanaka and J. Kindersberger, Tree initiation characteristics of epoxy resin, and epoxy/clay nanocomoposite, *IEEE Trans. DEI* Vol. **16**, No. 5, 1473–1480, October 2009.
45. G. Michael Danikas and T. Tanaka, Nanocomposites – a review of electrical treeing and breakdown, *IEEE Electr. Insul. M.* Vol. **25**, No. 4, 19–25, July/August 2009.

46. T. Tanaka, A Matsunawa, Y. Ohki, M. Kozako and S. Okabe, Treeing phenomena in epoxy/alumina nanocomposite and interpretation by a multi-core model, *Trans IEE J A* Vol. **126**, No. 11, 1128–1135, 2006.
47. Y. Chen, T. Imai, Y. Ohki and T. Tanaka, Tree initiation phenomena in nanostructured epoxy composites, *IEEE Trans. DEI* Vol. **17**, No. 5, 1509–1515, October 2010.
48. P. Maity, S. Basu and V. Parameswaran, Degradation of polymer dielectrics with nanometric metal-oxide fillers due to surface discharges, *IEEE Trans. DEI* Vol. **15**, No. 1, 52–62, 2008.
49. P. Maity, S. V. Kasisomayajula, V. Parameswaran, S. Basu and N. Gupta, Improvement in surface degradation properties of polymer composites due to pre-processes nanometric alumina fillers, *IEEE Trans. DEI* Vol. **15**, No. 1, 63–72, 2008.
50. S. Rätzke and J. Kindersberger, The role of the interphase on the resistance to high-voltage arcing and to tracking and erosion of silicone/SiO2 nanocomposite, *IEEE Trans DEI* Vol. **17**, No. 2, 607–614, 2010.
51. I. Ramirez, E. A. Cherney, S. Jayaram and M. Gauthier, Nanofilled silicone dielectrics prepared with surfactant for outdoor insulation applications, *Trans. IEEE DEI* No. 15, 228–235, 2008.
52. M. Kozako, N. Fuse, Y. Ohki, T. Okamoto and T. Tanaka, Surface degradation of polyamide nanocomposites caused by partial discharges using IEC (b) electrodes, *IEEE Trans. DEI* Vol. **11**, No. 5, 833–839, 2004.
53. M. Kozako, Y. Ohki, M. Kohtoh, S. Okabe and T. Tanaka, Preparation and various characteristics of epoxy/alumina nanocomposites, *IEE J. Trans. FM* Vol. **126**, No. 11, 1121–1127, 2006.
54. A. H. El-Hag, L. C. Simon, S. H. Jayaram and E. A. Cherney, Erosion resistance of nano-filled silicone rubber, *Trans IEEE DEI* Vol. **13**, No. 1, 122–128, 2006.
55. M. F. Fréchette, R. Y. Larocque, M. Trudeau, R. Veillette, R. Rioux, S. Pélissou, S. Besner, M. Javan, K. Cole, M. -T. Ton That, D. Desgagnés, J. Castellon, S. Agnel, A. Toureille and G. Platbrood, Nanostructured polymer microcomposites:a distinct class of insulating materials, *IEEE Trans. DEI* Vol. **15**, No. 1, 90–105, February 2008.
56. T. Tanaka, Y. Ohki, M. Ochi, M. Harad and T. Imai, Enhanced partial discharge resistance of epoxy/clay nanocomposite prepared by newly developed organic modification and solubilisation methods, *IEEE Trans. DEI* Vol. **15**, No. 1, 81–89, February 2008.
57. K. Ishimoto, E. Kanegae, Y. Ohki, T. Tanaka, Y. Sekiguchi, Y. Murata and C. Reddy, Superiority of dielectric properties of LDPE/MgO nanocomposites over microcomposites, *IEEE Trans. DEI* Vol. **16**, No. 6, 1735–1742, December 2009.
58. J. -W. Zha, Z. -M. Dang, H. -T. Song, Y. Yin and G. Chen, Dielectric properties and effect of electrical aging on space charge accumulation in polyimide/TiO2 nanocomposite films, *J. Appl. Phys*. Vol. **108**, 094113, 2010.
59. T. Takada, Y. Hayase, Y. Tanaka and T. Okamoto, Space charge trapping in electrical potential well caused by permanent and induced dipoles for LDPE/MgO nanocomposite, *IEEE Trans. DEI* Vol. **15**, No. 1, 152–160, February 2008.
60. R. J. Fleming, A. Ammala, P. S. Casey and S. B. Lang, Conductivity and space charge in LDPE containing nano- and micro-sized ZnO particles, *IEEE Trans. DEI* Vol. **15**, No. 1, 118–126, February 2008.

61. Y. Murakami, M. Nemoto, S. Okuzumi, S. Masuda, M. Nagao, N. Hozumi, Y. Sekiguchi and Y. Murata, DC conduction and electrical breakdown of MgO/LDPE nanocomposite, *IEEE Trans. DEI* Vol. **15**, No. 1, 33–39, February 2008.
62. X. Huang, P. Jiang and Y. Yin, Nanoparticle surface modification induced space charge suppression in linear low density polyethylene, *Appl. Phys. Lett.* Vol. **95**, 242905, 2009.
63. N. Fuse, Y. Ohki and T. Tanaka, Comparison of nano-structuration effects in polypropylene among four typical dielectric properties, *IEEE Trans. DEI* Vol. **17**, No. 3, 671–677, June 2010.
64. D. Fabiani, G. C. Montanari and L. Testa, Effect of aspect ratio and water contamination on the electric properties of nanostructured insulating materials, *IEEE Trans. DEI* Vol. **17**, No. 1, 221–230, February 2010.
65. H. Smaoui, L. E. L. Mir, H. Guermazi, S. Agnel and A. Toureille, Study of dielectric relaxations in zinc oxide-epoxy resin nanocomposites, *J. Alloys Compd.* Vol. **477**, 316–321, 2009.
66. M. Iijima, N. Sato, I. W. Lenggoro and H. Kamiya, Surface modification of BaTiO3 particles by silane coupling agents in different solvents and their effect on dielectric properties of BaTiO3/epoxy composites, *Colloids Surf. A: Physicochem. Eng. Aspects* Vol. **352,** 88–93, 2009.
67. M. Kurimoto, H. Okubo, K. Kato, M. Hanai, Y. Hoshina, M. Takei and N. Hayakawa, Permittivity characteristics of epoxy/alumina nanocomposite with high particle dispersibility by combining ultrasonic wave and centrifugal force, *IEEE Trans. DEI* Vol. **17**, No. 4, 1268–1275, 2010.
68. E. Tuncer, I. Sauers, D. R. James, A. R. Ellis, M. P. Paranthaman, T. Aytuğ, S. Sathyamurthy, K. L. More, J. Li and A. Goyal, Electrical properties of epoxy resin based nano-composites, *Nanotechnology* 18 025703 6, 2007.
69. R. Suresh, S. Vasudevan and K. V. Ramanathan, Dynamics of methylene chains in an intercalated surfactant bilayer by solid-state NMR spectroscopy, *Chem. Phys. Lett.* Vol. **371**, 118, 2003.
70. Q. H. Zeng, A. B. Yu, G. Q. Lu and R. K. Standish, Molecular dynamics simulation of the structural and dynamic properties of dioctadecyldimethyl ammoniums in organoclays, *J. Phys. Chem. B* Vol. **108**, 10025, 2004.
71. T. D. Fornes, D. L. Hunter and D. R.Paul Nylon-6, nanocomposites from alkylammonium-modified clay: the role of alkyl tails on exfoliation, *Macromolecules* Vol. **37**, 1793–1798, 2004.
72. H. -S. Lee, P. D. Fasulo, W. R. Rodgers and D. R. Paul, TPO based nanocomposites. Part 2. Thermal expansion behaviour, *Polymer* Vol. **47**, 3528, 2006.
73. L. Százdi, B. Pukánszky, E. Földes and B. Pukánszky, Possible mechanism of interaction among the components in MAPP modified layered silicate PP nanocomposites, *Polymer* Vol. **46**, 8001–8010, 2005.
74. N. Hasegawa and A. Usuki, Silicate layer exfoliation in polyolefin/clay nanocomposites based on maleic anhydride modified polyolefins and organophilic clay, *J. Appl. Polym. Sci.* Vol. **93**, 464–470, 2004.
75. M. Pramanik, S. K. Srivastava, B. K. Samantaray and A. K. Bhowmick, Rubber–clay nanocomposite by solution blending, *J. Appl. Polym. Sci.* Vol. **87**, 2216–2220, 2003.
76. W. Gianelli, G. Camino, N. T. Dintcheva, S. Lo Verso and F. P. La Mantia, EVA-montmorillonite nanocomposites: effect of processing conditions, *Macromol. Mater. Eng.* Vol. **287**, 238–244, 2004.

77. Y. Tang, Y. Hu, J. Wang, R. Zong, Z. Gui, Z. Chen, Y. Zhuang and W. Fan, Influence of organophilic clay and preparation methods on EVA/montmorillonite nanocomposites, *J. Appl. Polym. Sci.* Vol. **91**, 2416–2421, 2004.
78. H. Fisher, Polymer nanocomposites: from fundamental research to specific applications, *Mater. Sci. Eng.* C23, Vol. **23**, No. 6–8, 763–772, 2003.
79. J. Tudor, L. Willington, D. O'Hare and B. Royan, Intercalation of catalytically active metal complexes in phyllosilicates and their application as propene polymerisation catalysts, *Chem. Commun.* Vol. **2031,** 1996.
80. A. Y. A. Shin, L. C. Simon, J. B. P. Soares and G. Scholz, Polyethylene-clay hybrid nanocomposites: in-situ polymerization using bifunctional organic modifiers, *Polymer* Vol. **44**, 5317, 2003.
81. F. Ciardelli, S. Coiai, E. Passaglia, A. Pucci and G. Ruggeri1 Nanocomposites based on polyolefins and functional thermoplastic materials, *Polym. Int.* Vol. **57**, 805–836, 2008.
82. M. Martina and D. W. Hutmacher, Biodegradable polymers applied in tissue engineering research: a review, *Polym. Int.* Vol. **56**, 145, 2007.
83. Z. H. Huang and K.Y. Qiu, The effects of interactions on the properties of acrylic polymers/silica hybrid materials prepared by the in situ sol-gel process, *Polymer* Vol. **38**, 521, 1997.
84. P. Hajji, L. David, J. F. Gerard, J. P. Pascault and G. Vigier, *J. Polym. Sci. Part B: Polym. Phys.* Vol. **37**, 3172, 1999.
85. T. Ogoshi and Y. Chujo, Synthesis of organic–inorganic polymer hybrids utilizing amphiphilic solvent as a compatibilizer, *Bull. Chem. Soc. Jpn.* Vol. **76**, 1865, 2003.
86. C. -C. Chang, K. -H. Wei, Y. -L. Chang and W. -C. Chen, Synthesis of organic–inorganic polymer hybrids utilizing amphiphilic solvent as a compatibilizer, *J. Polym. Res.* Vol. **10**, 1, 2003.
87. Y. Wei, D. Yang, L. Tang and M. G. K. Hutchins, Synthesis, characterization, and properties of new polystyrene-SiO2 hybrid sol-gel materials, *J. Mater. Res.* Vol. **8**, 1143, 1993.
88. A. H. Yuwono, J. Xue, J. Wang, H. I. Elim, W. L. Ji, Ying and T. J. White, Transparent nanohybrids of nanocrystalline TiO2 in PMMA with unique nonlinear optical behavior, *J. Mater. Chem.* Vol. **13**, 1475, 2003.
89. M. Li, S. Zhou, B. You and L. Wu, Preparation and characterization of trialkoxysilane-containing acrylic resin/alumina hybrid materials, *Macromol. Mater. Eng.* Vol. **291**, 984, 2006.
90. L. Delattre, M. Roy and F. Babonneau, Design of homogeneous hybrid materials through a careful control of the synthetic procedure, *J. Sol–Gel Sci. Technol.* Vol. **8**, 567–570, 1997.
91. L. Matejka, J. Plestil and K. Dusek, Structure evolution in epoxy–silica hybrids: sol–gel process, *J. Non-Cryst. Solids* Vol. **226**, 114, 1998.
92. L. Matejka, K. Dusek, J. Plestil, J. Kriz and F. Lednicky, Formation and structure of the epoxy-silica hybrids, *Polymer* Vol. **40**, 171, 1998.
93. S. R. Davis, A. R. Brough and A. Atkinson, Formation of silica/epoxy hybrid network polymers, *J. Non-Cryst. Solids* Vol. **315**, 197, 2003.
94. A. Al-Mulla, Development and characterization of polyamide-10, 6/organoclay nanocomposites, *Int. J. Polym. Anal. Charact.* Vol. **14**, 540–550, 2009.
95. S. Tzavalas and V. G. Gregoriou, Infrared spectroscopy as a tool to monitor the extent of intercalation and exfoliation in polymer clay nanocomposites, *Vib. Spectro.* Vol. **51**, 39–43, 2009.

96. G. Gouadec and P. Colomban, Raman spectroscopy of nanomaterials: how spectra relate to disorder, particle size and mechanical properties, *Prog. Cryst. Growth Charact. Mater.* Vol. **53**, 1–56, 2007.
97. A. W. Musumeci, G. G. Silva, J. -W. Liu, W. N. Martens and E. R. Waclawik, Structure and conductivity of multi-walled carbon nanotube/poly(3-hexylthiophene) composite films, *Polymer* Vol. **48**, 1667–1678, 2007.
98. K. Jeon, L. Lumata, T. Tokumoto, E. Steven, J. Brooks and R.G. Alamo, Low electrical conductivity threshold and crystalline morphology of single-walled carbon nanotubes high density polyethylene nanocomposites characterized by SEM, Raman spectroscopy and AFM, *Polymer* Vol. **48**, 4751–4764, 2007.
99. J. W. Gilman, S. Bourbigot, J. R. Shields, M. Nyden, T. Kashiwagi, R. D. Davis, D. L. Vanderhart, W. Demory, C. A. Wilkie, A. B. Morgan, J. Harris and R. E. Lyon, High throughput methods for polymer nanocomposites research: extrusion, NMR characterization and flammability property screening, *J. Mater. Sci.* Vol. **38**, 4451–4460, 2003.
100. T. M. Arantes, K. V. Leao, M. I. B. Tavares, A. G. Ferreira, E. Longo and E. R. Camargo, NMR study of styrene-butadiene rubber (SBR) and TiO2 nanocomposites, *Polym. Test.* Vol. **28,** 490–494, 2009.
101. L. -M. Ai, W. Feng, J. Chen, Y. Liu and W. M. Cai, Evaluation of microstructure and photochromic behavior of polyvinyl alcohol nanocomposite films containing polyoxometalates, *Mater. Chem. Phys.* Vol. **109,** 131–136, 2008.
102. Y. Miwa, A. R. Drews and S. Schlick, Detection of the direct effect of clay on polymer dynamics: the case of spin-labeled poly(methyl acrylate)/clay nanocomposites studied by ESR, XRD, and DSC, *Macromolecules* Vol. **39**, 3304–3311, 2006.
103. Y. Li, B. Zhang and X. Pan, Preparation and characterization of PMMA–kaolinite intercalation composites, *Compos. Sci. Technol.* Vol. **68**, 1954–1961, 2008.
104. M. Zhang, P. Ding, L. Du and B. Qu, Structural characterization and related properties of EVA/ZnAl-LDH nanocomposites prepared by melt and solution intercalation, *Mater. Chem. Phys.* Vol. **109**, 206–211, 2008.
105. S. Hess, M. M. Demir, V. Yakutkin, S. Baluschev and G. Wegner, Investigation of oxygen permeation through composites of PMMA and surface-modified ZnO nanoparticles, *Macromol. Rapid Commun.* Vol. **30**, 394–401, 2009.
106. D. H. Kim, P. D.Fasulo, W. R. Rodgers and D. R. Paul, Effect of the ratio of maleated polypropylene to organoclay on the structure and properties of TPO-based nanocomposites. Part II: thermal expansion behavior, *Polymer* Vol. **49**, 2492–2506, 2008.
107. V. G. Ngo, C. Bressy, C. Leroux and A. Margaillan, Synthesis of hybrid TiO_2 nanoparticles with well-defined poly(methyl methacrylate) and poly(tert-butyldimethylsilyl methacrylate) via the RAFT process, *Polymer* Vol. **50**, 3095–3102, 2009.
108. G. Chen, C. Wu, W. Weng, D. Wu and W. Yan, Preparation of polystyrene/graphite nanosheet composite, *Polymer* Vol. **44**, 1781–1784, 2003.
109. G. Z. Li, L. Wang, H. Toghiani, T. L. Daultonc and C. U. Pittman Jr, Viscoelastic and mechanical properties of vinyl ester (VE)/multifunctional polyhedral oligomeric silsesquioxane (POSS) nanocomposites and multifunctional POSS-styrene copolymers, *Polymer* Vol. **43,** 4167–4176, 2002.

110. M. L.Cerrada, C. Serrano, M. Sánchez-Chaves, M. Fernández-García, F. Fernández-Martín, A. deAndrés, R. J. Jiménez Riobóo, A. Kubacka, M. Ferrer and M. Fernández-García, Self-sterilized EVOH-TiO2 nanocomposites: interface effects on biocidal properties, *Adv. Funct. Mater.* Vol. **18**, 1949–1960, 2008.
111. H. Zhou, Y. Chen, H. Fan, H. Shi, Z. Luo and B.Shi, Water vapor permeability of the polyurethane/TiO_2, nanohybrid membrane with temperature sensitivity, *J. Appl. Polym. Sci.* Vol. **109**, 3002–3007, 2008.
112. S. Filippi, C. Marazzato, P. Magagnini, A. Famular, P. Arosio and S. V. Meille, Structure and morphology of HDPE-g-MA/organoclay nanocomposites: effects of the preparation procedures, *Eur. Polym. J.* Vol. **44**, 987–1002, 2008.
113. R. H. Olly, Selective etching of polymeric materials, *Sci. Prog.*, Vol. **70**, No. 277 17–43 Part: 1 1986.
114. K. Wang, Y. Chen and Y. Zhang, Effects of organoclay platelets on morphology and mechanical properties in PTT/EPDM-g-MA/organoclay ternary nanocomposites, *Polymer* Vol. **49**, 3301–3309, 2008.
115. C. D. Green, A. S. Vaughan, G. R. Mitchell and T. Liu, Structure property relationships in polyethylene/montmorillonite nanodielectrics, *IEEE Trans. DEI* Vol. **15**, No. 1, 134–143, February 2008.
116. Y. Hu, L. Song, J. Xu, L. Yang, Z. Chen and W. Fan, Synthesis of polyurethane/clay intercalated nanocomposites, *Colloid Polym. Sci.* Vol. **279**, No. 8, 819–822, 2001.
117. S. -K. Yoon, B. -S. Byun, S. Lee and S. -H. Choi, Radiolytic synthesis of poly(styrene-co-divinylbenzene)–clay nanocomposite, *J. Ind. Eng. Chem.* Vol. **14**, No. 4, 417–422, 2008.
118. E. M. Moujahid, F. Leroux, M. Dubois and J. -P. Besse, In situ polymerisation of monomers in layered double hydroxides, *C. R. Chimie* Vol. **6**, No. 2, 259–264, 2003.
119. Y. Cai, Y. Hu, L. Song, L. Liu, Z. Wang, Z. Chen and W. Fan, Synthesis and characterization of thermoplastic polyurethane/ montmorillonite nanocomposites produced by reactive extrusion, *J. Mater. Sci.* Vol. **42**, No. 14, 5785–5790, 2007.
120. S. A. Gârea, H. Iovu, A. Nicolescu and C. Deleanu, A new strategy for polybenzoxazine–montmorillonite nanocomposites synthesis, *Polym. Test.* Vol. **28**, No. 3, 338–347, 2009.
121. Z. Sedláková, J. Plešti, J. Baldrian, M. Šlouf and P. Holub, Polymer-clay nanocomposites prepared via in situ emulsion polymerization, *Polym. Bull.* Vol. **63**, No. 3, 365–384, 2009.
122. H. Kim and C.W. Macosko, Morphology and properties of polyester/exfoliated graphite nanocomposites, *Macromolecules* Vol. **41**, No. 9, 3317–3327, 2008.
123. N. Jouault, P. Vallat, F. Dalmas, S. Said, J. Jestin and F. Boué, Well-dispersed fractal aggregates as filler in polymer-silica nanocomposites: long-range effects in rheology, *Macromolecules* Vol. **42**, No. 6, 2031–2040, 2009.
124. P. Schexnailder, E. Loizou, L. Porcar, P. Butler and G. Schmidt, Heterogeneity in nanocomposite hydrogels from poly(ethylene oxide) cross-linked with silicate nanoparticles, *Phys. Chem. Chem. Phys.* Vol. **11**, No. 15, 2760–2766, 2009.
125. D. R. Paul and L. M. Robeson, Polymer nanotechnology: nanocomposites, *Polymer* Vol. **49**, No. 15, 3187–3204, 2008.
126. B. Z. Jang and A. Zhamu, Processing of nanographene platelets (NGPs) and NGP nanocomposites: a review, *J. Mater. Sci.* Vol. **43**, No. 15, 5092–5101, 2008.
127. P. K. Sudeep and T. Emrick, Polymer-nanoparticle composites: preparative methods and electronically active materials, *Polym. Rev.* Vol. **47**, No. 2, 155–163, 2007.

128. F. Ciardelli, S. Coiai, E. Passaglia, A. Puccil and G. Ruggeri, Nanocomposites based on polyolefins and functional thermoplastic materials, *Polym. Int.* Vol. **57**, No. 6, 805–836, 2008.
129. G. Kickelbick, The search of a homogeneously dispersed material – the art of handling the organic polymer/metal oxide interface, *J. Sol-Gel Sci. Technol.* Vol. **46**, No. 3, 81–290, 2008.
130. T. Tanaka, G. C. Montanari and R. Mülhaupt, Polymer nanocomposites as dielectrics and electrical insulation-perspectives for processing technologies, material characterization and future applications, *IEEE Trans. DEI* Vol. **11**, No. 5, 763–783, 2004.
131. T. Tanaka, Dielectric nanocomposites with insulating properties, *IEEE Trans. DEI* Vol. **12**, No. 5, 914–928, October 2005.
132. Y. Cao, P. C. Irwin and K. Younsi, The future of nanodielectrics in the electrical power industry, *IEEE Trans. DEI* Vol. **11**, No. 5, 797–807, October 2004.
133. M. F. Fréchette, A. Vijh, L. Utracki, M. L. Trudeau, A. Sami, C. Laurent, P. Morshuis, T. Andritsch, R. Kochetov, A. Vaughan, É. David, J. Castellon, D. Fabiani, S. Gubanski, J. Kindersberger, C. Reed, A. Krivda, J. Fothergill, S. Dodd, F. Guastavino and H. Alamdari, Nanodielectrics – a panacea for solving all electrical insulation problems?, *Proc 2010 ICSD, IEEE*, Piscataway NJ, 130–158, July 2010.
134. T. Tanaka, M. Fréchette, A. Krivda, A. S. Vaughan, P. Morshuis, G. C. Montanari, Y. Tanaka, J. Castellon, S. Pélissou, J. Kindersberger, S. Gubanski, S. Sutton, J. -P. Mattmann, C. Reed, T. Shimizu, Polymer nanocomposites – fundamentals and possible applications to power sectors, CIGRE Working Group D1.24; Electra 02/2011, Vol. **254**, 68–73 and CIGRE Technical Brochure 451 – WG D1.24.
135. Shengtao Li, Guilai Yin, G. Chen, Jianying Li, Suna Bai, Lisheng Zhong, Yunxia Zhang and Qingquan Lei, Short-term breakdown and long-term failure in nanodielectrics: a review, *IEEE Trans. DEI* Vol. **17**, No. 5, 1523–1535, October 2010.
136. J. K.Nelsen (ed.), *Dielectric Polymer Nanocomposites*, Springer, New York, 2010.

9 Superconducting fault current limiters and power cables

W. HASSENZAHL, Advanced Energy Analysis, USA

DOI: 10.1533/9780857097378.2.242

Abstract: This chapter addresses the technology of superconducting power cables and superconducting fault current limiters (FCL). The first part of the chapter is a general summary of the electric power grid. This sets the stage for a discussion of the historical development of superconducting AC and DC cables, which begins with low temperature superconductor (LTS)-based systems. It then describes the present status of superconducting cable technology and summarizes some of the major programs around the world. Whereas power cables may be thought of as a replacement for conventional power system components, FCLs and fault current controllers (FCC) have no conventional equivalent in the power grid. Several conceptually different FCL designs are described. The status of their development and installation on the power grid are described.

Key words: electric power, power grid, power cable, AC power cable, DC power cable, fault current limiter, fault current controller, high-temperature superconductor (HTS), low temperature superconductor (LTS).

Note: This chapter is a revised and updated version of Chapter 8 'Superconducting fault current limiters and power cables' by W. Hassenzahl, originally published in *High temperature superconductors (HTS) for energy applications*, ed. Z. Melhem, Woodhead Publishing Limited, 2012, ISBN: 978-0-85709-012-6.

9.1 Introduction

9.1.1 Background

Shortly after superconductivity was discovered in 1911, many in the budding utility industry thought that it might provide a boon to the electric power industry, just as the development of undersea cables had increased world-wide communications in the late nineteenth century. That did not happen. In fact, the first half century after the discovery of superconductivity saw little progress in material characteristics that would allow them to compete with conventional materials. In particular, they did not have the ability to carry sufficient current to allow the design and construction of effective electric power devices. That changed in the early 1960s with the discovery of Nb-Ti and Nb_3Sn, and the subsequent development of conductors based on these materials. The discovery in 1986 of the HTSs caused a boom in interest in a variety of superconductor-based applications. After a period of hype, researchers settled down to a marathon of activities on superconducting materials and a variety of applications.

Today, some HTSs have current densities as great as 100 times that of copper, and have zero resistance in DC operation. These capabilities provide superconducting power applications several potential advantages over conventional devices, including higher efficiency, smaller size, and reduced weight. In addition, superconductors have the unique capability of changing between the superconducting state and the normal state under controlled conditions. This phase change, in which their resistivity can change by several orders of magnitude, provides an additional possibility for applications that are beyond the realm of conventional materials. As a result, today, there is an ever-increasing promise of real power applications based on superconducting materials.

9.1.2 The electric power grid

Although the electric power grid is seemingly familiar to many, few are aware of the level of complexity built into it and the critical need for stability and reliability. It is a very complicated interactive system with a variety of controls, checks, and balances that generally allow it to operate smoothly. A great deal more information is needed to understand why 'we, the users, can expect to flip a switch and see the lights go on' (Hassenzahl, 2005). An in-depth study of the electric power grid is a lifelong challenge. Here, we look at a few grid characteristics before moving on to HTS power applications.

It has been claimed that the North American Power Grid is the largest machine in the world. Power grids, and other large machines, function effectively because of controls, which are, in general, based on maintaining several parameters within small tolerances. One of the most critical parameters in the power system is the nominal frequency of transmitted electric power, which is usually 50 or 60 Hz. Figure 9.1 is a graphic modified from Eto *et al.*, (2010) of how various actions are taken to assure the proper frequency of the grid. Various mechanisms are in place to maintain that frequency over several temporal regimes. In the short term, seconds to minutes – and for small frequency variations – the power output of generators is adjusted. For greater frequency changes, a variety of approaches are used. One novel

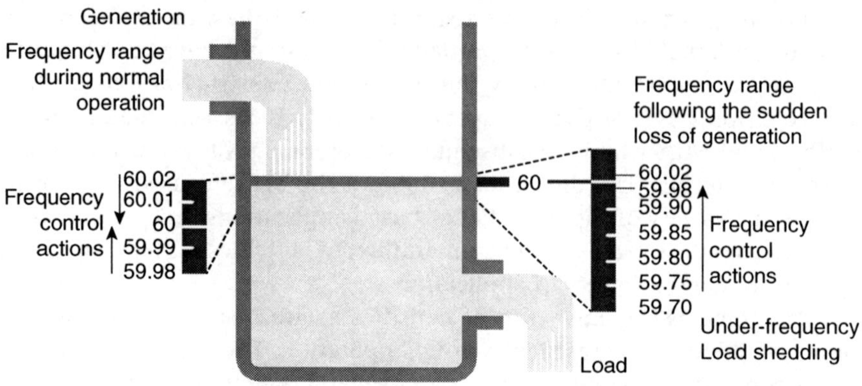

9.1 Simplified concept of frequency regulation and what controls are used to maintain frequency at the nominal value. The level of the 'water' represents the frequency. Normal load and generation variations are such that the nominal of 60 Hz frequency seldom varies by more than ± 0.02 Hz. Generators are controlled for adjustment. Should greater frequency variation occur, usually when generation is lost and the frequency decreases even more, additional controls are needed. These include using the spinning reserve, adding additional generation or using a fast response energy storage system. Ultimately, if the frequency drops below a prescribed level, the utility sheds load.

approach is to use rapid response flywheel storage energy systems. Because the possibility of loss of a generator or a line from a generator to the grid, government power regulating organizations require a certain fraction of the power generation that is online at any given time to be spinning, but not delivering power. This is referred to as spinning reserve and companies that can supply spinning reserve are paid a premium for their maintaining that capability. For extreme frequency adjustment, the utilities shed load via prearranged contracts with major users.

On a longer term, minutes to hours, the frequency is controlled by ramping generation up or down to accommodate the forecast load. This latter control is often referred to as dispatch or, somewhat more correctly, economic dispatch. The term economic is used because the utility prefers to use the most economi – and usually the most efficient – form of generation that is available at any given time. This dispatch mechanism requires some anticipation of the expected load. For example, in anticipation of the morning power demand, power generators that are not needed during evening and nighttime hours, and which require some time to reach full power, are carefully ramped up. Dispatch does not require a rapid response and, for many years was accomplished manually by engineers in a central control facility of each local utility. As computer controls became more effective, economic dispatch was taken over by electronic components, though often the generation and load mix remained the same. While mentioning dispatch, it is important to comment on renewables such as wind and solar. Neither is dispatchable in the traditional sense, and wind in particular can cause considerable upset to the normal operation of the grid where there is high penetration, i.e., where they produce a significant fraction of the power. Finally, on a daily basis, usually at a period of low load near midnight, the power output of a significant fraction of controllable generators is adjusted to achieve a small frequency error so that the total number of cycles during each day is precisely 216 000 for a 60 Hz system and 180 000 for a 50 Hz system.

Here we only address large power systems, but utility grids do not exist in many areas. Remote communities operate power systems without ties to an extended power grid (this is often referred to as an island, and the process is referred to as islanding). In addition, some large ships today require the equivalent of a small power grid for their operation. In some cases, even the propulsion system consists of an electric motor directly coupled to the propellers.

It is one thing to know about some characteristic of the power grid, but that does not provide us with a true understanding of what happens in the conversion of potential, chemical, solar, wind, wave, tidal, or nuclear energy into the power that is delivered to us for use on an almost instantaneous basis. To explore more deeply, we must follow the energy from the original source to the ultimate user.

Electrical energy is not available in nature. It must be produced by conversion from some other energy source, such as the thermal energy from the combustion of a fossil fuel, the potential energy in the water in a dam, etc. The method of electricity production from most energy sources is by converting heat to rotary mechanical motion. However, only a fraction of the initial energy is converted to rotary motion. The efficiency of that conversion is determined by various thermodynamic processes. In particular, most of our electricity comes from fossil and nuclear sources, which produce a high temperature. The efficiency of the production of electricity is determined by the upper temperature and the lower temperature in a turbine. This relationship is referred to as the Carnot efficiency (Moran and Shapiro, 2003). Generally, the efficiency of gas and steam turbines is in the range of 30–50%. Hydroelectric power generation is 85–90% efficient. The United States, Europe, and Japan use almost all the water available to produce electricity. Brazil and Paraguay share the 12.6 GW hydroelectric facility at the Itaipu dam on the Parana River. Brazil uses even more than their share of the electricity generated by purchasing some of Paraguay's share. China has recently developed the Three Gorges facility with a capacity of 18.2 GW. Each of these has several times the power capacity of the largest hydroelectric facilities in the USA, such as Boulder (or Hoover) Dam's 2.08 GW.

Additional sources of energy are wind and solar. They will eventually become important sources of electricity but, in the near term, they continue to have their own sets of barriers to large-scale implementation. They are both relatively inefficient. Commercial solar cells can convert 10–15% of the normally incident light into electricity. The filling factor of most solar configurations reduces that to between 8 and 10%. Note that some experimental cells have up to 40% conversion efficiencies. However, there is no clear path to their commercial production at a reasonable cost. Under optimum conditions, wind generators can convert about 35% of the energy that passes through them into electricity. However, they do not intercept a great fraction of the wind that is passing by, and the wind is intermittent.

In most cases, electricity is created in a rotating generator, and a transformer steps the voltage up to a level appropriate for long-distance transmission. Near the end of its journey to the user, the voltage is transformed to a level that is appropriate for local distribution systems, and is carried to the site of use by a distribution line or an underground cable. Along the energy flow path, the controls and feedback systems mentioned above ensure the near-continuous operation of the power grid. That is true even under many upset or fault conditions. For example, a large generator may fail or need to be taken out of service unexpectedly, or a significant load may drop from the grid. The former would tend to cause the frequency to decrease and the latter would cause it to increase. In addition, lightning may strike a power line or a transformer, trees may fall on power lines, overheated lines may sag during overload conditions and arc to ground, vandals may shoot the insulators on transmission lines, etc. These situations cause not only frequency variations, but significant disruption of power flow and service to small, and occasionally to large, areas.

Most generators are over 99% efficient in the conversion of rotational energy into electricity. The high efficiency is, in part, a result of their compact size, which is possible because they use iron and can operate with a small separation between the stationary and rotating parts. One result of this design is that generators work best at modest voltages – in the 10–50 kV range. Transformers have losses that are typically less than 0.5% of the energy that passes through them. Because of these two facts, for most of the twentieth century, AC was essentially the only widespread form of electricity. Today it is straightforward to convert AC electricity to DC electricity and vice versa. Thus the utility and the user have a choice of form of electricity used in any situation and in any part of the grid. High-voltage DC power lines are used in many situations to transmit large quantities of power over long distances.

Nearly 10% of the electrical energy is dissipated in the path from generator to user. Though this may seem to be a small fraction of the total, it is an enormous quantity of energy. Future technical improvements, including superconductivity, can hope to save only a portion of this 10%.

9.1.3 Superconductors in the power grid

In principle, superconductors could replace conventional conductors in the existing components. However, conventional power technology has become highly sophisticated over the past century, so that existing power system elements are effective, efficient, economical, and, most importantly, simple and easy to use. These attributes, combined with large-scale production, deliver economical functionality and present a significant challenge to the introduction of any new technology (Haldar and Abetti, 2011).

After type II superconductors became a practical reality in the 1960s, some early programs explored the power applications such as generators, cables, transformers, and motors. In the same period, new concepts with no conventional equivalent, such as flux pumps for converting power and charging high-current magnets, superconducting magnets for energy storage, and FCLs were suggested. Designing superconducting devices with the same capabilities as existing conventional systems is a tremendous challenge. In addition, only in the last decade have HTS materials become available that can operate at the temperature of liquid nitrogen. The requirement is that they support high-current densities and have the ability to operate in the magnetic fields found in power applications such as generators, motors, and Superconducting Magnetic Energy Storage (SMES).

Several other criteria must be met by superconductors to make them an effective part of most power applications. In particular, several mechanisms in the superconductors themselves and in the combined structure of a superconductor and normal metal convert electrical energy to heat. These include simple resistance, hysteresis losses, coupling currents, and eddy currents (Wilson, 1983). Whereas losses in conventional conductors are removed at near ambient temperatures by conduction or convection, losses in superconductors occur at cryogenic temperatures and must be removed by a refrigerator. Just as the Carnot efficiency determines the effectiveness of converting heat into rotational energy, it also determines the thermodynamic efficiency of removing heat from a cryogenic environment (Hassenzahl, 2006). In addition, refrigeration systems only function at a fraction of ideal. As a result, depending on the temperature at which the heat is generated and the total amount of heat, the room temperature electrical energy required to remove the heat may be 10–50 times as large as the thermal energy itself. The implication is that if efficiency is an important factor, the superconducting system must be 15–50 times as efficient as the conventional technology that it is replacing (Gouge *et al.*, 2002).

Nevertheless, a great deal of development and demonstration efforts are in the works today on a broad range of superconducting power technologies. Most of the programs have considerable government support, though there are some exceptions, such as a recent FCL installed by Nexans in Germany. The Nexans technology will be discussed in Section 9.2. As electrification increases, industrial, urban, and commercial areas require more and more power. The impact on the power grid is mainly seen as the need for greater and greater power flow along a particular corridor, greater power transfer in individual substations, etc. These factors are independent of any increase in overall power use. Superconducting power technologies such as transformers, cables, and FCLs will likely find use in niche situations where conventional solutions are inadequate. The fact that their enhanced performance can in some cases offset initial development costs will accelerate their use.

We can separate superconducting power applications into two groups. One includes those in which conventional materials are replaced by superconductors. The other includes novel technologies that are based on unique characteristics of superconductors. Cables, motors, generators, transformers, and flywheels form the first group and superconducting magnetic energy storage (SMES), FCLs, and FCCs are in the second. Additional applications of superconductivity are a superconducting substation (Hassenzahl, 2000; Zhang *et al.*, 2011) and integrated power transmission systems in which liquid hydrogen is transferred as a fuel and cools a superconducting cable (Bartlit *et al.*, 1972).

The IEEE has been instrumental in publications that address superconducting applications in various fields, including power. Two issues of the IEEE proceeding focused on superconductivity and had articles that show some of the historical developments and gives snapshots in August 1989 and in the October 2004 (Kirtley and Edeskuty, 1989; Hassenzahl *et al.*, 2004; Kalsi *et al.*, 2004). In addition, several articles on power applications of superconductors were published in the May, June, and August 2000 issues of the power engineering review (PER), at the time the official journal of the IEEE Power Engineering Society, and the July 1997 issue of the IEEE Spectrum was dedicated to developments of HTS applications for electric power systems.

9.1.4 What may the future hold?

Superconductivity power applications have great promise, but it seems that the challenges today are exceptional and will slow development for at least five reasons. The first is the cost of superconductor, which is several times that of copper. The second is the previously described issue of removing heat from a cryogenic environment. The third is the existing high level of development of conventional power components. The fourth is the need for large-scale production to achieve competitive costs. Both large quantities of superconductor and multiple instances of individual devices will be needed as step along the way to commercialization. Finally, the economic environment today is such that governments around the world are reducing their support for all but the most applied research.

Having heard the bad news, there is some hope because developments of superconducting materials continue to advance the technology, and there will be specific niche cases where the capabilities of the superconducting applications will overcome both the development costs for few-of-a-kind devices and the higher cost of materials. Thus, we can expect to see a few, perhaps even dozens, of superconducting FCLs, power cables, and transformers installed on utilities over the next decade. In addition, it is likely that some small generators will be installed in the nacelles of offshore wind power generators. The other power technologies will attend future breakthroughs.

9.2 Fault current limiters

9.2.1 An introduction to fault current limiters

Electric transmission and distribution systems are designed to operate in a variety of conditions. Normal operation has a fixed voltage at every location in the system and a nominal current that is usually in the range of a few hundred to a few thousand amperes. A variety of upset conditions can cause the current to increase to higher levels. For example, loss of a transformer in a substation typically requires one or more transformers in the same substation or nearby substations to operate at higher power levels. Transformers are designed conservatively and have overload ratings for operation at higher power levels for a certain period of time with little effect on operation or their functional life. During some emergency conditions, however, transformers are loaded to even higher levels, which causes the temperature to increase. Insulation life is reduced if excessive temperatures occur for long periods. Similarly, other components of the system are designed to accommodate the various off-normal conditions.

One upset condition that can have serious consequences is a fault, or short circuit. Faults can occur almost anywhere in the transmission and distribution system. Because they can be 10–50 times the rated current, they can cause major damage to various components. The damage is typically a result of: (a) increased forces, which are proportional to the square of the current, $F \propto I^2$, or (b) extreme heating when the fault continues for a long enough period. The temperature reached by any conductor in the system during a fault is given by:

$$T(t_f) = \int_0^{t_f} \frac{J^2 \rho}{C} dt, \qquad [9.1]$$

where T is the temperature, t is time, J is the current density in the conductor, ρ is its resistivity, and C its specific heat. The temperature can rise above the melting point of aluminum or even copper in a less than a second during some fault conditions. Note that this same equation is used in the design of fuses and in determining the maximum quench temperature of a superconductor.

Faults may occur between phases of a three-phase system, may be between one phase and ground, may be associated with a lightning strike, an induced short, etc. Fault duration, if unchecked, varies from a few cycles to much longer periods. No matter the source of the fault or the resulting current, each component of the grid is designed to withstand the mechanical and thermal loads for a few cycles. Early in the development of power systems, engineers came up with four simple lines of defense against faults. The first is to have a fusing element in the circuit that simply melts or vaporizes before other system components are damaged. The second is a circuit breaker, a device that is forcibly opened when the current exceeds a preset limit. The third is an explosive driven current interrupter. Fuses, circuit breakers, and explosive interrupters require a few cycles to operate. The fourth is the use of inductors or reactors that are permanently installed at critical locations in the grid.

As electrification expands, the number of power stations, other grid components, and interconnections on the existing system increase. The result vis-à-vis fault current is that it continues to increase in almost every transmission and distribution system. Technically, the reasons for the increase are two-fold; first, the increased number of interconnections reduces the effective system impedance, and second, the total power available in any area generally increases over time. Since the voltage is fixed, the lower impedance creates larger fault currents. It is not reasonable (nor economical) to replace older network elements just to deal with higher fault currents; thus, some method of limiting them is needed. In principle, an increase of the internal impedance of each element in the grid might accomplish this end. There are, however, two undesirable side effects of this type of solution. First, higher impedances cause increased energy loss, and second, fluctuating loads would require stronger control systems to avoid reduced overall grid stability.

9.2 Three reactors in elevated tanks (one is hidden from view) at the TIDD substation on the AEP power grid. The tanks for these reactors are live, i.e., one terminal of the reactor is tied to the case or tank. The TIDD substation was chosen by AEP, Zenergy, and the US Department of Energy (DOE) for the demonstration of a high-voltage superconducting FCL

Today, the electric utilities most frequently control the effects of fault currents, by either inserting reactors (inductors) in critical locations, or by interrupting the fault current with special switchgear. An example of reactors that are used is shown in Fig. 9.2. These reactors are passive devices that provide a voltage drop that is proportional to the current. At operating current that voltage is usually between 0.5 and 3.0% of the local system voltage. When a fault occurs and the current increases, the voltage across the reactor increases linearly with the current and may exceed half of the local system voltage; that is, during a fault, the impedance of the reactors is a large fraction of the impedance of the entire grid. The increased voltage drop causes the overall fault current to be reduced considerably. Steady state reactors have two disadvantages: first, they consume power on a continuous basis and thus reduce the efficiency of the power system, and second, the increased reactance decreases the responsiveness of the grid.

The second solution to controlling fault current is to install high-voltage circuit breakers that open when the current exceeds some predetermined value. Figure 9.3 shows some of the switchgear in use today. The interested reader may find further information on circuit breakers in Chapter 5 of this book, and in reference Dam (2009). Though the existing switchgear is effective today, the technology is near the theoretical limit of current interrupting capability. One issue is that the circuit must be broken in such a way that arcs, which are created as the switches are opened, can be quenched at the zero current point of the cycle. Often a combination of reactors and circuit breakers is used together. Note that circuit breakers alone have no effect on

9.3 Six circuit breakers at a high-voltage substation in California. They are dead-tank breakers, i.e., the tanks are at ground potential. These circuit breakers are oil-filled and operate independently. They open during a fault and reclose after a fixed number of cycles. They then reopen if the current remains high, i.e., if fault has not cleared.

maximum fault current during the first cycle, thus without an FCL, all grid components must be designed to accommodate the forces produced by the maximum fault current.

A device having a negligible impedance during normal operation and the ability to develop a considerably higher impedance within a very short time (less than a quarter of an AC cycle) after a fault occurs could limit the short-circuit current to a reasonable value. Ideally, such an FCL would operate passively and would return to its normal operational mode immediately after the fault clears. Note that electric power literature prior to the 1970s sometimes referred to reactors and circuit breakers as FCLs. Electric power utilities became interested in this issue as power levels were increasing in the 1970s. The first conference on the topic, which included superconductors as a possible component, was hosted by EPRI in 1976 (EPRI, 1977). The following section in this chapter describes some of the history of superconducting FCLs and the present status of their development.

It is of note that in addition to superconducting FCLs, there has been considerable effort on the use of silicon-based electronics to control current and power flow in various parts of the grid. There is a wide set of applications that use power electronics for this type of control. The general term is flexible alternating current transmission system (FACTS) (Hingorani and Gyugyi, 1999). The technology evolved from the power converters that are used for the connection of high-voltage direct current (HVDC) power cables and power lines to the AC grid. One specialized aspect of the technology is the development of silicon-based devices for the control of fault currents. This technology is often referred to as FCC. One of the superconducting technologies discussed below is an FCC.

9.2.2 Superconducting fault current limiters (SFCLs) and controllers

The first article with a proposal for the use of the phase change characteristics of practical superconductors for controlling fault currents in a power system was by Laquer (1968). Earlier, he provided the first description of superconducting switches and flux pumps for a variety of applications including charging superconducting magnets (Laquer, 1962a,b two patents; Laquer *et al.*, 1966). Since that seminal effort, hundreds of patents and thousands of papers have been written on superconducting FCLs. It is beyond the scope of this summary chapter to discuss a significant fraction of the concepts and the intricacies of the various designs. However, it is instructive to comment on some other early work and to include in the references some review papers that are specific to FCLs, or that include FCLs as a part of a general review of superconducting power applications.

Early developments of superconducting FCLs occurred long before the perovskite superconductors were discovered in 1986. In fact, most of the concepts that are the basis of ongoing developments were conceived prior to that discovery. One of the earliest approaches (Boenig and Paice, 1983; Rogers *et al.*, 1983) was to use a superconducting coil as a reactor that was electronically switched into the circuit when the current exceeded a predetermined level. The electronic controller operated rapidly and the coil could be in the circuit for small fractions of a cycle, as necessary. Multiple switches were installed so that the coil could be in the circuit twice during each cycle. The advantage of using a DC coil is that AC losses are avoided during normal operation and cryogenic refrigeration requirements are small. The first FCC constructed and tested in a utility environment was based on this concept (Leung *et al.*, 1997). Development began in the early 1990s and two devices were fabricated using HTS BSCCO-2223. The first, a 2.2 kV, 2.4 kA_{rms} device was completed in 1995 and was successfully tested at the Southern California Edison (SCE) Center Test Substation in Norwalk, CA. A second, larger device, specified at 26 MVA and designed to operate at 12.5 kV, 1.2 kA_{rms}, was tested at the same facility in 1999. This device was designed so that it would be transportable, as shown in Fig. 9.4. Arcing occurred in the connections between one of the three superconducting coils and the connection to a bushing during full voltage tests. The device was subsequently repaired by the staff at the Los Alamos National Laboratory and one phase was successfully tested at a Los Alamos substation where

9.4 A transportable fault current controller at the substation in Los Alamos. The picture was taken during testing that occurred after repair of the damage caused by electrical arcing during the initial tests at SCE. (*Source*: Los Alamos National Laboratory.)

the active voltage was 13.7 kV (Waynert *et al.*, 2003). Other work on this concept, and several variations thereof, has been carried out by various researchers; One example is Hoshino *et al.* (2001). However, no other large devices have been constructed. This technology's loss of favor is mainly due to the physical size of the device, the amount of superconducting material needed, and the fact that it is not passive, i.e., it requires a feedback and control system to operate.

Today, the functionality of all commercial and pre-commercial superconducting FCLs depends on a phase change in either a superconducting or in a magnetic material. Perhaps the most significant factor in this choice is that relying on a phase change to accomplish current control allows the devices to operate passively. Over the past decade, several general papers on FCLs describe how they function, discuss the dozen or so ways in which they can be used in the grid, and give a snapshot of the status of the technology (Noe and Oswald, 1999; Nagata *et al.*, 2001; Hongesombut *et al.*, 2003; Hassenzahl *et al.*, 2004; Noe and Steurer, 2007; Young and Hassenzahl, 2009; Hassenzahl and Young, 2010). Here we focus on three FCL technologies and present some information on existing and planned installations.

Three conceptually different types of superconducting fault current limiter (SFCL) are under development in various laboratories and industries around the world. These are best described as resistive, shielded core, and saturated (or saturable) core FCLs. The latter two types, which rely on a ferromagnetic core, are sometimes referred to collectively as inductive type FCLs. However, because the way they operate is so different, they are treated separately here. The three types are described briefly in Table 9.1, which is adapted from reference Xiao (2009).

Table 9.1 General characteristics of three SFCL concepts

Technology	Losses	Triggering	Recovery	Size/weight	Distortion
Resistive SFCL	Hysteretic (depends on HTS materials).	Passive	HTS conductor must be re-cooled.	Potential to be small, because HTS performs limiting action.	Only during first cycle.
Shielded core SFCL	Hysteretic (amount depends on HTS materials).	Passive	Faster than resistive, but re-cooling required.	Large and heavy due to iron core and windings.	Only during first cycle.
Saturable core SFCL	Continuous power for DC coil that saturates the iron core, and Joule heating in copper coils.	Passive	Immediate.	Large and heavy due to iron core and conventional windings.	Some, caused by nonlinear characteristic of iron core.

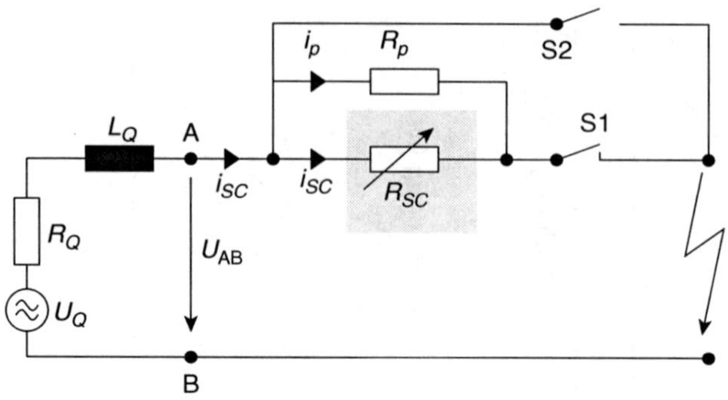

9.5 Schematic of a resistive SFCL showing the superconductor as a variable resistance. There are several variations on this design that have been used by various manufacturers.

9.2.3 Resistive SFCLs

Figure 9.5 shows the principle of the resistive SFCL. The current limiting element consists of a superconductor, which can be wire, tape, or bulk material that provides a path R_{SC} for the current. A parallel resistor R_P may be in the form of a sheet of normal material in contact with the superconductor, or may be a separate element that has no physical contact with the superconductor. During normal operation, the resistance of the superconductor is zero, it carries the entire operating current and the voltage drop is zero. In the event of a fault in which the current increases over some threshold, the superconductor must rapidly transition to the normal state over its entire volume in a fraction of a cycle. Completely quenching the superconductor avoids hot spots and assures sharing, particularly if multiple parallel superconducting elements are used. Once the superconductor quenches, its resistance increases by several orders of magnitude and the normal material begins to carry the majority of the current. Even though the overall current is greater than the normal operating current, the superconductor current $i_{SC}(t)$ decreases sufficiently so that the superconductor is not damaged:

$$i_{SC}(t) = i_{ac}(t)\frac{R_p}{R_p + R_{SC}(t)} \qquad [9.2]$$

The normally closed switch, S1, may be included as part of the FCL or it may already be a part of the grid. Its operation will depend on the overall capabilities of the FCL. If the FCL can recover under load, then S1 will remain closed after the fault. If the FCL requires some time to recover, then switch S2 can be closed and switch S1 opened until the FCL is ready to operate again.

The passive transition of the superconductor and the high normal resistance causes the actual current to be considerably less than the prospective fault current. The value of the limited fault current depends on R_p and the impedance of the load. The ratio of the prospective fault current to the resulting fault current can be as much as ten, and depends on the design of the FCL. The action protects the power grid and decreases the maximum current that is seen by the circuit breakers.

Several groups have developed variations on the theme shown in Fig. 9.3. Termed hybrid resistive FCLs, they may include a separate, fast switch in series with the superconducting element. This switch quickly isolates the superconductor after most of the current has transitioned to the shunt element, thereby allowing the superconducting element to begin the recovery cycle while the limiting action is sustained by the shunt. Other possibilities are to install a reactor in parallel with the FCL. The current automatically transfers to the reactor when the superconductor quenches.

9.2.4 Shielded core SFCLs

The shielded core FCL combines the phase change of a superconductor with the transformer effect of a ferromagnetic core. In a sense, it may be considered a variation of the resistive FCL with the HTS electrically and cryogenically isolated from the high-voltage grid, as shown in Fig. 9.6. An electromagnetic connection is made between the line and the HTS element through mutual coupling of a high-voltage AC coil and a closed circuit superconducting coil. In essence, this FCL is a transformer in which the secondary is a shunted HTS element. During a fault, the increasing current in the superconductor causes the HTS element to quench. In the resistive state the shunted transformer effect is lost, and the iron core and the high-voltage AC coil form a reactor that reduces the fault current. Following a quench,

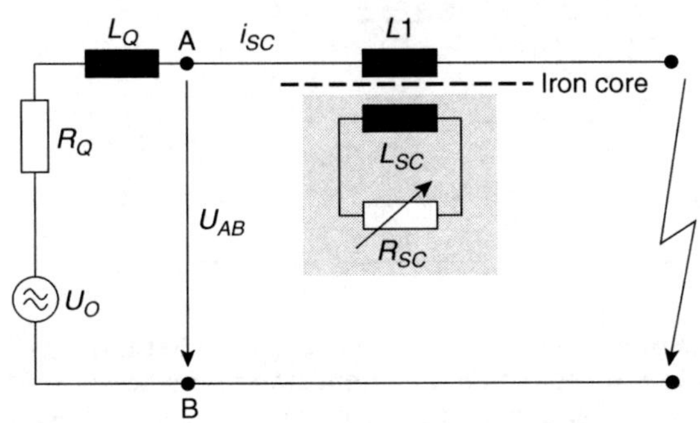

9.6 Schematic of a shielded core SFCL showing that the high-voltage circuit is isolated from the cryogenic environment.

only a small current flows in the superconducting coil. Thus, recovery can be quite rapid once a fault has cleared.

9.2.5 Saturable core SFCL

The saturable core SFCL is conceptually quite different from other types of SFCLs. Rather than using a phase transition in the superconductive material, it relies on a phase change in an iron core. The behavior of the magnetic properties of iron is dynamic, in that it can change from the saturated state to the unsaturated state or vice versa almost instantaneously. The exact time for a transition depends on a variety of factors including material characteristics and the format of the core, which is usually made of many thin sheets of a specialty steel. Because of the extensive development of transformers over the past century, the transition time can be measured in microseconds.

There are two independent iron cores and two AC windings for each phase. Figure 9.7 is an example of the concept for one phase of a three-phase system. The AC windings use conventional conductors that are wrapped around the core to form an inductance in series with the AC line. Both coil and core assemblies are positioned inside a superconducting coil that saturates the iron. As shown in the figure, the two conventional coils are in series and are wrapped with opposite helicities. Because the current alternates in the coils, during one halfcycle the field in one of the coils bucks the superconducting field and the other boosts the superconducting field. During the next half cycle, the condition reverses.

The number of ampere-turns in the superconducting coil must be sufficient to maintain iron saturation so long as the AC current is at or below

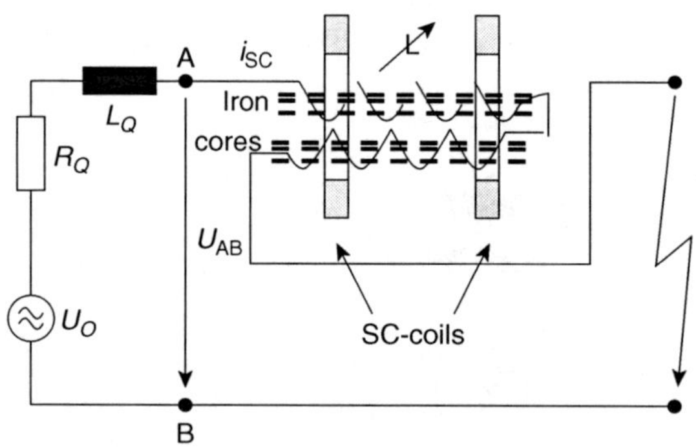

9.7 Schematic of a one phase of a saturated core SFCL showing the two high-voltage coils that are connected to the power grid and superconducting solenoids that saturate the iron cores. There are several variations on this design that have been used by various manufacturers.

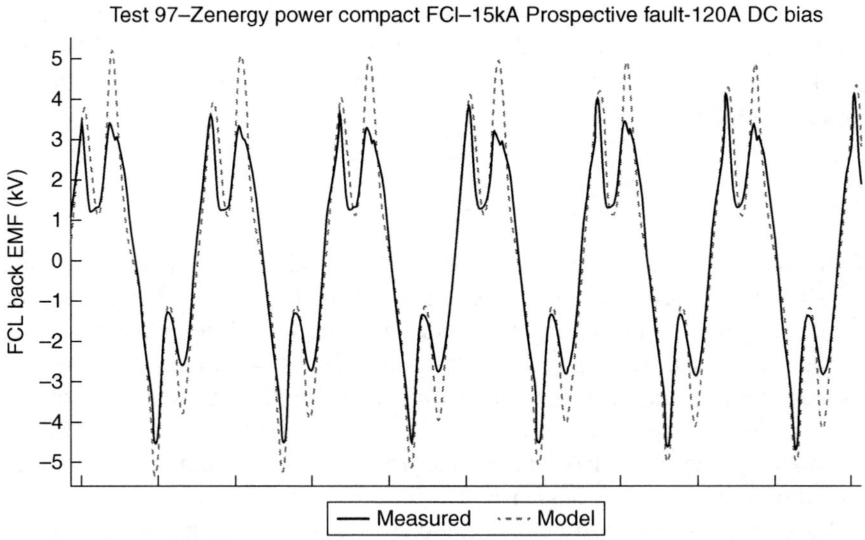

9.8 Results of a test of one phase of a saturated core SFCL during a fault. The abscissa is time, with the large divisions representing 10 ms. Both measured and calculated voltages are shown. The back emf of an FCL is a measure of its current limiting capability.

the nominal operating values. At any instant this is only an issue in the coil whose field opposes the field induced by the superconducting coil. When a high-current fault occurs and the current in the AC coil exceeds a critical level during a part of a cycle, the iron in the coil that is in opposition falls out of saturation. While the current is above that critical level the device appears as a reactor and has a significant voltage drop, or back electromotive force (emf), which causes the fault current to be decreased. Because the value of the inductance is quite variable on a millisecond basis, the waveform of the voltage across the device is not a simple sinusoid. The results of a test of a model saturated core FCL during a fault are shown in Fig. 9.8. The abscissa on the plot is time, with each large division representing 10 ms. The two curves show the measured voltage and the voltage calculated in a model of the device. The differences are not significant in terms of device performance.

The geometry shown in Fig. 9.7 is the simplest form that allows an easy understanding of the concept. Practical considerations in the design of the SC coil and the high-voltage AC coils have led to other designs. Early designs used a single superconducting coil to saturate the iron in all three phases. Several devices of this design have been installed and tested at low voltage as a prototype on utility systems. This geometry is often referred to as the 'Spider' design.

The important advantages of the saturated core SFCL are (1) the superconducting coil is fully isolated from the high voltage of the power grid, and (2) the high-voltage coils and their containment are based on conventional transformer design.

9.2.6 Commercial activities

There have been many pre-commercial SFCL activities. Up to 2011 they were mostly centered in the US. However, at the beginning of 2013 most of the developmental effort on FCLs and all of the installations have been in Europe and Asia. Nexans is the premier developer of FCLs in Europe and has installed several devices. The initial work by Nexans was on resistive SFCLs, and they were the first to fabricate and install a commercial SFCL (Dommerque *et al.*, 2010). The device described in this paper received no government support for design, construction, or installation.

A European consortium was formed to design, build, and test a versatile resistive-type SFCL. The consortium has a total of 15 partners, including five utilities. Industrial participants include: Nexans SuperConductors-Germany; Karlsruhe Institute of Technology-Germany; École PolytechniqueFédérale de Lausanne-Switzerland; Ricerca Sistema Energetio-Italy; and Institut Néel-G2Elab-CRETA-France. The utility companies include: a2a-Italy; endesa-Spain; VorwegGehen and RWE Group-Germany; Východoslovenskáenergetikaa.s.-Slovakia.

The SFCL will be tested at two sites to evaluate its flexibility, reliability, and efficiency. In addition to the design, building, and testing of a system, the project will also investigate the economics of the SFCL when used in diverse applications. Participants hope that this project will provide the first steps toward realization of standardized medium voltage SFCL.

The SFCL design is based on rare earth barium copper oxide (REBCO) tapes that meet the specifications of the two host utilities. The tape selected is produced by Furukawa/SuperPower (Schenectady N.Y. USA).

A bifilar arrangement (to reduce inductance) of five parallel conductors of about 16 m length each forms the basic component. This allows an operating current of 1005 A rms without the need of having parallel conducting paths. Two layers of polyimide insulation are wound between adjacent conductors to avoid local contact and the resulting AC losses. The maximum voltage across each component is 800 V rms. Twelve components are connected in series and mechanically coupled to produce a module ('stack'). Three such modules, one for each AC phase, are installed in a single vacuum enclosure. The three-phase rating of the SFCL is 24 kV and 1 kA.

Over the past decade, Nexans has manufactured and deployed several resistive SFCLs using a different superconductor technology. That design

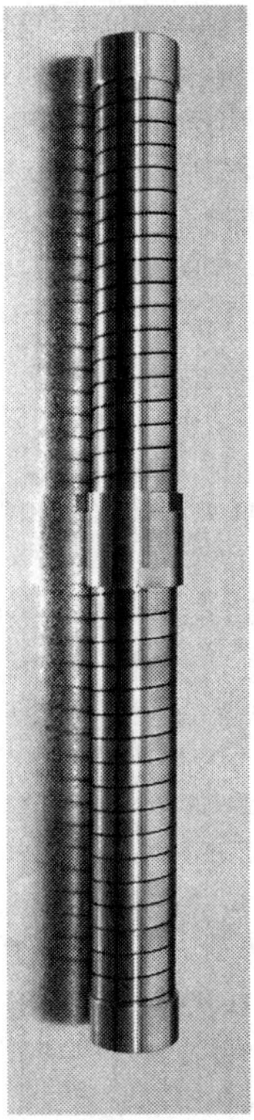

9.9 A Nexans HTS SFCL element produced by a Melt Casting Process using BSCCO-2212. The typical element is somewhat less than 1 m long and has a diameter of about 6 cm.

uses tubes made of bulk BSCCO-2212 (barium, strontium, calcium, copper, oxygen). Each tube is formed into a coil by having grooves cut in a helical pattern on its surface, as shown in Fig. 9.9. These tubes are manufactured using a sophisticated melting, casting, and processing (MCP) technology. The number, size, and arrangement of these elements in a given FCL are determined by the operating voltage and current at the installed site, and by the expected and allowable fault currents.

The bulk superconducting material is soldered on the inside of a normal conducting metal tube and equipped with suitable contacts at each end. The assembly is then machined to form a spiral coil and the resulting element is insulated. The spiral allows a 1 m long element to have an effective superconducting length of several meters. The normal conducting bypass is necessary for protection against hot regions, which is especially important in the case of low current faults where an entire element might not transition to the normal state. A critical part of the element fabrication is to assure complete electrical contact between superconductor and the normal metal over the entire length of the spiral. A critical parameter in the overall SFCL design is the voltage drop per centimeter of length along the superconductor. The choice of this value is a trade-off among several factors including reliability and overall cost, and can vary from one installation to another.

A large number (a few tens) of such elements are connected in series. This structure has a given current and voltage capability. If the current requirements for the FCL are greater than the capability of individual elements, then two or more may be connected in series as needed. A concept of the assembly of one phase of the resistive SFCL is shown in Fig. 9.10. This figure shows the fabrication of a module consisting of many superconducting elements, and its installation into a cryostat.

Recently, the FCL manufacturer, Zenergy Power (Zenergy), a multinational company with subsidiaries in Germany, the USA, and Australia, was purchased by Applied Superconductivity Limited (ASL), a British company that has installed FCLs from several manufacturers, including Zenergy.

ASL will continue the programs initiated by Zenergy to the extent possible, meaning mostly in Europe, their focus being saturated core SFCL technology. ASL installed an SFCL at a CE Electric UK substation near Scunthorpe, England in 2012. They plan to continue the SFCL effort with a higher voltage (33 kV) installation in Sheffield. Effort on a transmission voltage system at the American electric power (AEP) TIDD substation in Ohio, which was shown in Fig. 9.3, has been discontinued.

The ASL/Zenergy SFCL design has moved through several stages, progressing from some initial concepts for modest devices to more robust systems that can operate at transmission level voltages of 130 kV or greater. A model SFCL employing four copper DC coils is shown under test at the Lane Cove facility in Sydney, Australia in Fig. 9.11. The similarity with the concept shown in Fig. 9.7 is clear.

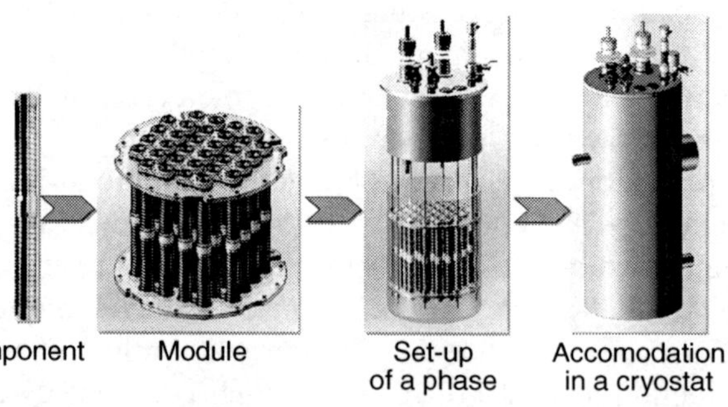

9.10 Assembly of multiple elements in series and parallel as needed to match system voltage and current requirements. Also shown are the set-up and installation of a single phase into a cryostat.

9.11 A test set-up for a model Zenergy SFCL. The coils that saturate the iron are conventional and the heat that is generated is such that each test can only last a few seconds. Full height of the model is about 2 m.

9.12 Tank for the ASL/Zenergy SFCL at CE Electric. Three high-voltage coil sets and iron cores, which are based on transformer technology, are installed in the tank, which is filled with transformer oil.

The oil tank for CE Electric installation is shown in Fig. 9.12, and one of the two superconducting coils is shown in Fig. 9.13. Installation of this device should be completed before publication of this book. Because of the

9.13 One of the two ASL/Zenergy superconducting magnets for the SFCL at CE electric. The coil uses a BSCCO-2223 from American Superconductor. It can produce a magnetic field over 2 T and fully saturates the iron cores in the tank shown in the previous slide. This picture shows power leads for the coil and the flanges where two refrigerators will be installed.

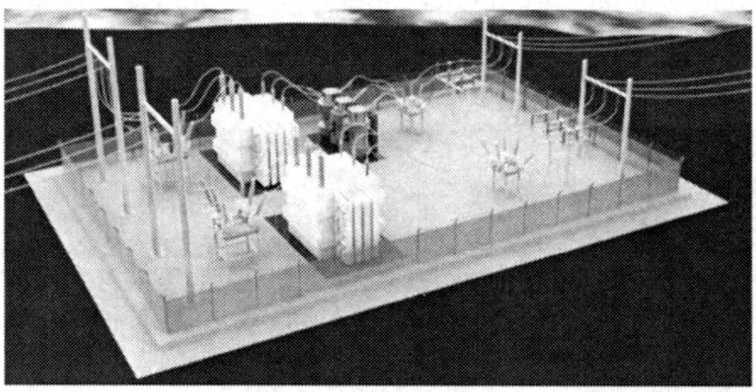

9.14 An artist's concept of an the three single phase SFCLs that were to be installed at AEP's TIDD substation. Though approximately 3 m tall, they are dwarfed by the other components of the substation, only part of which is shown.

high voltage, it will use three single phase SFCLs, each in a separate oil tank. Figure 9.14 is an artist's concept of the installed device. For size comparison, each of the three units will be approximately the same size as the circuit breakers shown in Fig. 9.4.

There are several other SFCL projects worldwide. The list and device details are too extensive to include in this chapter. However, EPRI's recently completed 'Technology Watch' (Eckroad, 2012) includes details of SFCL and Superconducting Power Cable activity worldwide.

9.3 Superconducting power cables

9.3.1 An introduction to conventional cables

Conventional underground cables are used for power transmission and distribution in a variety of situations. Urban underground and undersea power cables are the most highly developed and widely distributed of all cable applications. The most important drivers for these and other cable applications are rights-of-way, environmental concerns, and effectiveness in niche markets. Undersea power cables were first used in locations where islands are near a mainland and where cables coupled with use of large and efficient power plants were more effective and reliable than local, island-based power plants, which needed fuel on a continuous basis. The urban use of power cables has a longer history and is typical in all cities having high population density, particularly in industrialized countries. Cable technology for three-phase power systems is highly developed. In general, the length of single cables is limited by total shipping weight and shipping spool diameter. Figure 9.15 shows a cable in place on a spool and Fig. 9.16 shows the transport of three such spools on a highway between the manufacturing facility and the site of installation. The rationale for the historical development of various conventional cable designs, their installation, and the critical issue of splicing two cables together are described in detail in the EPRI publication referred to as the Green Book (EPRI, 2007).

The key to the effectiveness of cables compared to elevated power lines is that the voltage breakdown capability (V/m) of the insulators used for

9.15 A typical high-voltage power cable made for underground installation. Drum diameter and length are approximately 3 and 4 m, respectively. Shipping weight, in this case about 60 t, and dimensions determine the maximum length of cables of a given diameter. Usually cable lengths are 700–2000 m.

9.16 Three drums similar to that shown in Fig. 9.15 during transit to an installation site. Three, generally identical cables are needed to form a three-phase power system.

cables is 100–1000 times greater than that of air. Thus, the power flow along underground cables in a restricted corridor can be considerably greater than is possible with overhead lines along the same right-of-way. A further advantage is that the stray electric field from a three-phase underground cable system is negligible compared to that of overhead power lines.

Both overhead power lines and underground cables transmitting AC electric power are affected by the inductances among their conductors, and between the conductors and ground. In all cases, the inductance causes the relative phase of the current and the voltage to shift as the distance from the power source increases. The real power, P, that can be used is equal to the product of the current, the voltage, and the cosine of the phase between them; $P = I_{rms}V_{rms}\cos(\alpha)$.

The phase shift α depends on the geometry of the lines or cables, and also depends on the magnitude of the current. Electric utilities install capacitors along AC lines and cables to oppose the effect of the inductance. One disadvantage of AC cables compared to overhead AC lines is their higher inductance, which requires phase correction every 30 or so km, instead of 500 km for overhead power lines.

Superconducting AC cables having the same cable diameter as conventional cables will be able to carry much greater currents, and, depending on the design, will have a reduced inductance, so that the phase variation with length will be reduced. Thus, they will have the capability of carrying more power in a given corridor than conventional cables. Most conventional AC cables are buried and have significant losses. Thus, since cooling in the underground environment is limited, they tend to operate at elevated temperatures when the loads are high. Superconducting cables may be more efficient than conventional cables, even considering the refrigeration load, they may have longer lives, and the life cycle cost may be less. In the not too distant future, superconducting cables will, in some niche applications, replace both overhead lines and conventional cables.

9.3.2 History of superconducting power cables

Before addressing various programs that are in effect today, it is of interest to review some of the more significant developments in the history of superconducting power cables. It is of note that before the early work on superconducting cables, there were several programs on the development of cryoresistive cables. In particular, in the 1960s, teams in the US, France, Germany, and Russia considered the possibility of using the decreased resistivity of high purity aluminum at cryogenic temperatures to produce a more efficient power cable. The rationale for this approach was that the resistivity of both aluminum and copper decrease more rapidly than the efficiency of refrigeration so that there would be a net increase in overall efficiency of cables using conventional conductors and operating at lower temperature. Depending on the magnetic field and the purity of the material, the resistivity of copper at 4 K can be less than 0.5% of its value at room temperature, and the resistivity for aluminum can be as low as 0.1% of its value at room temperature. This was a strong driver for the development of cryogenic power cables. Though there is not space to discuss cryogenics here, developments in the industrial gas industry in the 1940s and 1950s led to the availability of large capacity, highly efficient, and economical cryogenic refrigerators in the 1960s. It is that effort that has provided an enabling technology for the development of superconducting power cables.

Although it was generally thought that superconducting power cables would be very efficient, it was not until the late 1960s that Long and Nottaro (1971) and others carried out the experiments needed to show that the losses (including the refrigeration power requirement) in superconducting power cables could in fact be lower than those of cooled normal conductors. Making these early loss measurements was a challenge, and the researchers in this area had to develop both cryogenic and electronic equipment as well as establishing measurement procedures in order to definitively determine the actual losses. Some of those measurements and many other early observations are described in the Proceedings of the 1968 Summer Study on

Superconducting Devices and Accelerators held at Brookhaven National Laboratory (BNL, 1968). Note that the emphasis of that study was particle accelerators for physics experiments. Several reports of the time, including Hammel and Rogers (1970) and Hammel *et al.* (1970) summarized work in that area. Particle accelerator technology was instrumental in the improvement of the characteristics of superconducting materials more than any other technology until the invention and commercial development of magnetic resonance imaging.

Many attendees at the 1968 study were already considering power applications of various types. Aggressive development of superconducting power cables had begun a few years earlier when Nb-Ti and Nb_3Sn became available on a near commercial scale. It was also in the 1967–1969 timeframe that multi-filamentary Nb-Ti superconductors became available. Prior to that time, however, Klaudy (1967) in Austria had determined many of the criteria needed for the development of superconducting AC cables. Early work on model and prototype superconducting cables began in Europe (Moisson and Leroux, 1971; Klaudy *et al.*, 1981) and later in the US (Laquer *et al.*, 1977; Forsyth and Thomas, 1986) and in Russia (Dolgosheyev *et al.*, 1979). Significant development programs existed for AC and DC superconducting cables. These programs addressed the significant issues of superconducting cable technology, and the research in the 1970s and early 1980s is generally applicable to many of the technical issues associated with HTS power cables today. The most critical issue that led to their disfavor was need for an extensive refrigeration system to maintain the operating temperature. The utilities were concerned with operating and maintenance (O&M) costs and other issues of a superconducting cable with an extensive refrigeration system that would not be completely under utility control (Hassenzahl, 2006).

The USA had the largest power cable efforts, which included Brookhaven National Laboratory's AC cable program (Forsyth and Thomas, 1986) and the Los Alamos Scientific Laboratory's DC cable program (Laquer *et al.*, 1977). At the time, electric power use was increasing by up to 10% per year, and it was generally thought that energy costs would go down as more and more nuclear reactors were built and grouped in farms with 10 or more GW capacity. History has shown us differently, but the need for long-distance high-power transmission was already affecting several corridors. In particular, a 1.4 GW DC line termed the Pacific DC Intertie had been completed in 1970. It moved power from the hydro resources in the Columbia basin (specifically the Celilo dam) to Southern California. Recognizing the effectiveness of DC power transmission and the extended hydro resources in places such as Siberia and the James Bay region of Canada, (Garwin and Matisoo, 1967) proposed superconducting DC cables with the capability of carrying 10 GW to transmit power from these areas to load centers in more populated areas.

The discovery of materials that were superconducting above 70 K in 1986 led to renewed interest in superconducting power cables. Since the refrigeration issue had turned out to be the show-stopper for LTS cables, the fact that HTS-based power cables would require a factor of ten or so less refrigeration power encouraged many cable manufacturers and utilities to embrace the new materials. Though capital and operating costs are less, O&M is still an issue and was the subject of an electric utility workshop (EPRI, 2004). Though the refrigeration requirements for HTS cables are less than for LTS cables, there are several barriers to the wide-scale implementation of HTS cable technology. The two most important advantages are: (1) it is more expensive per unit of current carried and (2) the perovskite materials resemble mica and are brittle rather than robust like Nb-Ti. The latter material is malleable and has mechanical characteristics that in some regards are better than steel.

In many locations AC power cables are pulled into ducts, which are installed in multiples of three. In some cases the utilities plan for the future expansion by having additional holes that can be filled with additional cables when power demand increases. Because superconductors can carry orders of magnitude more current than conventional conductors of the same cross-section, a likely use for HTS AC cables will be their installation as additional or retrofit cables in some circuits where the existing power cables are operating at or near their safe limit.

9.3.3 Types of superconducting power cables

Superconducting power cable designs are, for the most part, modifications of the designs that have been used for over a century for conventional cables. There are, however, some adaptations to take advantage of the characteristics of superconductors. Many of those adaptations were proposed during the development of LTS cables (Laquer *et al.*, 1977; Forsyth and Thomas, 1986).The most important differences take advantage of some of the characteristics of superconductivity.

The simplest, and the design first used for superconducting cables, used a superconductor in a cryostat with a conventional electrical insulation on the outside. The HTS wires are wrapped around a copper core and are in a tube filled with flowing coolant, typically liquid nitrogen. This design is shown in Fig. 9.17 and is referred to as a warm dielectric design. The copper at the center is a protection element that carries the current for short periods when the critical current of the superconductor is exceeded. It uses the least amount of HTS conductor for a given level of power transfer. However, it has a high inductance and an external magnetic field. If the three phases are in close proximity, the magnetic field produced by one phase causes hysteresis losses in the other phases.

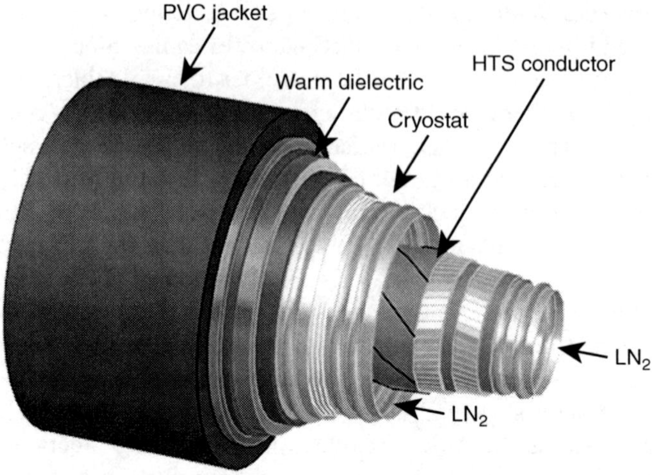

9.17 One configuration of a single phase warm dielectric cable. The coolant flows in one direction on the inside of the flexible core tube and in the other direction outside of the superconductor. This arrangement may have a significant counterflow of heat between the two fluid paths, which limits the length of cables made of this design. A conventional extruded dielectric is used on the outside of the cable. Not shown is a copper layer that is a protective element when the critical capabilities of the superconductor are exceeded.

Because of the high losses associated with the warm dielectric cable, a cable with two layers of superconductor, with the outer layer acting as a magnetic shield for adjacent cables was proposed. The two concentric HTS layers are insulated from each other by a cold dielectric as shown in concept in Fig. 9.18. This cold dielectric design has been used for most superconducting cable installations. Liquid nitrogen coolant may contact both layers, providing both cooling and dielectric insulation between the center conductor layer and the outer shield layer. Compared to the warm dielectric design, it has a lower inductance, a higher current carrying capacity, reduced AC losses, and very low stray magnetic fields. It also uses more superconductor than either of the other designs.

Perhaps the most effective AC cable design is the three-conductor concentric triplex design as shown in Fig. 9.19. Because each of the three layers of insulation must be fairly thin for the cable to be flexible, it seems to be most effective for low to medium voltage levels. The design has many advantages because it has no stray magnetic fields, a very low inductance, and it uses less superconducting material than the other designs. However, it does require special wound insulation that is permeated with the cryogenic coolant.

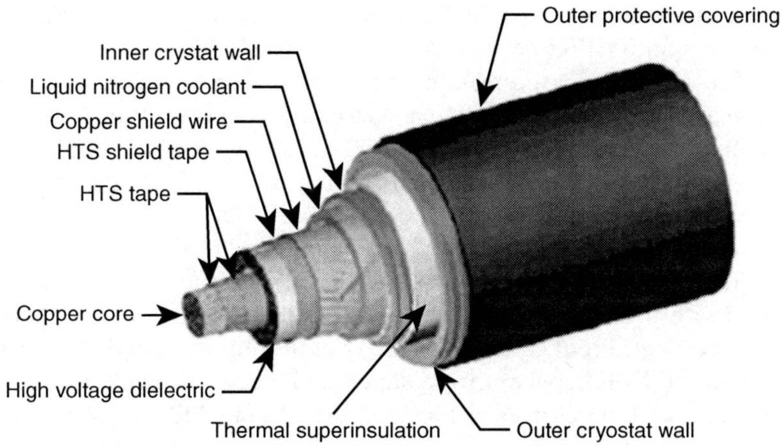

9.18 One phase of a cold dielectric cable system. Because the dielectric is cold it must be a flexible structure, such as a wound tape that is permeated with the liquid nitrogen.

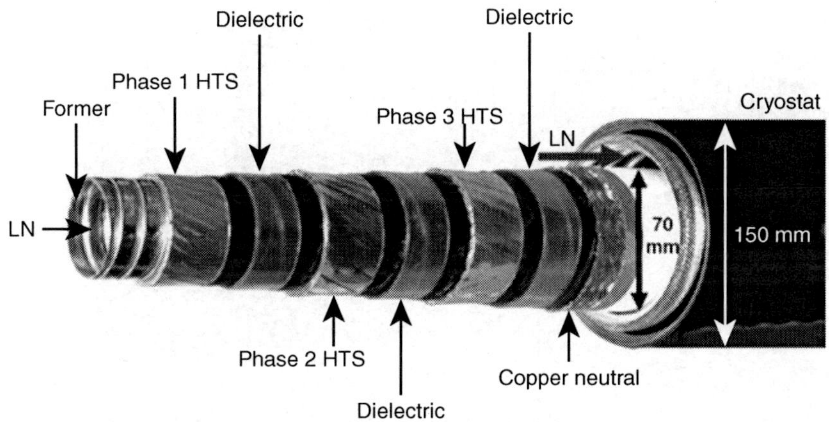

9.19 A Triaxial three-phase cable showing the three layers of HTS tape and the three layers of electrical insulation. This particular cable is a section of the cable installed by Southwire at AEP's Bixby substation.

There has been little interest in HTS DC cables because the power levels and lengths necessary to make a DC power system affordable are well above the capabilities of the manufacturing capabilities of HTS materials at the present time. To explain, the cost of the converter stations at the terminals of a DC line may equal the cost of a hundred or more kilometers of the line itself. Similarly, the cooling requirements of the terminations of a high-power, high-current DC cable, are greater than the cooling requirements of several kilometers of cable. Thus the topic was discussed, but not addressed in detail until EPRI began a detailed study in 2006.

The developing interest in renewables such as wind and solar have increased the possibility of the need for massive power transfers, e.g., greater than 10 GW over distances of 1000 km or more. This requirement must be compared to the AC cables that carry at most 250 MW and are generally less than 1 km in length. Most of those HTS AC cables use many short cryostats that are permanently evacuated. An installation of 1000s of kilometers must use a different approach because of the reliability issues associated with 10 000 cryostats that must work in concert. Avoiding this difficulty requires a different concept for the overall cable system design. Figure 9.20 shows the EPRI concept for a superconducting DC cable that is designed to transmit 10 GW over distances greater than 1000 km (Hassenzahl *et al.*,

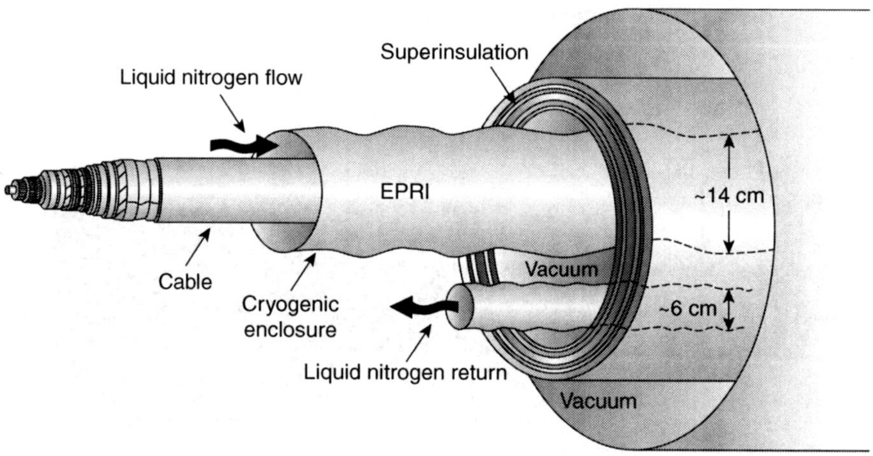

9.20 A superconducting DC cable designed to carry 10 GW over distances greater than 1000 km. The cable is vacuum insulated and uses multilayer insulation to keep the total heat input to the coolant to less than 1 W per meter of length. Two flow paths for the cryogen are used to allow long-distance operation without a counterflow heat exchange effect. The outer wall can be made of a conventional steel pipe of the type that is being used for high-pressure gas pipelines. The cable shown inside the cryogenic enclosure contains two layers of superconductor and two protective layers of copper. Insulation between the two layers of superconductor is rated at the peak value of the system with safety margins. A thinner layer of insulation is placed between the outer layer of superconductor, which is nominally at ground potential and an outer corrugated steel shield. Reproduced with kind permission from EPRI.

2009). Refrigeration stations are required every 10–20 km. Liquid nitrogen below 70 K and at a pressure of 20–30 atmospheres seems to be the ideal coolant. Because of the long length, elevation changes may be several hundred meters along the path between refrigerators, which can lead to pressure changes of more than 20 atmospheres. The high pressure was chosen to accommodate these effects.

Not all power systems that are in close proximity operate at the same frequency, and even if they are at the same frequency they need not be fully synchronous. Japan uses both 50 and 60 Hz power in different parts of the country. The Itaipu dam produces power at both 50 and 60 Hz to satisfy users in Brazil and Paraguay. The United States Power grid is separated into three distinct control areas, western, eastern, and Texas. These areas all operate at 60 Hz, but are not synchronized. As a result, where it is necessary to transfer power from one area to another, it must be converted to DC and then back again into AC. The three control areas abut in the south-eastern part of New Mexico. The Tres Amigas facility has been proposed to interconnect the three areas, as shown in Fig. 9.21. Because the land requirement for the 5 GW power converters is so large, the proposal is to separate them by about 10 km. A DC line or cable is needed for the interconnection and the consortium developing Tres Amigas is considering superconducting DC cables.

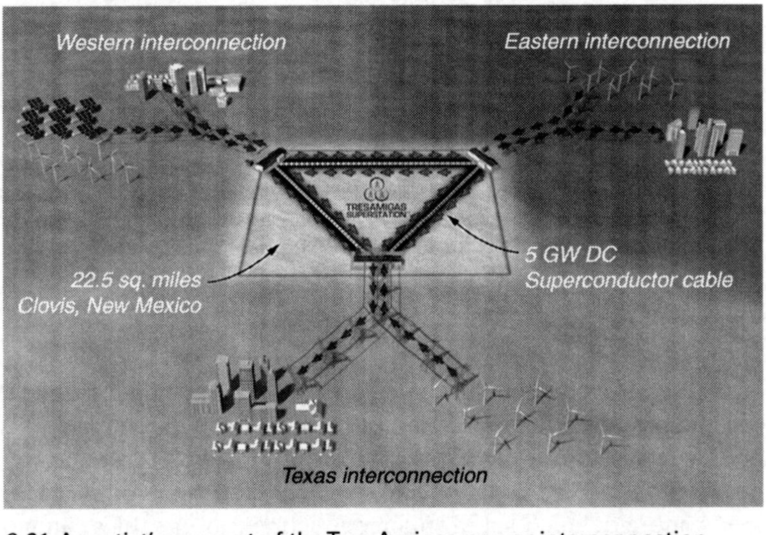

9.21 An artist's concept of the Tres Amigas power interconnection between the three power areas of the United States. Tres Amigas will be located in southeastern New Mexico.

9.22 Terminations for the three phases of the LIPA cable produced by Nexans and AMSC. The installation is on Long Island in New York. The inset shows a cross-section of the cold dielectric cable.

9.23 The first long term superconducting cable installation at the Southwire facility. Southwire is a producer of copper conductors for transformers, cables, etc. This three-phase cable installation delivered power to the plant for over 26000 h.

9.3.4 Installed superconducting power cables

Several superconducting power cables have been installed around the world over the past decade or so. Some of them have operated for a considerable time. A few of those cables are shown in Figs 9.22–9.24. The captions on those figures indicate location and some characteristics of the cable installation. A more detailed list of superconducting power cables can be found in reference Eckroad (2012).

9.24 Test installation of the three phases of the Changtong warm dielectric HTS cable at the Changtong Cable plant in Baiyin, China. This cable is now installed at the 12.5 kV substation in Baiyin city, Gansu province. (*Source*: Courtesy Changtong Cable.)

9.4 Conclusion

Superconductivity power applications have been under development since the 1960s. Early analyses and progress with power applications using LTSs provided a basis for work during the last two decades on power applications based on HTSs. There has been sufficient progress in several of the technologies that they are capable of satisfying many of the electric power utilities technical requirements. Unfortunately, the cost of superconducting materials has not decreased to a sufficient level where they can be economically competitive with conventional technologies. However, there are a significant number of niche applications where cost is not so important as performance. For example, superconducting cables will be used in areas where increased power must be delivered in critical corridors to urban areas where energy use is growing rapidly and environmental and other concerns do not allow the installation of additional power cables or transmission lines. Another niche application is the use of superconducting FCLs to avoid upgrading existing equipment. Several FCLs are under development and they show promise in several demonstration projects around the world. Not mentioned in the text above is the recent suggestion that MgB_2 may be an ideal material for near term development of FCLs that will operate at distribution level voltages (15–50 kV).

9.5 References

Bartlit J, Edeskuty F and Hammel E (1972), 'Multiple Use of Cryogenic Fluid Transfer Lines', Proc. Int. Cryogenic Eng. Conf., Eindhoven, May 24–26, 1972, pp. 176–179.

Boenig H and Paice D (1983), 'Fault current limiter using a superconducting coil', *IEEE Transactions on Magnetics*, Vol. **19**, No. 3, pp. 1054–1058, May 1983. BNL (1968), *Summer Study on Superconducting Devices and Accelerators* Brookhaven National Laboratory Report BNL-50155 C-55 Vol. **1**, **2**, and **3**, 1968.

Dam Q (2009), *Preserving Circuit Breaker Adequacy in Electric Power Systems*, Lambert Academic Publishing, Saarbrücken, Germany.

Dolgosheyev P, Peshkov I, Svalov G, Bortnik I, Karapazyuk V, Kubarev L, Panov A, Petrovsky Y and Turkot V (1979), 'Design and first stage test of 50-meter flexible superconducting cable', *IEEE Transactions on Magnetics*, Vol. **15**, No. 1, pp. 150–154, 1979.

Dommerque R, Boch J, Krämer S, Hobl A, Böhm R, Bludau M, Bock J, Klaus D, Piereder H, Wilson A, Krüger T, Pfeiffer G, Pfeiffer K and Elschner S (2010), 'First commercial medium voltage superconducting fault current limiters – production, test and installation', *Superconductor Science And Technology*, (SuST) Vol. **23**, 034020, 2010.

Eckroad S (2012), 'Superconducting Power Equipment: Technology Watch 2012', EPRI report 1024190, EPRI, 2012.

EPRI (1977), New Concepts in Fault Current Limiters and Power Circuit Breakers EPRI Symposium Proceedings, EL-276-SR, April, 1977, Palo Alto, CA.

EPRI (2006), *Underground Transmission Systems Reference Book,* EPRI product number 1012334, December 2006.

Eto J H, Undrill J, Mackin P, Daschmans R, Williams B, Haney B, Hunt R, Ellis J, Illian H, Martinez C, O'Malley M, Coughlin K and LaCommare K H (2010), 'Use of Frequency Response Metrics to Assess the Planning and Operating Requirements for Reliable Integration of Variable Renewable Generation', LBNL Report LBNL-4142E, 2010.

Forsyth E and Thomas R (1986), 'Performance summary of the Brookhaven superconducting power transmission system,' *Cryogenics*, Vol. **26**, No. 11, pp. 599–614, November 1986.

Garwin L and Matisoo J (1967), 'Superconducting lines for the transmission of large amounts of electrical power over great distances,' *Proceedings of the IEEE*, Vol. **55**, No. 4, pp. 538–548.

Gouge M, Demko J, McConnell B, and Pfotenhauer J (2002), 'Cryogenics Assessment Report', ORNL Report.

Hammel E and Rogers J (1970), 'Cryogenics and nuclear physics – Part I,' *Cryogenics*, Vol. **10**, pp. 5–13, June 1970.

Hammel E, Rogers J and Hassenzahl W (1970), 'Cryogenics and nuclear physics-Part II,' *Cryogenics*, pp. 186–195, June 1970.

Haldar P and Abetti P (2011), 'Superconductivity's First Century', *IEEE Spectrum*, Vol. **48**, March 2011.

Hassenzahl W (2000), 'The All Superconducting Substation: A Comparison with a Conventional Substation,' EPRI Tech. Rep. TR-1000915, December 2000.

Hassenzahl W Hazelton D, Johnson B, Komarek P, Noe M and Reis C (2004), 'Power applications of superconductivity, *Proceedings of the IEEE*, Vol. **92**, No. 10, pp. 1655–1674, October 2004.

Hassenzahl W (2005), 'Electricity storage: Believe in it', *IEEE Power and Energy Magazine*, Vol. **3**, pp. 20–21 March/April, 2005.

Hassenzahl W (2006), 'Cryogenics: A Utility Primer', EPRI report 1010897.

Hassenzahl W, Gregory B, Nilsson S, Daneshpooy A, Grant P and Eckroad S (2009) 'A Superconducting DC Cable', EPRI Report 1020458, December 2009.

Hassenzahl W and Young M (2010), 'Superconducting Power Equipment: Technology Watch 2010', EPRI report 1019995, December, 2010.

Hingorani N and Gyugyi L (1999), *Understanding FACTS: Concepts and Technology of Flexible AC Transmission Systems*, Wiley-IEEE Press, December 1999. ISBN 978-0-7803-3455-7.

Hongesombut K, Mitani Y and Tsuji K (2003), 'Optimal location assignment and design of superconducting fault current limiters applied to loop power systems', *IEEE Transactions on Applied Superconductivity*, Vol. **13**, No. 2, pp. 1828–1831, March 2003.

Hoshino T, Salim K M, Nishikawa M, Muta I and Nakamura T (2001), 'DC reactor effect on bridge type superconducting fault current limiter during load increasing', *IEEE Transactions on Applied Superconductivity*, Vol. **11**, No. 1, pp. 1944–1947, June 2001.

Kalsi S, Weeber K, Takesue H, Lewis C, Neumueller H and Blaugher R (2004), 'Development status of rotating machines employing superconducting field windings', *Proceedings of the IEEE*, Vol. **92**, No. 10, pp. 1688–1704, October 2004.

Kirtley J and Edeskuty F (1989), 'Application of superconductors to motors, generators, and flywheels', *Proceedings of the IEEE*, Vol. **77**, No. 8, pp. 1143–1155, August 1989.

Klaudy P (1967);'Some remarks on cryogenic cables', *Advances in Cryogenic Engineering*, Vol. **11**, pp. 684–693, 1967.

Klaudy P, Gerhold I, Beck A, Rohner P, Scheffler E and Ziemek G (1981), 'First field trials of a superconducting power cable within the power grid of a public utility', *IEEE Transaction on Magnetics*, Vol. **17**, pp. 153–156, 1981.

Laquer L (1962a), 'Superconductive electric switch', US patent 3145284, filed 2 October 1962.

Laquer L (1962b), 'Incremental electrical method and apparatus for energizing high current superconducting magnets', US patent 3150291, filed 2 October 1962.

Laquer, Carroll J and Hammel E (1966), 'Automatic superconducting pump', US patent 3414777, filed 1 June 1966.

Laquer L (1968), 'Automatic Superconducting Switches', *Journal of Applied Physics*, Vol. **39**, No. 6, pp. 2639–2640, May 1968.

Laquer H, Dean J and Chowdhuri P (1977), 'Electrical, cryogenic, and systems design of a dc superconducting transmission line', *IEEE Transactions on Magnetics*, Vol. **17**, pp. 182–187, 1977.

Leung E, Rodriguez I, Albert G, Burley B, Dew M, Gurrola P, Madura D, Miyata M, Muehleman K, Nguyen L, Pidcoe S, Ahmed S, Dishaw G, Nieto C, Kersenbaum I, Gamble B, Russo C, Boenig H, Peterson D, Motowildo L and Haldar P, 'High temperature superconducting fault current limiter development', *IEEE*

Transactions on Applied Superconductivity, Vol. **7** No. 2, pp. 985–988, June 1997.

Long H and Nottaro J (1971), 'Design features of ac superconducting cables', *Journal of Applied Physics*, Vol. **42**, No. 1, pp. 155–162, 1971.

Moisson F and Leroux J (1971),'Development of a superconducting cable for transmission of high electric power', *Journal of Applied Physics*, Vol. **42**, No. 1, p 154, 1971.

Moran M and Shapiro H (2003), *Fundamentals of Engineering Thermodynamics*, Wiley Text Books, 5th edition, June 2003, ch. 2.

Nagata M, Tanaka K and Tanaguchi H (2001), 'FCL location in large scale power system', *IEEE Transactions on Applied Superconductivity*, Vol. **11**, No. 1 pp. 2489–2494, March 2001.

Noe M and Oswald B (1999), 'Technical and economical benefits of superconducting fault current limiters in power systems', *IEEE Transactions on Applied Superconductivity*, Vol. **9**, No. 2, pp. 1347–1350, June 1999.

Noe M and Steurer M (2007), 'High-temperature superconductor fault current limiters: Concepts, applications, and development status,' *Superconductor Science and Technology*, Vol. **20**, pp. R15–R29, 2007.

Rogers J, Boenig H, Chowdhuri O, Schermer R, Wollan J and Weldon D (1983), 'Superconducting fault current limiter and inductor design', *IEEE Transactions on Magnetics*, Vol. **19**, No. 3, pp. 1051–1053, May 1983.

Waynert J, Boenig H, Mielke C, Willis J and Burley B (2003), 'Restoration and testing of an hts fault current controller', *IEEE Transactions on Applied Superconductivity*, Vol. **13**, No. 2, pp. 1984–1987, June 2003.

Wilson M (1983), *Superconducting Magnets*, Oxford, Oxford University Press.

Xiao L (2009), Chinese academy of Sciences private communication 'A 35 kV, 1.5 kA spider type of FCL is installed at the Puji substation in China', November 2009.

Young M and Hassenzahl W (2009), 'Superconducting Fault Current Limiters: Technology Watch 2009', EPRI report 1017793, December 2009.

Zhang G, Dai S, Song N, Guo W, Zhang J, Zhang D, Zhang Z, Zhu Z, Xiao L and Lin L,' (2011), The Construction Progress of a High Temperature Superconducting Power Substati on in China', *Applied Superconductivity, IEEE Transactions*, Vol. **21**, No. 3, pp 2824–2827, June 2011.

Part III

Electricity storage technologies

10

Techno-economic analysis of electricity storage systems

J. OBERSCHMIDT, M. KLOBASA and
F. GENOESE, Fraunhofer Institute for Systems
and Innovation Research, Germany

DOI: 10.1533/9780857097378.3.281

Abstract: Against the background of an increasing share of intermittent renewable energies in power generation, in the medium to long term, electricity storage could be a key technology of strategic importance with regard to grid stability, grid extension and security of supply. In this context, this chapter analyses the economic feasibility of applying electricity storage technologies in energy markets. Additionally, resource requirements and environmental aspects, as well as health and safety issues, resulting from electricity storage are examined. The focus is on stationary storage technologies with the potential of balancing fluctuations due to intermittent renewable energies, i.e. allowing storing electricity over time periods of several hours, days or weeks. Conventional technologies, such as pumped hydro-energy storage and lead acid (LA) batteries, as well as innovative alternatives such as advanced adiabatic (AA) compressed air energy storage (CAES) and redox flow batteries (RFB), which are still under development, are considered. Hydrogen storage, lithium ion batteries and sodium sulphur batteries are also considered, representing further alternatives widely discussed today.

Key words: electricity storage, intermittent renewable energies, economic analysis, environmental assessment.

10.1 Introduction

Electricity demand is subject to significant daily, weekly and seasonal fluctuations. Supply needs to adapt to these fluctuations. In this context, electricity storage fulfils important functions within energy systems, in particular in ensuring stability and reliability of supply. Due to the ongoing integration of non-dispatchable renewable energies, such as wind and solar based generation, the amount of intermittent electricity fed into the grid is increasing. This causes a growing imbalance of supply and demand, temporally as well as geographically, which results in increased efforts for grid management as well as system balancing. Furthermore, a trend towards decentralized electricity supply based on renewables and combined heat and power increases the need for storage to balance supply and demand on a decentralized level. Therefore, it is expected that electricity storage technologies will play an increasingly important role over the coming years (Baker, 2008; Bouffard and Kirschen, 2008; Bünger *et al.*, 2009; Gatzen, 2008; Tester *et al.*, 2005).

Storage technologies can help to better match disposable reserves with demand, and also to increase flexibility in grid operation. In the medium to long term, electricity storage could be a key technology of strategic importance with regard to grid stability, grid extension and security of supply. Alongside other flexibility options such as load management, electricity storage technologies could contribute to avoiding cost-intensive reserve capacities and grid extension. In this context it is important to understand the impacts of electricity storage on power systems. In the long term only those technologies will be applied that are economically feasible. Thus, the economic viability of different possibilities for electricity storage is in the centre of further development. Furthermore, in view of climate change and the increasing scarcity of natural resources, environmental concerns are becoming ever more important. Against this background, this chapter analyses the economic feasibility of applying electricity storage technologies in energy markets with an increasing share of intermittent renewable energies. Additionally, resource requirements, environmental aspects as well as health and safety issues resulting from electricity storage are examined. The focus is on stationary storage technologies with the potential of balancing fluctuations due to intermittent renewable energies, i.e. allowing storing electricity over time periods of several hours, days or weeks. Conventional technologies, such as pumped hydro-energy storage and LA batteries, as well as innovative alternatives such as AA-CAES and RFB, which are still under development, are considered. Hydrogen storage, lithium ion batteries and sodium sulphur batteries are also considered, representing further alternatives widely discussed today.

10.2 Economic issues and analysis

In the following, the possibilities of applying selected electricity storage technologies in energy markets and the potentially resulting economic benefits are analysed. As a basis, important technical characteristics and costs of the selected electricity storage technologies are summarized, followed by a description of general conditions on energy markets under which they could be applied. Finally, the possible revenues of applying electricity storage technologies are presented based on modelling results.

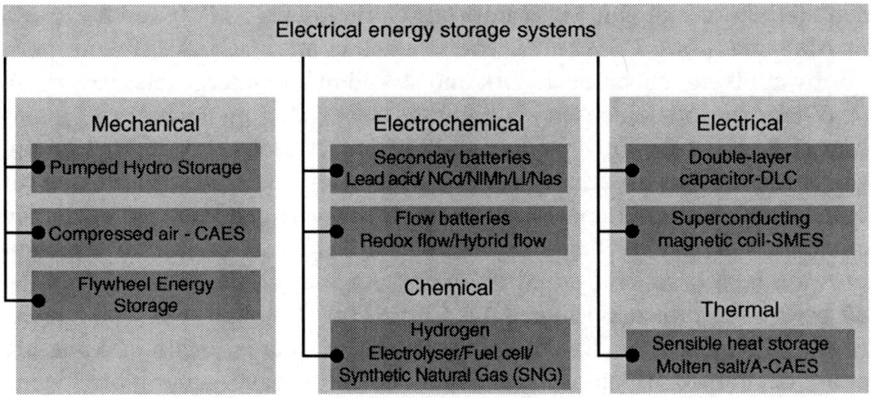

10.1 Categorization of energy storage systems. (*Source*: IEC, 2011.)

10.2.1 Technical characteristics and costs of electricity storage

Electricity storage technologies can be categorized based on the physical principle used, distinguishing mechanical, electrochemical and electromagnetic storage (Baker, 2008; Gatzen, 2008). A widely used approach for classifying energy storage systems is based on the form of energy used (Fig. 10.1). The area of application for specific electricity storage technologies is restricted by physical, technical and economic restrictions. Storage technologies such as superconducting magnetic electricity storage (SMES) and super capacitors (SuperCaps) are usually not suitable for storing large amounts of energy over long time periods of several hours or days. They are therefore not considered further in this article.

Mechanical storage technologies include pumped hydro, compressed air systems and flywheels (Baker, 2008). Because for **flywheels**, high losses occur if energy is stored over periods longer than a few minutes, they do not seem to be suitable for balancing longer term fluctuations due to intermittent renewables and are therefore not the focus of this article (Baker, 2008; Bünger *et al.*, 2009; Gatzen, 2008; Oertel, 2008). In **pumped hydro storage plants**, electricity is used to pump water from a lower reservoir to an upper reservoir in times of low demand. To generate electricity in peak-load periods, the water flows downhill through a turbine driving a generator. Efficiencies are high, ranging from 75–80% and the plants can be activated within a few seconds. However, the number of suitable sites to build up additional pumped hydro storage plants is limited due to their particular topographic requirements.

In principle, **CAES** can also be used to store energy over several days or weeks. However, these systems carry high costs, and worldwide only two installations are known, i.e. Huntorf in Germany (321 MW) and McIntosh in Alabama, USA (110 MW). There, air is compressed in off-peak periods by applying mechanical work and stored in an underground cavern. In peak-load periods, electricity is generated by expanding the air in a gas turbine (Bünger *et al.*, 2009; Oeding and Oswald, 2004; Oertel, 2008). Usually, gas is required as an additional fuel to heat up the cold, compressed air beforehand. To avoid consumption of natural gas and increase efficiency, adiabatic systems are being developed, with the aim of storing the compression heat in an additional thermal storage and using it to heat up the air before the expansion stage (AA CAES) (cf. Bullough *et al.*, 2004; Jakiel *et al.*, 2007; Zunft *et al.*, 2006). Furthermore, smaller scale systems are being developed which are not dependent on special geographic formations because they use storage vessels such as steel cylinders to store the air (Baker, 2008; Oertel, 2008).

Electrochemical and chemical storage systems include batteries and hydrogen technologies. A wide variety of different battery types is available or under development, with efficiencies ranging from 70% to 95% (Mulder *et al.*, 2010). In general, batteries convert the chemical energy contained in their active materials directly into electric energy via an electrochemical oxidation-reduction (redox). LA **batteries** are based on mature and established technology. They are advantageous in terms of initial investment and efficiency. However, cycle life and energy density are rather low (Oertel, 2008; Wietschel *et al.*, 2010). **Lithium ion batteries** are achieving higher energy densities, therefore are the focus of current research, e.g. in the context of electric vehicle development. Disadvantages include lower robustness and easily flammable materials. High temperature batteries, such as **sodium sulphur (NaS) batteries,** can reach efficiencies from 70–80%. So far they have not been applied widely, due to high costs. Other battery types, which are not the focus of this article, include further developments of the above, such as lithium sulphur (LiS) or sodium nickel chloride (NaNiCl, also known as ZEBRA) systems (Baker, 2008; Gatzen, 2008; Oertel, 2008; Wietschel *et al.*, 2010). In general, batteries can only provide higher power ratings based on serial connection. Therefore, increasing power ratings results in higher costs to the same order of magnitude.

Another type of electrochemical storage is **RFB**. They differ from conventional batteries because the reactants (electrolytes, e.g. vanadium, zinc bromide, polysulphide) storing the electricity come from containers outside the electrochemical cell. They flow through the cell, absorbing energy when it is being charged and delivering energy when it is being discharged. The reactants are prevented from mixing within the electrochemical cell by the ion selective membrane or a micro-porous separator. The available power is determined by the size of the stack (surface area of electrodes and number of cells), whereas the capacity is determined by the volume of electrolyte, i.e. the size of the external storage tanks. Therefore, power and capacity can be scaled independently from each other. The most common RFB system today is the all-vanadium system, where both electrolytes contain only vanadium species at different valence states. Advantages over other types of RFBs include higher energy density, low electrolyte costs and lower demands regarding membrane electrolyte crossover (Tassin *et al.*, 2003). However, to be able to offer competitive solutions with a storage volume of several hours, costs of RFB still need to be reduced significantly (Baker, 2008; Gatzen, 2008).

Another possibility for electrochemical storage is based on the production of **hydrogen** using electricity, storage of the hydrogen produced and its re-electrification. Different technologies are available or under development for this purpose. To generate hydrogen, electrolysis (electrolytic water decomposition) is regarded as the best method (Boulanger *et al.*, 2003). For storage, pressurized tanks are most commonly used. For larger scale systems, underground caverns could also be used. In order to generate electricity based on hydrogen, thermal processes (combustion to drive a generator) or electrochemical fuel cells can be used. In larger scale applications, the use of a combined cycle gas turbine (H_2-CCGT) is expected. Generally, the overall efficiency of hydrogen storage systems is rather low, at around 30%, and costs of the technologies are high. Still, hydrogen storage is regarded as one of the most promising alternatives, in particular if large amounts of energy need to be stored, due to its high energy density.

Even though it is technically possible to store electricity with the technologies described above, most of them are still very expensive, or under development and not yet commercially available. However, economic efficiency is crucial for energy technologies to be applied on the market. Therefore, the following economic analysis reveals the conditions for competitive application of the selected storage technologies in energy markets. The profitability of electricity storage on the one hand depends on the costs of the technologies applied and, on the other hand, on the earning potentials in energy markets. Costs are determined by the initial investment for the converter and for the storage unit, as well as operation and maintenance (O&M) requirements. Furthermore, variable costs during operation result from purchasing electricity, i.e. power input costs depending on electricity prices at the time of charging, which are determined by the electricity markets (cf. Section 10.2.2). Table 10.1 shows an overview of costs of the selected electricity storage systems alongside important technical characteristics which can also influence economic profitability. These parameters and costs provide the basis for the evaluation of energy storage applications in power trading presented in the following section.

Table 10.1 Technical characteristics and cost data for electricity storage technologies

Technology (alphabetical)	Size (MW)	Size (MWh)	Specific investment (€/kW)	Cycle efficiency (%)	Cycle life (number)	Calendar life (years)	Energy density (Wh/l)	Energy density (Wh/kg)
AA-CAES	250	2500	1143	70		20–30	15–5	
H_2-CCGT	250	2500	1650	37		15	170–190[1]	
LA	25	250	1190	78	1400	7	10–400	75–300
Li	25	250	1284	84	1500	8	200–400	75–200
NaS	25	250	2200	69	4500	15–20	170	120
PHES	250	2500	730[2]	80		40	0 7	
RFB	25	250	2012	70	9000	30	20–30	

Note: AA-CAES = advanced adiabatic compressed air energy storage; CCGT = combined cycle gas turbine; LA = lead acid battery; Li = lithium ion battery; NaS – sodium sulphur battery, PHES = pumped hydro-energy storage; RFB = redox flow battery.

[1] Gaseous H_2 in salt caverns.

[2] Investment for pumped hydro storage can vary widely depending on the specific characteristics of the project. Spahic *et al.* (2006) give 600 €/kW for Goldisthal in Germany. Steffen (2012) quotes 1048 €/kW on average for projects in Germany and Deane *et al.* 960 €/kW on average for projects in Europe. As specific investment for PHES can also be higher (for example, EPRI (2010) mention 2500–4300 $/kW) the value used here represents the lower range of current projects.

Sources: Deane *et al.*, 2010; Honsel, 2007; Spahic *et al.*, 2006; Steffen, 2012; Stieler, 2007; Wietschel *et al.*, 2010.

10.2.2 Application of electricity storage in power trading

Different business cases and application strategies are possible for electricity storage systems. For example, **wind energy plant operators** could use storage to level out deviations from the planned output capacity due to forecasting errors. This could be profitable in the future if wind plant operators are obligated to announce their feed-in in advance (as are the operators of conventional power plants) or for power trading at the electricity stock exchange if potential revenues are higher than feed-in tariffs or other incentives guaranteed by political support instruments as the renewable energy act (Erneuerbare Energien Gesetz or EEG) in Germany. **Large industrial electricity consumers** could use storage technologies to avoid peak loads and thus to avoid high energy prices for peak loads, if a defined maximum load is exceeded. Furthermore, **distribution system operators** can use storage to guarantee uninterruptible power supply. For distribution system operators, electricity storages can moreover be applied to optimize the operation of the distribution grid and thus to better preserve the operating equipment within the grid. In this context, electricity storage could also help to avoid grid extension or reinforcement by homogenizing usage of up-stream grids. Power plant operators could become storage power plant operators as well, and operate on ancillary service markets especially on reserve markets. In times when the storage power capacity is not contracted on the reserve markets, it could be bid on day-ahead spot markets as well.

The profitability of electricity storage application in energy markets, apart from the necessary investments, depends on the spread between the selling prices that can be achieved at the time of discharging and the purchasing price paid at the time of charging. Due to energy losses during charging, storage and discharging, the profitability also depends on the overall cycle efficiency of the technology. Furthermore, the economic feasibility of electricity storage applications depends on the technology's life-time in years and its cycle life, which in turn depend on the operations strategy (number of cycles per period, full load hours, charging conditions). Because storage technologies can be regulated easily with fast reaction times, they can be operated on different reserve markets where further revenues can be realized.

In many countries a large share of total electricity generation and demand is traded on an hourly basis on the spot market, which is executed every day as a day-ahead market (Sensfuß, 2008). Typically day and night price spreads arise due to lower electricity demand at night time. In countries with a large share of electric heating systems, and therefore a higher electricity demand at night, this price spread can be reduced significantly. However, today's storage power plants (basically pumped hydro storage) operate on these price differences, and load at night time and generate peak power at day time. Adjustments to the amounts of energy traded on the spot market are made on the intraday market. Prices on the intraday market are close to spot market prices.

Due to imperfect forecasts and power station blackouts, deviations to the planned day-ahead and intraday trade need to be balanced. The prices of balancing power are determined on the reserve power market. Prices on the reserve power market can deviate significantly from those on the spot market. The price spread between positive and negative reserve power is much higher than between day and night electricity spot prices. Reserve capacity is needed to provide system services in order to ensure system stability. The transmission system operators (TSO) are responsible for the stability of the electricity system. Total reserve capacity amounts to considerable volumes not available on the spot market. Three kinds of reserve capacity can be distinguished:

1. Primary reserve or spinning reserve is the first reserve used to stabilize the electricity system and has to be able to operate at full capacity within 30 s. It is only used for time periods of less than 15 min before

it is substituted for secondary reserve. The primary reserve is activated automatically and mostly provided by power plants which are already in operation. For example, in Germany the primary reserve market is organized as a tender for all four transmission zones together, where the capacity is contracted 1 month ahead for a period of 1 month. Several technical qualifications have to be met in order to be able to take part in the primary reserve market, so that participation is restricted to only a few players. Primary reserve is paid for by a capacity price only. Due to the high technical restrictions, storage systems will not be operated as primary reserve in the near future.

2. Secondary reserve has to be at full capacity within a period of 30–300 s, and replaces primary reserve. The use of secondary reserve is also determined automatically. For the purchase of this system service, a tender for 1 month is used by the grid operators, similar to primary reserve. The number of players is limited by technical restrictions. In addition to the capacity price, a price per MWh is also paid for any secondary reserve capacity that is called. The price spread between positive and negative secondary reserve energy is very high in most countries. From a technical point of view, many storage systems would be able to take part in the secondary reserve markets.
3. Minute reserve or tertiary reserve is used to free secondary reserve. It is activated manually and has to be available within 15 min. Usually, minute reserve remains activated for up to 1 h. However, this period can be extended to several hours in the case of serious grid imbalance. The transmission grid operator purchases minute reserve in a day-ahead tender. There are lower technical requirements compared to primary and secondary reserve, so that there are more players on this market. Similar to secondary reserve, minute reserve is paid for by a capacity price and an additional price per MWh. The price spread between positive and negative minute reserve energy is typically smaller than for secondary reserve, but still much higher than on the spot market. The only German CAES power plant in Huntorf operates in this market.

For short-term trading (intraday and balancing power) prices are determined within one price area which can exist of one or more control areas. If the sum of generation is higher than consumption, in general, the additional amount of electricity available can only be sold for prices below the spot market price. If consumption is higher than the planned generation, the missing amount of electricity generally has to be purchased at higher prices than the prices on the spot market.

Bulk electricity price arbitrage on the spot market is an important source of revenues for energy storages. It involves the purchase of inexpensive electricity available during periods when the demand for electricity is low and/

or the supply from renewable energy is high. In these times the storage unit is charged, so that the low priced energy can be used or sold at a later time when the price for electricity is high due to high demand and/or renewable energy supply is low. For historical market prices in Germany, these revenues are calculated using an optimization model.[1] In the optimization, the revenues for a certain time frame (i.e. 1 year) are maximized while price distribution is assumed to be known ('perfect foresight'). The storage is operated whenever a positive profit margin can be achieved. The storage unit is not operated when the price arbitrage between low-price and peak-price hours is too low, so that efficiency losses cannot be compensated. For this analysis, the optimization horizon was set to 1 year. While this allows the possibility that load can not only be shifted between the hours of a day but also between multiple days, such shifting seldom occurs, since the storage volumes considered in this analysis are in the range of full load *hours* and not *days*. Typically, the reservoir is loaded at night and discharged during peak hours. One important assumption is that electricity storage operations have no influence on the electricity price. Therefore, and because of the perfect foresight of market prices, the revenues calculated represent theoretical maximum revenues.

Figures 10.2 and 10.3 show possible revenues that could be achieved by applying electricity storage technologies for power trading on the German spot market EEX for the years 2008 and 2009. It also shows the annual costs for O&M and the annuity of capital costs. The battery technologies (RFB, NaS, Li-Ion, LA) are assumed to be configured with 25 MW of output capacity and 250 MWh of storage volume, while the remaining, large-scale storage options (pumped hydro-energy storage, PHES, AA-CAES, H_2-CCGT) are assumed to be configured with 250 MW of output capacity and 2500 MWh of storage volume. This means that all storage systems are able to provide 10 h of full load output. In absolute terms, the 250 MW systems generate more income than the 25 MW. Therefore, values relative to the annual costs (O&M costs and the annuity of capital costs) are also displayed.

The years 2008 and 2009 have been very diverse market-wise. Being the year of the economic crisis, 2009 showed a relatively low electricity demand and therefore low spot market prices (⊘ = 38.85 €/MWh), whereas in 2008 record peak prices for crude oil and high energy carrier prices in general could be observed which, in consequence, led to high spot market prices (⊘ = 65.76 €/MWh). While the average price level is an indicator for revenues – i.e. higher prices lead to higher revenues – storage revenues mainly depend on the distribution of spot market prices: a high price spread between peak and off-peak prices is needed for storage units to be able to operate.

[1] The corresponding optimization problem is described in Hartel *et al.* (2010).

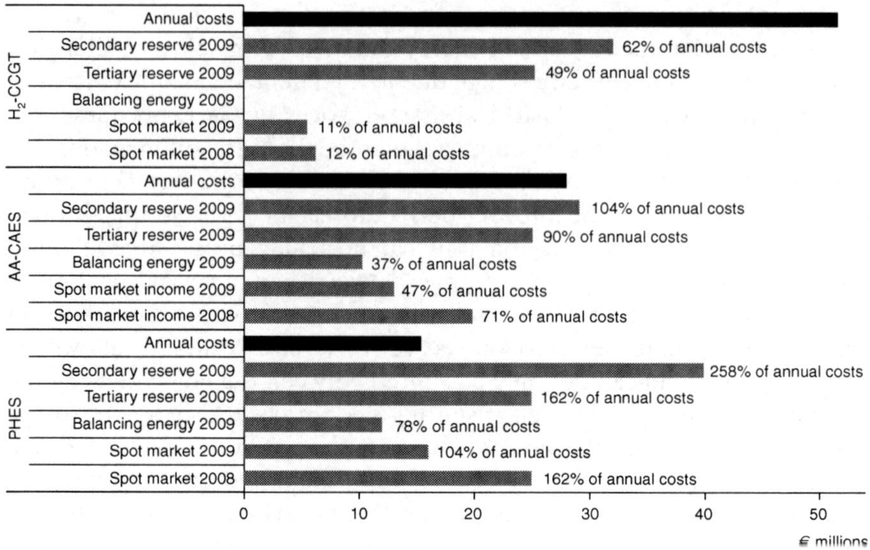

10.2 Potential revenues and annual costs (large-scale systems). (*Source*: Fraunhofer ISI.)

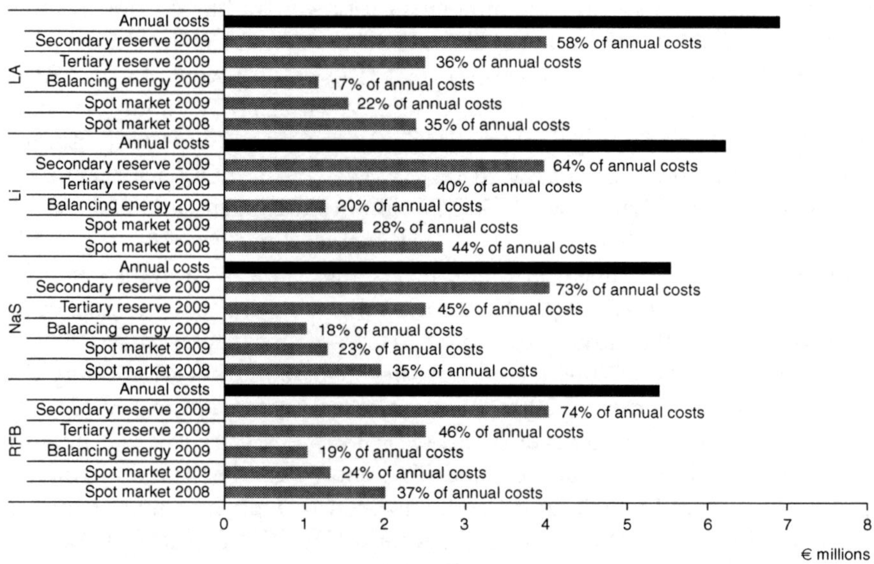

10.3 Potential revenues and annual costs (small-scale systems). (*Source*: Fraunhofer ISI.)

Table 10.2 Annual discharge utilisation

	2008 (hours)	2009 (hours)
Lead-Acid	2679	2644
Lithium-Ion	3041	3015
Sodium-Sulfur	2170	2122
Redox-Flow	2247	2207
Hydrogen CCGT	526	562
Adiabatic Compressed Air	2239	2191
Pumped Hydro	2824	2789

Source: Fraunhofer ISI.

The optimization results show that the year 2008 was more profitable than 2009 for all storage technologies. These results can be understood by looking at the price distributions: for example, in 2009 electricity prices above 100 €/MWh were achieved in only 45 h, whereas in 2008 there were 892 h above this price. Prices below 10 €/MWh could be observed in 144 h in 2008 and in 362 h in 2009. While there would have been more opportunities to charge a storage unit at low prices in 2009, there were far more peak prices in 2008 than in 2009, which could have been utilized to sell the stored electricity and generate a higher income.

The optimization results also show that at current costs none of the storage technologies available would generate enough income to cover the total annual costs (with the exception of pumped hydro). Most notably, the costs of the H_2-CCGT system exceed its possible spot market income by a factor of 8–9. This is a direct effect of the low round-trip efficiency of the H_2 system, which leaves fewer opportunities to exploit the price spread between peak and off-peak prices. Hydrogen-based storage units could be more useful for long-term storage (weeks to months) since they offer a much higher energy density when compared to PHES resulting in lower energy-specific capital costs. Short-term storage based on hydrogen, however, is unlikely to become economically feasible even if substantial cost reductions could be achieved. The AA-CAES technology performs worse than the PHES, and its income is below the annual costs but is still in a promising range.

Among the 25 MW systems, the Li system performs best, although having the second highest annual costs. This is a result of the high round-trip efficiency of the Li system, ensuring a high utilization. The utilization performance of all storage technologies is summarized in Table 10.2.

Other potential revenue streams for electricity storage systems result from avoiding the use of balancing power, or from directly acting on reserve markets.

The combination of fluctuating generators with storage systems can increase forecast accuracy and therefore reduce the need for balancing

Table 10.3 Price data for secondary and tertiary reserve in Germany in 2009

		capacity price (€/MW per day)		price per MWh (€/MWh)	
		2008	2009	2008	2009
Positive	Secondary reserve	150.07	126.53	142.71	136.86
	Tertiary reserve	128.51	58.41	99.50	109.98
Negative	Secondary reserve	84.39	115.67	4.12	–20.30
	Tertiary reserve	62.30	173.59	0.19	–6.10

Source: Fraunhofer ISI.

power for the renewable energy sources (RES) operator, if they are responsible for balancing. Utilization of such a storage system (25% of installed wind capacity, 10 full load hours storage capacity) is estimated to be 1400 h discharging, based on forecasted and real wind-park generation data. Balancing energy prices are taken from the four German control areas in 2009. Typically, wind power forecast deviations are not fully correlated with deviations in the control areas. In 2009 the correlation factor was between 0.18 and 0.25 in the different German control areas. The average electricity price for charging the storage systems was 28 €/MWh, and for discharging 70 €/MWh.

Revenues from reserve markets are estimated for secondary and tertiary reserve with price data of the German reserve market for 2009 (Table 10.3). Utilization for discharging is set to average activation values for secondary respectively tertiary reserve capacity for all storage technologies. This was about 440 h for secondary and 130 h for tertiary reserve in Germany in 2009.

Revenues are highest in the secondary reserve market as it has higher capacity prices, utilization and prices per MWh than the minute reserve market. Only the capacity price for negative tertiary reserve was higher than for secondary reserve in 2009, but this could not compensate the revenue difference between secondary and tertiary reserve. Both secondary and tertiary reserve markets offer higher revenues than the case of avoiding balancing energy. Although utilization of the storage system is then higher, this could not compensate for the larger price spread and the capacity payments in the reserve markets.

10.3 Environmental aspects of electricity storage

With a view to sustainable energy supply, not only economic profitability should be considered for a thorough analysis of electricity storage systems, but also environmental issues. Electricity storage systems can contribute to an improved environmental performance of energy supply indirectly by supporting the integration and increased utilization of fluctuating renewable energy sources. During operation storage systems hardly produce any direct CO_2 emissions and usually no additional fuel is needed (an exception is non-adiabatic CAES, where co-firing of natural gas is required). In this regard, electricity storage is advantageous in environmental terms compared to peak-load plants such as gas turbines or diesel generators. On the other hand, it has to be recognized that storage always results in a loss of electricity and additional energy demand depending on the round-trip efficiency of the specific system. Furthermore the manufacture, use and disposal of electricity storage systems can cause additional environmental impacts as well as health and safety hazards. These depend on the type of technology used and are interrelated with the resources and materials needed to manufacture and operate the respective electricity storage system. Thus, the benefits and drawbacks of electricity storage should be analysed considering the whole life cycle of the system. Against this background, in this section resource requirements, health and safety issues, as well as environmental impacts of electricity storage are examined.

10.3.1 Resource requirements

Efficient resource and material use have become increasingly important over recent years due to environmental as well as economic considerations. On the one hand, material costs usually account for the largest part of total production costs. On the other hand, the extraction, processing and refinement of resources cause negative environmental impacts. Furthermore, attention should be paid to potential supply risks, in particular for 'vulnerable' raw materials. The latter are highly important for the economy, but are characterized by their concentration in only a few countries, with low political stability (Angerer *et al.*, 2009; Frondel *et al.*, 2007).

Considering the materials required, **pumped hydro-energy storage** as well as CAES and **H_2-CCGT systems** seem to be unproblematic, because the bulk of resources used are standard materials, such as steel and concrete. Therefore, the development of these technologies seems not to be limited by critical raw material requirements, but rather by the availability of adequate geographic sites. For **pumped hydro storage** a sufficient topographic gradient is required, while for **CAES** and **hydrogen storage** suitable caverns are needed, preferably close to the place where the electricity is produced. Different types of subterranean excavations are suitable to store air, for example, salt caverns, aquifers, closed mines, exhausted oil or natural gas fields (Oertel, 2008; Wietschel *et al.*, 2010).

Resource requirements for **electrochemical** storage are presumably more critical in terms of vulnerability and potential hazards. Therefore, this section focuses on battery technologies. The electrodes of **batteries** are usually made of metals or metal oxides, which are placed in an electrolyte solution. A huge number of theoretically possible combinations of electrode materials and electrolytes exist, but in practice for rechargeable batteries only a few systems are applied (Stieler, 2007). In a charged state, the cathodes of LA **batteries** consist of elementary lead (Pb) and the anodes of lead (IV) oxide (PbO_2), placed in diluted sulphuric acid as the electrolyte. In a discharged state both electrode plates are covered by a lead salt (lead (II) sulphate $PbSO_4$), while the electrolyte basically becomes water. The housing of LA batteries is usually made of plastic (Oertel, 2008). The global consumption of lead is increasing, with more than half of the worldwide production used for electrodes in LA batteries. However, the availability of lead is regarded as unproblematic. Also, diluted sulphuric acid has been produced on a large-scale for a long time and is widely available, i.e. there are no critical resource requirements for LA batteries (Wietschel *et al.*, 2010).

For **lithium ion batteries** a liquid, anhydrous **electrolyte** is used which contains conducting salt. The liquid component is a mixture of organic solvents, which is often composed of carbonates (e.g. propylene carbonate, ethylene carbonate, dimethyl carbonate or diethyl carbonate). Different mixtures are possible depending on the requirements regarding the characteristics of the cell. The solvents need to be chemically pure, in particular free of water. Thus, elaborate purification and dehydration are applied before use. The conducting salt nowadays usually consists of $LiPF_6$ (lithium hexa fluorine phosphate). Other substances, such as lithium perchlorate ($LiClO_4$) and lithium borofluoride ($LiBF_4$), can also be used as salt (Oertel, 2008; Wietschel *et al.*, 2010). The **negative electrode (anode) of lithium ion batteries** is usually made of graphite. As an alternative, lithium metals, lithium alloys, metal oxides (e.g. SnO, SiO_2) or amorphous carbon can also be used. The safety characteristics of graphite and amorphous carbon are similar, but with graphite significantly higher storage capacity can be achieved. For the **positive electrode (cathode)** different metal oxides with varying properties can be used. The **cathode** of most of the lithium ion batteries produced to date are of the cobalt type ($LiCoO_2$), which enable high energy intensity. Recently, manufacturers have begun making batteries with cathodes that include transition metals such as nickel and manganese, in addition to cobalt. With LiNiCo higher energy density can be achieved. Manganese cathodes ($LiMn_2O_4$) allow higher intensity of current. Additionally, copper and aluminium are needed for the electrodes. As separators between the electrodes, thin micro-porous polyolefin films are used (Angerer *et al.*, 2009; Stieler, 2007; Wietschel *et al.*, 2010).

Inside a cylindrical **NaS battery cell** an electrode made of sodium can be found. The electrode is enclosed in a metal casing for safety reasons. The outside consists of beta-alumina as a solid electrolyte, which is enclosed by a sulphur electrode (Pohl and Kriebs, 2006; Simon *et al.*, 2006). Sodium (Na) and sulphur (S) are liquid at operating temperatures of 270–350°C. During discharge sodium ions migrate from the inside to the outside of the cell, where NaS compounds develop. With ongoing discharge the amount of sodium in these compounds increases (Na_2S_3, Na_2S_4, Na_2S_5). Due to high operating temperatures, adequate insulation materials are needed. However, no critical resource requirements are known for NaS batteries (Wietschel *et al.*, 2010). None of the main materials needed for NaS batteries is regarded as being critical in terms of its availability.

The electrodes of RFBs are usually made of graphite and carbon. In many cases the membrane is made of Nafion-sulphated polytetrafluoroethylene (PTFE). Instead of Nafion, some systems use polystyrene sulphuric acid ($C_8H_8O_3S$) for the membrane. Depending on the type of redox flow battery, the electrolyte contains different substances. In the electrolyte of the **vanadium/vanadium system** the oxidation states of vanadium V^{2+}, V^{3+}, V^{4+} and V^{5+} are dissolved in sulphuric acid (H_2SO_4). Some systems contain ruthenium and niobium to speed up the reaction. For stabilization of the electrolyte, phosphoric acid (H_3PO_4) is added, and sometimes also small amounts of ammonium phosphate. The electrolyte **of vanadium/bromide systems** contains vanadium (III) bromide (VBr_3) or vanadium (II) chloride (VCl_3) dissolved in hydrochloric acid (HCl). For stabilization of the electrolyte hydrogen bromide (HBr), sodium bromide (NaBr), potassium bromide (KBr) or mixtures of these substances are used. The electrolyte of **polysulphide/bromide systems** consists of sodium sulphide (Na_2S), sodium poly sulphides (Na_2S_x), sulphur (S) and sodium hydroxide (NaOH). For **iron/chromium** RFB iron (II) chloride ($FeCl_2$) and chromium (III) chloride ($CrCl_3$), dissolved in HCl are required for the electrolyte. Table 10.4 summarizes the substances required for battery systems.

Specific resource requirements of potentially critical substances are summarized in Table 10.5, alongside global production capacities and known reserves. For example, literature data regarding the lithium content of **lithium ion batteries** ranges between 50 and 300 g of lithium per kilowatt hour. The share of cobalt amounts to around 60% in $LiCoO_2$ cathodes, making up around 22 wt% of the battery system. Due to growing markets for lithium ion batteries with expected growth rates of around 5–7% until 2030, the total demand for the materials required will increase correspondingly. While the increasing demand for lithium is regarded rather unproblematic, cobalt markets are seen to be more critical. Angerer *et al.* (2009) estimate that even if cobalt substitutes (such as phosphates) come to play an increasingly important role in lithium battery production, the cobalt market will still be affected considerably by growing battery markets. On the other hand, as recycling of used batteries is extended, fostered by national and European legislation (e.g. European Battery Directive 2006/6/EC,

Table 10.4 Substances required for batteries

Storage System	Component	Substances
LA	Electrodes	Elementary lead (Pb) Lead (II) oxide (PbO_2) Lead (II) sulphate ($PbSO_4$)
	Electrolyte	Diluted sulphuric acid (H_2SO_4)
Li	Electrodes	Graphite (alternatively, e.g., amorphous carbon, lithium metals, lithium alloys, metal oxides, e.g. SnO, SiO_2) $LiCoO_2$ (alternatively, e.g., LiNiCo, $LiMn_2O_4$) Copper, aluminium
	Separator	Micro porous polyolefin films
	Electrolyte	Organic solvents, e.g. propylene carbonate, ethylene carbonate, dimethyl carbonate, diethyl carbonate Conducting salt, e.g. $LiPF_6$,$LiClO_4$, $LiBF_4$
NaS	Electrodes	Sodium (Na) Sulphur (S) Sodium sulphur compounds (Na_2S_5, Na_2S_4, Na_2S_3)
	Electrolyte	Beta-alumina
	Housing	Insulation materials
RFB (general)	Electrodes	Graphite Carbon
	Membrane	Nafion (sulphated poly tetra fluorine ethylene PTFE) (alternatively: polystyrene sulphuric acid $C_8H_8O_3S$)
RFB (vanadium/ vanadium)	Electrolyte	Vanadium (V^{2+}, V^{3+}, V^{4+}, V^{5+}) Phosphoric acid (H_3PO_4) Sulphuric acid (H_2SO_4)
RFB (vanadium/ bromide)	Electrolyte	Vanadium (V^{3+}, V^{4+}) vanadium (III) bromide (VBr_3) (alternatively: vanadium (II) chloride VCl_3) Hydrogen bromide (HBr) (alternatively: sodium bromide NaBr, potassium bromide KBr, mixtures) Hydrochloric acid (HCl)
RFB (zinc/ bromide)	Electrolyte	Zinc (II) bromide ($ZnBr_2$) Zinc (II) chloride ($ZnCl_2$) Potassium chloride (KCl) Potassium bromide (KBr)
RFB (polysulfide/ bromide)	Electrolyte	Sodium sulphide (Na_2S) Sulphur (S) Sodium hydroxide (NaOH) Sodium bromide (NaBr)
RFB (iron/ chromium)	Electrolyte	Iron (II) chloride ($FeCL_2$) Chromium (III) chloride ($CrCl_3$) Hydrochloric acid (HCl)

Source: Fraunhofer ISI.

Table 10.5 Specific resource requirements

Storage System	Substance	Specific resource requirements	Global production 2006	Global production 2007	Reserves[1]	Resources
		g/kWh	10^3 t	10^3 t	10^6 t	10^6 t
Li	Li	50–300	23.5	25	4.1	> 13.76
	Co	[2]				
RFB (V/V)	V	2920	55.7	58.6	13	> 68
RFB (V/Br)	V	2720	55.7	58.6	13	> 68
RFB (Zn/Br)	Zn	1360	10 000	10 500	180	1900
RFB (Fe/Cr)	Fe	3470	865 000	940 000	73 000	n.a.
	Cr	3230	17.5	18	> 500	> 10 000

[1] Share of resources which can currently be produced economically.
[2] 60% of cathode; 22 weight% of battery system.
Sources: Angerer *et al.*, 2009; USGS, 2008; Fraunhofer ISI.

Waste Electrical and Electronic Equipment (WEEE) Directive 2002/96/EC), the demand for primary resources could be reduced. While in 2003 around 8% of cobalt used in rechargeable batteries was recycled, in 2012 recycling rates of 12.5%, and in 2016 22.5%, are expected for lithium ion batteries (Angerer *et al.*, 2009).

10.3.2 Health and safety issues

Due to the production, operation and disposal of electricity storage systems different health and safety hazards can arise. A comparative assessment of health and safety issues related to different electricity storage technologies based on expert judgements is provided by Jossen and Protogeropolousos (2004) and Sauer (2002). Expert judgements were made considering different categories (normal and abnormal system operation as well as normal and abnormal system production) and aggregated afterwards. A summary of results is shown in Fig. 10.4. As can be seen, health and safety risks are rated similarly for LA, NiCd, NiZn and metal–air batteries as well as for compressed air storage in the medium to lower risk range. For SuperCaps as well as for RFB lower risks are assumed, while for lithium ion systems health and safety risks are rated rather higher.

The health and safety risks associated with the storage systems are in many cases related to the materials used. For example, lead contained in **LA**

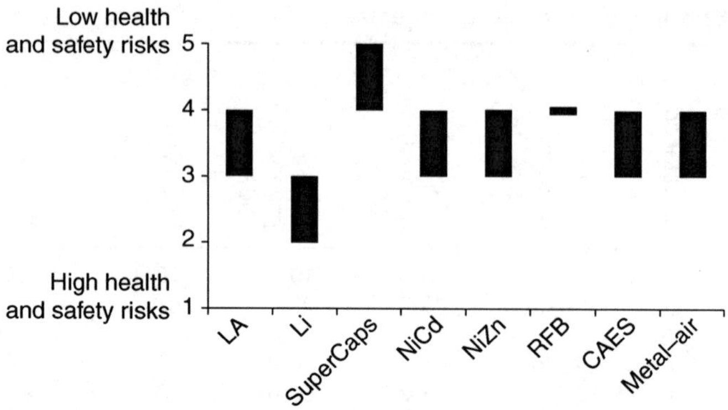

10.4 Expert judgements regarding health and safety risks of electricity storage technologies. (*Source*: based on Jossen and Protogeropolousos, 2004; Sauer, 2002.)

batteries is a heavy metal causing health and environmental hazards (Daniel and Pappis, 2008; Lindhqvist, 2010). Furthermore, in the case of accidents, leaking of sulphuric acid is probable, which can cause heavy chemical burns and is rated hazardous to water. Therefore, precautionary measures have to be taken for handling sulphuric acid. Further safety requirements include measures to avoid explosions, protection against high voltages, protection against possible short circuits and transport regulations. To avoid explosion risks due to higher concentrations of H_2 (explosive limit at concentrations of > 4%) sufficient aeration is required. Today, LA batteries are usually supplied with valve regulation (VRLA: valve regulated LA). Safety instructions for LA batteries are described in the European standard EN 50272. However, conventional LA **batteries** are generally adjudicated high safety, even in the event of misuse (Oertel, 2008).

Lithium ion batteries are generally less robust compared to other battery types, due to high energy densities causing higher accident hazards. The battery consists of easily flammable materials which react violently with water in case of failure or damage. $LiCoO_2$, at overcharging, reacts to cobalt oxide, which is very reactive, resulting in temperatures of up to 500°C with severe reaction of the cell components ('thermal runaway'). Potential consequences of accidents increase with increasing battery size and energy content. In the case of serial connection of several cells, the risks increase correspondingly. To keep voltage, current and temperature within secure margins, it is necessary to include a protection circuit. However, safe and failure-free operation can usually be provided for with adequate and professionally installed safety devices within the cell or the battery stack (Oertel, 2008; Wietschel *et al.*, 2010).

Concerning the substances contained in **NaS batteries,** it has to be remembered that Na_2S_x is water soluble and flammable, and hazardous to health and to the environment, and therefore it needs to be labelled as toxic, and it is also rated hazardous to water. Furthermore, sulphur is flammable and, if it is liquid and in contact with air, sulphur dioxide can develop, which is very toxic (GESTIS, 2008; RÖMPP, 2008).

Graphite and carbon used for the electrodes of RFB do not exhibit any serious hazards. Nafion used for the membrane can develop vapours, if it is heated too much, which can cause fever but no permanent harm. For disposal, Nafion is usually buried at landfill sites, as it does not release toxic substances nor does it contain any useful substances for recovery. Combustion should be avoided, because it results in release of toxic substances such as sulphur dioxide and hydrogen fluoride. Polystyrene sulphuric acid, which can be used as an alternative material for the membrane, can irritate skin and respiratory systems and cause chemical burns. Therefore, adequate labelling is mandatory (GESTIS, 2008; RÖMPP, 2008).

In case of leakage of the electrolyte of **vanadium/vanadium RFBs,** V^{2+} and V^{3+} would transform to V^{4+} and V^{5+}, which are stable under ambient conditions. V^{4+} and V^{5+} are ligated quickly in soil and do develop complexes which are poorly water soluble. The electrolyte is produced based on V_2O_5, which is hazardous to health and to the environment, a mutagen, and has to be labelled as toxic. Furthermore, ignition hazards exist for V_2O_5 together with calcium, sulphur, water, lithium and chlorine trifluorine. The electrolyte solution is usually based on H_2SO_4, which causes heavy chemical burns and is rated low hazard to waters. At end of life, the vanadium can be removed from the electrolyte and recycled, while H_2SO_4 neutralized and disposed of. Ruthenium and niobium, which are sometimes contained to speed up the reaction, are both easily flammable as a powder. In contact with air, explosion risks exist for ruthenium. The stabilizing agent H_3PO_4, together with nitro-methane, results in explosion risks. It is rated caustic, hazardous to health and low hazardous to water (GESTIS, 2008; RÖMPP, 2008).

The electrolyte **of vanadium/bromide systems** is produced by dissolving vanadium oxide (V_2O_5) in hydrogen bromide with the addition of HCl. VBr_3 together with water reacts to hydrogen bromide, which can result in strongly exothermic reactions, heating up and development of dangerous gases and vapours. Toxic decomposition products can develop. Furthermore, VBr_3 has to be labelled caustic and severely hazardous to water. VCl_3, which can be used as an alternative, shows similar properties and is even rated carcinogenic. For stabilization of the electrolyte HBr, NaBr or KBr is used. HBr, as well as KBr, are rated low hazardous to water. While HBr is also hazardous to health and needs to be labelled, neither NaBr nor KBr requires any labelling. NaBr develops flammable gases together with alkaline metals and bromide trifluoride. For KBr, strongly exothermic reactions can result, together with heavy metals and salts, oxidizing agents and mercuric salts. Intake of very high doses can result in chronic bromine poisoning (GESTIS, 2008; RÖMPP, 2008).

Sodium sulphide (Na_2S) contained in **polysulphide/bromide systems** oxidizes easily in contact with air, which can result in self-ignition. Dangerous gases and vapours can develop in contact with acids (e.g. hydrogen sulphide). Furthermore, Na_2S is rated hazardous to the environment and acute and chronic health hazards exist. The properties of Na_2S_x and sulphur have already been discussed for NaS batteries. NaOH is also part of the electrolyte and can cause heavy chemical burns. Together with water or flammable substances, ignition hazards exist and flammable vapours or gases can result (GESTIS, 2008; RÖMPP, 2008).

$FeCl_2$ as well as $CRCl_3$, both part of the electrolyte of **iron/chromium RFBs**, can cause acute and chronic health hazards. Together with alkaline metals and ethylene oxide, $FeCl_2$ can heat up and strong exothermic reactions can arise. Also, $CrCl_3$ together with fluorine and lithium results in explosion risks. $CrCl_3$ and $FeCl_2$ are dissolved in HCl, which can cause chemical burns, and in contact with air reacts in building caustic acid fumes. As a catalyst, thallium can also be found in the electrolyte, which is rated highly toxic as well as dangerous to health and the environment. At charging, thallium can precipitate from thallium chloride dissolution and condense at the anode. Thallium chloride is also very toxic, as well as dangerous to health and the environment. Furthermore, it is rated highly hazardous to water and reacts strongly exothermically with fluorine and potassium (GESTIS, 2008; RÖMPP, 2008).

10.3.3 Environmental impact assessment

To date, only a few environmental impact assessment studies of electricity storage technologies exist, considering the whole life cycle of the product including production, use and end of life. For example, Rydh (1999) compares conventional LA batteries and vanadium/vanadium RFBs based on life-cycle assessment. Overall the results show that the environmental impact of the RFB is lower than that of the LA battery. According to Rydh (1999), this is due to lower primary energy demand during operation, higher cycle life, and better recycling possibilities for the vanadium RFB. Another comparative environmental assessment of electricity storage can be found in Denholm and Kulcinski (2004), where a vanadium/vanadium RFB, a polysulfide/bromide RFB, pumped hydro-energy storage and CAES are compared. The environmental impact categories considered include cumulative energy demand and greenhouse gas (GHG) emissions resulting from building up and operating the electricity storage plants. The emissions

Table 10.6 Cumulative energy demand and greenhouse gas emissions of electricity storage (example)

	Round-trip efficiency	Cumulative energy demand		Greenhouse gas emissions	
		Manufacture (GJ/MWh storage capacity)	Use (GJ/GWh)	Manufacture (tons of CO_2-equ./ MWh storage capacity)	Use (tons of CO_2-equ./ GWh)
PHES	78%	373	25.8	35.7	1.8
CAES	71%	266	5210	19	288
Polysulfide RFB	70%	1755	54	125	4
Vanadium RFB	70%	2253	45	161	3.3

Source: Denholm and Kulcinski, 2004.

taken into account include indirect emissions based on the power input as well as emissions resulting from production and operation. The results of the evaluation are summarized in Table 10.6. As can be seen, PHES has the lowest cumulative energy demand during production as well as operation. The cumulative energy demand to produce RFBs is significantly higher than for PHES and CAES. This can be traced back to the energy demand for extraction of the required resources. The vanadium RFB shows the highest cumulative energy demand for production. On the other hand, during operation, CAES exhibits considerably higher cumulative energy demand than the other three alternatives shown, in particular caused by the required co-firing of natural gas in non-adiabatic conventional CAES plants.

The lowest GHG emissions during production are caused by CAES, while for operation GHG emissions are highest. Furthermore, as an additional environmental impact of **CAES,** emissions of compression waste heat to the atmosphere should also be considered (Wietschel *et al.*, 2010). GHG emissions caused by the operation of RFBs and PHES are considerably lower than CAES, because no natural gas firing is needed. GHG emissions caused during production of RFB systems are again higher than for PHES and CAES. Even though the comparison shown suggests low environmental impacts for PHES in terms of GHG emissions and energy demand, it should also be considered that the installation of **PHES** can have severe impacts on adjacent ecosystems, and issues of landscape protection should not be neglected (Oertel, 2008; Wietschel *et al.*, 2010). However, this kind of impacts is not easily measurable.

As an example, further possible environmental impacts of an adiabatic CAES system, a hydrogen system and a vanadium redox flow battery are compared based on own research and calculations. Production and operation phases are taken into account. The environmental impacts analysed include CO_2 emissions, cumulative energy demand, land usage and noise emissions, which are regarded as some of the most important environmental impact categories for electricity storage (cf. Oberschmidt and Klobasa, 2008). For better comparability, the same output power (2 MW), storage capacity and full load hours (1400 h per year) are assumed for all of the technologies considered. However, the storage systems vary widely in terms of their round-trip efficiency (assumptions: 70% for RFB, 55% for CAES, 35% for H_2 system). For the production phase, the material input required is the basis for the evaluation of energy demand and CO_2 emissions, while for the operation phase, the loss of electricity due to storage is valuated. As can be seen in Figs 10.5 and 10.6, the main impact occurs during the operation phase due to additional energy demand/CO_2 emissions associated with the electricity losses caused by storage. Thus, the impacts are closely interrelated with the round-trip efficiency of the energy storage and also with the underlying electricity mix assumed (current German electricity mix for this example). Therefore, in this comparison RFB performs best, followed by CAES and the H_2 system, which has the lowest efficiency. However, environmental impact assessment results could vary significantly due to the specific assumptions for technical characteristics, operation strategies and underlying electricity mix.

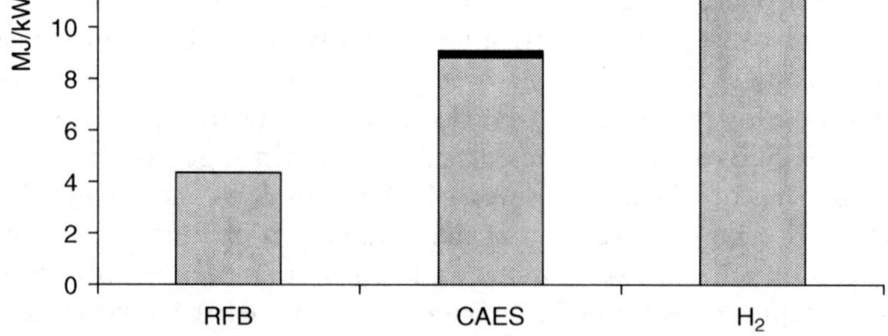

10.5 Cumulative energy demand of electricity storage (example). (*Source*: Fraunhofer ISI.)

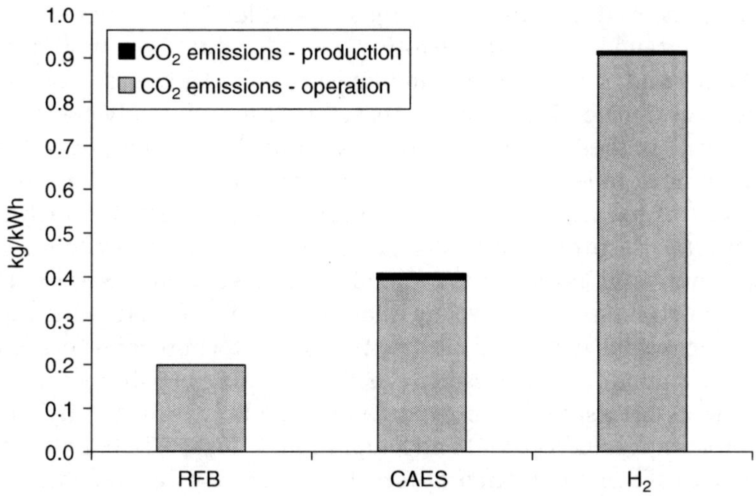

10.6 CO_2 emissions due to electricity storage (example). (*Source*: Fraunhofer ISI.)

Table 10.7 Noise emissions and land usage of electricity storage (example)

	RFB	CAES	H_2
Noise emissions (dBA)	50	75	78
Land usage (m^2*a/kWh_{out})	214	3.57	159

Source: Fraunhofer ISI.

The comparison of land usage and noise emissions is shown in Table 10.7. Regarding noise emissions, the RFB battery performs better than the CAES and H_2 system. Specific land usage is considerably higher for the RFB and the H_2 system compared to the CAES plant. The highest specific land usage is due to the RFB battery.

Further environmental issues can arise at the **end of life** of electricity storage technologies, particularly if toxic materials are used. For example, heavy metals used in **batteries** can cause considerable problems in waste management (Lindhqvist, 2010). Therefore, within the European Union recycling of batteries with certain amounts of mercury, lead or cadmium is mandatory. Disposal together with municipal solid waste is prohibited. Manufacturers are obligated to label their products (European Battery Directive 2006/6/EC). For instance, **LA batteries** cause problems in waste management due to the toxic properties of lead. These have been known for thousands of years, resulting in strict environmental regulations (Daniel and Pappis, 2008; Lindhqvist, 2010). However, because in LA batteries the metallic components are almost exclusively made from lead, nearly all of the battery is completely recyclable. In any case, LA batteries do not need to be disposed of together with municipal solid waste, but be treated as hazardous waste, even though (Oertel, 2008). Further critical properties of problematic substances used in battery systems, including environmental hazards, have already been discussed in Section 10.3.2.

Opportunities for recycling play an important role with regard the environmental impact of electricity storage technologies at their end of life. A comparative assessment of the recycling capability for electricity storage systems based on expert judgements is provided by Jossen and Protogeropolousos (2004) and Sauer (2002) (Fig. 10.7). For all of the technologies shown, some kind of recycling technology is available at present. For LA and NiCd batteries, as well as SuperCaps and CAES, recycling technologies are commercially available with possible recycling rates of more than 80%. For Li, NiZn and RFB batteries, recycling seems more problematic. However, it is expected that recycling possibilities for Li

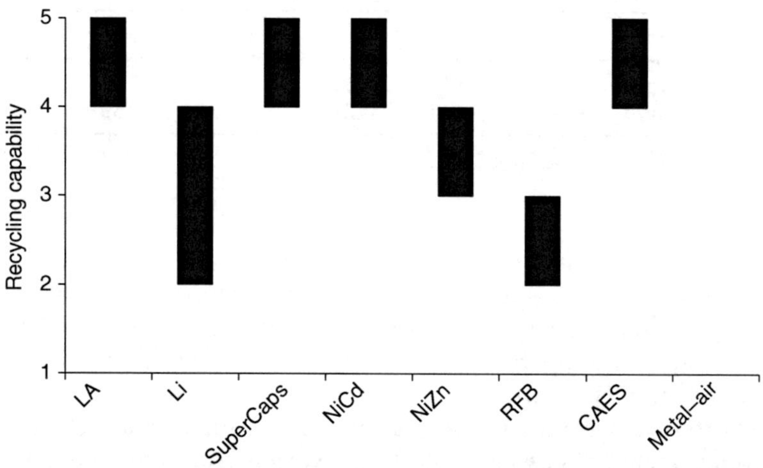

10.7 Expert judgements regarding the recycling capability of electricity storage systems. *Note*: 5 = recycling technology commercially available, with > 80% material recycling; 4 = recycling technology commercially available, with < 80% material recycling or technology available on pilot scale; 3 = technology demonstrated at lab scale; 2 = only ideas for recycling technology, not tested yet; 1 = no technology available at present. (*Source*: based on Jossen and Protogeropolousos, 2004; Sauer, 2002.)

batteries will be improved in the future, shown by the upper range of the expert judgements.

10.4 Challenges and future trends

If a strong diffusion of storage technologies should happen, further cost reduction will be necessary in the future. Until now only hydro pump storage power plants can be operated profitable under today's market conditions. Some technologies are close to be profitable like CAES. For existing battery technologies further improvements are necessary, but the expectations for redox flow, lithium and natrium sulphur batteries are promising.

Two main drivers for higher price spreads can be seen already today. A higher share of renewables will bring time periods with very low market prices. In periods with low generation from renewables market prices are expected to rise due to higher fuel and CO_2 prices. These periods might be limited, forcing storage systems to be profitable and with short utilization times. The second driver for higher price spreads is the increased deviation from day-ahead generation planning. In this case, storage systems have to compete with shorter planning horizons for power plant operation and with other flexibility options such as demand response technologies. Especially in reserve markets, the market volume will be limited and the implementation of new flexibility options will also have a negative impact on the average price level. In the past, both price trends could be observed, e.g. a strong increase of negative reserve capacity prices, but also a strong decrease for positive reserve capacity prices.

Regarding resources, much of current research focuses on the improvement of materials and new material development to be used in electrochemical storage systems. For example, the use of nano-materials could help to improve the storage capacity (Oertel, 2008; Stieler, 2007). For lithium ion batteries, currently various materials are being tested or are under development. Iron phosphate cathodes ($LiFePO_4$) are being tested, because they are advantageous in economic and environmental terms. Furthermore, they allow high intensity of current. Other materials, such as nickel oxides and mixtures of different metal oxides, are also currently being tested (e.g. $LiNiO_2$, $Li(N_xCo_yMn_z)O_2$). Further strategies to improve the properties of electrode materials include blending of $LiCoO_2$ with good electronic and lithium ion conductors, such as blends of $LiCoO_2$ and $LiRuO_3$ (Angerer *et al.*, 2009; Stieler, 2007). Further developments include **lithium polymer and lithium titanate** batteries. Significant improvements are expected for the electrolyte in the future; for example, newly developed separators made of ceramics, which are more robust in case of thermal or mechanical stress and safer in mechanical handling. Improvements are also expected with additives for existing electrolytes, as well as for newly developed electrolytes which are not flammable (Oertel, 2008).

Regarding resources for CAES, in particular, materials for heat storage in non-adiabatic systems still need to be further developed. Further future trends concern the underground systems discussed for pumped hydro storage plants. Another future option are coastal sites for pumped hydro storage. However, saline water imposes additional requirements on the components and materials used. One salt water plant is currently installed in Japan to test technical and environmental aspects (Oertel, 2008).

10.5 Conclusion

This chapter analyses the economic and environmental impacts of electricity storage systems against the background of the ongoing integration of intermittent renewable energies. The focus is on stationary electricity storage, in particular pumped hydro storage systems, (adiabatic) CAES, hydrogen storage systems, LA batteries, lithium batteries, NaS batteries, as well as RFBs. Different application strategies in energy markets are described. Possible economic benefits based on various application strategies are presented based on modelling results and compared to costs. Furthermore, critical resource requirements as well as health and safety issues and environmental impacts of electricity storage are analysed. The results of the economic analysis show that costs in general need to be reduced significantly for electricity storage systems in order to offer competitive solutions. Regarding environmental impacts, each technology has specific advantages and drawbacks, while for a comprehensive analysis the whole life cycle as well as the specific framework conditions must always be considered.

10.6 References

Angerer G, Erdmann L, Marscheider-Weidemann F, Scharp M, Lüllmann A, Handke V and Marwede M (2009), *Rohstoffe für Zukunftstechnologien*, Stuttgart, Fraunhofer IRB.

Baker J (2008), 'New technology and possible advances in energy storage', *Energy Policy*, **36** (12), 4368–4373.

Bouffard F and Kirschen D S (2008), 'Centralised and distributed electricity systems', *Energy Policy*, **36** (12), 4504–4508.

Boulanger P and Perrin M (2003), Storage Technology Report Electrolyser, Hydrogen storage and fuel cell, INVESTIRE-Network.

Bullough C, Gatzen C, Jakiel C, Koller M, Nowi A and Zunft S (2004), 'Advanced adiabatic compressed air energy storage for the integration of wind energy', *Proceedings of the European Wind Energy Conference, EWEC 2004*, 22–25 November 2004, London.

Bünger U, Crotogino F, Donadai S, Gatzen C, Glaunsinger W, Kleinmaier M, Könemund M, Landinger H, Lebioda T J, Leonhard W, Sauer D, Weber H, Wenzel A, Wolf E, Woyke W and Zunft S (2009), *Energiespeicher in Stromversorgungssystemen mit hohem Anteil erneuerbarer Energieträger. Be-deutung, Stand der Technik und Handlungsbedarf*, Frankfurt, Energietechnische Gesell-schaft im VDE (ETG).

Daniel S E and Pappis C P (2008), 'Application of LCIA and comparison of different EOL scenarios: The case of used lead-acid batteries', *Resources, Conservation and Recycling*, **52**, 883–895.

Deane J, O Gallachoir B, McKeogh E (2010), 'Techno-economic review of existing and new pumped hydro energy storage plant', *Renewable and Sustainable Energy Reviews*, **14**, 1293–1302.

Denholm P and Kulcinski G L (2004), 'Life cycle energy requirements and greenhouse gas emissions from large scale energy storage systems', *Energy Conversion and Management*, **45**, 2153–2172.

Electric Power Research Institute EPRI (2010), Electricity Energy Storage Technology Options: A White Paper Primer on Applications, Costs, and Benefits, 1020676. Palo Alto, CA: EPRI, December 2010.

Frondel M, Grösche P, Huchtemann D, Oberheitmann A, Peters J, Angerer G, Sartorius C, Buchholz P and Wagner M (2007), *Trends der Angebots- und Nachfragesituation bei mineralischen Rohstoffen*, Bundesministeriums für Wirtschaft und Technologie (BMWi).

Gatzen C (2008), The economics of power storage: theory and empirical analysis for Central Europe, München, Oldenburg Industrieverlag.

GESTIS (2008), GESTIS-Stoffdatenbank, Institut für Arbeitsschutz der Deutschen Gesetzlichen Unfallversicherung, BGIA. Available from http://www.dguv.de/bgia/de/gestis/stoffdb/index.jsp (Accessed August 2008).

Hartel R, Keles D, Genoese M, Möst D and Fichtner W (2010), 'Optimierter Einsatz von adiabaten und diabaten Druckluftspeichern', *11th Symposium Energy Innovation (EnInnov2010)*, 10–12 February 2010, Graz, Austria.

Honsel G (2007), 'Wind auf Vorrat', *Technology Review*, August 2007, 70–71.

Jakiel C, Zunft S and Nowi A (2007), 'Adiabatic compressed air energy storage plants for efficient peak load power supply from wind energy. The European project AA-CAES', *International Journal of Energy Technology and Policy*, **5** (3), 296–306.

Jossen A and Protogeropolousos C (2004), Existing data. Synthesis of the performance of storage technologies for given use, INVESTIRE-Network.

Lindhqvist T (2010), 'Policies for waste batteries – learning from experience', *Journal of Industrial Ecology*, **14** (4), 537–540.

Mulder G, De Ridder F and Six D (2010), 'Electricity storage for grid-connected household dwellings with PV panels', *Solar Energy*, **84**, 1284–1293.

Oberschmidt J and Klobasa M (2008), 'Economical and technical evaluation of energy storage systems', *Third International Renewable Energy Storage Conference (IRES 2008)*, Berlin: 24–25 November 2008.

Oeding D and Oswald B R (2004), *Elektrische Kraftwerke und Netze*, Berlin, Springer.

Oertel D (2008), *Energiespeicher – Stand und Perspektiven*, Büro für Technikfolgen-Abschätzung beim Deutschen Bundestag (TAB).

Pohl C and Kriebs K (2006), *Wirtschaftliche Einsatzmöglichkeiten der NaS-Batterie*, Bingen, Institut für Innovation, Transfer und Beratung GmbH.

RÖMPP (2008), Thieme RÖMPP Online, Verlag Thieme Chemistry. Available from http://www.roempp.com/prod/ (Accessed August 2008).

Rydh C J (1999), 'Environmental assessment of vanadium redox and lead-acid batteries for stationary energy storage', *Journal of Power Sources*, **80**, 21–29.

Sauer D U (2002), *Technical criteria and specifications*, INVESTIRE-Network.

Sensfuß F (2008), Assessment of the impact of renewable electricity generation on the German electricity sector. An agent based simulation approach, Düsseldorf, VDI-Verlag.

Simon R, Pohl C and Kriebs K (2006), Einsatzmöglichkeit einer NaS-Batterie für die Regenerativstromversorgung am Beispiel der Gemeinde Bruchmühlbach, Bingen, Institut für Innovation, Transfer und Beratung GmbH.

Spahic E, Balzer G, Münch W and Hellmich B (2006), 'Speichermöglichkeiten der Windenergie', *Energiewirtschatfliche Tagesfragen*, **105** (25), 46–50.

Steffen B (2012), 'Prospects for pumped-hydro storage in Germany', *Energy Policy*, **45**, June 2012, 420–429.

Stieler W (2007), 'Optimierte Vielfalt', *Technology Review*, August 2007, 64–69.

Tassin N (2003), *Redox Systems Report*, INVESTIRE-Network.

Tester J W, Drake E M, Driscoll M J, Golay M W and Peters W A (2005), *Sustainable Energy: Choosing Among Options*, Cambridge, MIT Press.

USGS (2008), *Mineral Commodity Summaries*, U.S. Geological Survey.

Wietschel M, Arens M, Dötsch C, Herkel S, Krewitt W, Markewitz P, Möst D and Scheufen M (2010), *Energietechnologien 2050 – Schwerpunkte für Forschung und Entwicklung. Technologiebericht*, Stuttgart: Fraunhofer IRB.

Zunft S, Jakiel C, Koller M and Bullough C (2006), 'Adiabatic compressed air energy storage for the grid integration of wind power', *Sixth International Workshop on Large-Scale Integration of Wind Power and Transmission Networks for Offshore Windfarms*, Delft, 26–28 October 2006.

11

Nickel-based batteries: materials and chemistry

P-J. TSAI and S. L. I. CHAN, University of New South Wales, Australia

DOI: 10.1533/9780857097378.3.309

Abstract: This chapter provides a comprehensive review on Nickel-based batteries, where nickel hydroxide electrodes are utilised as positive plates in these batteries. An example is the popular nickel/metal hydride batteries, which are one of the most important power sources for a wide range of electronic devices. The chapter first gives a brief history of these batteries, the fundamental electrochemistry behind this system. It will detail the construction of the batteries, including the active materials and electrolyte. The chapter will compare the performances, the advantages and limitations of different types of battery in this system. New materials to be applied in this system, the current development and the future trend of these Ni-based batteries will be discussed.

Key words: Ni-based batteries, Ni-Fe, Ni-Cd, Ni-H_2, Ni-Zn, Ni-MH, secondary batteries, NiOOH, anode, cathode.

11.1 Introduction

Nickel-based batteries, including nickel-iron, nickel-cadmium, nickel-zinc, nickel-hydrogen, and nickel-metal hydride batteries, are similar in the way that nickel hydroxide electrodes are utilised as positive plates in the systems. As strong alkaline solutions are generally used as electrolyte for these systems, they are also called alkaline secondary batteries. Ni-based batteries have been, and still are the most important power sources for a wide range of electronic devices.

The present chapter offers a comprehensive review on the past and present available Ni-based battery systems, including the fundamental electrochemistry behind this system, active materials and electrolyte, the performances, and the advantages and disadvantages for each different type of battery. The current popular and novel materials that have potential in the applications of Ni-based batteries will be discussed, followed by the challenges and the future trend of the batteries.

On 11 March 1899, Jüngner in Sweden patented the very first idea of utilising an electrolyte that remains the same throughout the charging and discharging of the battery. That is, the electrolyte itself does not react chemically with the electrodes, while conducting the ions between electrodes satisfactorily. As a result, the amount of electrolyte required for the batteries can be reduced significantly, and hence a reduction in the weight of the batteries. During the period of late 1890s to early 1900s, Michalowski from Germany (Michalowski, 1899), Jüngner from Sweden (Jungner, 1901), and Edison from the United States (Edison, 1901) have successively patented Ni-Zn, Ni-Cd, and Ni-Fe battery systems, respectively. The Ni-Zn system, however, was not commercialised until the 1930s, when an Irish chemist, Drumm, used it in electric trains. In the 1960s and 1970s, Ni-Cd system was the primary power source for a majority of spacecraft and satellites, owing to its superior properties in cycle life capability and robustness. Despite the fact that Ni-H_2 batteries had been discovered and patented in the 1950s, it was not until the 1970s when Ni-H_2 batteries slowly found their way to replacing Ni-Cd systems for aerospace applications. An alternative battery system, Ni-MH, had been introduced to replace the frequently-used Ni-Cd batteries in the 1980s. The primitive metal hydride, $LaNi_5$ alloy, was discovered in Dutch Philips Research Laboratories in the late 1960s, where it was proven that this intermetallic compound was capable of absorbing a reversibly large amount of hydrogen. However, it took nearly 20 years before the commercialisation of the Ni-MH batteries. Among the five Ni-based battery systems, only sealed Ni-Cd and Ni-MH batteries are still present in the market.

Nickel electrodes on the other hand, were developed based on pocket-plate technology and initiated in Sweden, Germany, and the United States in the late 1890s (Shukla *et al.*, 2001). The structure of pocket-plate technology involved pelletising the nickel hydroxide with conductive additive and binder. The pellet was then enfolded in a perforated nickel-plated steel sheet to serve as a current collector. Pocket-plate batteries were the oldest and least expensive type, with a very reliable and long-life cell design that could tolerate severe mechanical and electrical abuse. The advancement of the tubular-plate structure in nickel electrodes took place in 1908, where the electrode's durability was improved efficaciously by restricting the mechanical forces induced from the expansion of the active material. This design typically involved the assembly of perforated nickel-plate steel tubes filled with compacted layers of the nickel hydroxide and conductive additive, into a frame comprising these tubes in parallel (Shukla *et al.*, 2001). The complicated design made the processing laborious and therefore cost-ineffective and, for these reasons, tubular-plate structure is no longer in production (Falk and Salkind, 1986).

The next revolution in nickel electrodes was the development of sintered-plate technology, which was invented by Pflider in 1928 (Berndt, 1998). As it literally means, the loose nickel particles are transformed into a coherent body at a temperature just below the melting point of nickel in a reducing atmosphere. In sintered electrodes, a porous sintered plaque is impregnated with nickel hydroxide active material by either chemical or electrochemical approaches (Shukla *et al.*, 2001). The electrochemical impregnation is considered to be more effective than the chemical approach in terms of material loading into porous sintered substrate, which would consequently lead to a better material utilisation and durability.

The fabrication of sintered plaque can either be a dry-powder or wet-slurry process. The latter is commercially favoured, as electrodes fabricated via this approach possess high porosity, large surface area, and high electrical conductivity in combination with good mechanical strength. The higher degree of utilisation in terms of faradaic efficiency of 90%, as compared to 60% for the pocket-plate electrodes, is attributed to a better material loading in the porous structure of sintered substrate (Shukla *et al.*, 2001). Up till now, more than 50% of batteries based on nickel electrodes being currently manufactured use sintered electrodes as the positive plate (Shukla *et al.*, 2001).

The gap between the superior but expensive and size limited (< 100 Ah/cell) sintered battery, and the low cost but bulky and heavy pocket-plate battery, was filled in the 1980s by the development of fibre plate batteries, and later the plastic bonded electrode batteries (Dahlen, 2003). The fibre plate Ni-Cd batteries were developed primarily for electrical vehicle applications, and are today available for general industrial applications.

Many researches have been devoted to improving the nickel electrode's performances by maximising the active material loading and energy density. Foam-nickel electrodes, which have a similar working mechanism to sintered-plate electrodes, were introduced in the mid 1980s. Pasted nickel electrodes are prepared by smearing the spherical nickel hydroxide active material densely on highly-porous nickel foam through the mechanical and physical impregnation process. These electrodes have an exceptional high energy density and low production cost compared to those of conventional sintered nickel electrodes, and are therefore widely available in Ni-Zn (Taucher-Mautner and Kordesch, 2003, 2004), and Ni-MH batteries (Chen *et al.*, 2003; Lv *et al.*, 2004).

The nickel foam substrate used in pasted electrodes is generally produced by nickel-plating a porous synthetic material (e.g. polyurethane or acrylic fibre) followed by pyrolysis of the plastic material. The as-synthesised nickel foam substrate has a three-dimensional porous texture, with a typical volumetric porosity of 97% and a pore size of 600 μm (Olurin *et al.*, 2003). The high porosity and large pore size of nickel foam are beneficial for loading the nickel hydroxide, which in turn would result in a higher packing density of the active material. It would, however, also increase the internal resistance between the substrate and nickel hydroxide particles, as well as among the nickel hydroxide particles themselves. Therefore, pasted nickel electrodes

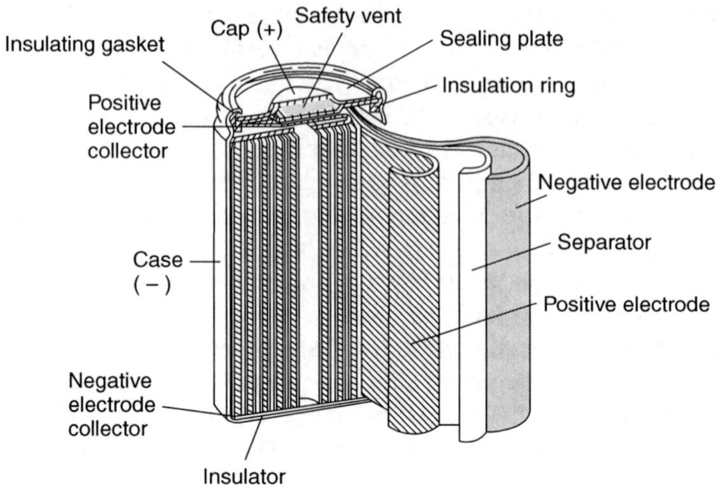

11.1 Structure of a cylindrical Ni-MH battery (Taniguchi *et al.*, 2001).

are advantageous in attaining a greater capacity and energy density; but are inferior to sintered electrodes in terms of the electrical conductivity and high rate capability. Sintered nickel electrodes generally have energy densities of 450–500 mAh/cm^3, whereas the value of 700 mAh/cm^3 is obtained for pasted electrodes.

11.1.1 Structure of the battery

Prior to the beginning of the detailed coverage of Ni-based systems, it is of great importance to understand the general structure of a battery. Though the detailed structure of the battery varies between the types of batteries, the essential parts are similar: the positive and negative electrodes separated by a separator, electrolyte and a casing. The structure of a current popular cylindrical Ni-MH battery is illustrated in Fig. 11.1. In this battery, both positive (nickel electrode) and negative electrodes are coiled and separated by the separator. The battery design should consider the optimisation of the reaction area of the electrodes, reduction of resistance for current collection, and improvement in electrolyte composition to obtain high power characteristics. The electrodes, separator, and electrolyte are contained in steel case; the sealing plate is equipped with a valve to prevent bursting when the internal pressure increases from overcharge, short circuit or reverse charge situation.

11.2 Nickel hydroxide electrode

With the ever-increasing demand for higher energy for multitudinous applications, batteries with high energy density, power capability, long cycle stability, light weight, and low cost are in desperate need. Tremendous research has therefore been devoted in the past decades in attempting to understand the mechanism, and increase the electrochemical performance, of the nickel electrodes. This section covers the fundamental understanding of the electrochemical reaction of the nickel electrode, followed by methods for improving the performance of the electrode, and concludes with a final remark on current and future perspectives.

11.2.1 Redox reactions during charge/discharge process

As nickel hydroxide is employed as the positive electrode for all five types of nickel battery systems, it is of great importance to understand its working mechanisms and related properties. The charge/discharge reactions of the nickel electrode have been expressed as follows (Watanabe and Kumagai, 1997; Jain *et al.*, 1998):

$$\mathrm{NiOOH} + \mathrm{H_2O} + e^- \underset{\text{charge}}{\overset{\text{discharge}}{\rightleftarrows}} \mathrm{Ni(OH)_2} + \mathrm{OH^-} \quad (E^0 = 0.49\ \mathrm{V}\ vs\ \mathrm{SHE}) \tag{11.1}$$

The reversible electrode potential (E_{rev}) for the nickel electrode can then be expressed by the Nernst Equation as follows:

$$E_{rev} = 0.49 - 0.059 \log\left(\frac{a[\mathrm{Ni(OH)_2}]a[\mathrm{OH^-}]}{a[\mathrm{NiOOH}]a[\mathrm{H_2O}]}\right) \tag{11.2}$$

As the charge/discharge reaction expressed in Equation [11.1] involves an equivalent diffusion of hydrogen ions (protons) through the solid-state lattices of $Ni(OH)_2$ and NiOOH, it would inevitably induce a continuous change in the composition of the active material between the fully discharged nickel hydroxide and fully charged oxyhydroxide. Equation [11.1] may therefore be written as:

$$\mathrm{NiOOH} + \mathrm{H_2O} + e^- \underset{\text{charge}}{\overset{\text{discharge}}{\rightleftarrows}} \mathrm{Ni(OH)_2} \tag{11.3}$$

11.2.2 Phase transformation of nickel hydroxides

Nickel hydroxide $Ni(OH)_2$, the cathodic active material in nickel electrodes, has been studied by many workers. There are several reviews of work in the field (Oliva *et al.*, 1982; Halpert; McBreen, 1990), and an updated review by Song and Chan (2009). Before 1966, the progress of understanding the reaction in nickel electrodes has been relatively slow, owing mainly to the complex nature of the reactions involved in the Ni-based batteries. Much effort has been devoted to understanding the items that are trivial in most of the other batteries, such as overall reaction, determination of the open-circuit potential, and the oxidation state of the charge material. The major advance was made by Bode *et al.* in 1966, where the detailed overall reaction of the nickel electrode was proposed for the first time. It was pointed out that both the charged (NiOOH) and discharged ($Ni(OH)_2$) materials could exist in two forms. One form of the discharged material, namely β-$Ni(OH)_2$, is anhydrous and has a layered brucite ($Mg(OH)_2$) structure; whereas the other form, designated as α-$Ni(OH)_2$, is hydrated and has intercalated water between brucite like layers.

Oxidation of β-$Ni(OH)_2$ during charging produces β-NiOOH, and oxidation of α-$Ni(OH)_2$ produces γ-NiOOH. The formation of β-$Ni(OH)_2$ and α-$Ni(OH)_2$ takes place when β-NiOOH and γ-NiOOH are discharged, respectively. Furthermore, α-$Ni(OH)_2$ can dehydrate and recrystallise in the concentrated alkaline electrolyte to form β-$Ni(OH)_2$ when aged; and that β-NiOOH could be converted to γ-NiOOH when the electrode is overcharged. The overall reaction scheme can be clearly illustrated by the Bode diagram, as shown in Fig. 11.2 (Wehrens-Dijksma and Notten, 2006). The chemical structure, degree of hydration, and morphology of these various forms of nickel hydroxides are distinct from each other. The two reaction schemes are often referred to as the β/β and the α/γ cycles.

11.2.3 β/β Redox model for nickel electrodes

Conventional nickel hydroxide electrodes are designed to operate on the β/β cycle, with the aim to accommodate the volume changes during cycling, and to ensure that adequate electronic conductivity is provided to yield high utilisation of the active material during discharging. The β/β cycle is generally favoured as the volume expansion associated during cycling is less than any other forms.

Among the different polymorphic modifications of nickel hydroxide, β-$Ni(OH)_2$ is widely adopted as the preferable active material in positive electrode in all nickel-based secondary batteries, owing to its high stability in strong alkaline electrolyte (Song *et al.*, 2002). β-$Ni(OH)_2$ shows a good

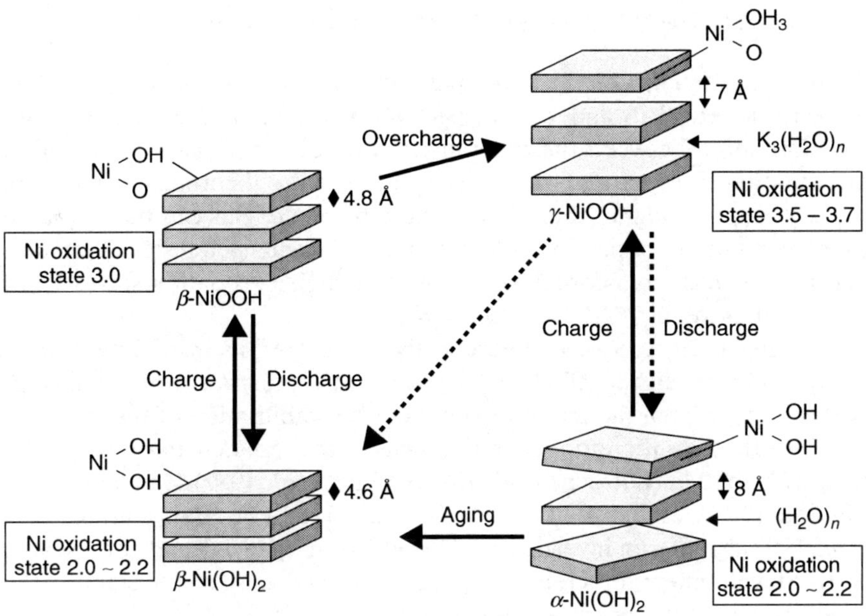

11.2 Bode diagram illustrating the phase transformation of nickel hydroxide/oxyhydroxides with various Ni oxidation states (Wehrens-Dijksma and Notten, 2006).

reversibility when charged to form β-NiOOH, which has a similar layered structure. When overcharged, however, β-NiOOH is then converted to γ-NiOOH; this phase is undesirable as it involves a large volumetric change and would result in swelling of the nickel electrode and drying of electrolyte in the separator. Consequently, the formation of γ-NiOOH considerably damages the nickel electrode and causes immature cell failure. Furthermore, it was reported that the formation of γ-NiOOH is responsible for the memory effect in alkaline batteries (Sato *et al.*, 2001).

These drawbacks provoked the study by Oshitani and his co-workers of suppressing the formation of γ-NiOOH in the nickel electrode by means of elemental additives (Oshitani *et al.*, 1986). The addition of cobalt and cadmium together was found to be most effective in suppressing the formation of γ-NiOOH. The partial substitution of zinc for nickel was found to have a similar effect by having a larger ionic radius than nickel in the β-$Ni(OH)_2$ solid solution, which would cause a distortion in the crystal lattice (Yuasa and Ikoma, 2006). In addition to the drawbacks of the β/β couple, the limited theoretical capacity of 289 mAh/g, which corresponds to the exchange of 1e$^-$ per Ni atom, has been the bottleneck to the higher performance of the nickel electrodes.

11.2.4 α/γ Redox model for nickel electrodes

As illustrated in Fig. 11.2, the lattice parameter c in α-$Ni(OH)_2$ and γ-NiOOH is nearly identical (8 and 7 Å, respectively). This allows reversible phase transformation between the two phases without experiencing mechanical deformation or constraints. It is envisaged that the theoretical capacity in the α/γ couple is higher than that of the β/β couple. This can be attributed to the fact that more than one electron may be exchanged per Ni atom during the α/γ phase transformation, owing to a higher oxidation state (3.5 or higher) of nickel in γ-NiOOH (Liu *et al.*, 2009).

Though the α/γ couple is theoretically favourable, its instability in thermodynamics in strong alkaline solution and rapid transformation to β-$Ni(OH)_2$ have been the major drawbacks. The stabilisation of the structure of α-$Ni(OH)_2$ has therefore been the focus of the research. Partial substitutions of Ni by Mn (Latroche *et al.*, 1995; Colinet *et al.*, 1987), Zn and Al (Van Mal *et al.*, 1973; Meli *et al.*, 1995; Wu *et al.*, 2003) and Fe (Balasubramaniam *et al.*, 1993) have been investigated. Despite more than a decade of research on the development for a reliable approach to synthesise and to stabilise the α-$Ni(OH)_2$, it still remains a major scientific challenge.

11.2.5 Improvement of electrochemical performances of nickel hydroxide

Despite the fact that the nickel hydroxide electrode used in Ni-based systems is fundamentally the same as that used by Edison a century ago, the electrode widely used nowadays is more complicated. Electrode characteristics, including capacity, energy density, high rate capability, cell duration and material utilisation are continuously being improved to satisfy the ever-increasing demand for energy. Two scientific challenges related to nickel hydroxide are its chemical stability and the catalytic properties towards oxygen evolution reaction (Vidotti *et al.*, 2009).

The implementation of high-density spherical nickel hydroxide in commercial nickel electrodes had been established in 1990s, where it was found that spherical nickel hydroxide powder with a particle size distribution from several microns to tens of microns has a high filling density and superior flow characteristics, optimising pasting conditions (Sakai *et al.*, 1990; Reisner *et al.*, 1997). This spherical nickel hydroxide is synthesised in a precipitation process in which metal salts, such as nickel sulphate, are reacted with sodium hydroxide in the presence of ammonia.

Though the abovementioned physical properties of spherical nickel hydroxide are beneficial to the loading of the active material and energy density, its electrochemical properties are inferior (Song *et al.*, 2005). Thus, considerable interest has been devoted to improving the electrochemical activity of nickel hydroxide. Modifications of nickel electrode can be divided into two main categories. Firstly, modifying the nickel hydroxide with additives; secondly, modifying the nickel hydroxide by means of mechanochemical processing, such as ball-milling; and finally, controlling the microstructure of nickel hydroxide by changing the preparation techniques and synthesis parameters. These three modifications are discussed in the following sections.

Additive/doping

In general, adding additives in nickel electrodes is the most popular and effective approach to improving the electrochemical properties of the nickel electrode. The purpose of incorporating additives in nickel hydroxide materials and electrodes is to (Song and Chan, 2009):

- Refine the microstructure and increase the solid-state proton diffusion ability of the active material, for instance, Co (Watanabe *et al.*, 1996), Al (Liu *et al.*, 2008), and other metals.
- Stabilise the structure of the α-type active material by partial substitutions for Ni in nickel hydroxide, for example, Al (Kamath *et al.*, 1994; Chen *et al.*, 2005, 2009), trivalent cations Co with Al and Cr (Dixit and Vishnu Kamath, 1995), Mn (Morishita *et al.*, 2008) and V (Avendano *et al.*, 2005).
- Lower the oxidation potential of nickel hydroxide and improve the redox reversibility of Ni(II)/Ni(III) – Co was proven an effective element for this purpose (Corrigan and Bendert, 1989; Unates *et al.*, 1992).
- Improve the charging efficiency and suppress the oxygen evolution side reaction by separating the redox couples OH^-/O_2 and Ni(II)/Ni(III), for examples, Ca, Mg (Zhu *et al.*, 1995), Cd, Co and Zn (Provazi *et al.*, 2001).
- Enhance the mechanical properties and increase the long-term stability by avoiding the so-called γ-NiOOH effect, for instance, Co, Fe, Al, Zn etc., (Oshitani *et al.*, 1986) – more references are given in Section 11.2.4.
- Improve the electrical conductivity and enhance the high rate capability of the nickel electrode, for example, Co (Pralong *et al.*, 2001), Zn (Yuan *et al.*, 1999), nickel, Co compound (Yunchang *et al.*, 1995) and carbonaceous materials (Lv *et al.*, 2004).
- Enhance the performance at elevated temperatures by suppressing premature oxygen evolution on charge, for example, $Y(OH)_3$ (Cheng *et al.*, 2005), $Yb(OH)_3$ (He *et al.*, 2006), $Ba(OH)_2$ and $Co(OH)_2$ (Shaoan *et al.*, 1998) and combination of Co with Ca compounds (Yuan *et al.*, 1998).

Cobalt is undoubtedly the most popular element among the various additives used in improving the performance of the nickel hydroxide electrodes. It has been added to the nickel electrode in different forms, such as metal powders, CoO, $Co(OH)_2$ and $Na_{0.60}CoO_2$. The cobalt will dissolve and reprecipitate on the surface of nickel hydroxide when added to the nickel electrode, and form highly conductive and stable cobalt oxyhydroxides (H_xCoO_2) during the activation charge. This conductive network ensures good electrical conductivity between the active material and the substrate, as well as among the active material particles. It has also been found that cobalt can efficaciously improve the protonic conductivity of $Ni(OH)_2$, increase the oxygen evolution potential, delay the mechanical failure of the electrode, lower the oxidation potential of $Ni(OH)_2$, improve the redox reversibility of Ni(II)/Ni(III), inhibit the formation of γ-NiOOH, and suppress the electrode swelling.

One concern associated with the use of Co is its high cost; the incorporation of Co would mean an increase in material costs. Therefore, efforts have been devoted to replace Co with Al (Kamath *et al.*, 1994; Hu and Noréus, 2003), Zn (Tessier *et al.*, 2000), Mn (Guerlou-Demourgues *et al.*, 1994), Ca (Yuan *et al.*, 1998), and other metals (Demourgues-Guerlou and Delmas, 1993; Zhu *et al.*, 1995; Bardé *et al.*, 2006). However, just like many other techniques, a trade-off is generally inevitably associated with the doping of additives in nickel hydroxide. For example, doping Ca or rare earth additives is accompanied by a loss of specific power and cycle life (Yuan *et al.*, 1998). Besides, some of these metals are expensive and environmentally unfriendly, thus the following two methods have also been considered.

Ball-milling

Numerous studies have shown that better electrochemical properties, such as proton diffusion rate and concentration polarisation of proton during discharge/recharge (Watanabe *et al.*, 1995), can be realised for active material with reduced crystalline size. This has brought about the physical modifications of nickel hydroxide. Mechanochemical processing of nickel hydroxide has received considerable interest for the preparation of the active material, as it is simple, highly efficient, and cost-effective.

High energy ball-milling (HEBM) is one of the most popular techniques for modifying the physical structure of materials. It has been extensively used for many applications, including hydrogen storage alloys in negative electrode of the Ni-MH batteries (Stubicar *et al.*, 2001; Abrashev *et al.*, 2010), and electrode materials for Li-ion batteries (Machida *et al.*, 2005; Zhang *et al.*, 2005b; Park *et al.*, 2006; Hassoun *et al.*, 2007). It is generally recognised that the milling process could decrease the particle and crystallite size of materials, and induce continuous formation of structural defects through the cycling deformation of large crystallites.

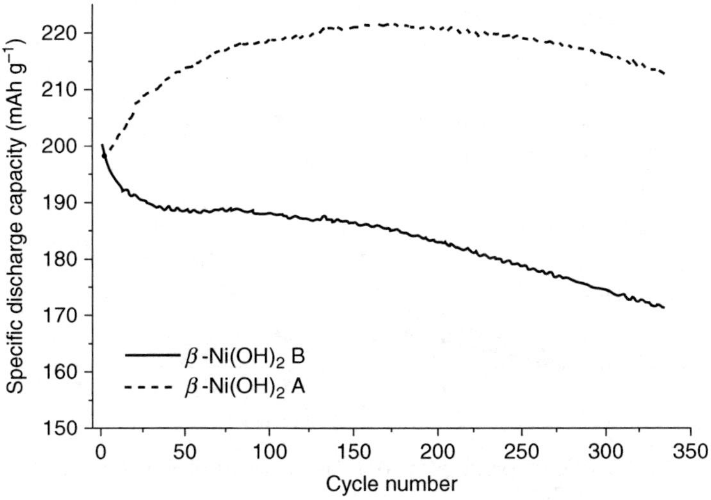

11.3 The cycle life of β-$Ni(OH)_2$ at 0.5 C rate (Chen *et al.*, 2003).

Chen *et al.* (2003) have modified the $Ni(OH)_2$ by HEBM, and it was found that the ball-milled $Ni(OH)_2$ as active material was advantageous to electrochemical properties of β-$Ni(OH)_2$, such as specific discharge capacity, discharge potential, and cycle performance. The cycle stability at 0.5 C rate of the β-$Ni(OH)_2$ before and after (abbreviated in Fig. 11.3 as β-$Ni(OH)_2$B and β-$Ni(OH)_2$ A, respectively) HEBM are shown in Fig. 11.3. The maximum discharge capacity for as-synthesised and milled samples was 200 and 225 mAh/g, respectively. After 335 repetitive charge/discharge cycles, 85.5 and 99.5% of their maximum discharge capacity was retained for as-synthesised and milled samples, respectively.

The electrochemical performance of the electrodes was further analysed by electrochemical impedance spectroscopy (EIS), and the measurements of β-$Ni(OH)_2$B and β-$Ni(OH)_2$ A at 100% state of charge (SoC) (defined as the ratio of the available capacity of an electrode to its maximum attainable capacity) after 335 cycles are illustrated in Fig. 11.4. This is a typical EIS plot where a semicircle at high frequency region and a linear line at low frequency region are observed. It is clear that the resistance associated with charge transfer and proton diffusion was much smaller in the ball-milled sample as compared to that of the as-synthesised one. Consequently, the electrochemical and diffusion polarisation of the electrode will decrease, which will lead to the increase of the discharge potential and the decrease of the charge potential.

Later, Song *et al.* employed a low cost normal ball-milling (NBM) process on spherical β-$Ni(OH)_2$ as a means to alter the microstructure of pasted nickel electrodes and to improve the distribution of the active material in

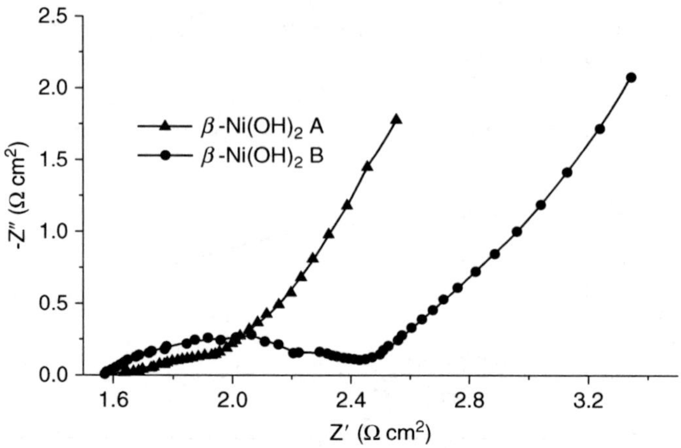

11.4 Complex plane plots of β-$Ni(OH)_2$ samples at 100% SoC (Chen *et al.*, 2003).

the porous electrode (Song *et al.*, 2006). The electrode was prepared by mixing 8 wt.% of $Ni(OH)_2$ powder that was milled for 120 h with 92 wt.% of spherical $Ni(OH)_2$ powder as the active material. The electrodes with (electrode B) and without (electrode A) the addition of milled $Ni(OH)_2$ were subjected to charge and discharge at 0.2 and 0.5 C rate, and the results are illustrated in Fig. 11.5. It shows that the top-of-charge voltage of electrode B was lower than that of electrode A, implying that the former has a better chargeability and a lower intrinsic resistance. The specific discharge capacity of electrode B was larger than that of electrode A, and the discharge plateau of the former electrode was also higher and flatter than that of the latter. The CV results, as shown in Fig. 11.6, suggest better reaction reversibility and electrochemical reactivity, as well as a higher active material utilisation in the electrode with 8 wt.% of milled $Ni(OH)_2$.

11.2.6 Synthesis of nanostructured nickel hydroxide

As discussed earlier, reducing the crystallite size of the nickel hydroxide was proven to be an efficient approach to improving the electrochemical activity and charge/discharge performance of the material. In addition, nanostructural nickel hydroxide particles have a magnificently wide surface area which is capable of absorbing a large amount of water molecules on the surface of the particles. This surface water is believed to improve the wettability of the nickel hydroxide particles and, as a result, represents an enhancement in proton diffusion within the active material during the charge/discharge process, and a better material utilisation of the electrode (Audemer *et al.*, 1997) can be achieved. In the same year, (Reisner *et al.*,

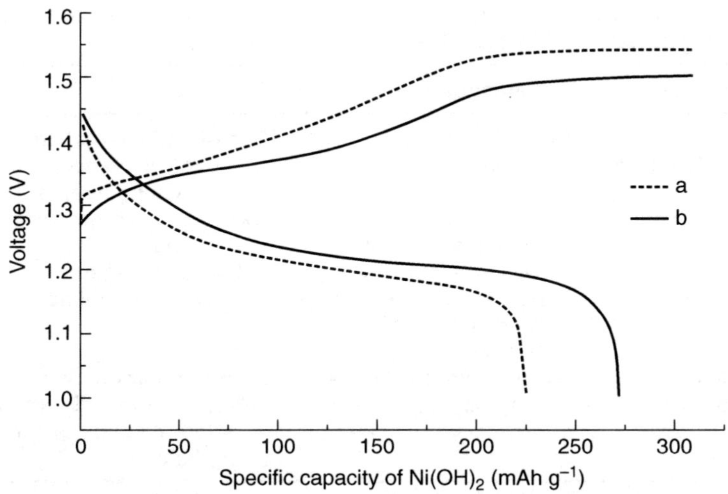

11.5 Typical charge (0.2 C rate) and discharge (0.5 C rate) curves for pasted nickel electrodes: (a) without and (b) with an addition of 8 wt.% ball-milled $Ni(OH)_2$ powder (Song *et al.*, 2006).

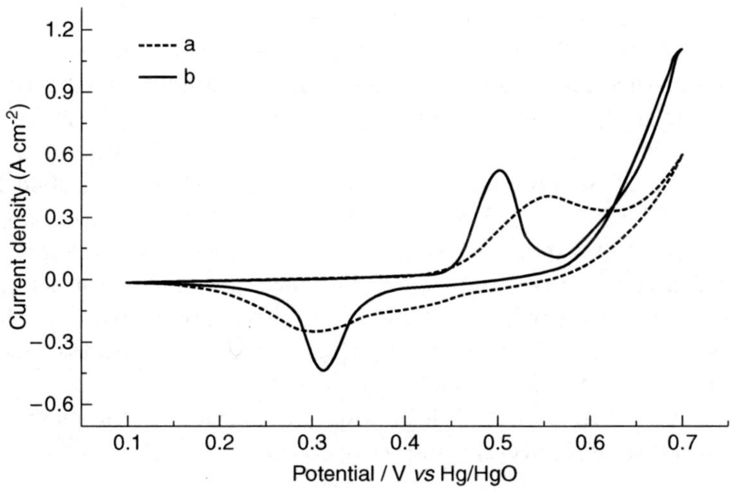

11.6 Cyclic voltammograms for pasted nickel electrodes: (a) without and (b) with an addition of 8 wt.% ball-milled $Ni(OH)_2$ powder (Song *et al.*, 2006).

1997) also reported the development of the nanostructured β-$Ni(OH)_2$ with a mixture of nanofibres and nanoparticles, which was expected to yield at least a 20% improvement in cathode energy content for rechargeable batteries. Later, it was found that the optimum particle size for β-$Ni(OH)_2$ sample to deliver the largest capacity was 25 nm. Supportive information has

Table 11.1 Pros and cons associated with several methods for the synthesis of nanostructured nickel hydroxides

Techniques	Advantages	Disadvantages
Solution precipitation	Simple and easy to control	Use of surfactant/dispersants, organic solvents or complexing agents
Hydrothermal	Simple and cost-effective, no tedious washing involved	Use of environmental unfriendly urea as hydrolytic agent
Template	Tunable geometry and morphology of nickel hydroxide	Removal of templates
Solid-state reaction	No surfactant/solvents, eco-friendly, high yields, simple reaction	High temperature is required, homogenous distribution of different solid reactants may be difficult

Source: Song and Chan, 2009.

been acquired by a more recent study (Kiani *et al.*, 2010). All these have provoked increasing interest in the development of nanostructured nickel hydroxide for its application in Ni-based batteries.

Numerous methods have therefore been developed for the synthesis of nanostructured nickel hydroxide. These methods can be categorised into four groups, namely solution precipitation (Han *et al.*, 2003, 2005; Hu *et al.*, 2006), hydrothermal (Jayalakshmi *et al.*, 2005; Orikasa *et al.*, 2007; Sakai *et al.*, 2010), template (Rahman *et al.*, 2004; Duan *et al.*, 2006; Peng and Shen, 2007; Cheng and Hwang, 2009), and solid-state reaction approaches. As the detailed experimental procedures for each of these approaches has been described elsewhere (see the references above, and also see (Song and Chan, 2009)), they will be excluded from this chapter. However, the advantages and disadvantages in relation to each of the synthesis methods are tabulated in Table 11.1 for easy comparison.

The following sections will provide an overview of the electrochemical performance of β-Ni(OH)$_2$ and α-Ni(OH)$_2$.

Electrochemical performance of nanostructured β-Ni(OH)$_2$ powder

As suggested by many reports, nanosized β-Ni(OH)$_2$ powder exhibits better redox reversibility, smaller reaction resistance, lower polarisation, and better charge/discharge properties than commercial spherical β-Ni(OH)$_2$ powder. Zhou and Zhou (2005) synthesised β-Ni(OH)$_2$ with particle size ranging from 10 to 40 nm via aqueous solution precipitation reaction, aided by ultrasonication. The charge and discharge curves at the ninth and thirteenth

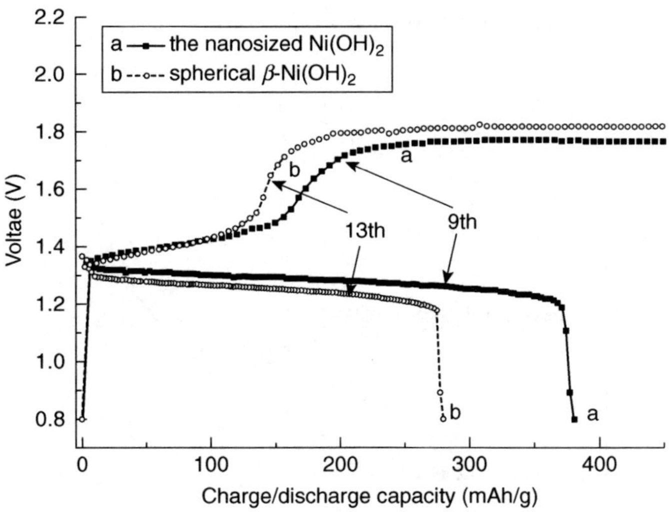

11.7 The charge/discharge curves of (a) the nanosized $Ni(OH)_2$ and (b) spherical β-$Ni(OH)_2$ electrode, at a charge/discharge rate of 0.1 C (Zhou and Zhou, 2005).

cycles for the nanosized and micron-sized spherical $Ni(OH)_2$ are displayed in Fig. 11.7. The maximum specific discharge capacity for nanosized $Ni(OH)_2$ was 381 mAh/g, which was much higher than the theoretical capacity of 289 mAh/g for β-phase $Ni(OH)_2$. The formation of a considerable amount of stable α-$Ni(OH)_2$, which has been confirmed by X-Ray diffraction (XRD) analysis, accounted for the exceptional high capacity.

In the same year, Han's group (Han *et al.*, 2005) investigated the electrochemical behaviour of nickel hydroxide electrodes with and without 8 wt.% of nanometer β-$Ni(OH)_2$. Based on the cyclic voltammograms as displayed in Fig. 11.8, it was confirmed that the electrode reaction of nickel hydroxide was controlled by proton diffusion, and the proton diffusion coefficients for nanometer β-$Ni(OH)_2$ and spherical $Ni(OH)_2$ were 1.93×10^{-11} cm²/s and 5.50×10^{-13} cm²/s, respectively. In addition to the charge/discharge characteristics of the electrodes, it was found that the cathode discharge specific capacity of the Nano-E (electrode with nanometerβ-$Ni(OH)_2$ addition) was about 10% larger than that of the Micro-E (electrode with only spherical β-$Ni(OH)_2$), as evident in Fig. 11.9. The curves further demonstrate that a higher discharge plateau, chargeability, discharge specific capacity, lower polarisation, and faster electrode reaction in Nano-E electrode were observed in comparison to those of the Micro-E. These improvements were attributed to the higher oxidation state of Ni in Nano-E, which offered more vacancies for proton diffusions (e.g. higher conversion rate between Ni^{2+} and Ni^{3+}).

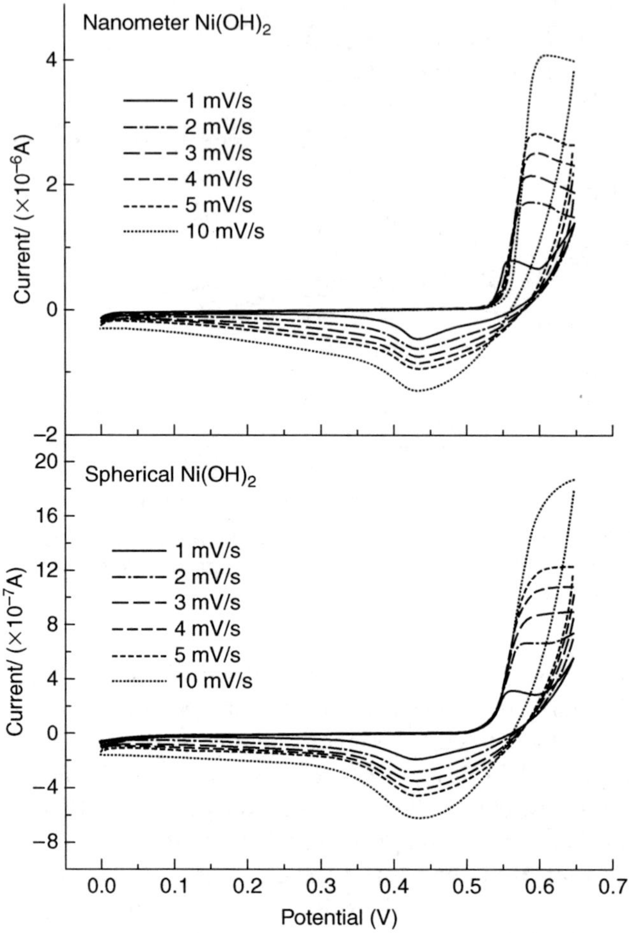

11.8 Cyclic voltammograms for nanometer $Ni(OH)_2$ and spherical $Ni(OH)_2$ under different scanning rates (Han *et al.*, 2005).

Electrochemical performance of nanostructured α-$Ni(OH)_2$ powder

The other focus of research has been on the improvement of the electrochemical properties of nickel electrode by using nanostructured α-$Ni(OH)_2$ powder. Jayalakshmi *et al.* (2005) studied the electrochemical performance of the phase-pure α-$Ni(OH)_2$ synthesised by a hydrothermal method via CV, and the results are illustrated in Fig. 11.10. The nanoparticles of α-$Ni(OH)_2$ were immobilised on a paraffin impregnated graphite (PIGE), and the shift of the anodic current peak towards positive potentials confirmed that the α-$Ni(OH)_2$was thermodynamically unstable or, in other words, the conversion to β-phase was energetically favourable. The peak

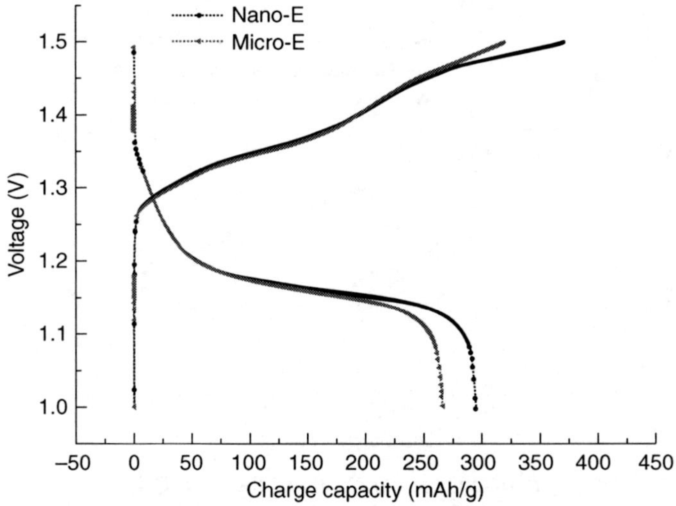

11.9 Charge/discharge curves for Nano-E and Micro-E at 0.2 C, with a limiting charge voltage of 1.5 V and a cut-off discharge voltage of 1.0 V (Han *et al.*, 2005).

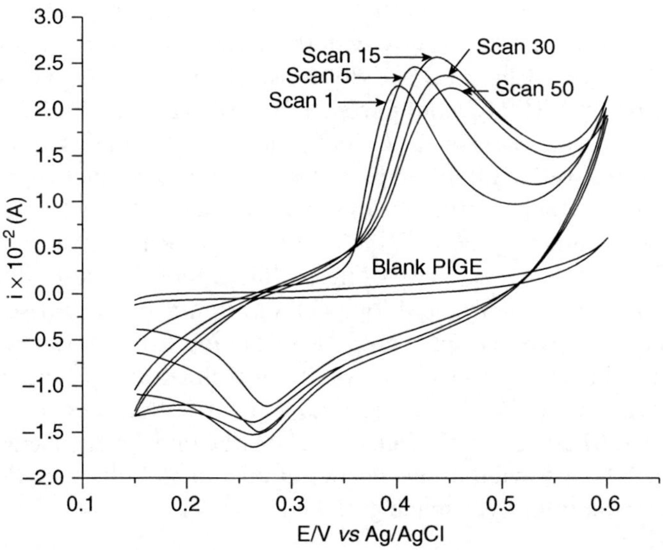

11.10 Cyclic voltammograms of immobilised α-$Ni(OH)_2$ on a PIGE in 1.0 M NaOH solutions; continuous cycling up to 50 scans (Jayalakshmi *et al.*, 2005).

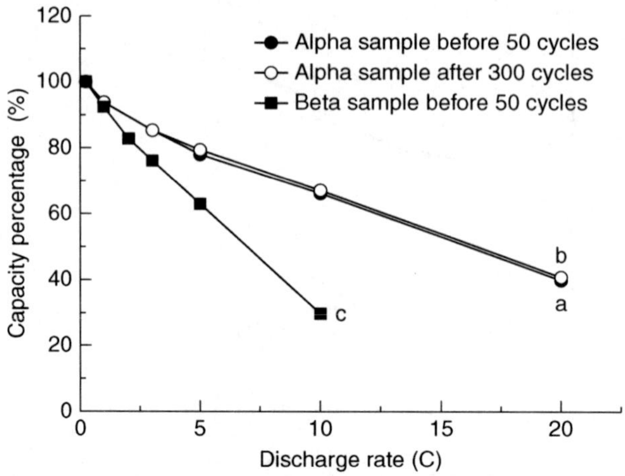

11.11 High rate discharge capability of the α-Ni(OH)$_2$ electrode synthesised by hydrothermal treatment at 165°C for 100 h (a) before, (b) after 300 cycles, and (c) β-Ni(OH)$_2$ electrodes (Hu *et al.*, 2006).

shift between first and thirtieth cycles further suggests the existence of ageing of the nanoparticles in alkaline solution. This ageing, which has been assigned to change in water activity in the solid and solution phases, would decrease the electrochemical activity. Since then, research focus has been on stabilising α-Ni(OH)$_2$ nanoparticles by doping with foreign elements. Hu and his co-workers (Hu *et al.*, 2006) prepared Al-stabilised α-Ni(OH)$_2$ with nanosized, well-crystallised particles via a hydrothermal approach, and the electrochemical performance of the as-synthesised powder was characterised. The results showed that the α-Ni(OH)$_2$ powder exhibited not only a high electrochemical capacity of up to 400 mAh/g, but also an excellent rate-capacity performance, and long cell durability. A comparison of high rate discharge capability between β-Ni(OH)$_2$ and α-Ni(OH)$_2$ prepared by hydrothermal treatment at 160°C for 100 h is illustrated in Fig. 11.11. The capacity retained at 10 C discharge current for cycled α-Ni(OH)$_2$ and β-Ni(OH)$_2$ was 67 and 29%, respectively. The large interlayer spacing of 7.9 Å in the α-Ni(OH)$_2$ powder allowed a better proton mobility than β-Ni(OH)$_2$ powder with an interlayer spacing of 4.6 Å.

11.2.7 Present status and future challenges

Among many other factors, the performance of nickel-based secondary batteries is dependent on the electrochemical behaviour of the nickel electrodes. The overall electrochemical properties of nickel electrodes are governed by the microstructure, textural characteristics, and physicochemical properties of the nickel hydroxide active material. Advances so far include modifying the active material via mechanochemical processing, doping of additive elements, and the synthesis of nanosized nickel hydroxides. Despite discrepancies in the literature, it has been demonstrated that exceptionally high discharge capacity, long durability, and high rate discharge capability for nickel electrodes can be obtained by introducing nanosized nickel hydroxide. However, the distribution of nanosized $Ni(OH)_2$ particles as-pasted onto electrode is generally poor between the $Ni(OH)_2$ particles and the conductive additive, and among the $Ni(OH)_2$ particles. This can subsequently deteriorate the electrochemical performance of the electrode owing to poor contact between particles. The challenges in nanostructured nickel hydroxide are therefore:

- to develop an efficient and simple method with high yield and low cost for the synthesis of nanostructured nickel hydroxide;
- to be able to control the phases, compositions, crystallinity, dimensions, and morphology of the nanostructured nickel hydroxide, as a means to tailoring the electrochemical performance of nanostructured nickel hydroxide;
- to find suitable conductive additives and an efficient, yet effective approach for electrode fabrication.

Little has been reported on the application of nickel hydroxide thin-film electrodes in rechargeable alkaline batteries, mainly because of the limited active material loading and energy density. Apart from the enhancement in the active material loading and energy density, the adhesion between the film and the substrate is also a critical factor to be overcome for nanostructured nickel hydroxide thin films to be commercially applicable.

11.3 Nickel-iron systems

The nickel-iron (Ni-Fe) battery was developed by Edison from the USA and Jüngner from Sweden in 1901, using nickel oxyhydroxide at the positive electrode and iron at the negative electrode. The porous separators, such as polyvinyl chloride, polyethylene, polyamide or polypropylene, are used to separate the electrodes. Nickel-iron batteries had early industrial application as traction batteries in forklift trucks, and in mine and railway locomotives. The interest in this battery decreased during the 1960s but revived again in 1975, particularly for electric vehicle (EV) applications (Eagle-Picher Industries, 1980).

Iron electrodes, unlike other electrode materials such as cadmium, lead and zinc, are environmentally benign. Besides, iron electrodes are both mechanically and electrically robust, meaning they have a high tolerance of

Table 11.2 A comparison between the specific energies of Ni-Fe and Pb/acid batteries at two discharge rates

Rate (W/kg)	Ni-Fe (Wh/kg)	Pb/acid (Wh/kg)
20	54	36
40	50	26

Source: Dell, 2000.

abuse (overcharge, overdischarge, and short-circuiting). Unlike Ni-Cd batteries, Ni-Fe batteries do not suffer from memory effect. The Ni-Fe batteries are famous for their long cycle life even under abusive usage. Vijayamohanan *et al.* (1991) reported that under normal conditions of use, the Ni-Fe batteries when discharged at 80% depth-of-discharge (DoD) can be cycled up to 2000 cycles, which is equivalent to a calendar life of up to 20 years. Another attraction of this battery is its high specific energy (1.5–2 times higher than that of lead/acid batteries) with exceptionally good discharge rates (Dell, 2000), as shown in Table 11.2.

11.3.1 Electrochemistry of Ni-Fe batteries

The charge/discharge reactions of the battery are (Shukla *et al.*, 1994, 2001):

$$2NiOOH + Fe + 2H_2O \underset{\text{charge}}{\overset{\text{discharge}}{\rightleftarrows}} 2Ni(OH)_2 + Fe(OH)_2 \quad (E_{cell} = 1.37\ V) \tag{11.4}$$

Under deep discharge, an Ni-Fe cell with a negative-limited configuration will undergo a further discharge reaction at a potential that is lower than the first step represented by Equation [11.4], i.e.

$$2NiOOH + Fe(OH)_2 \underset{\text{charge}}{\overset{\text{discharge}}{\rightleftarrows}} 2Ni(OH)_2 + FeOOH \quad (E_{cell} = 1.05\ V) \tag{11.5}$$

The cell reaction in alkaline electrolyte is highly reversible, particularly if the discharge reaction is restricted to the first step. The battery's charge/discharge cycle life is dependent on the reversibility of the two electrodes. The electrode stack is kept immersed in an alkaline electrolyte of 30 wt.% aqueous potassium hydroxide (KOH). In general, cell terminals and links are made of nickel-plated mild steel. Vents for battery cells are designed to prevent spillage and carbonation, while allowing the escape of gases produced in the cell.

The charge/discharge reactions at the negative electrode of a Ni-Fe battery take place in two steps as follows:

$$\mathrm{Fe} + 2\mathrm{OH}^- \underset{\text{charge}}{\overset{\text{discharge}}{\rightleftarrows}} \mathrm{Fe(OH)_2} + 2\mathrm{e}^- \quad (\boldsymbol{E}^0 = -0.88 \text{ V } vs \text{ SHE}) \qquad [11.6]$$

$$\mathrm{Fe(OH)_2} + \mathrm{OH}^- \underset{\text{charge}}{\overset{\text{discharge}}{\rightleftarrows}} \mathrm{FeOOH} + \mathrm{H_2O} + \mathrm{e}^- \quad (\boldsymbol{E}^0 = -0.56 \text{ V } vs \text{ SHE}) \qquad [11.7]$$

where E^0 stands for the standard electrode potential and standard hydrogen electrode is the standard hydrogen electrode. Kabanov and Leikis (Kabanov and Leikis, 1946) suggested that Equation [11.6] occurs in two steps, and the charge/discharge data associated with the steps for an iron electrode are illustrated in Fig. 11.12. First, iron is oxidised to $HFeO_2^-$ ions, then a loose porous $Fe(OH)_2$, weakly bound to the metal, precipitates from solution:

$$\mathrm{Fe} + 3\mathrm{OH}^- \underset{\text{charge}}{\overset{\text{discharge}}{\rightleftarrows}} \mathrm{HFeO_2^-} + \mathrm{H_2O} + \mathrm{e}^- \quad (E^0 = -0.75 \text{ V}) \qquad [11.8]$$

The reaction is followed by:

$$\mathrm{HFeO_2^-} + \mathrm{H_2O} \underset{\text{charge}}{\overset{\text{discharge}}{\rightleftarrows}} \mathrm{Fe(OH)_2} + \mathrm{OH}^- \quad (\Delta G^0_{298} = -24.7 \text{ V}) \qquad [11.9]$$

When the cell is subjected to prolonged discharge, the active δ-FeOOH is being continuously changed to iron hydroxide, similar to that of the nickel hydroxide positive electrode. Proton diffusion is associated with the electrode reaction between the solid lattices of $Fe(OH)_2$ and δ-FeOOH. The solid-state transformation of $Fe(OH)_2$ to δ-FeOOH in bulk indicates that the mechanism in the second discharge step (Equation [11.7]) is homogenous.

In the same solution, the open-circuit potential of a charged alkaline iron electrode is always more negative than the hydrogen electrode reaction ($E^0 = -0.88$ V vs SHE). As the hydrogen overvoltage is very low, the iron is thermodynamically unstable and suffers from corrosion through the evolution of hydrogen during charging according to the reaction:

$$2\mathrm{H_2O} + 2\mathrm{e}^- \rightarrow \mathrm{H_2} + 2\mathrm{OH}^- \quad (E^0 = -0.83 \text{ V } vs \text{ SHE}) \qquad [11.10]$$

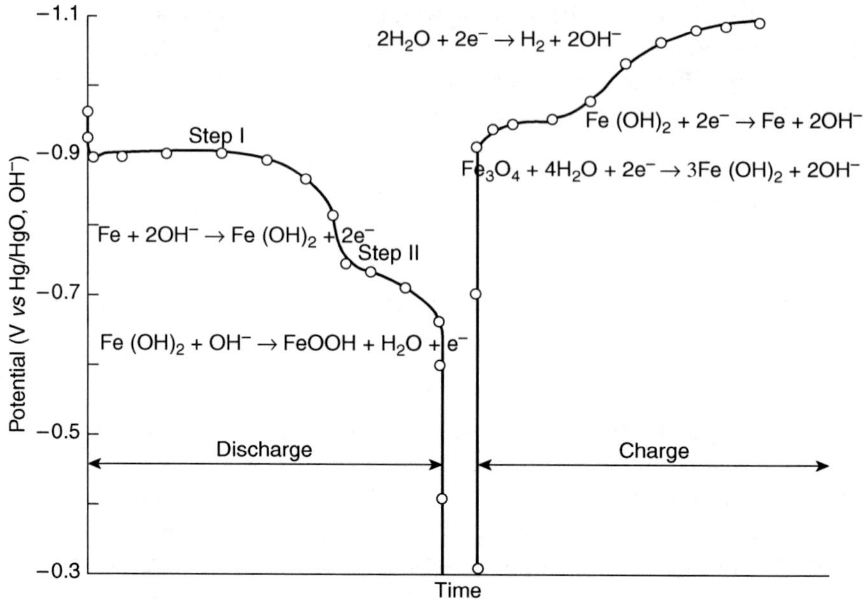

11.12 Typical charge/discharge data for an iron electrode in 6 M KOH (Vijayamohanan *et al.*, 1991; Shukla and Hariprakash, 2009b).

The oxygen reduction reaction can be expected as a result of the dissolved oxygen in alkaline solution:

$$O_2 + 2H_2O + 4e^- \rightarrow 4OH^- \quad (E^0 = -0.41 \text{ V } vs \text{ SHE}) \qquad [11.11]$$

As a result of these reactions, iron electrodes are prone to self-discharge at a daily rate at 298 K of 1–2% of their nominal capacity (Chakkaravarthy *et al.*, 1991). In addition, the hydrogen evolution reaction can compete with the discharge reaction, which would result in low charge efficiency of the electrodes (Shukla *et al.*, 1994).

11.3.2 Solid-state chemistry of iron electrodes

The electrochemical performances of the iron electrodes are in direct relation to the active material's phase composition, surface morphology and crystallinity. The changes in the morphology and volume of the active mass influence the porosity, electrical conductivity, and mechanical strength of the electrodes. It is therefore of critical importance to understand the interrelation between various solid-state parameters. Uniformly dispersed grains of the active mass are preferred for electrochemical reactions to proceed symmetrically. The change in lattice parameters of the active mass upon electrode cycling should be suppressed to preserve the integrity of the electrodes, which is critical for the long-term performance of the electrodes with increasing cycles. A shift in the orientation of crystallites can also influence electrode performance, as the electrochemical reaction would proceed asymmetrically (Shukla and Hariprakash, 2009b).

The crystal chemistry data for iron, magnetite (Fe_3O_4), iron hydroxide, and iron oxyhydroxide (FeOOH) can provide insight into the understanding of the structural changes associated with the iron electrodes when in operation. The raw material has a body-centred cubic (bcc) structure. Magnetite is a ferromagnetic mixed-valence 3d transition metal oxide that has an inverse spinel structure (Zhang *et al.*, 2005a), a unit cell containing 32 O^{2-}, 8 Fe^{2+} and 16 Fe^{3+} ions. Iron hydroxide has a CdI_2-type layered structure, with each OH^- ion forming three bonds to iron atoms in its own layer, while joining to three OH^- ions in the adjacent layer. Iron oxyhydroxide exists in four different forms:

1. α-FeOOH (goethite) has a hexagonal, close-packed lattice, wherein OH^- ions are hydrogen-bonded to O^{2-} ions.
2. β-FeOOH has a bcc lattice with similar structure to α-MnO_2; presumably, β-FeOOH is not a pure oxyhydroxide, as it is stable only in the presence of certain interstitial impurities.
3. γ-FeOOH has a cubic closest packing (ccp) lattice with hydrogen bonding. Each iron atom is surrounded by a distorted octahedral group of oxygen atoms that are linked together to form corrugated layers with hydrogen bonding between the OH^- groups.
4. In δ-FeOOH, Fe^{3+} ions are randomly distributed over a hexagonal closest packing (hcp) lattice with Fe^{3+} ions in octahedral sites similar to iron hydroxide with an additional 20% of Fe^{3+} ions in tetrahedral sites.

11.3.3 Performance of Ni-Fe batteries

The discharge curves for commercial cells at different rates at 25°C are given in Fig. 11.13 (Shukla *et al.*, 1994). As can be seen in this figure, nominal voltage of Ni-Fe cells could vary from approximately 1.23 V at the *C*/8 rate to 0.85 V at the *C*/1 rate. The open-circuit voltage and the nominal voltage at discharge rates between *C*/10 and *C*/100 lie in the range of 1.3–1.4 V. The discharge curves are relatively flat with change in the cell voltage at the *C*/8 rate, for instance, 1.32 and 1.15 V at 10 and 90% DoD, respectively. Constant-voltage charging on Ni-Fe batteries is not desirable as it may cause thermal runaway that can severely damage the batteries. Galvanostatic charging set

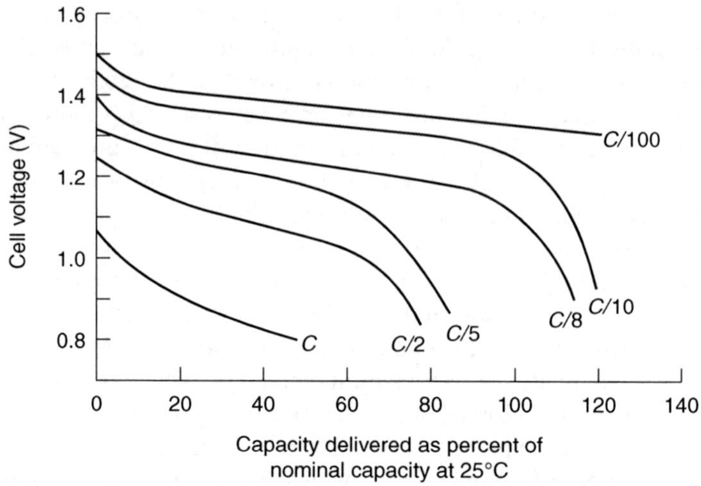

11.13 Typical discharge curves for a nickel/iron cell at 25°C. Numbers on curves are discharge rates with *C* as the current in A numerically equal to the nominal Ah capacity of the cell (Shukla *et al.*, 1994).

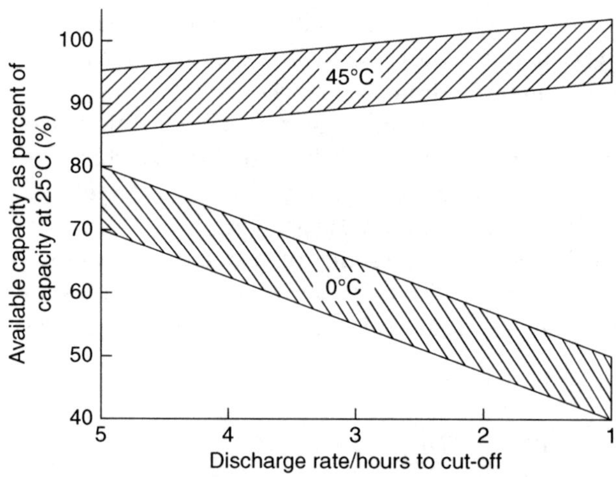

11.14 Discharge capacity of Ni-Fe cell at low and high temperatures as a function of rate of discharge. The range of the shaded area is affected by the battery packs' size, type and number of cells (Shukla *et al.*, 1994).

at 1.7 V is recommended to control the cell temperature, as the cell temperature will increase when the cell is almost fully charged. In addition to the rate of discharge, the operational temperature was also found to affect the discharge capacity of this type of battery, as displayed in Fig. 11.14 (Shukla *et al.*, 1994). This explains the limited applications of this battery system for high discharge rate at low temperatures.

As previously discussed in Section 11.1, one of the major deficiencies related to Ni-Fe batteries is their high self-discharge rate. The rate of discharge can be as high as 8–10% of the nominal capacity per day at an operational temperature of 40°C (Shukla *et al.*, 1994). It is therefore highly recommended that the Ni-Fe batteries are recharged once every two days at this operating temperature to ensure 80% of the nominal cell capacity is available. A longer interval between each recharge is permitted when the operating temperature is at 20°C or below. Both the self-discharge and cycle life behaviours of Ni-Fe batteries deteriorate at higher operating temperature; this is mainly attributed to the faster corrosion rate on iron electrode at higher temperature. Thus, efforts have been devoted to improving the performance of the Ni-Fe batteries by suppressing the rate of hydrogen evolution and corrosion.

11.3.4 Improvement of electrochemical performances of Ni-Fe batteries

To overcome the aforementioned limitations of the iron electrode induced by low over-potential for hydrogen evolution, a number of additives have been incorporated in both the electrolyte and the iron negative electrode during fabrication. Hills (1965) reported that the stabilisation of iron electrodes in pure 45% KOH electrolyte yielded efficiencies equivalent to those obtained with lithiated 20% KOH. He concluded that the increased efficiency in the LiOH-aided KOH electrolyte was attributable to the delay in the onset of passivity. In addition, the incorporation of Na_2S in alkaline electrolyte was found to improve the performance of iron oxide electrodes, such as maximum capacity, by increasing the value of open-circuit potential (Periasamy *et al.*, 1996).

Besides copper and mercury, which are generally used in iron electrode to increase the hydrogen overvoltage (Linden, 1984), sulphide is one of the most popular additives used to improve the performance of the iron electrodes. Based on the several previous studies (Rozentsveig and Shcherbakova, 1961; Rozentsveig *et al.*, 1962; Teplinskaya *et al.*, 1964; Cerny and Micka, 1989; Vijayamohanan *et al.*, 1990), it is inferred that sulphide additives could affect the behaviour of alkaline iron electrodes in either of the following ways: (a) by causing a substantial increase in the anodic current density prior to passivation and subsequently permitting higher discharge rates; (b) by increasing the bulk electrical conductivity of the electrodes; (c) by enhancing the rate of the $Fe/Fe(OH)_2$ reaction; (d) by increasing the solubility of iron compounds; and (e) by modifying the electrode texture and morphology. The report by Vijayamohanan *et al.*, 1990 reviewed the effect of grain refinement (Salvarezza *et al.*, 1982) and depassivation models (Micka

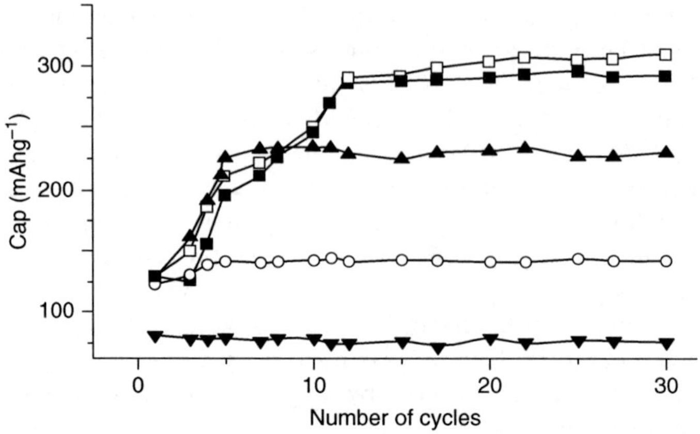

11.15 Discharge capacity of iron electrodes in the absence and presence of FeS or PbS as a function of charge/discharge cycle. The values were obtained in 6 M KOH + 0.33 M LiOH solution at I_d = 12.9 mA/g; where (□) 1 wt.% PbS, (■) 1 wt.% FeS, (▲) 0.5 wt.% PbS, (○) in absence of additive, and (▼) 2 wt.% PbS (Souza *et al.*, 2004).

and Zábranský, 1987), and confirmed that the incorporation of sulphide can improve the conductivity of bulk electrode.

The effect of FeS and PbS on capacity and self-discharge of an iron porous electrode used in a Fe-Ni battery was investigated by Souza *et al.*, (2004). The discharge capacity of electrodes with various amounts of FeS and PbS additions over the first 30 cycles has been illustrated in Fig. 11.15. A significant increase in capacity in electrodes with 1 wt.% FeS and PbS additions was observed with respect to the additive-free electrode; this was ascribed to the incorporation of sulphide ions in the $Fe(OH)_2$ film, which caused a distortion in film structure, leading to an increase in ionic conductivity. This resulted in a thicker film being produced, and subsequently raising the discharge time. From Fig. 11.15, it can also be observed that the capacity of the electrode containing 1 wt.% PbS was higher than that of the electrode with 1 wt.% FeS, indicating that in addition to the sulphite ion, the metal present in the additive also affected the electrode capacity. Furthermore, the optimum content of the PbS addition was found to be 1 wt.%; an excess addition of PbS can cause a decrease in capacity due to the formation of PbO or $Pb(OH)_2$.

Fig. 11.16 displays the self-discharge behaviour of the electrodes with and without 1 wt.% of PbS and FeS additions. Clearly, both FeS and PbS can retard the degree of self-discharge, and this effect was most significant in the presence of PbS. As the hydrogen evolution over-potential of Pb is higher in comparison to the Fe over-potential, by introducing PbS or FeS to raise the

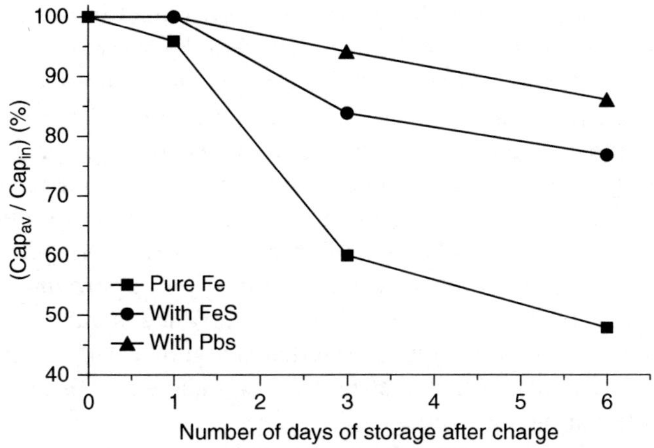

11.16 Self-discharge behaviour of iron electrodes, in the absence and presence of additives 1 wt.% PbS, and 1 wt.% FeS, at the end of different periods of storage at 30°C. The values were obtained in 6 M KOH + 0.33 M LiOH solution at I_d = 12.9 mA/g; where (■) in absence of additive, (•) 1 wt.% FeS, and (▲) 1 wt.% PbS (Souza *et al.*, 2004).

over-potential of the iron electrode, the self-discharge rate of iron electrode can be significantly reduced.

11.3.5 Current status and future challenges

Research and development on Ni-Fe batteries has mainly been conducted by Eagle-Picher Industries (EPI) in the United States, under a cost-shared contract with the Department of Energy (DoE). Since the late 1970s, EPI has focused on the development of high-performance batteries using sintered nickel and iron electrodes. Significant improvements on the battery performance, cell duration, and cost have been achieved. The capacity of the EPI battery, even after powering an EV for over 45 000 km of operation, was reported to maintain its original rated capacity (Shukla and Hariprakash, 2009c). However, this work was discontinued, as the focus has been shifted towards nickel/metal hydride batteries.

The advanced Ni-Fe batteries, which are generally used in EVs, utilise a superior electrode-grid design employing fibre metal that provides intimate contact with the iron active material throughout the porous structure (Chakkaravarthy *et al.*, 1991). With this design, the specific energy of 40 and 60 Wh/kg with cycle lives of about 1000 and 2000 cycles at 100 and 80% DoD can be achieved (Shukla and Hariprakash, 2009c).

As previously stated, though Ni-Fe batteries do not suffer from memory effect, their corrosion-induced high self-discharge has been the key drawback in the development of Ni-Fe batteries. The improvement in overall performance of the Ni-Fe batteries has been achieved in the past decades substantially by electrocatalysis of the iron electrode reaction. Although complete suppression of hydrogen evolution seems to be a great challenge, efforts have been made to realise sealed Ni-Fe batteries by employing efficient hydrogen–oxygen recombinant catalysts in the headspace of each cell. The performance characteristics of such sealed Ni-Fe batteries were found to be comparable to the vented cells of similar size. However, the performance at sub-zero temperature is very poor; for instance, only approximately 10% of the nominal capacity was obtained at –20°C, which is attributed to the limited solubility of the reaction intermediates together with the increased resistance and viscosity of the electrolyte in conjunction with the retarded reaction kinetics at the electrodes.

In the sealed Ni-Fe batteries, the maximum temperature in the catalyst-bed (catalyst on a porous mat) can reach 65°C at *C* rate charging at 25°C, which is ascribed to the exothermic hydrogen–oxygen recombination reaction (Shukla and Hariprakash, 2009c). Major difficulties associated with the development of commercial-grade sealed Ni-Fe batteries include engineering design in suppressing the temperature of catalyst-bed to below 60°C and excessive internal heating of the cells during charge/discharge operations.

Although sealed and maintenance-free batteries are desirable for most applications, Ni-Fe batteries seem to be difficult to fit into these criteria as compared to current popular nickel/cadmium (Ni-Cd), nickel/metal hydride (Ni-MH), and Li-ion batteries. Nevertheless, Ni-Fe batteries are attractive for applications such as forklift trucks and open-pit mining vehicles, and for meeting certain power requirements in offshore platforms and petrochemical industries.

11.4 Nickel-cadmium systems

Ni-Cd cell utilises nickel hydroxide as the positive active material, a mixture of cadmium and iron as the negative electrode material, and an aqueous alkaline OH as an electrolyte. This type of battery has been developed in different ways to produce a wide range of commercial secondary batteries, including sealed and maintenance-free cells with capacities ranging between 10 mAh and 100 Ah. Vented standby power units with capacities over 1000 Ah, as well as cranking batteries that are capable of delivering peak currents as high as 8000 A, are also available (Shukla *et al.*, 2009).

The operating voltage of Ni-Cd cell is 1.20 V, which is slightly less than that of Ni-Fe (1.25 V), and the mass of cadmium is higher than that of iron; these together give a specific energy of approximately 40–60 Wh/kg. Though the specific energy is lower, the high rate and low-temperature performances are better than those of lead/acid. More superior properties, such as stable discharge voltage, long cycle life, continuous overcharge capability, high power performance, quick discharging and charging, low maintenance requirement and excellent reliability, have made Ni-Cd the most widely used secondary battery after lead/acid.

11.4.1 Operating principle of sealed Ni-Cd batteries

During discharge of an Ni-Cd battery, there is a homogeneous solid-state mechanism through proton transfer between nickel (Ni^{3+}) hydroxide (charged active material) and nickelous (Ni^{2+}) hydroxide (discharged material). On the other hand, cadmium (charged active material) is transformed to cadmium hydroxide (discharged material) by a heterogeneous dissolution–precipitation process at the negative plate. The net charge/discharge reactions at the negative plate are:

$$Cd + 2OH^- \underset{\text{charge}}{\overset{\text{discharge}}{\rightleftarrows}} Cd(OH)_2 + 2e^- \quad (E^\circ = -0.81\ V) \qquad [11.12]$$

And the overall cell reaction is therefore expressed as:

$$2NiOOH + Cd + 2H_2O \underset{\text{charge}}{\overset{\text{discharge}}{\rightleftarrows}} 2Ni(OH)_2 + Cd(OH)_2 \quad (V_{cell} = 1.30\ V) \qquad [11.13]$$

Fig. 11.17 illustrates the operating principle of a sealed Ni-Cd battery. Under deep discharge conditions, the gaseous hydrogen that is evolved at the positive electrode, owing to the inevitable difference in the storage capacities of series-connected cells in a battery, would be consumed at the positive electrode. Thus, repeated overdischarge can cause internal pressure build-up and lead to bursting of the cell.

Ni-Cd batteries are designed to be positive limited, to ensure that a sealed Ni-Cd cell is properly operational under various operating conditions. This approach ensures that only oxygen evolution would occur under normal operating conditions, and the oxygen would subsequently diffuse to the cadmium electrode and combine with active cadmium to form cadmium hydroxide according to the reaction:

$$Cd + \frac{1}{2}O_2 + H_2O \rightarrow Cd(OH)_2 \qquad [11.14]$$

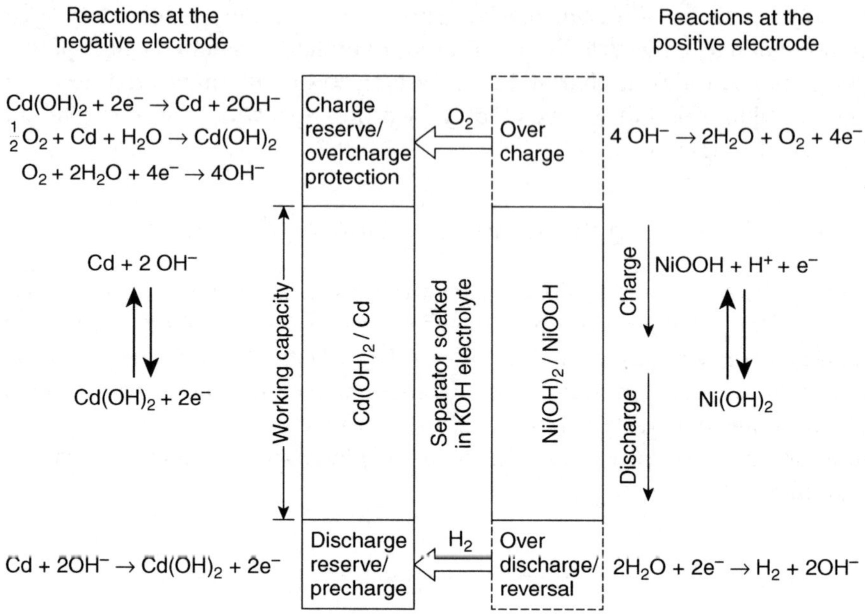

11.17 Operating principle of a sealed Ni-Cd battery (Shukla *et al.*, 2001).

The as-produced cadmium hydroxide from Equation [11.14] is converted to active cadmium during charging of the cell. Such a recombination of gaseous oxygen, produced on the nickel positive plate and on the cadmium electrode, is called the oxygen recombination cycle, as illustrated in Fig. 11.17. In order to realise an effective oxygen recombination cycle, the negative-to-positive electrode capacity ratio is generally controlled between 1.5 and 2.

11.4.2 Crystal chemistry of cadmium hydroxide

The cadmium negative electrode, when undergoes a reduction reaction, is accompanied by a complex combination of solid-state and dissolution–precipitation processes. There are three different forms of cadmium hydroxide: the α, β, and γ varieties. β-$Cd(OH)_2$ is the most stable phase with one molecule per unit cell among the three forms. The γ-$Cd(OH)_2$, on the other hand, is much more active than the β-phase (easily charged) as it has four molecules per unit cell, and is dominant at low temperatures.

Both α-$Cd(OH)_2$ and γ-$Cd(OH)_2$ have a hexagonal-layered lattice system of the brucite type (C6 type), where the latter is well-crystallised with defined lattice parameters (a = 3.496 Å and c = 4.702 Å). However, α-$Cd(OH)_2$ is poorly crystalline (Shukla and Hariprakash, 2009a). It is unstable in KOH electrolyte and tends to convert to β-$Cd(OH)_2$. Perfect crystalline, stoichiometric β-Cd(OH)$_2$ is electrochemically inactive. On the contrary, γ-Cd(OH)$_2$ possesses a crystal structure entirely different from that of the brucite type. The γ-phase is monoclinic, with four molecules per unit cell with lattice parameters of a = 5.67 Å, b = 10.25 Å, and c = 3.41 Å.

11.4.3 Active materials and their fabrication approaches for cadmium electrodes

Two synthesis methods are generally employed for preparing the active materials for pocket-type cadmium plates, namely coprecipitation (Kazarinov *et al.*, 2001) and dry-mixing (Negeevich *et al.*, 1990). In sintered-type Ni-Cd cells, cadmium electrodes are synthesised by an impregnation method – a process by which porous sintered plaques are filled with active material (Kalaignan *et al.*, 1996). Among chemical impregnation processes, polarisation, thermal decomposition and cadmium formate procedures have reached a commercial stage for the precipitation of finely-divided cadmium, cadmium hydroxide, or cadmium oxide in the pores of the negative plaques.

Cadmium alone as active material in the negative electrode has a tendency of losing capacity; this is because of the agglomerated cadmium particles within the core of the electrode fail to take part in the electrochemical reactions. Thus, compounds such as iron, which are found capable of maintaining the cadmium material in the finely-divided state, have become indispensable in the making of the cadmium electrode.

11.4.4 Performance of Ni-Cd batteries

The performance of Ni-Cd batteries is dependent on numerous factors: type of cell in the battery, cell construction, manufacturing process and operating temperature, charge/discharge rates, the age of the cells and, most direct of all, the performance of the negative cadmium electrode. In a sealed Ni-Cd system, the nominal voltage is around 1.2 V at room temperature, which is comparable to the majority of the Ni-based battery systems, for example, Ni-Fe and Ni-MH; the specific energy and energy density are 40–60 Wh/kg and 50–150 Wh/L, respectively (Cattaneo and Riegel, 2009). These values are higher than for lead/acid and Ni-Fe battery systems, but lower than for Ni-MH batteries.

The cycle life of Ni-Cd batteries can reach 1000 cycles in 100% DoD at room temperature. This superior cycling performance is linked to the excellent stability of active materials, and consequently to the very limited variation of the internal resistance, which remains practically constant as a function of the state of charge and cycling life (Bernard, 2009).

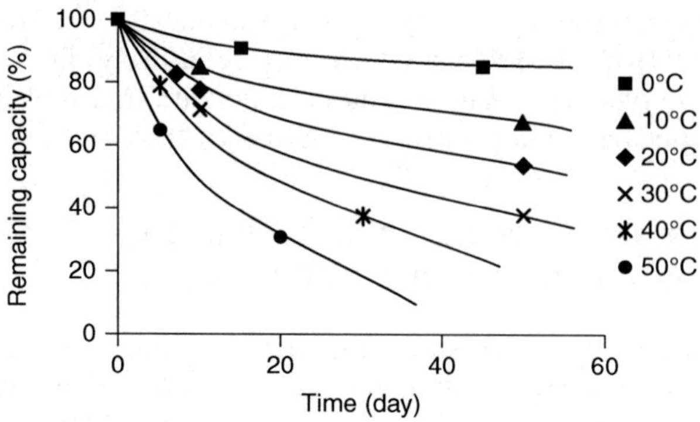

11.18 Rate of self-discharge vs temperature for sealed Ni-Cd batteries (Bernard, 2009).

The Ni-Cd batteries lose reversibly some capacity during storage through the self-discharge mechanism. Figure 11.18 illustrates the rate of capacity loss in sealed Ni-Cd batteries at different temperatures. It clearly shows that the rate of self-discharge is non-linear. Initial self-discharge rate is significant (20% in 7 days at 40°C), and gradually slows down. This suggests that the rate of self-discharge is likely to be correlated to two mechanisms as described below.

In the early period of storage, the instability of the charged positive electrode is responsible for the loss of capacity. When a positive electrode is fully charged and left in open-circuit, a decrease in the oxidation state of the positive electrode is observed, and oxygen gas evolution takes place on the positive electrode. Subsequently, reduction of the oxygen on the cadmium electrode takes place according to the Equation [11.12]. The observed potential of the positive nickel electrode is a mixed potential of both nickel reaction and oxygen reaction, and it determines the self-discharge current. Consequently, the oxidation state of a charged positive electrode will experience a fast decrease initially and will be nearly stable when the self-discharge current becomes nearly negligible. In the later period of storage, the rate of self-discharge is governed by the amount of nitrogen impurities in the electrolyte. Nitrite ions, which are reduced from nitrate ions, diffuse back to the charged positive electrode, where they are oxidised to nitrate. This nitrate/nitrite shuttle reaction equally discharges the positive and negative electrodes.

As mentioned, the overall performance of the Ni-Cd battery is dependent on the behaviour of the negative cadmium electrode, therefore considerable efforts have been devoted to improving the properties of the cadmium electrode. Some of the key issues associated with cadmium electrodes are as follows:

- The involvement of dissolved ion intermediate (see Equation [11.12]) in the mechanism of the cadmium electrode reaction can lead to recrystallisation and redistribution of cadmium hydroxide active material during charge/discharge cycling (Sathyanarayana, 1985).The former will result in electrode passivation, and the latter will cause the formation of metallic cadmium dendrite during charging; this will deteriorate the capacity over cycling and eventually lead to cell failure.
- The low hydrogen overvoltage of the cadmium electrode will in turn cause hydrogen evolution as a normal overcharge reaction in vented cells and as a side reaction in sealed cells. The consequence of the hydrogen evolution in the latter can lead to high pressure and result in cell failure (Shukla and Hariprakash, 2009a).
- The graphite loss and swelling in the pocket-type positive electrode, which is evidently seen as a second low voltage plateau in the discharge curve, will result in capacity loss in Ni-Cd pocket-plate batteries (Ahlberg *et al.*, 2000).
- Cadmium active material is expensive and eco-unfriendly.

In-depth investigation on the causes of failure in sealed Ni-Cd batteries are described elsewhere (Gross, 1971). Enormous research has focused on improving the performance of the Ni-Cd cells by introducing metal (Tamil Selvan *et al.*, 1990), organic (Munshi *et al.*, 1985; Munshi *et al.*, 1988), or inorganic additives (Kalaignan *et al.*, 1996) into cadmium electrodes, and cell design (Britting, 1984; Miller, 1986; Miller, 1987; Petchjatuporn *et al.*, 2008). Simulation works have also been performed on Ni-Cd batteries for aerospace applications (Montalenti and Stangerup, 1977; Ratnakumar *et al.*, 1996).

Selvan *et al.* (Tamil Selvan *et al.*, 1990) studied the high- and low-rate performance characteristics of a cadmium electrode with different percentages of iron added during impregnation. A significant improvement in open-circuit potential of the iron-containing electrodes was observed at both high- and low-rate discharge. The capacity output of the electrodes as a function of iron content at different rates of discharge is displayed in Fig. 11.19. Clearly, electrode with 20 wt.% of iron addition to the cadmium electrodes performed better than the other electrodes; excessive iron addition (>20 wt.%) can degrade the performance of the cadmium electrode.

The charging characteristics of cadmium electrodes containing iron, as shown in Fig. 11.20, demonstrate that the charging potential lies between −800 and −930 mV, and that the end-of-charge potential is between −975 and −1075 mV, which is the charging potential range for a cadmium electrode. From this observation, it is acknowledged that the iron does not take part in the reaction of the cadmium electrode, but rather provides an increase in the active surface area. A higher surface area can improve the rate of

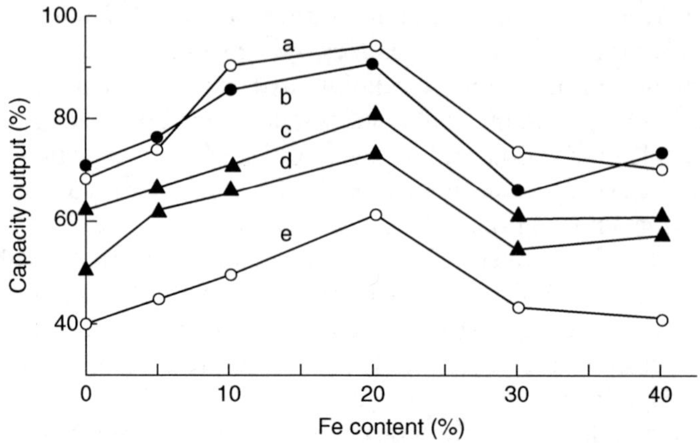

11.19 Capacity output of Cd electrode vs iron content at: (a) 0.1 *C*, (b) 0.2 *C*, (c) 1 *C*, (d) *3 C*, and (e) 5 *C* (Tamil Selvan *et al.*, 1990).

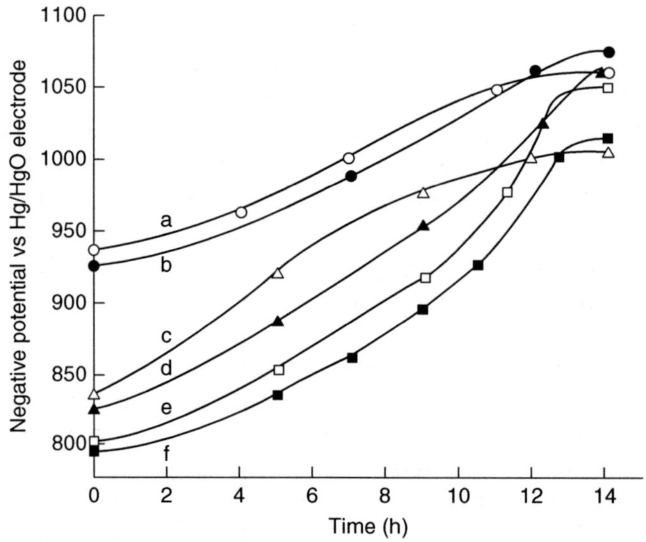

11.20 Charging characteristics of Cd electrode with iron additions: (a) 0 wt.%, (b) 5 wt.%, (c) 10 wt.%, (d) 20 wt.%, (e) 30 wt.%, and (f) 40 wt.% (Tamil Selvan *et al.*, 1990).

dissolution, and the soluble species required for the charging process can be easily supplied to the electrode and subsequently give rise to a significant higher charging efficiency. The capacity decrease for electrodes with excessive iron addition (> 20 wt.%) was attributed to a decrease in the amount of cadmium available for reaction.

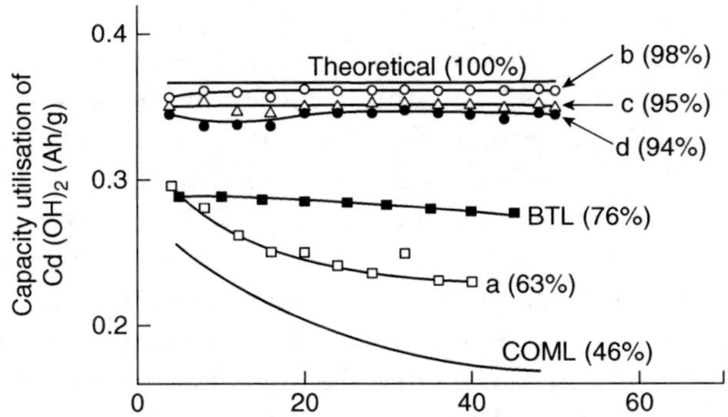

11.21 Dependence of capacity utilisation of sintered-plate $Cd/Cd(OH)_2$ electrode on the number of charge/discharge cycles with 100% DoD at room temperature. (a) – electrodes prepared without any additive, (b) and (c) – electrodes prepared with a coating of ethyl cellulose by different methods, (d) – electrodes prepared with a coating of ethyl cellulose by both (b) and (c) (Sathyanarayana, 1985).

Another approach has been demonstrated to be effective in retarding the growth of metal dendrite and suppressing the migration of active material (Sathyanarayana, 1985). A colloidal surfactant, namely ethyl cellulose, was incorporated into the cadmium electrode during the fabrication so as to form a monolayer envelope on the cadmium hydroxide crystallites whose growth rate on ageing was then automatically retarded. The faradaic efficiency of cadmium electrodes with and without the surfactant additive is shown in Fig. 11.21. An appreciable improvement in ethyl cellulose-treated cadmium electrodes (b–d) was observed. These electrodes showed an excellent faradaic efficiency, which was within 3% of the theoretical value even after 50 cycles of deep discharge between cycles. In other words, the addition of ethyl cellulose during the fabrication of electrodes could idealise the utilisation efficiency of the active material and minimise the capacity fade-out of the cadmium electrodes. The author concluded that the colloid layer thus formed on the cadmium hydroxide crystallites prevented the growth of these crystallites, and subsequently migration.

Another drawback associated with nickel-cadmium system is the infamous 'memory effect'. The memory effect can be defined as the apparent reduction in cell capacity at a fixed cut-off voltage. The true cause of the memory effect is still unclear. Nevertheless, the view previously held was that it was related to growth of Cd or $Cd(OH)_2$, causing ohmic suppression (Pensabene and Gould, 1976). In the same work, authors found that the crystals are larger in the negative materials of 'memory' (problematic) cells than that of the normal cells. Larger crystals have a smaller surface area and will therefore be more difficult to discharge. Another important cause of the memory effect was discovered by Crompton and discussed in more detail by Berndt (Berndt, 1997; Crompton, 2000b). It has been found that cadmium intermetallic compounds were formed with the nickel substrate of sintered plates. Compounds such as Ni_2Cd_5 or Ni_5Cd_{21} build up under the operating conditions that cause the memory effect. Approximately 150 mV of the voltage drop would take place if a complete discharge process were performed using these cadmium compounds. The set-up of the memory effect can be exceptionally difficult, and the results can be hard to reproduce. Therefore, it can be seen that the memory effect is not a simple phenomenon, but is established by a rather complex process. Nevertheless, it can be stated that true memory is restricted to cells with sintered negative plates and involves both crystalline effects and reactions between cadmium and nickel (McDowall, 2003). A simple cure for this complex phenomenon is to discharge the battery completely before recharging it normally.

11.4.5 Final remarks and challenges

Sealed Ni-Cd batteries have an inherently good durability. However, their cycle life in service is strongly dependent on the type of application. Failure behaviours such as low discharge voltage, high charge voltage, loss of capacity, high self-discharge rate, internal shorts, open-circuiting, excessive oxygen pressure, generation of hydrogen, and loss of electrolyte have been characterised.

Service temperature is the most important parameter in affecting the performance of the Ni-Cd batteries, with the longest battery life between 5 and 15°C (Shukla *et al.*, 2009). At lower operating temperature, though the capacity degradation is slow, the hydrogen evolution can occur, especially at high charge rates. On the other hand, the accelerated cadmium migration, crystal growth, and enhanced solubility of cadmium in aqueous KOH electrolyte at high temperature can accelerate the battery wear-out. In addition, nylon separators also deteriorate faster at high temperature on reacting with the electrolyte; this creates carbonate, dilutes the electrolyte, and causes cadmium hydroxide to be reduced to cadmium with loss of some of the overcharge protection.

Despite substantial improvement in Ni-Cd batteries over the last few decades, problems such as hydrogen evolution when cadmium electrode is overcharged, cell reversal, memory effect, dendritic shorting, and environmental concern over cadmium still remain. The last has undoubtedly made the largest negative impact on the development of Ni-Cd batteries. The high toxicity and the environmental concern associated with inappropriate disposal of Ni-Cd batteries have posed threats to the health of humans; short-term ingestion can cause flu-like illness, long-term ingestion can cause lung cancer, and human ingestion of up to 9 g of cadmium can be lethal (Boreiko, 2009). Nevertheless, Ni-Cd batteries dominated the market in secondary batteries until the 1990s. Since then, the Ni-Cd batteries have been rapidly replaced by Ni-MH and lithium-ion batteries.

11.5 Nickel-hydrogen systems

In the beginning of the space era, Ni-Cd secondary batteries were employed for powering the majority of spacecraft, satellites, and space probes, owing to their superior properties such as long cycle life, high robustness, and relatively simple charge-control mechanism. The late 1970s saw the development of the nickel-hydrogen (Ni-H_2) battery system established and employed in US Navy's Navigation technology satellite-2 (NTS-2) (Dunlop and Stockel, 1977; Stockel *et al.*, 1980).

Ni-H_2 is a sealed secondary battery that incorporates a fuel cell with a conventional electrode. The fuel-like electrode is negative and uses hydrogen as the negative active material. Similar to other Ni-based batteries, the positive electrode is the nickel electrode, which uses nickel hydroxide as the active material. The lightweight nature of the hydrogen gas electrode allows the Ni-H_2 cell to have exceptional high gravimetric energy density, but its volumetric energy density is lower than for other nickel-based batteries. In addition to its high gravimetric energy density, more superior properties such as high DoD capacity (up to 80%), good reliability, high resistance to high rate overcharging and cell reversibility and long lifetime (more than 10 years) have been the selling points for Ni-H_2 batteries. Ni-H_2 batteries have now been used in over 800 satellites, with good reliability (Borthomieu and Bernard, 2009).

11.5.1 Electrochemistry of Ni-H_2 batteries

An Ni-H_2 cell is composed of a nickel positive electrode and a negative platinum catalyst electrode, which permits the electrochemical oxidation of the hydrogen gas as in a fuel cell. A cell case in the form of a vessel is therefore required to handle changes in gas pressure during charging and discharging.

The active material of the positive electrode, namely $Ni(OH)_2$, is converted to NiOOH upon charging through a simple deprotonation mechanism, and transformed back to $Ni(OH)_2$ when discharged (see Chapter 12 for detailed explanation). Accordingly, the reaction in the nickel positive electrode is expressed as in Equation [11.1]. The hydrogen electrode, on the other hand, oxidises hydrogen to water during discharge and reforms the gas during charge:

$$H_2 + 2OH^- \underset{\text{charge}}{\overset{\text{discharge}}{\rightleftarrows}} 2H_2O + 2e^- \quad [11.15]$$

The overall reaction of the cell can thus be expressed as:

$$2NiOOH + H_2 \underset{\text{charge}}{\overset{\text{discharge}}{\rightleftarrows}} 2Ni(OH)_2 \quad [11.16]$$

It can be depicted from the overall reaction that upon charging, nickel oxyhydroxide is formed at the positive electrode, and hydrogen is produced at the negative. As the cell is within a sealed container, the hydrogen pressure increases from 65 to 70 atm (Borthomieu and Bernard, 2009). During overcharge, oxygen evolution is expected to take place at the positive electrode:

$$4OH^- \rightarrow 2H_2O + 4e^- + O_2 \quad [11.17]$$

During both charge and overcharge, oxygen is recombined at the negative electrode:

$$2H_2O + 2e^- \rightarrow 2OH^- + H_2 \quad [11.18]$$

$$O_2 + 2H_2 \rightarrow 2H_2O \quad [11.19]$$

Therefore, there is a pressure stabilisation, the kinetics of which depends mainly on the charge rate and temperature. The recombination reaction (Equation [11.18]) is an exothermic reaction, which is accompanied by heat generation. This heat can cause a microexplosion, which is also termed popping, as a result of spontaneous reaction of oxygen and hydrogen when concentrations reach the combustible range. The occurrence of popping is highly undesirable, as it can severely damage the cell and lead to short circuits. Highly permeable separators, which allow faster oxygen diffusion through the separators, are required for suppressing the oxygen bubble formation that leads to popping.

11.5.2 Types of $Ni\text{-}H_2$ batteries

The $Ni\text{-}H_2$ battery has been developed in two versions, and is distinguished by the pressure. In the high pressure version, the Pt/H_2 is employed as the negative electrode, where the H_2 released during charge remains in the gas phase and the gas pressure rises to a maximum of 40 atm, falling to 2 atm on discharge. In the recently developed low pressure version, the conventional Pt/H_2 electrode is replaced by a metallic hydride, for instance $LaNi_5H_6$. This metal hydride electrode discharges to the alloy $LaNi_5$, which is electrochemically rehydrided on recharge. The ambient hydrogen pressure of the electrode corresponds to the dissociation pressure of the alloy hydride, typically 1–2 atm at 20°C. However, the $Ni\text{-}H_2$ battery with metal hydride as the negative electrode suffers from a progressive loss of capacity on cycling, as a result of irreversible oxidation processes, but this deficiency has been largely overcome in the later design (Markin and Dell, 1981).

11.5.3 Negative electrode

The negative electrode is a consequence of fuel cell technology. It comprises a Teflon-bonded, platinum black catalyst supported on a photo-etched nickel grid. A membrane is pressed on the back of the grid. The hydrophobic nature of this membrane allows hydrogen to react efficiently with the platinum at the three-phase (solid platinum, liquid electrolyte, and hydrogen gas) boundary. The photo-etched nickel grid is generally used as the current collector to avoid cutting edges that may cause electrode shorts during stack assembly. The mixture paste of platinum black and Teflon solution is deposited into the holes of the grid prior to drying and sintering at low temperature. The quantity of the platinum black commonly used is close to 7–20 mg/cm^2.

Markin and Dell (1981) demonstrated the fabrication of metal hydride negative electrode by mixing small quantity of $LaNi_5$ with binder and pasted onto Ni grids. The active materials incorporated in the making of the electrode include AB_2 Laves type alloy (Moriwaki *et al.*, 1989) and AB_5 hexagonal close-packed alloy (Iwakura *et al.*, 1988).

11.5.4 Separator

The properties of the separator in $Ni\text{-}H_2$ batteries have great influence on overall battery performance. A suitable separator must not only act as an electronic insulator between the electrodes, and be chemically stable in concentrated alkaline electrolyte, but also (Borthomieu and Bernard, 2009):

- have high electrolyte retention characteristics based on its pore size and wettability properties;
- be able to prevent the transport of active substance particles (i.e. dendrites) from one electrode to the other;
- be resistant to the oxidising potential of the nickel electrodes and to oxygen liberated at this electrode during the latter portion of the recharge cycle;
- be dimensionally stable to maintain proper compression in the stack assembly throughout cell life; and
- manage the oxygen transportation from the surface of the nickel positive electrode to minimise the probability of popping from occurring.

These requirements are indispensable for improving the performance of the Ni-H_2 cell, and are the main factors for the lack of candidates available for the Ni-H_2 cells. A considerable number of studies have been performed to obtain separators with the required characteristics of thickness, oxygen bubble pressure, density, compression, and electrolyte retention. To date, at least three types of separators material have been used for Ni-H_2 cells, namely asbestos, Zircar, and polyamide. Detailed explanation is available elsewhere (Borthomieu and Bernard, 2009).

11.5.5 Electrolyte

Two key factors associated with electrolyte for Ni-H_2 batteries are known to be influential for battery performance: electrolyte concentration and quantity. The standard concentration of KOH presently employed in Ni-H_2 cells is 31 wt.%. However, Lim and Verzwyvelt (Lim and Verzwyvelt, 1988) pointed out that the cycle life of the Ni-H_2 cell for Low Earth Orbit (LEO) satellites can be extended by a factor of 5–10 simply by lowering the KOH concentration from 31 to 26 wt.%. The improved performance at lower electrolyte concentration can be explained on the basis that β/β nickel electrode reaction (see Section 11.2.3) is favoured at lower KOH concentration, whereas the α and γ phases (see Section 11.2.4) are promoted at higher concentration. The latter reaction is accompanied by a large lattice expansion of the active material (30% increases in volume), and subsequently increases the electrode swelling, which is the primary cause for the inferior cycle life performance at deep DoDs. On the other hand, the lower electrolyte concentration leads to a slight decrease in capacity, poorer low-temperature performance, and higher freezing temperatures with fully discharged cells.

Adequate quantity of KOH in the Ni-H_2 battery was proven necessary for enhancing performance (Thaller and Zimmerman, 1996). In a battery with insufficient electrolyte, the reserve is incapable of coping with the increase in positive electrode porosity (owing to swelling) over lifetime; consequently separator dry-out takes place, which eventually leads to an increase in battery impedance, capacity deterioration, and lower heat dissipation. Conversely, the porosity may not be sufficient to obtain a good transfer of oxygen gas in the battery with excessive electrolyte, thereby causing popping and short circuits within the electrodes. The adequate quantity of electrolyte in a battery is dependent on the battery design, and can range from 1.8 to 2.2 g/Ah (the electrolyte quantity is often quoted in terms of grams per ampere-hour of rated capacity).

11.5.6 Battery designs

As this type of battery has been used extensively in satellites, most of the battery design depends on the satellite platform, the power requirements, and the type of mission. For example, the power control unit of the satellite determines the number of cells according to the voltage range, and the required power sets the cell capacity; the allowable volume for the batteries is dependent on the type of platform. The following maximum DoDs are specified for Geosynchronous Earth Orbit (GEO) and LEO applications (Borthomieu and Bernard, 2009):

- GEO: 10–15 years from 50% to 80% DoD, and
- LEO: 3–7 years from 20% to 40% DoD.

In general the design of Ni-H_2 batteries utilises a base plate with holes/cavities to position the cells. Anodised aluminium is used for fabricating the base plate, and the cells are supported with cylindrical mounting sleeves made from aluminium or graphite/epoxy. A layer of room temperature vulcanised silicone rubber and adhesive (Solithane) serves three purposes (Borthomieu and Bernard, 2009) namely:

1. allowing the cell to be restrained mechanically during the launch vibration,
2. insulating the cell electrically from the battery structure, and
3. conducting the heat generated by the cell to the base plate (Borthomieu and Bernard, 2009).

Therefore, important parameters of the layer are conductivity, thickness, and interface characteristics. The sleeve design must be capable of withstanding the stress resulting from the pressure vessel expansion as the cell is charged

11.22 Design of SAFT Ni-H_2 battery (Borthomieu and Bernard, 2009).

and the pressure increases. The sleeves are then secured to the base plate by screws, as shown in Fig. 11.22 for a SAFT battery design. This typical design ensures there are two electrical insulation barriers from the cell to the base plate. The electrical connections are established with copper or aluminium wires, and are attached to the terminal studs. A power connector allows the battery to be charged and discharged. A good overview of the design, development, and application of Ni-H_2 batteries has been provided in a paper by Thaller and Zimmerman (2003).

11.5.7 Performance of Ni-H_2 batteries

Ni-H_2 secondary batteries are popular for their long cycle life and high specific energy. The LEO cycle life test results of some commercially available Ni-H_2 batteries cells are tabulated in Table 11.3 (Crompton, 2000a). For a LEO cycle regime at 40% DoD, all cells were operative without experiencing cell failure, even after 15 000 cycles.

The performance and characteristics of three commercially available Ni-H_2 cells are listed in Table 11.4 (Crompton, 2000a). As expected, the capacity decreased with increasing temperature. For the Superbird battery the capacity at 35°C was only about 57% of the capacity at 0°C, at 30°C about 67%, at 20°C about 81%, and at 10°C about 94%. For the Spacenet cells, the capacity at 20°C on average was approximately 84% of the capacity at 0°C, and 97% at 10°C.

Though Ni-H_2 batteries have the longest cycle life among the Ni-based batteries, premature failure of these batteries has been reported, and in one instance the failure was attributed to the rapid reaction of oxygen with hydrogen (Fuhr, 1987). In addition to the premature failure, the loss in capacity on storage has been a serious problem for spacecraft applications. The early Ni-H_2 batteries, which were fabricated by Hughes Aircraft Company using the hydrogen precharge, can lose up to 20–40% of their capacity after a period of storage longer than two to three weeks. This capacity loss was reversible, and most of the capacity loss can be recovered by either cycling the batteries at high rate in LEO with an 80% DoD or by subjecting the batteries to an extended period of trickle charge or periodic recharge. Manzo *et al.* (Manzo, 1990) found that the capacity loss on storage varied with regard to hydrogen pressure, storage temperature, electrode manufacturing processes, and cobalt content in nickel electrode.

Lim and Stadnick (1989) proposed an effective method for relieving the self-discharge phenomenon in Hughes-built Ni-H_2 batteries by precharging the nickel electrode. The capacity changes of cells with different precharges on storage are illustrated in Fig. 11.23. Batteries with a nickel precharge of 6.6% of the rated capacity, and H_2 precharges of 2.8% (0 atm), and 12.3% (3.4 atm) are included in each series. The capacity decreased steadily with storage without an indication of stabilising even after 120 days for the latter two hydrogen precharged cells. The rate of capacity loss was neither quantitatively correlated to the amount of H_2 precharge, nor reproducible with

Table 11.3 Summary of LEO cycle life test results for IPV Ni-H_2 battery cells

Manufacturer	No. cells	Capacity (A h)	Cycle regime	DoD (%)	T (°C)	Cycles	Status	Test laboratory
Hughes	3	50	LEO	40	10	19 000	No failures test continuing	GE Astro Space
Yardney	5	50	LEO	40	10	18 000	No failures test continuing	Martin Marietta
GE (now Hawker Energy Inc.)	16	50	LEO	40	10	15 968	No failures test continuing	Martin Marietta
Eagle-Picher	16	50	LEO	40	10	16 912	No failures test continuing	Martin Marietta

Source: Crompton, 2000.

Table 11.4 Characteristics of three types of Ni-H_2 cells and batteries

	Spacenet comset design +	Intelstat V comset design	Superbird (similar to Airforce Hughes design)
Individual cells			
Rated capacity (ah)	40	30	83
Weight (kg)	117	0.89	1.87
Capacity (ah) at 0°C	50	35	–
Capacity (ah) at 20°C	42	32	92
Discharge voltage (V)	1.25	1.25	1.24
Specific energy at 20°C (Wh/kg)	44.6	44.8	60.9
	27 cell batteries		
Battery weight (kg)	32.6	30.1	64.5
Energy at 10°C (Wh)	1328	1174	3448
Specific energy at 10°C (Wh/kg)	40.7	39	53

Source: Crompton, 2000.

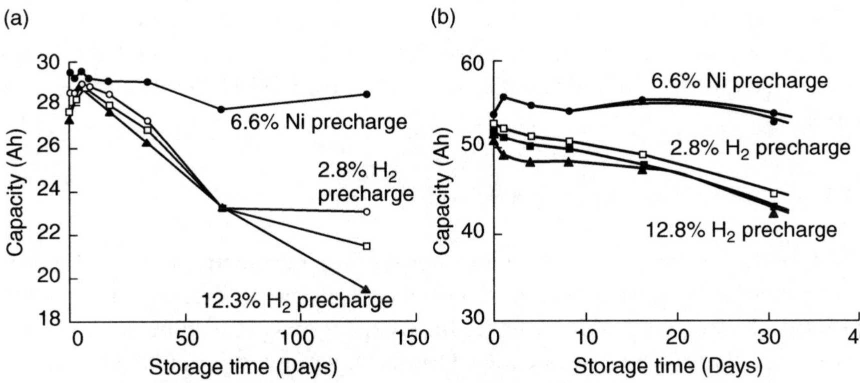

11.23 Effect of precharge on capacity changes with open-circuit storage time at 20°C of (a) 25 A h, and (b) 50 A h Ni-H_2 cells. Capacities were measured by discharge to 1.0 V after charging for 18 h at *C*/10 rate (Lim and Stadnick, 1989).

different cell fabrication lots. In addition, the presence of hydrogen after discharging, and its reaction with cobalt, are two possible causes responsible for the loss of capacity during storage. The latter was consistent with Zimmerman *et al.*'s finding (Zimmerman and Seaver, 1990). Nickel precharge has the advantage of low operation pressure, therefore possible long life because the cell pressure increased gradually with cycling. They concluded that the amount of nickel precharge has to be sufficient to use up all H_2 when the cell is fully discharged before storage. Later, Visintin and his co-workers (Visintin *et al.*, 1995) investigated the self-discharge kinetic of Ni-H_2 batteries, and concluded that the rate of self-discharge was determined by diffusion of the dissolved hydrogen in electrolyte.

As aforementioned, the cell life of a Ni-H_2 battery is affected by the performance of the separator. The asbestos diaphragms, which have been employed as separators in early Ni-H_2 batteries, suffer from corrosion at high temperature (> 100°C); they are also highly toxic and can cause a range of lung diseases. Last but not least, the availability of the basic separator materials is becoming an increasing problem.

Alternative separators, for instance Zircar and nylon, have therefore been in development for replacing the asbestos. However, Zircar (zirconium oxide cloth) has the drawback of being expensive and delicate from the assembly point of view. Studies have been attempting to replace the asbestos separator with Zirfon, which is a porous composite separator material composed of a polysulphone matrix and ZrO_2 (for increased wettability) fabricated via film-casting technique (Vermeiren *et al.*, 1998). The cycle life performance of the batteries using Zirfon was superior to those using the standard separator. The improved cycle life performance was attributed to better electrolyte retention and wetting of the electrode. A mathematical model was proposed to characterise the self-discharge of a nickel oxyhydroxide electrode in a hydrogen environment for a Ni-H_2 battery (Mao and White, 1991).

11.5.8 Final remarks on Ni-H_2 batteries

Ni-H_2 batteries have higher specific energy, longer cycle life, and good inherent protection against overcharge and overdischarge as compared to Ni-Cd batteries. Therefore, Ni-H_2 batteries have captured a great share of the space battery market in the last decades. However, a conventional Ni-H_2 battery relies on high pressure hydrogen to provide the necessary hydrogen capacity to match the positive electrode capacity. The indispensable pressure vessels required to seal the battery lead to high materials cost and complicated vessel sealing. In addition, many other mechanical components are needed for the battery assembly, and the volumetric density of the battery is therefore significantly decreased.

An attempt has been made to tackle the problems by fabricating a microfabricated nickel-hydrogen battery via thick film printing technique (Tam and Wainright, 2007). Though an improvement in the cycle percent utilisation was observed, the total cell impedance was increased as a result of poor oxygen recombination/water redistribution.

With the rapid rise of lithium-ion batteries in the recent years, the disappearance of Ni-H_2 batteries for space applications can be foreseen in the near future. Most satellite manufacturers have already modified their satellite power systems to suit lithium-ion batteries.

11.6 Nickel-zinc systems

The nickel-zinc (Ni-Zn) secondary battery was discussed as early as 1899 in a German patent credited to Michalowski (1899). Despite over a century of history on the Ni-Zn batteries, progress on this technology has been slow. It was not until recently that important advancements have made Ni-Zn batteries viable and competitive with other commercially available secondary batteries.

The Ni-Zn batteries have a high open-circuit cell voltage of about 1.75 V, which results in a very favourable energy density compared to that of Ni-Cd or lead/acid. In addition, zinc is relatively inexpensive and, in the absence of mercury additive, is environmentally benign. The properties of Ni-Zn batteries, such as energy density of 55–85 Wh/kg, power density of 140–200 W/kg, and a self-discharge rate of less than 0.8% per day, have attracted great interest for a wide range of applications, including cordless power tools, cordless telephones, digital cameras, and light EV sectors.

Several versions of Ni-Zn systems have been patented and developed; the version patented by Edison and Drumm in the 1930s for electric trains showed good performance, but suffered from short lifetime. As a matter of fact, the poor cycle life performance of this battery system (150–200 cycles maximum) is responsible for the slow development of Ni-Zn batteries.

11.6.1 Operating principles

Zinc is the 24th most abundant element in the Earth's crust. It is widely adopted as battery material because of its low potential (giving rise to a high cell potential), excellent reversibility (rapid kinetics), compatibility with aqueous electrolytes, low equivalent weight, high specific capacity and volumetric capacity density, low cost, low toxicity, and ease of handling. However, zinc electrodes are commonly accepted for primary batteries, but have found rather limited use in secondary batteries as they have serious issues regarding short or unpredictable lifetime when subjected to charge/discharge cycling. The redistribution of zinc active material (so-called shape change) and the development of deleterious zinc electrode morphologies (dendrites, filamentary growths, nodules) during recharge are responsible for the short lifetime of the Ni-Zn cells. These behaviours are affected by two important characteristics of zinc: (1) its high solubility in alkaline electrolyte (as the zincate ion $Zn(OH)_4^{2-}$), and (2) its fast electrochemical kinetics (Cairns, 2009).

The reaction of Ni-Zn cells is based on dissolution-precipitation reactions at the Zn electrode:

$$Zn + 4OH^- \underset{\text{charge}}{\overset{\text{discharge}}{\rightleftarrows}} Zn(OH)_4^{2-} + 2e^- \quad [11.20]$$

$$Zn(OH)_4^{2-} \underset{\text{charge}}{\overset{\text{discharge}}{\rightleftarrows}} ZnO + 2OH^- + H_2O \quad [11.21]$$

The reaction expressed in Equation [11.20] is the electrochemical reaction that gives rise to the electrical current and generates the soluble zincate ion. This zincate ion gradually undergoes the reaction expressed in Equation [11.21], which results in the precipitation of solid zinc oxide. Each of these two reactions is reversed during recharge. Combining this with the reaction at the nickel electrode, which is described in Equation [11.1], gives the overall cell reaction:

$$Zn + 2NiOOH + H_2O \underset{\text{charge}}{\overset{\text{discharge}}{\rightleftarrows}} ZnO + 2Ni(OH)_2 \quad [11.22]$$

This reaction corresponds to a theoretical specific energy of 326 Wh/kg. A working cell with about one-quarter of the theoretical value or in the range of 60–70 Wh/kg has been realised.

The zinc electrode is inherently thermodynamically unstable in aqueous electrolytes as its reversible potential is well below that of the hydrogen electrode. Therefore, there is always a driving force that favours the dissolution of zinc and the evolution of hydrogen, as expressed by:

$$Zn + H_2O \underset{\text{charge}}{\overset{\text{discharge}}{\rightleftarrows}} ZnO + H_2 \quad [11.23]$$

Heavy metals, such as lead and indium, have been added to the zinc electrode to minimise the rate of reaction and hydrogen evolution. The latter two additives have been found effective in minimising the hydrogen evolution, so this is no longer a significant problem.

11.6.2 Electrolyte

Similar to other Ni-based batteries, Ni-Zn batteries employ KOH as the electrolyte. Maximum conductivity of the electrolyte is obtained when it contains 30 wt.% of KOH. Alkaline solution is known to react with carbon dioxide in air to produce carbonate ions (CO_3^{2-}) in the electrolyte, therefore exposure of the electrolyte to air should be avoided.

As zinc is soluble in KOH, it can migrate in solution and deposit in a location distant from its origin. Zinc redistribution, which is largely uneven on the surface of the electrode after a number of cycles, is the main factor limiting the cycle life of zinc electrodes. As expected, this deficiency can be reduced by lowering the solubility of the zincate ion in the electrolyte. Nichols *et al.* demonstrated the zinc solubility in alkaline solution with various additives, as shown in Fig. 11.24 (Nichols *et al.*, 1985). Though the results show that the added anion has no significant effect on the ZnO solubility, the solubility appears to be strongly dependent on the concentration of hydroxide-ion in the electrolyte. Lower concentrations of KOH will reduce the favour zincate solubility.

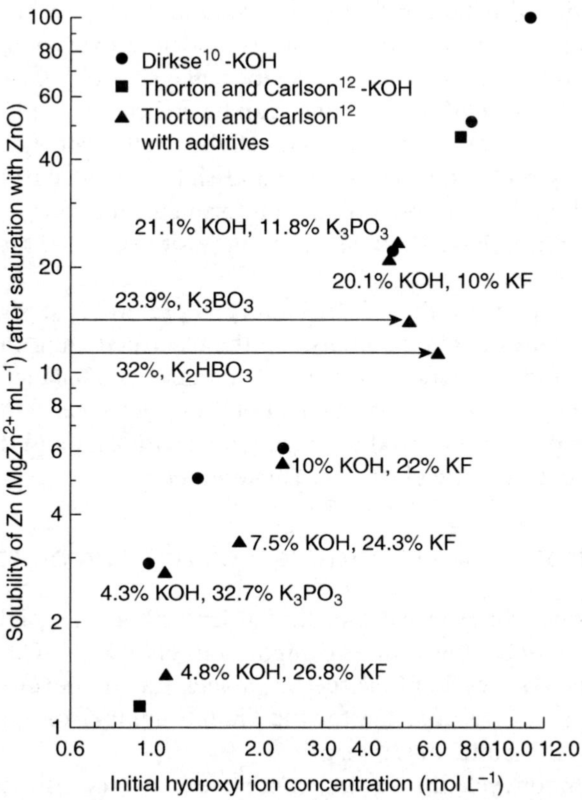

11.24 Solubility of ZnO in alkaline electrolytes with F-, BO_3^{3-}, and PO_4^{3-} compared to solubility of ZnO in KOH electrolytes. Note that the solubility rises as the square of the hydroxide concentration (Nichols *et al.*, 1985).

The effect of V_2O_5, ZnO, PbO, and $(NH_4)_2CS$ in electrolyte on the reversibility of the redox couple was examined by potentiodynamic polarisation and triangular potential sweep voltammetric techniques (Shivkumar *et al.*, 1995). V_2O_5 and PbO are not ideal, as the former stimulates corrosion through a reduction in the cathodic oxygen overvoltage, and the latter induces lead deposition and subsequently reduces the open-circuit voltage. The additives ZnO and $(NH_4)_2CS$ are found to be beneficial for improving the discharge capability and cycle life of the cell by efficaciously reducing the corrosion and passivation of the electrode.

11.6.3 Battery design

An adequate battery design is of great importance in handling the rapid kinetics of the zinc electrode and the solubility of zincate ion. In addition, the current density distribution tends to be non-uniform as a result of an exceptionally high exchange current density and low overvoltage at the zinc electrode, which exacerbates the shape change problem associated with Ni-Zn cells. The rate of shape change can be greatly reduced by proper design of the current collector. The conductivity of the current collector must be extremely high, and the location and shape of the current input/output tabs are important in providing a uniform current density distribution. It is preferable to have the tabs at the opposite edges of the positive and negative electrodes.

As previously explained, the Ni-Zn battery is preferably designed as a sealed cell to prevent the reaction between the electrolyte and the carbon dioxide in the air and to preserve the state of charge (SoC) balance between the positive and negative electrodes. In light of the sealed design, the oxygen evolved from the positive electrode is recombined with zinc, and thus a balance of the SoC between the electrodes is maintained.

11.6.4 Electrochemical performances of Ni-Zn batteries

The main attractive characteristics of the Ni-Zn cell are the high specific power and the relatively low cost as compared to Ni-MH, Ni-Cd, and lithium-ion batteries. Poor cycle life is the bottleneck for the development of Ni-Zn batteries, and is responsible for the limited number of applications and suppliers for this battery system.

Regardless of poor cycle life, several companies have recently developed Ni-Zn battery systems for high power applications, examples being sub-C cell by PowerGenix, and cell with an ion-exchange polymer electrolyte by Xellerion. The discharge performance of the PowerGenix cell is presented in Fig. 11.25 (Cairns, 2009). The discharge medium voltage decreased with

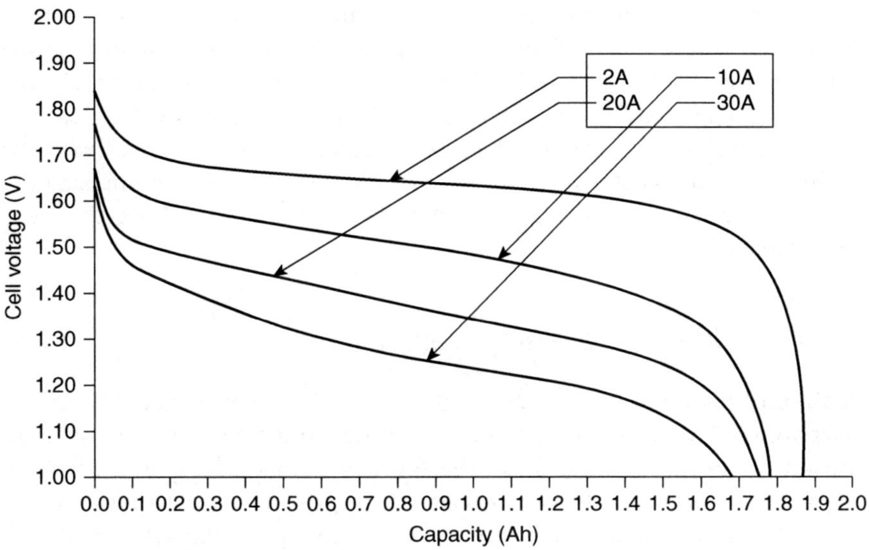

11.25 Discharge curves for the PowerGenix sub-C cell at different discharge rates (Cairns, 2009).

increasing discharge current, with only a subtle decrease in capacity when the battery was discharged at high current, indicating the excellent high rate discharge performance of the battery.

Considerable research effort has been spent on exploring ways of either eliminating or minimising the effect of two failure mechanisms associated with the poor cycle life of Ni-Zn cells: namely the redistribution of active material (shape change), and the formation of unwanted Zn electrode morphologies (dendrites, etc.) during recharging. The research can be divided into five distinct categories:

1. Addition of additives to the electrode.
2. Addition of additives to the electrolyte (as previously discussed).
3. Development and improvement of separators.
4. Miscellaneous techniques.
5. Synthesis and morphology alteration of active materials.

11.6.5 Improvement of electrochemical performance of Ni-Zn batteries

The introduction of additives to the zinc electrode as a means to improve the cycle life of the Ni-Zn has been extensively investigated. In the early stages of research, mercury was a common additive to minimise the corrosion in zinc active material, as the hydrogen overvoltage of mercury is higher than that of zinc. Mercury was found to be extremely effective in inhibiting the self-discharge of the zinc anode (Gregory *et al.*, 1972). Unfortunately, the effect of amalgamation was also detrimental to the cycling performance of the zinc anodes; furthermore, the rate of shape change increases with the addition of mercury. A substitute additive that is capable of minimising the rate of self-discharge while maintaining or reducing that of shape change is therefore essential.

11.6.6 Additives to electrodes

McBreen and Gannon (McBreen and Gannon, 1983) investigated the effect of various additives in pasted zinc electrodes and found that heavy metals, such as Pb, Cd, Tl, and In in the form of their oxide/hydroxides were the most successful additives in terms of improving the current distribution of the electrode and reducing the shape change of the active material. Another series of additives which has received considerable attention was that of the alkaline-earth-metal oxides or hydroxides. Of these additives, calcium hydroxide was found to be the most promising for improving the cycle life of the zinc electrodes. The formation of insoluble calcium zincate is beneficial for minimising the migration of zincate away from the electrode, hence decreasing the shape change (Gagnon, 1986). One shortcoming associated with the calcium hydroxide is that it reduces the zinc utilisation; however, optimum efficiency can be obtained by controlling the molar ratio of $Ca(OH)_2$ to ZnO to greater than 0.5–1.0 (Gagnon and Wang, 1987).

Organic additives such as polytetrafluoroethylene (Duffield *et al.*, 1985), poly(vinyl alcohol) (Poa and Lee, 1979), and poly(ethylene) (Cenek *et al.*, 1977) have also been extensively investigated. The main advantages of these organic additives are mainly mechanical – for example, binding the electrode – and also structural – by providing a stable network to retain the zinc active material, thus slowing the shape change. As these polymer additives are inherently non-conductive, high concentration (> 10%) addition can result in a reduction in electrode cycle life and capacity (Hampson and McNeil, 1985).

The main synthesis approach of mixing the abovementioned additives with ZnO is by physical mixing. However, such a process cannot provide homogenous mixture or sufficient contact between ZnO and additives. Zhang *et al.* (2008b) proposed a novel approach for preparing ZnO with conductive-ceramic nanocomposite using a homogeneous precipitation method. Figure 11.26 displays the morphologies of the nanocomposite with 9 wt.% conductive-ceramic, and it can be seen that ZnO nanorods with an average size of about 200 nm were distributed uniformly.

11.26 SEM images of the ZnO/conductive-ceramic nanocomposite synthesised by homogeneous precipitation method at (a) low, and (b) high magnification (Zhang *et al.*, 2008b).

The variation of specific discharge capacity with cycle number of pure nanosized ZnO, and ZnO/conductive-ceramic nanocomposites, is illustrated in Fig. 11.27. The specific discharge capacity and capacity retention rate after 50 charge/discharge cycles were found to be 513 mAh/g and 52.2%, respectively. Electrodes with added conductive-ceramic nanocomposite exhibited higher discharge capacity and better cycle life; the beneficial effect of this additive was most profound for electrode with 14 wt.% addition of conductive ceramics (Zhang *et al.*, 2008b). A maximum discharge capacity of 659 mAh/g and capacity retention rate of 99.5% were obtained. The improved performance in these electrodes with additive was ascribed to uniform redistribution of active material on the surface of the electrode, and good preservation of the initial shape of the active material. Consequently, the

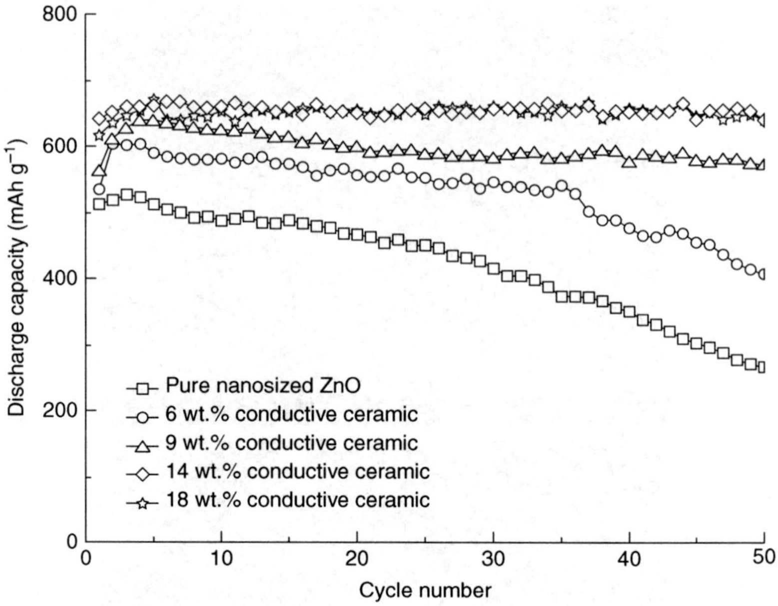

11.27 Discharge capacity as a function of cycle life for electrodes with pure nanosized ZnO and ZnO/conductive-ceramic nanocomposites at 0.2 C rate (Zhang *et al.*, 2008b).

unwanted dendrite growth and shape change of the zinc electrode can be effectively inhibited.

11.6.7 Development and improvement of separators

Cycle life performance of the Ni-Zn cell appears to be related to separator's mass transport properties; also, the lower the electrolytic resistivity and higher water permeability, the longer is the cycle life. The correct selection of separator is therefore vitally important to the successful operation of a battery. According to Bass *et al.* for Ni-Zn batteries, the separator should be (Bass *et al.*, 1991):

- resistant to degradation by either the electrolyte and/or active materials,
- low in electrical resistance and high in ionic conductivity,
- effective in preventing particle migration between electrodes,
- high in wettability by the electrolyte,
- mechanically strong and flexible enough to withstand the battery fabrication, and
- preferably inexpensive.

There are two basic kinds of separators employed in Ni-Zn batteries, namely membranous and microporous (Lundquist, 1983). The former are those in which ionic transport occurs through the interaction of the hydrophilic groups attached to the polymer with the ionic groups in the electrolyte; whereas the ionic transport in the latter occurs by diffusion through discrete pores. It was found by these authors that the rate of shape change was greater in batteries with membrane separators than those with microporous separators. Microporous separators with small pore size of 300 Å or below in diameter have superior mass transfer characteristics, and thus beneficial to the cycle life of the batteries.

The development of separators for the Ni-Zn cell has progressed through the selection of cellophane, grafted and radiation cross-linked polyalkenes (Hsu and Sheibley, 1982; Sheibley *et al.*, 1983) and inorganic separators (Sheibley and Manzo, 1980). Cellophane was first chosen but it was unstable in alkaline solution. The radiation polymers consist of a film of the polyalkene which is cross-linked to give a uniform three-dimensional structure. Radiation polymers have the advantage that they hinder the growth of any penetrating species such as dendrite bridging. The grafting and radiating of the polymer in the film increase its conductivity and oxidation resistance to alkaline electrolyte. However, these separators do not swell upon wetting, thus allowing more space for the electrodes to expand on cycling. This is detrimental to the batteries, as shape change is encouraged (Bass *et al.*, 1991). Further research in improvement in the performance of Flexel cellophane film and Viskase fibre-reinforced tubular sausage casing polyvinyl alcohol (PVA) film as separators in alkaline primary and secondary cells with zinc anode has been carried out (Lewis *et al.*, 2001). The results show both separators exhibit similar discharge performance up to 50 cycles, with no cells lost to shorting for the latter. Most important of all, the available volume generated by the reduction in separation wet swollen thickness for energy density increases through the use of just one layer of tubular SC with PVA. This additional volume available in the cell case will allow at least a 25% increase in volumetric energy density.

Rigid separators were firstly made from aluminosilicate compositions, but it was not long before the flexible separators were found to have greater use and appeal. As compared with cells with cellulose separators, the rigid separators give increased cycle life, whereas the flexible ones give slightly less. However, these inorganic separators for silver/zinc cells were mainly developed for space and military applications, and they were prohibitively expensive for everyday civilian uses. At present, the use of combined separator systems has been the major research direction. Future work on the separators should include the coating of metallic-nickel or copolymer on the separators, so as to improve the slightly poorer cycle life of these separators.

11.6.8 Miscellaneous techniques for improving the cycling life of Ni-Zn batteries.

An alternative approach to improving the cycle life of the Ni-Zn cells, which has received a considerable amount of attention (Bennion, 1980; Chin and Venkatesh, 1981), is by altering the charge current profile. This is achievable through various methods, for example, pulsing the current, superimposing alternating current (a.c.) on direct current (d.c.), periodically reversing the current, and multi-component pulse charging. In the first method, a high current peak is introduced to create a high overvoltage on the surface of the Zn electrode, which can then activate a large number of nucleation sites for Zn redeposition. Thus the dendrite formation can be inhibited with increased cycle life. The rest period following the high current allows local zinc concentration gradients to relax by diffusion into the depleted diffusion layer, and consequently a more uniform deposit is achieved by electrodeposition from a replenished diffusion layer.

The effect of pulsed d.c. current on the electrodeposition of zinc was investigated (Arouete *et al.*, 1969), and the authors concluded that the nature of the deposit with pulsed charging depended on the current density, the amount of charge passing through the electrolytic cell, and the on and off time. Katz *et al.* (Katz *et al.*, 1988) examined the effect of various pulse-charging regimes on the cycle life and performance of Ni-Zn cells. Using a pulse-charging regime of 30 ms*on*/90 ms*off* with 16 mA/cm^2 peak current density, the cycle life of the batteries can be improved two- to three-fold as compared to those charged under constant current.

Another modification on the charging regime is the superimposition of an a.c. on a constant d.c. waveform, as was demonstrated by Chin and Venkatesh (Chin and Venkatesh, 1981). They concluded that with superimposed a.c., the rate of nucleation was greatly enhanced during the cathodic deposition reaction as a result of more uniform zinc deposition on the electrode.

The final method of modifying the charging waveform was to use a multi-component pulse current (Binder and Kordesch, 1986). This consisted of a deposition current, a dissolution current, and a rest period. Though the electrodeposition using a multi-component pulse current was found to be beneficial for the plating of zinc, the required time for charging was much longer than that using constant current. In the case of the optimised pulse current, it can be four times longer (Bass *et al.*, 1991). This is obviously undesirable.

11.6.9 Synthesis and morphology alteration of active materials

One of the more recent approaches for improving the cycle life of the Ni-Zn cell was by altering the synthesis method for calcium zincate. Preparation of calcium zincate has been well-studied since 1980s owing to its effect on improving the performances of zinc electrode. Several researchers reported the preparation of calcium zincate by chemical coprecipitation under strong alkali solution, such as KOH (Wang and Wainwright, 1986; Wang, 1990; Zhang *et al.*, 2001), or by ball-milling (Yang *et al.*, 2004). However, all these methods have their vital shortcomings: (a) the calcium zincate prepared by chemical coprecipitation requires repetitive washing in distilled water, which is a tedious process; and (b) the quantity of product prepared by milling method is small and is therefore unsatisfactory for mass production. Wang *et al.* (Wang *et al.*, 2008) synthesised calcium zincate by solid-phase synthesis method without the use of strong alkali. This synthesis method yielded calcium zincate with well-crystallised tetragonal particles, and when employed as active material in the zinc electrode, the battery exhibited good cycle life and a high discharge platform.

Wu *et al.* (2009) modified ZnO with silver by a facile silver mirror reaction process so as to inhibit the dissolution of ZnO into the alkaline electrolyte. Figure 11.28 presents the cycle performance between batteries with unmodified and with Ag-modified anodes. Both electrodes have comparable maximum discharge capacity of about 530 mAh/g; however, the capacity fading rate was 4.85 mAh/cycle for the battery with unmodified anode, and 0.89 mAh/cycle for the Ag-modified anode, representing a better cycle life for the latter. The improvement of Ag-modified ZnO on the discharge performance and cycle life can be ascribed to better electrical contact between

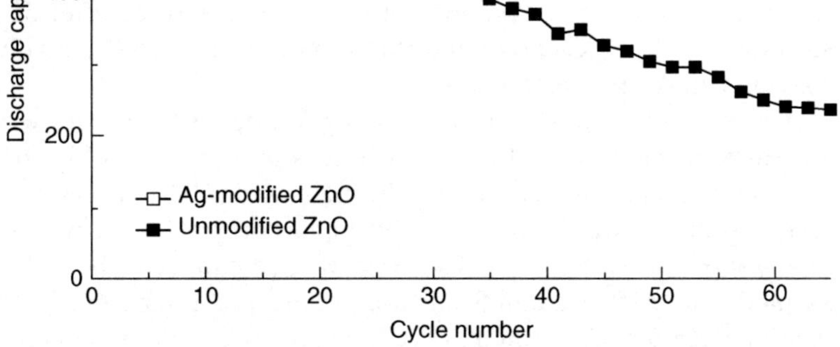

11.28 Cycle life performance of Ni-Zn cells with Ag-modified ZnO and unmodified ZnO anodes at 0.2 C charge/discharge rate (Wu *et al.*, 2009).

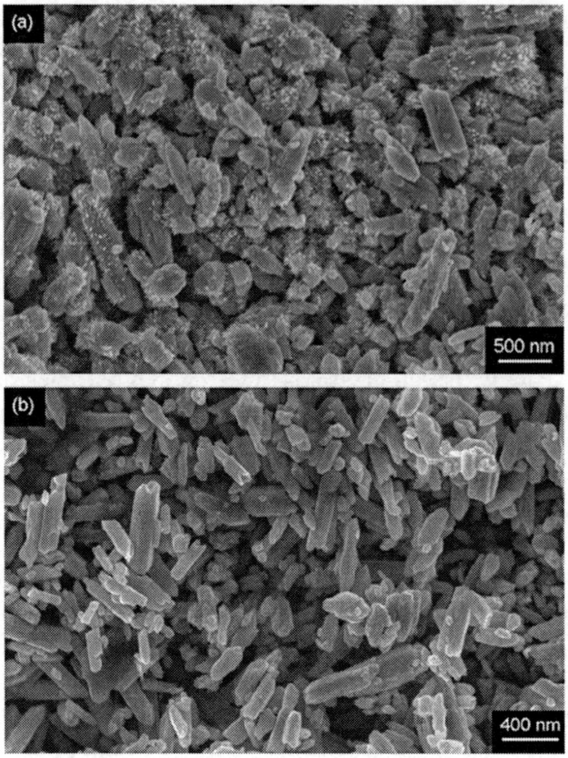

11.29 SEM images of (a) unmodified ZnO, and (b) Ag-modified ZnO anodes after 31 cycles (Wu *et al.*, 2009).

ZnO/ZnO and ZnO/Ni foam with the presence of Ag, and hence facilitating electron transfer in the ZnO electrodes. Second, the Ag nanoparticles modified on the ZnO surface can reduce the contact surface of the ZnO and electrolyte, suppress the dissolution of ZnO, and subsequently inhibit the formation of dendrite. This statement is supported by the SEM images showing the surface morphology of the anode electrode after 31 cycles, as shown in Fig. 11.29. Clearly, dendrite formation has been effectively reduced for cell with Ag-modified anode.

Another novel approach to inhibiting the dendritic growth was demonstrated by Yang *et al.* (2010), in which the size and morphology of the ZnO material was carefully controlled. In their work, ZnO nanowires with perfect crystallinity were synthesised by a hydrothermal route without the use of any substrate or template. The cycle life of a battery with ZnO nanowires as anode material was significantly improved as compared to that with conventional ZnO, as illustrated in Fig. 11.30. In addition to improved cycle life, the battery with ZnO nanowires also shows excellent discharge capacity,

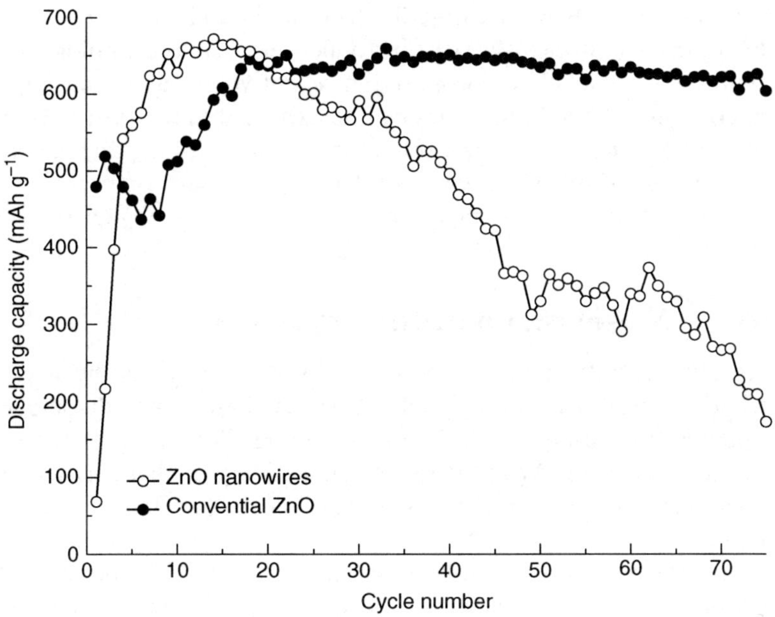

11.30 Electrochemical cycle behaviour of cells with ZnO nanowires and the conventional ZnO at 0.2 C charging/discharging rate (Yang *et al.*, 2010).

higher discharge voltage, and lower charge voltage. The absence of dendrite in the ZnO nanowire electrode was the main factor for the improved electrochemical performance of the battery.

11.6.10 Final remarks and future challenges

Regardless of the recent advances in extending the cycle life of the Ni-Zn batteries, the current status is still far from satisfactory. Preferably, the cycle life of the batteries should be prolonged to 1000 cycles or more for a greater market penetration. Nevertheless, the exceptionally high specific energy, specific power, and high nominal cell voltage of the Ni-Zn cells are attractive for applications such as power tools, electric bicycles and scooters, electric lawnmowers, emergency batteries for automobiles, and power sources for camping.

The paramount factors which limit the cycle life of the Ni-Zn batteries are the redistribution of active material (shape change) and the formation of unwanted Zn electrode morphologies (dendrites) during recharge. Improvements can be made via careful control of the electrodes, electrolyte, separator, cell design, and formation and charging. The performance of the zinc electrode is undoubtedly the dominant element affecting the cycle life of these batteries. With the introduction of nanotechnology and various synthesis methods been proposed in recent years, significant improvement in cycle life of Ni-Zn batteries can be expected in the future. However, for Ni-Zn batteries to find their role in the future energy devices, the commercial success of this battery system has to take place quickly as the performance and cost of the competing systems are improving every year.

11.7 Nickel-metal hydride systems

An Ni-MH battery utilises hydrogen storage alloys as the negative electrode material. The commercialised Ni-MH batteries in the late 1980s utilised mischmetal-based AB_5 hydride-forming alloys as active material in the negative electrode. With ever-increasing energy demand, new intermetallic compounds have been developed, leading to a promising future in these batteries.

Ni-MH batteries are in high volume commercial production for small portable battery applications, achieving a worldwide production of over 1 billion batteries per year (Fetcenko *et al.*, 2007). The driving force for such a rapid growth of Ni-MH is both technical and environmental. In addition to the explosive growth of Ni-MH batteries in portable electronic devices, they have also become one of the dominant advanced battery technologies for EV and hybrid electric vehicle (HEV) applications owing to their reasonably high volumetric and mass energy density, improved high rate discharge capability, good durability, high resistance to overdischarging, and low cost.

Unlike Li-ion batteries, Ni-MH batteries are free from dendrite formation, meaning the possibility of short circuit and overheating is greatly minimised. In addition to the sealed Ni-MH batteries, prismatic design with superior packaging can be realised. Despite the higher price in comparison to lead-based batteries, Ni-MH battery systems have been mainly adopted for those applications where lead acid batteries are inferior: continuous discharge power capability, recharge rate and service life.

11.7.1 Operating principle

An Ni-MH battery comprises a hydride-forming MH electrode, and a nickel electrode electrically insulated by the separator. Both separator and electrodes are impregnated with a strong alkaline solution (generally 6M KOH) that provides for the ionic conductivity between the two electrodes. A schematic depicting the operating principle of a sealed secondary Ni-MH cell is shown in Fig. 11.31 (Shukla *et al.*, 2001). The electrochemical reaction in the

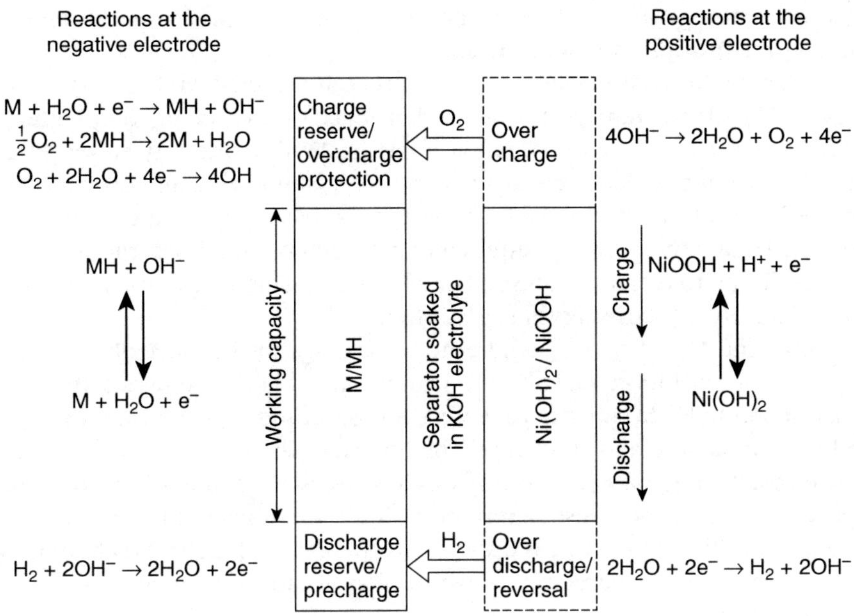

11.31 Operating principle of a sealed Ni-MH cell (Shukla *et al.*, 2001).

nickel positive electrode is presented in Equation [11.1], and the reaction in the MH negative electrode is presented as follows:

$$MH + OH^- \underset{\text{charge}}{\overset{\text{discharge}}{\rightleftarrows}} M + H_2O + e^- \qquad [11.24]$$

As before, nickel hydroxide $Ni(OH)_2$ at the positive electrode is oxidised to nickel oxyhydroxide NiOOH on charging, and reduced back to nickel hydroxide during discharging. At the negative electrode on the other hand, reduction of water produces atomic, adsorbed hydrogen that diffuses into the lattice of the intermetallic alloy to form a metal hydride during charging, and a reverse reaction takes place during discharging. It is noteworthy that the hydrogen stored in the MH electrode is in equilibrium with that in the gas phase. Accordingly, the overall cell reaction can be presented as follows:

$$NiOOH + MH \underset{\text{charge}}{\overset{\text{discharge}}{\rightleftarrows}} Ni(OH)_2 + M \quad (E_{cell} = 1.32\ V\ vs\ SHE) \qquad [11.25]$$

The cathodic reaction in an Ni-MH battery proceeds based on a homogenous solid-state mechanism via proton transfer between $Ni(OH)_2$ and the hydrogen storage alloy, whereas the anodic reaction proceeds through a dissolution–precipitation mechanism.

This charge/discharge mechanism differentiates the Ni-MH battery from the Ni-Cd battery. Many performance deficiencies observed in other battery systems, such as changes in crystallography, mechanical integrity, and surface morphology of the electrode as a result of recrystallisation and dissolution, and reduced electrical conductivity in the oxidised state, are inhibited in the Ni-MH system allowing better compact assembly, longer durability, and the possibility to realise an all solid-state battery using a proton-conducting solid electrolyte (Hariprakash *et al.*, 2009).

Unlike the Ni-Cd battery where no net change in the electrolyte quantity and concentration was observed over charge/discharge cycles, the cell reaction in the Ni-MH battery involves the generation and consumption of water upon charge and discharge. This prevents the cell from drying, and hence results in good gas recombination, good high- and low-temperature operations, and good resistance to cycle life limitations induced by corrosion and swelling. However, as opposed to Ni-Cd batteries, Ni-MH batteries have a faster self-discharge rate, lower electrical capacity, and are less tolerant to overcharging.

In nickel oxide electrode (cathode), the oxygen evolution takes place as a parallel and competing reaction owing to the fact that nickel oxide electrode is thermodynamically unstable. The parasitic reaction during cell charge is described as Equation [11.17], and on the negative electrode, which will increase the battery's internal pressure:

$$MH + MH \rightarrow 2M + H_2 \uparrow \qquad [11.26]$$

The Ni-MH battery is designed to be positively-limited with negative-to-positive capacity ratio varying between 1.5 and 2, as to ensure the battery can be operative under varying conditions. Typically, the discharge reserve is kept at about 20% of the positive capacity. The reaction described in Equation [11.17] depicts that the oxygen evolution occurs at the positive electrode during overcharging, which diffuses to the metal hydride electrode and combines to form water. The reaction on the negative electrode was described in Equation [11.24], and combining these two reactions (Equation [11.17] and [11.24] will give:

$$4M + 2H_2O \rightarrow 4MH + O_2 \qquad [11.27]$$

Under deep discharging, hydrogen evolution occurs at the positive electrode as a result of inevitable differences in storage capacities of series-connected cells in a Ni-MH battery, and it is oxidised to water at the MH electrode. Accordingly, there are recombination mechanisms for both hydrogen and oxygen gases evolved during overdischarging and overcharging, respectively, which facilitate sealed operation of the Ni-MH cells. The following reactions occur during the cell overcharge:

$$\frac{1}{2}H_2 + OH^- \rightarrow H_2O + e^- \qquad [11.28]$$

and

$$H_2O + e^- \rightarrow \frac{1}{2}H_2 + OH^- \qquad [11.29]$$

11.7.2 Hydride-forming electrode materials

Different types of hydride-forming compounds as active material for the MH electrode have been investigated extensively. They can be classified into five major categories, namely AB_5, AB_2, AB, A_2B_2, and A_2B. Table 11.5 summarises the crystal structure and metals available for each of the MH system. The 'A' in these MHs is composed of hydride-forming rare earth metals, whereas the 'B' is made up of non-hydride-forming supportive metals. Examples of some popular metal hydrides are $LaNi_5$/$CeNi_5$ for AB_5, ZrV_2/$ZrMn_2$ for AB_2, TiFe for AB, Y_2Ni_7 for A_2B_7, and Mg_2Ni-Ti_2Ni for A_2B. In this section, only AB_5 and AB_2 metal hydrides will be discussed in detail as they have been extensively studied and commercialised; other metal hydrides including non-stoichiometric metal hydrides (Hu, 1998; Shu *et al.*, 2003; Li *et al.*, 2010a), AB_{3-4}(Di *et al.*, 2000; Li *et al.*, 2010c), magnesium (Rongeat *et al.*, 2006; He *et al.*, 2008), and Ti-V-based alloys (Guiose *et al.*, 2009) are described elsewhere and are therefore excluded from this chapter.

AB_5-type compounds

The first employment of intermetallic compound as negative electrode for Ni-MH batteries was $LaNi_5$, which can be dated back to 1970s. The $LaNi_5$ alloy exhibits favourable properties, such as fast activation, reasonable theoretical capacity of 372 mAh/g (Boonstra and Bernards, 1990), and good reversibility, which have made it the dominant material in negative electrode for Ni-MH batteries over the last two decades. However, the major concern of this alloy is its poor cycle life, induced by the decomposition of

Table 11.5 Families of intermetallic compounds

Intermetallic compounds	Examples	A	B	Structure
AB_5	$LaNi_5$, $YCoH_3$	Group III (including rare earths and Th)	Group VIII	Haucke phases, hexagonal
AB_2	ZrV_2, $ZrMn_2$	Group III, rare earth or Group IV metal	Group VIII (Group II, IV, VI or VII)	Laves phase, hexagonal or cubic
AB	TiFe, ZrNi	Group IV or rare earth	Group VIII	Cubic, CsCl type
A_2B_7	Y_2Ni_7, Th_2Fe_7	At least one rare earth, and also includes Mg	Include at least Ni, the atomic ratio of x to y is between 1:2 and 1:5 (A_xB_y)	Hexagonal, Ce_2Ni_7 type
A_2B	Mg_2Ni, Ti_2Ni	Group IV or Group IIA	Group VIII	Cubic, $MoSi_2$- or Ti_2Ni-type

the alloys into lanthanum hydroxide $La(OH)_3$ and nickel particles in concentrated KOH solution, and gradual alloy pulverisation. Only 12% of its initial capacity remained after 400 cycles (Willems and Buschow, 1987). The driving force for the $LaNi_5$ decomposition in aqueous solution is the strong affinity of $LaNi_5$ for water. The free enthalpy change of the corrosion reaction is –472 kJ (mol $LaNi_5)^{-1}$, which is almost four times larger than the formation enthalpy of $LaNi_5$.

Willems *et al.* (1987) demonstrated the cycle life of this alloy can be significantly improved by substituting Ni with Co. In addition, they also found that the residual capacity after cycling is inversely proportional to the volume expansion $\Delta V/V$ (measured after exposure to hydrogen gas at 150 atm) of the alloy. The improved performance was ascribed to a decrease in volume expansion accompanying the hydride formation, which prevents the diffusion of lanthanum to the surface area.

The effect of a series of substitutes for Ni on the cycle life of the negative electrode was investigated by Sakai and his co-workers (Sakai *et al.*, 1990). The effectiveness of substituting elements to improve the cycle life increased in the order: Mn, Ni, Cu, Cr, Al, and Co. A large storage capacity of the alloy is generally accompanied by a large volume expansion ratio, which leads to an increase in the internal stress. Under stress, the pulverising rate depends on the mechanical strength of the alloy. The more tenacious alloys, such as

$LaNi_{2.5}Co_{2.5}$, have lower pulverising rates. Continued pulverisation of the alloy causes a larger increase in the oxidised surface area, leading to decay in storage capacity. The depth of the oxidised layer which internally protects the alloy depends on the composition of the alloy. Thus, alloys containing such elements for instance as aluminium, silicon, titanium, and zirconium seem to form a more protective oxide layer on the alloy surface, which prevents further internal oxidation of the alloy.

Though the addition of Co has been demonstrated to be beneficial in improving the cycle life and corrosion resistance of the negative electrode, it remains the most costly element of the alloy. An evaluation of the necessity of Co was performed by Latroche *et al.* (Latroche *et al.*, 1999), in which the influence of Co content (5 and 10%) in Mm-based (Mm = Mischmetal) AB_5 alloys on its electrochemical behaviour by *in situ* neutron diffraction was studied. In addition to the well-established one-step phase transformation of the $LaNi_5$ alloy from α to β upon charging as observed in the 5% Co sample, the 10% Co sample exhibited a two-step phase transformation ($\alpha \rightarrow \gamma$ and $\gamma \rightarrow \beta$) through the formation of a metastable intermediate γ phase. The presence of γ phase in the Co-rich sample relaxes the large strain that was observed in the one-step $\alpha \rightarrow \beta$ transformation in the 5% Co sample, consequently limiting the decrepitation (pulverisation) of the alloy.

The expensive rare earth La has been replaced by other less expensive metals, for example Ce. The cycle life of the negative electrode of an Ni-MH battery can be significantly improved by substituting La with 20% Ce ($La_{0.8}Ce_{0.2}B_5$) (Adzic *et al.*, 1995). The formation of a protective layer of CeO_2 on the alloy surface was found to retard the corrosion rate effectively and hence increase the cycle life of the electrode. Other metals such as Nd and Pr also improve the cycle life in a similar fashion. However, the trade-off for the improved cycle life by substituting La with these metals is a decrease in capacity as a result of decreased cell volume of the alloy. This issue can be addressed by optimising the ratio between the La and the substituting metal, and alternatively by substituting Ni with metal that has larger atomic radius than Ni, for example Al (Bliznakov *et al.*, 2008).

AB_2-type compounds

Hydride-forming compounds with increased weight capacity with respect to AB_5 compounds have been explored in order to meet the demand of high-capacity accumulators for portable devices and (hybrid) EVs. The Laves phases ZrV_2, $ZrCr_2$, and $ZrMn_2$ can be hydrogenated up to 3.6 H per formula unit, which represents a capacity of 500 mAh/g (Joubert *et al.*, 1996). However, the AB_2 compounds suffer from several problems such as difficulties in preparing single-phase compounds, passivation and slow activation, and corrosion (Notten and Latroche, 2009), the last two being the most significant. Substantial research thus has been focused on these topics.

The slow activation of the AB_2 alloys is related to passivation or surface corrosion of the alloy particles in concentrated electrolyte solution. Cycles required for activation can be as high as 50 for $Zr(V_xNi_{1-x})_2$ alloy electrodes (Züttel *et al.*, 1994). A number of techniques have been employed in order to address this issue; these include surface treatments, incorporation of highly electrocatalytic alloy or a pure metal, and secondary phase precipitation. An oxidation surface treatment has been proven effective in improving the activation of some WaTi-Zr-V alloy electrodes from 45 cycles to only 3 cycles (Sawa *et al.*, 1989). The oxidation treatment was performed by annealing the electrode under low pressure oxygen (20–100 mm Hg) at a temperature of 250–300°C. Wakao and his co-workers (Wakao *et al.*, 1991) reduced the activation time for Zr-V-Ni alloy electrode by immersing the electrode in KOH solution for up to 5 days. Züttel *et al.* (1994) investigated the influence of different chemical etchings of the alloy surfaces and found that the best electrochemical performance was obtained with hydrofluoric acid treatment.

The effects of fluorination on AB_2 alloys on their electrodes and battery performances have been examined by Li *et al.* (2002). The battery using fluorinated AB_2 alloys displayed better activation properties and longer cycle life at 0.2C as compared with the battery with untreated AB_2 alloys. This improvement was ascribed to the elimination of Zr and Mn oxides and the deposition of $Ni(OH)_2$ on the alloy surface after fluorination treatment. The deposited $Ni(OH)_2$ could be converted to metallic Ni during battery charging, and serve as an electrochemical catalytic agent and subsequently speed up the activation reaction.

The gradual decrease in hydrogen storage capacity of the metal hydride during repeated cycling is attributed to the alloy deterioration of the alloy powder and dissolution of the constituent species (Ovshinsky *et al.*, 1993). Kim *et al.* (1998) studied the effect of Cr or Mn substitution with Ni on the cycle life of the Zr-Ti-V-Ni electrodes. They found the substitution of Ni by Cr increases the cycle life of the electrode, due to the formation of a V-Cr solid solution which inhibits vanadium dissolution into alkaline electrolyte, while the substitution of Mn for Ni increases the discharge capacity, but shortens the cycle life as a result of the formation of highly corrosive V-Mn phase. Furthermore, a Zr to Ti ratio of 3 to 1 is found to have the most stable cycle life; this can be attributed to an increase in the corrosion resistance of the alloy due to the formation of less corrosive V-Mn phase.

Indeed, the electrochemical performance of the electrodes can be significantly improved by optimising both the A- and B-side composition. Elements that can increase the number of hydrogen atoms stored per metal atom include Mg, Ti, V, Zr, Nb, and La; elements that can adjust the metal-hydrogen bond strength (hence stabilise or destabilise the alloy) include V,

Mn, and Zr; catalytic elements to facilitate charge and discharge reaction rate and gas recombination include Al, Mn, Co, Fe, and Ni; and elements that can impart desirable surface properties such as oxidation and corrosion resistance include Cr, Mo, and W (Ovshinsky *et al.*, 1993). Compounds with up to eight elements are commonly used in AB_5 alloys with improved performance, and perhaps for the same reasons, the larger the number of elements involved in the alloy composition, the better are its electrochemical properties and its resistance to surface oxidation. In other words, electrochemical performance of the MH electrode can be engineered upon optimisation of the alloy composition as a means to pursue the ever-increasing demand in energy.

11.7.3 Electrolyte

Similar to other Ni-based batteries, 6M of KOH solution has been widely adopted as the electrolyte for Ni-MH batteries. As this concentrated solution is very aggressive to the alloys, several methods have been introduced to improve the performance of the electrodes. Wang *et al.* (2002) found the cycle life of the La-based AB_5 alloy electrode can be significantly extended by introducing 0.5M ZnO to the 6M of KOH solution. The underpotential deposited Zn on the surface of the alloy inhibited the formation of $La(OH)_3$, increased the conductivity and/or retarded the disintegration process, all of which are beneficial to the cycle life performance of the electrodes.

A recent study was performed on the effect of KOH concentration on the performance of Ni-MH batteries (Khaldi *et al.*, 2009). Cells with 1M and 8M electrolyte concentration have been examined for their electrochemical performance. The results suggest the cell with low KOH concentration has higher maximum capacity, better cycle stability, lower polarisation after activation, and better corrosion resistance. It demonstrated the possibility of utilising a less aggressive electrolyte (lower concentration) for improving the performance of the Ni-MH batteries.

The electrochemical performance of the MH electrode has been investigated in 6M KOH electrolytes with different compositions (Wang *et al.*, 2009, 2010). Electrodes have better durability in electrolyte with either saturated $Al_2(SO_4)_3$ or $MnSO_4$ addition, as compared with that in conventional KOH electrolyte. The improved performance was attributed to lower charge transfer resistance at the interface between electrode and the electrolyte in these electrolytes.

Solid polymer electrolytes, as opposed to conventional concentrated KOH solution, are ionic conductive solutions of salts in a polymeric solvent, typically poly(ethylene oxide) (PEO). The diversity and the complexity of the current research in the field of polymer electrolytes can be explained by their attractive advantages over the conventional solid or liquid electrolytes. For example, problems such as chemical and electrochemical instability, corrosion and gas emission, can be avoided in such polymer-based systems, therefore the cycle life of the cell can be extended. However, the major drawback for this polymer electrolyte is the smaller current density to be drawn because of low ionic conductivity of the electrolyte. This issue was later addressed by Iwakura *et al.* (Iwakura *et al.*, 2002), where a high water-absorbing capacity, high water-holding capacity, high gel strength crossed-linked poly(acrylic acid) was introduced. The polymer gel electrolyte prepared from potassium salt of cross-linked poly(acrylic acid) and a KOH aqueous solution was found to have high ionic conductivity that was almost comparable to the conventional KOH aqueous solution. The solid electrolyte N,N-dimethylpyrrolidoium hydroxide (PIIOH) with polymer poly(tetramethyl ammonium acrylate) (PTMA) was investigated as an electrolyte in Ni-MH cells. The cell with electrolyte polymer-PIIOH showed a discharge capacity of 142 mAh/g at 25 °C (Wang *et al.*, 2005).

11.7.4 Electrochemical performance of Ni-MH batteries

Ni-MH batteries of all common sizes are now commercially available in the market. Panasonic has produced some sub-AAA size Ni-MH batteries, while prismatic cells with capacity up to 250 Ah are manufactured by Ovonic Battery Company in United States and Gold Peak Industries in Hong Kong for EV application. An Ni-MH cell can deliver a capacity that is about two times that of an equivalently sized Ni-Cd cell. From the initial product introduced in 1991 of cylindrical cells having an energy of 54 Wh/kg, today's small consumer cells with a specific energy over 100 Wh/kg and energy densities above 300 Wh/L are now available (Pistoia, 2009). The power associated with the commercial Ni-MH batteries has increased from less than 200 to 1200 W/kg, and up to 2000 W/kg at a development level (Fetcenko, 2005). The advancement of Ni-MH batteries from 1991 to 2005 has been tabulated in Table 11.6.

Under mild conditions, for example, charge/discharge at 0.2C, at a temperature of 20°C, and under limited overcharge, a standard Ni-MH battery can deliver in excess of 500 cycles (up to about 1000–1200) before its capacity decreases to 80% of the initial value. In HEV applications, more than 100 000 cycles are possible under high current pulses and SoC between 2 and 10%.The performance of the Ni-MH batteries is, however, temperature dependent. It has been ascertained that the best charging efficiency and cell durability were obtained at temperatures ranging from 10 to 30°C. Cycle life deterioration of the cells can be as high as 60% if exposed continuously over this temperature. Nevertheless, in today's batteries, excellent power can be

Table 11.6 Advance in specific energy and energy density values in Ni-MH batteries

Year	Specific energy ($Wh\ kg^{-1}$)	Energy density ($Wh\ L^{-1}$)	Ni–MH AA cell capacity (mAh)
1991	54	190	1100
1993	70	235	1400
1996	80	255	1600
2000	92	300	1900
2002	95	345	2100
2003	102	385	2300
2005	107	428	2600

Source: Fetcenko *et al.*, 2007.

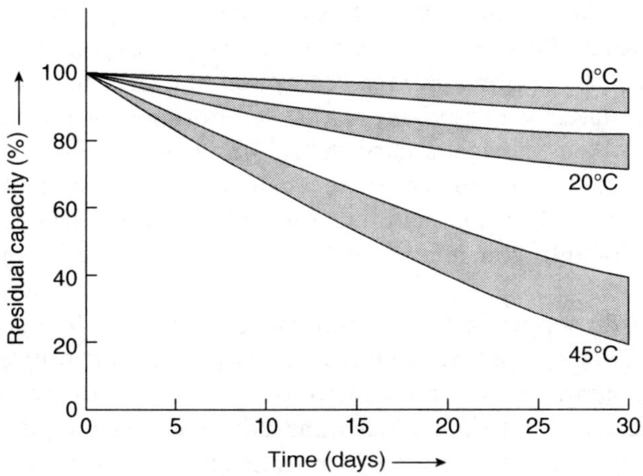

11.32 Self-discharge characteristics of an Ni-MH battery at various temperatures. The grey regions indicate the spread between different batteries (Notten, 1994).

provided at cold temperatures of –30°C and high capacity of over 90% can be achieved at 70°C (Fetcenko *et al.*, 2007).

Ni-MH batteries, similar to Ni-Cd and lithium-ion batteries, lose their stored energy under open-circuit conditions to a certain extent. For an Ni-MH battery, the daily self-discharge rate is typically 1% of the nominal storage capacity. However, the rate is strongly dependent on the external conditions, such as SoC and temperature. Figure 11.32 illustrates that the self-discharge rate increases greatly with increasing temperatures.

As the applications of Ni-MH batteries are greatly limited by the high self-discharge rate, considerable efforts have been devoted to overcome, or at least, to minimise the rate of self-discharge. Ikoma *et al.* (Ikoma *et al.*,

1996) suggested the following five possible factors that are responsible for the self-discharge behaviour in Ni-MH batteries:

1. self-decomposition of positive active material (NiOOH);
2. shuttle reactions of nitrate ions in the positive electrode;
3. reduction of a NiOOH positive electrode by hydrogen gas coming from the metal hydride negative electrode or oxidation of the negative electrode by oxygen gas evolved from the positive electrode;
4. shuttle reactions of organic impurities dissolved in an alkaline electrolyte solution from a separator;
5. shuttle reactions of metal ions dissolved in the alkaline electrolyte solution from the hydrogen storage alloy.

In addition, Iwakura *et al.* (1989) classified self-discharge mechanism of Ni-MH batteries into two types, namely reversible and irreversible capacity loss. The former was attributed to the desorption of hydrogen from the MH anode, and the latter was ascribed to the deterioration of the hydrogen-absorbing alloy. As a short summary, the self-discharge rate of the Ni-MH battery is expected to be improved through proper selection of separator, electrolyte, and electrode materials for both electrodes as to minimise the formation of nitrogen impurities and the degradation of the active materials.

Li *et al.* (2010b) suppressed the self-discharge rate of the Ni-MH battery from 35% to 25% by employing acrylic acid grafted polypropylene (PP) separator; this separator can trap the impurities more effectively and hence suppress the detrimental shuttle reactions. Alternatively, microencapsulation of the alloy with a thin layer of Cu can suppress the hydrogen diffusion in MH electrode and hence lower the rate of self-discharge (Feng and Northwood, 2005).

The discharge performance of Ni-MH batteries when discharged at higher current is typically much worse than with lower current. The dependence of high rate discharge capability of AA-sized Eneloop Ni-MH batteries on discharge current density is illustrated in Figure. 11.33. As expected, that the discharge capacity and plateau voltage significantly decreased with increasing C rate. Improvement is essential to enhance its feasibility in high power applications.

High rate discharge performance of Ni-MH batteries is controlled by the rate of charge transfer at the alloy/electrolyte interface, and the rate of hydrogen diffusion in the alloy. Over the last few years, additives have been incorporated in either positive (Wu *et al.*, 2006; Zhang *et al.*, 2008c) or negative (Zhang *et al.*, 2008a) electrode as a means to improve the high rate discharge capability. Compared with traditional carbon materials, carbon nanotubes (CNT) possess higher electron transfer rate, and are therefore

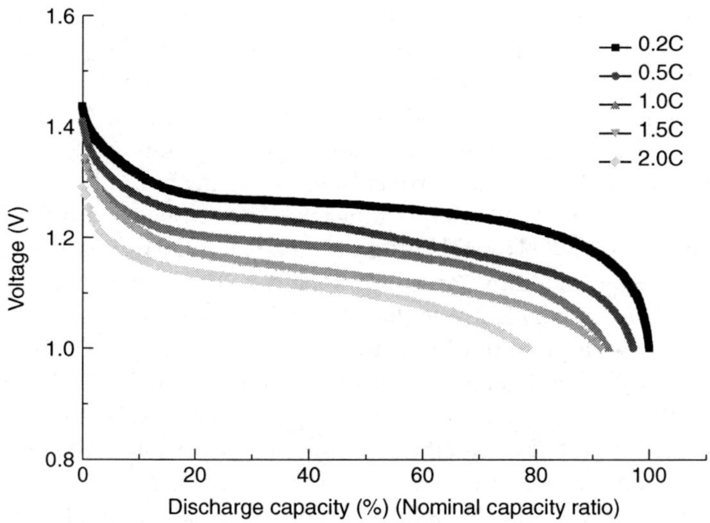

11.33 Discharge characteristic of AA-sized Eneloop Ni-MH batteries discharged at different C rates.

widely adopted as effective additives in improving the electrochemical properties of the electrodes. Zhang *et al.* (Zhang *et al.*, 2008a) demonstrated that the high rate discharge and cycle life performance of AB_5 electrodes were significantly improved by mixing with 0.8 wt.% of CNTs.

The durability of an Ni-MH battery is highly affected by the performance to the MH electrode. In general, the charge/discharge cycle life shortens due to the formation of hydroxides and oxides on the fresh surface. The creation of fresh surface proceeds progressively through alloy pulverisation. The improvements can be made with the following strategies: (1) partial replacement of alloy components by foreign metals such as cobalt; (2) amorphisation of the alloy; (3) microencapsulation of the alloy surface with foreign metals such as copper or nickel; (4) others, including doping and alloy fabrication methods.

Hydrogen storage alloys are subjected to pulverisation during charging and discharging owing to repeated lattice expansion and shrinkage as a result of hydrogen absorption and desorption. As aforementioned, metals such as Co, Al, Si, Sn or Ge for Ni can improve the cycle life of the MH electrode by suppressing the degree of volume change or by the formation of a surface oxide film and subsequently inhibit the pulverisation or corrosion from taking place.

Amorphous alloys when compared with their crystalline counterparts show a longer cycle life, but lower discharge capacity. It was found that amorphous MNi_2 (M = La, Ce, Pr) alloys showed exceptional low discharge capacity of less than 100 mAh/g at room temperature, but had a long cycle life. The grain size and crystal structure of the alloy can be manipulated by controlling the solidification rate by the linear velocity of the copper wheel in melt-spinning technique. High solidification rates can produce alloys with smaller grains containing nanocrystalline and amorphous phases, resulting in good cycle stability. Long-term ball-milling is an alternative technique for preparing amorphous alloys with long cycle life (Rongeat and Roué, 2005).

The cycle life of an Ni-MH battery can be extended by coating a thin protective layer on the alloy surface. Boonstra *et al.* (1989) demonstrated the cycle life of an $LaNi_5$ electrode that can be improved by coating the surface of the alloy with platinum black to protect the alloy from being oxidised with the oxygen evolved at the positive electrode. Mixing of $LaNi_5$ and its ternary alloys with a large amount of copper powder can also increase the cycle life. However, these methods seem to be undesirable from a viewpoint of energy density and cost.

An encapsulated Mm-Ni_5 alloy powder covered with an electroless copper layer was used as the negative electrode material in an Ni-MH battery (Raju *et al.*, 2009). The coated Cu layer induced an increase in electrocatalytic activity for hydrogen evolution reaction and reduced the over-potential of charge/discharging, resulting in a remarkable increase in both electrode capacity and cycle life. Co encapsulated electrodes were also found to have higher capacity and superior cycle life than the bare alloy (Durairajan *et al.*, 1999).

Other approaches to improving the cycle life of a Ni-MH battery include the doping of additives, such as CNTs, alkaline treatment (Ikoma *et al.*, 1999), incorporation of non-stoichiometric alloys (Notten and Latroche, 2009), and alloy fabrication techniques (Anik *et al.*, 2009). The relationship of the ending capacity and cycle number for an $LaNi_5$ electrode with various multi-walled carbon nanotube (MWNT) addition in an Ni-MH battery is shown in Fig. 11.34. The capacity was higher for battery with MWNTs addition than that without MWNT; also, the cycle life of the batteries is remarkably improved with 0.8 wt.% of MWNT addition (sample B). The improved cycle life was inferred to the addition of MWNT preventing the inner electric resistance from rising in the battery.

11.7.5 Final remarks and future challenges

Ni-MH batteries seem to have widespread commercial viability and significant opportunity for improvement among the various nickel-based batteries. Ni-MH batteries appear to be the technology of choice for applications such as HEV, emerging EVs, and fuel cell EVs. With the advances of Ni-MH batteries over the last decades, prototype Ni-MH batteries of specific

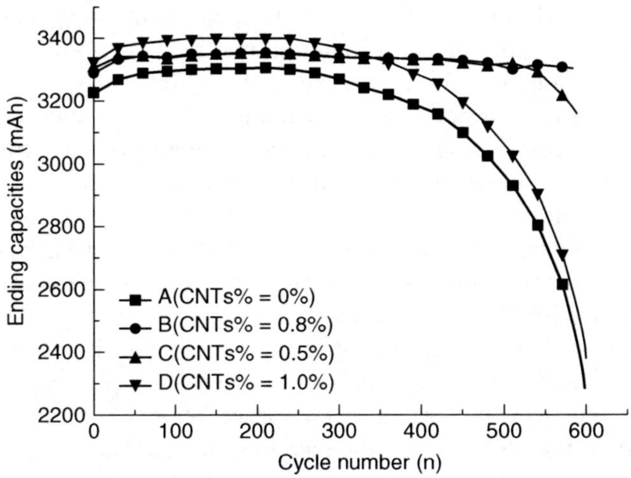

11.34 The relationship curve of the cycle number and its ending capacity of battery samples with different MWNTs contents in the negative electrodes under 2C-rate discharge conditions (Zhang *et al.*, 2008a).

energy over 100 Wh/kg have been established, while the specific power has advanced from 150 to over 1000 W/kg.

In addition to the essential performance targets (energy, power, cycle life, and operating temperatures), flexible vehicle packaging, easy application to series and series/parallel strings, safety, maintenance-free operation, fast and less expensive charging, and environmentally acceptable and recyclable materials have established the eminence of Ni-MH batteries. Despite the continuous advances in the performance of Ni-MH batteries, common problems such as capacity deterioration upon cycling induced by alloy pulverisation and corrosion, and high self-discharge as a result of active material decomposition, still require further improvement. While Ni-MH batteries still have the advantage of being safe to use, the competition from low cost lithium-ion batteries with high performance is posing a great threat to replace Ni-MH batteries. Unless a further leap in the advancement of Ni-MH battery is seen in the near future, both in terms of electrochemical performance and cost, the lithium-ion battery is expected to gradually replace the Ni-MH battery in many applications.

11.8 Conclusion

This chapter reviewed the history, current status, and future trend of the Ni-based batteries. The characteristics of these Ni-based battery systems have been tabulated in Table 11.7. Ni-Fe batteries have relatively long cycle life as compared with other Ni-based batteries, but their low specific energy and energy density, and high self-discharge rate have been the bottlenecks for their development and greater market penetration.

Ni-Cd batteries have superior electrochemical characteristics in terms of long cycle life, and reasonable specific energy and energy density, which have made them the second most widely used secondary batteries after lead/acid. However, severe memory effect, dendritic shorting and hydrogen evolution when cadmium electrode is overcharged are problematic for battery performance. The detrimental environmental effect of Cd also makes this system non-competitive with Ni-MH batteries.

Ni-H_2 batteries have exceptionally long service life, high gravimetric energy density, high robustness, and a good inherent protection against overcharging and overdischarging as compared to Ni-Cd batteries, which have made them the technology of choice in space applications. The drawbacks related to this type of battery are the poor volumetric energy density and high self-discharge rate, which have made them inferior to lithium-ion batteries.

Among various Ni-based batteries, Ni-Zn batteries possess the highest nominal voltage. In addition, the Zn electrode is environmentally benign and inexpensive, and has therefore attracted a considerable amount of research interest. Despite efforts having been devoted to improving the cycle life of the Zn electrode, this still remains the major obstacle to its wider development and applications.

Among all Ni-based batteries, Ni-MH batteries appear to have more success in terms of both development and commercialisation, particularly in HEVs. They exhibit long durability, high compatibility to alkaline battery systems, exceptionally high specific energy and energy density, high resistance to rapid charging and discharging, and are safe to operate. The characteristics of Ni-based secondary batteries have been tabulated in Table 11.7. The performance of the MH electrode is, however, limited by the inevitable alloy pulverisation and corrosion. Improvements have been made in the

Table 11.7 Characteristics of Ni-based secondary batteries

System units	Nominal voltage (V)	Cycle life (cycles)	Specific energy (Wh/kg)	Energy density (Wh/L)	Self-discharge (%/month) at 20°C
Ni-Fe	1.25	2000–3000	20–25	30	25–30
Ni-Cd	1.2	> 1200	40–60	50–150	20
Ni-H_2	1.25	> 2000	40–75	65–80	60
Ni-Zn	1.75	300–600	60–75	100	15–20
Ni-MH	1.2	800–1200	50–80	140–300	15–30

Source: Pistoia, 2009.

past that have had some success in reducing them, but it is still far from satisfactory. The characteristics of recent and current popular rechargeable batteries, namely Ni-Cd, Ni-MH and Li-ion, for portable applications have been compared and the results are tabulated in Table 11.8 (Rashidi *et al.*, 2009). Though electrochemical properties of Li-ion batteries, such as

Table 11.8 Characteristics of rechargeable batteries for portable applications

Battery type	Ni-Cd	Ni-MH	Li-ion
Gravimetric energy density (Wh/kg)	40–60	30–80	90
Volumetric energy density (Wh/L)	180	140	210
Nominal cell voltage (V)	1.25	1.25	3.6
Equivalent series resistant (Ω)[a]	Extremely low	Extremely low	High
Cycle life (to 80% of initial capacity)	300–500	300–500	300–600
Self-discharging at 20°C (%/month)[b]	20–30	15–20	5–10
Typical slow charge time (h)	12–36	4–10	Does not tolerate slow charge time after fully charged (charging by constant voltage only)
Typical fast charge time (h)	1	0.25–1	1.5
High rate discharge capacity (C)	10	2–3	1–1.5
Cost comparison	Least expensive	More expensive than Ni-Cd, but less than Li-ion	Most expensive
Most common/ severe degradation mechanisms (reliability)	High current overcharge, cell polarity reversal (during discharging)	High current overcharge, cell polarity reversal (during discharging)	Accidentally shortening the battery
Operating temperature (°C)	Discharging: 10–40 Optimum operating temperature: 25	Discharging: 10–40 Optimum operating temperature: 25	Discharging: –20–60 Optimum operating temperature: 25
Compatibility to alkaline batteries	Yes	Yes	No

Source: Gray and Smith, 2009; Rashidi *et al.*, 2009.

gravimetric and volumetric energy density, nominal cell voltage, and rate of self-discharge are superior to the Ni-Cd and Ni-MH batteries, their poor high rate discharge capability and high cost have been restricting factors for their wider spread in the market. In addition, the Li-ion battery is not compatible with devices which require alkaline battery systems, for example, in EVs.

With the several secondary Ni-based battery systems reviewed here, the ultimate goal for ongoing research worldwide is to meet the upsurge in demand for exploring environmental-friendly power sources with long endurance, high energy density, high power capability, smaller in size, and low cost in electronic devices. Undoubtedly, finding a battery system meeting these requirements will be the research focus for the following years, and revolutionary technology in batteries can be foreseen upon success.

11.9 References

http://industrial.panasonic.com/www-data/pdf/ACK4000/ACK4000PE2.pdf (Accessed 7 July 2013).

Abrashev, B., Spassov, T., Bliznakov, S. and Popov, A. (2010). Microstructure and electrochemical hydriding/dehydriding properties of ball-milled TiFe-based alloys. *International Journal of Hydrogen Energy,* **35**, 6332–6337.

Adzic, G. D., Johnson, J. R., Reilly, J. J., Mcbreen, J., Mukerjee, S., Kumar, M. P. S., Zhang, W. and Srinivasan, S. (1995). Cerium content and cycle life of multicomponent AB[sub 5] hydride electrodes. *Journal of the Electrochemical Society,* **142**, 3429–3433.

Ahlberg, E., Palmqvist, U., Simic, N. and Sjövall, R. (2000). Capacity loss in Ni-Cd pocket plate batteries. The origin of the second voltage plateau. *Journal of Power Sources,* **85**, 245–253.

Anik, M., Gasan, H., Topcu, S., Akay, I. and Aydinbeyli, N. (2009). Electrochemical hydrogen storage characteristics of Mg1.5Al0.5-xZrxNi (x = 0, 0.1, 0.2, 0.3, 0.4, 0.5) alloys synthesized by mechanical alloying. *International Journal of Hydrogen Energy,* **34**, 2692–2700.

Arouete, S., Blurton, K. F. and Oswin, H. G. (1969). Controlled current deposition of zinc from alkaline solution. *Journal of the Electrochemical Society,* **116**, 166–169.

Audemer, A., Delahaye, A., Farhi, R., Sac-Epee, N. and Tarascon, J. M. (1997). Electrochemical and Raman studies of Beta-type nickel hydroxides Ni[sub 1 – x]Co[sub x](OH)[sub 2] electrode materials. *Journal of the Electrochemical Society,* **144**, 2614–2620.

Avendano, E., Azens, A., Niklasson, G. A. and Granqvist, C. G. (2005). Proton diffusion and electrochromism in hydrated NiO[sub y] and Ni[sub 1 – x]V[sub x] O[sub y] thin films. *Journal of the Electrochemical Society,* **152**, F203–F212.

Balasubramaniam, R., Mungole, M. N. and Rai, K. N. (1993). Hydriding properties of MmNi5 system with aluminium, manganese and tin substitutions. *Journal of Alloys and Compounds,* **196**, 63–70.

Bard, F., Palac, N., M. R., Beaudoin, B., Christian, P. A. and Tarascon, J. M. (2006). Cationic substitution in [gamma]-type nickel (oxi)hydroxides as a means to prevent self-discharge in Ni/Zn primary batteries. *Journal of Power Sources*, **160**, 733–743.

Bass, K., Mitchell, P. J., Wilcox, G. D. and Smith, J. (1991). Methods for the reduction of shape change and dendritic growth in zinc-based secondary cells. *Journal of Power Sources*, **35**, 333–351.

Bennion, D. N. (1980). Review of membrane separators and zinc-nickel oxide battery development. *Other Information: Portions are Illegible in Microfiche Products*. http://www.osti.gov/energycitations/product.biblio.jsp?osti_id=6137574.

Bernard, P. (2009). Secondary batteries – nickel systems | nickel-cadmium: sealed. *In: Encyclopedia of Electrochemical Power Sources*, X00FC and Rgen, G. (eds.). Amsterdam: Elsevier.

Berndt, D. (1997). *Maintenance-Free Batteries* (Second Edition). Hoboken, New Jersey, USA John Wiley & Sons, Inc.

Berndt, D. July (1998). Development of battery. *Varta Special Report*. http://company.varta.com/eng/content/presse/download/sp_report_4e.pdf (Accessed 23 August 2013).

Binder, L. and Kordesch, K. (1986). Electrodeposition of zinc using a multi-component pulse current. *Electrochimica Acta*, **31**, 255–262.

Bliznakov, S., Lefterova, E., Dimitrov, N., Petrov, K. and Popov, A. (2008). A study of the Al content impact on the properties of MmNi4.4-xCo0.6Alx alloys as precursors for negative electrodes in NiMH batteries. *Journal of Power Sources*, **176**, 381–386.

Bode, H., Dehmelt, K. and Witte, J. (1966). Zur kenntnis der nickelhydroxidelektrode--I. Über das nickel (II)-hydroxidhydrat. *Electrochimica Acta*, **11**, 1079–1087.

Boonstra, A. H. and Bernards, T. M. N. (1990). The effect of the electrolyte on the degradation process of LaNi5 electrodes. *Journal of the Less Common Metals*, **161**, 355–368.

Boonstra, A. H., Lippits, G. J. M. and Bernards, T. N. M. (1989). Degradation processes in a LaNi5 electrode. *Journal of the Less Common Metals*, **155**, 119–131.

Boreiko, C. J. (2009). Safety | materials toxicity. *In: Encyclopedia of Electrochemical Power Sources*, X00FC and Rgen, G. (eds.). Amsterdam: Elsevier.

Borthomieu, Y. and Bernard, P. (2009). Secondary batteries – nickel systems | nickel-hydrogen. *In: Encyclopedia of Electrochemical Power Sources*, X00FC and Rgen, G. (eds.). Amsterdam: Elsevier.

Britting, J. R and Britting, A. O. (1984). Design, development, performance, and reconditioning of Ni-Cd batteries using polypropylene separators. *Journal of Power Sources*, **12**, 305–316.

Cairns, E. J. (2009). Secondary batteries – nickel systems | nickel-zinc. *In: Encyclopedia of Electrochemical Power Sources*, X00FC and Rgen, G. (eds.). Amsterdam: Elsevier.

Cattaneo, E. and Riegel, B. (2009). Chemistry, electrochemistry, and electrochemical applications | nickel. *In: Encyclopedia of Electrochemical Power Sources*, X00FC and Rgen, G. (eds.). Amsterdam: Elsevier.

Cenek, M., Kouril, O., Sandera, J. and Calabek, M. (1977). New possibilities in the technology of electrodes for alkaline accumulators. *Power Sources*, **6**, 215.

Cern, J. and Micka, K. (1989). Voltammetric study of an iron electrode in alkaline electrolytes. *Journal of Power Sources*, **25**, 111–122.

Chakkaravarthy, C., Periasamy, P., Jegannathan, S. and Vasu, K. I. (1991). The nickel/iron battery. *Journal of Power Sources*, **35**, 21–35.

Chen, H., Wang, J. M., Pan, T., Xiao, H. M., Zhang, J. Q. and Cao, C. N. (2003). Effects of high-energy ball milling (HEBM) on the structure and electrochemical performance of nickel hydroxide. *International Journal of Hydrogen Energy,* **28**, 119–124.

Chen, H., Wang, J. M., Pan, T., Zhao, Y. L., Zhang, J. Q. and Cao, C. N. (2005). The structure and electrochemical performance of spherical Al-substituted [alpha]-Ni(OH)2 for alkaline rechargeable batteries. *Journal of Power Sources,* **143**, 243–255.

Cheng, F. -Y., Chen, J. and Shen, P. -W. (2005). Y(OH)3-coated Ni(OH)2 tube as the positive-electrode materials of alkaline rechargeable batteries. *Journal of Power Sources,* **150**, 255–260.

Cheng, M. Y. and Hwang, B. J. (2009). Control of uniform nanostructured alpha-Ni(OH)(2) with self-assembly sodium dodecyl sulfate templates. *Journal of Colloid and Interface Science,* **337**, 265–271. 10.1016/j.jcis.2009.05.008.

Chin, D. T. and Venkatesh, S. (1981). A-C modulation of a rotating zinc electrode in an acid zinc-chloride solution. *Journal of the Electrochemical Society,* **128**, 1439–1442. 10.1149/1.2127658.

Choi, B., Lee, S., Iizuka, M., Otsuji, Y., Fushimi, C. and Tsutsumi, A. (2009). Al-doped & alpha;-nickel hydroxide electrode: Addition of Co and effect of Al ion in electrolyte. *Journal of Chemical Engineering of Japan,* **42**, 452–456.

Colinet, C., Pasturel, A., Percheron-Guégan, A. and Achard, J. C. (1987). Enthalpies of formation and hydrogenation of La(Ni(1-x)Cox)5 compounds. *Journal of the Less Common Metals,* **134**, 109–122.

Corrigan, D. A. and Bendert, R. M. (1989). Effect of coprecipitated metal ions on the electrochemistry of nickel hydroxide thin films: Cyclic voltammetry in 1M KOH. *Journal of the Electrochemical Society,* **136**, 723–728.

Crompton, T. R. (2000a). Nickel batteries. *Battery Reference Book* (Third Edition). Oxford: Newnes.

Crompton, T. R. (2000b). Secondary batteries. *Battery Reference Book* (Third Edition). Oxford: Newnes.

Dahlen, M. (2003). Investigation on Storage Technologies for INtermittent Renewable Energies: Evaluation and recommended R&D strategy. INVESTIRE-NETWORK.

Dell, R. M. (2000). Batteries: Fifty years of materials development. *Solid State Ionics,* **134**, 139–158.

Demourgues-Guerlou, L. and Delmas, C. (1993). Structure and properties of precipitated nickel-iron hydroxides. *Journal of Power Sources,* **45**, 281–289. 10.1016/0378-7753(93)80017-j.

Di, Z., Yamamoto, T., Inui, H. and Yamaguchi, M. (2000). Characterization of stacking faults on basal planes in intermetallic compounds La5Ni19 and La2Ni7. *Intermetallics,* **8**, 391–397.

Dixit, M. and Vishnu Kamath, P. (1995). Electrosynthesis and stabilization of [alpha]-cobalt hydroxide in the presence of trivalent cations. *Journal of Power Sources,* **56**, 97–100.

Duan, G. T., Cai, W. P., Luo, Y. Y., Li, Z. G. and Lei, Y. (2006). Hierarchical structured Ni nanoring and hollow sphere arrays by morphology inheritance based on ordered through-pore template and electrodeposition. *Journal of Physical Chemistry B,* **110**, 15729–15733. 10.1021/jp062255q.

Duffield, A., Mitchell, P. J., Hampson, N. A., Kumar, N. and Shield, D. W. (1985). A rotating-disk study on Teflon-bonded porous zinc electrodes. *Journal of Power Sources,* **15**, 93–100.

Dunlop, J. D. and Stockel, J. F. (1977). Orbital performance of NTS-2 nickel-hydrogen battery. *Comsat Technical Review,* **7**, 639–660.

Durairajan, A., Haran, B. S., Popov, B. N. and White, R. E. (1999). Cycle life and utilization studies on cobalt microencapsulated AB5 type metal hydride. *Journal of Power Sources,* **83**, 114–120.

Eagle-Picher Industries, I. (1980). Research, development and demonstration of a nickel-iron battery for electric vehicle propulsion. *Journal of Power Sources,* **5**, 325–325.

Edison, T. A. (1901). Reversible galvanic battery. *US Pat,* 678,722.

Falk, S. U. and Salkind, A. J. (1986). Alkaline storage batteries. *Electrochemical Society Series.* New York, Wiley.

Feng, F. and Northwood, D. O. (2005). Self-discharge characteristics of a metal hydride electrode for Ni-MH rechargeable batteries. *International Journal of Hydrogen Energy,* **30**, 1367–1370.

Fetcenko, M. A. (2005). 22nd International Seminar & Exhibit on Primary and Secondary Batteries. Ft. Lauderdale, FL, 14–17 March 2005.

Fetcenko, M. A., Ovshinsky, S. R., Reichman, B., Young, K., Fierro, C., Koch, J., Zallen, A., Mays, W. and Ouchi, T. (2007). Recent advances in NiMH battery technology. *Journal of Power Sources,* **165**, 544–551.

Fuhr, K. H. (1987). Failure analysis of 3.5 inch, 50 A h nickel-hydrogen cells undergoing low earth orbit testing. *Proc. 22nd Intersociety Energy Conversion Engineering Conference,* Philadelphia, PA, AIAA, New York, 10–14 August 1987.

Gagnon, E. G. (1986). Effects of KOH concentration on the shape change and cycle life of Zn/NiOOH cells. *Journal of the Electrochemical Society,* **133**, 1989–1995.

Gagnon, E. G. and Wang, Y. -M. (1987). Pasted-rolled zinc electrodes containing calcium hydroxide for use in Zn/NiOOH cells. *Journal of the Electrochemical Society,* **134**, 2091–2096.

Gray, F. M. and Smith, M. J. (2009). Secondary batteries – lithium rechargeable systems | lithium polymer batteries. *In: Encyclopedia of Electrochemical Power Sources,* X00FC and Rgen, G. (eds.). Amsterdam: Elsevier.

Gregory, D. P., Jones, P. C. and Redfearn, D. P. (1972). The corrosion of zinc anodes in aqueous alkaline electrolytes. *Journal of the Electrochemical Society,* **119**, 1288–1292.

Gross, S. (1971). Causes of failure in sealed nickel-cadmium batteries. *Energy Conversion,* **11**, 39–42, 43–45.

Guerlou-Demourgues, L., Denage, C. and Delmas, C. (1994). New manganese-substituted nickel hydroxides: Part 1. Crystal chemistry and physical characterization. *Journal of Power Sources,* **52**, 269–274. 10.1016/0378-7753(94)02023-x.

Guiose, B., Cuevas, F., Camps, B., Leroy, E. and Percheron-Guégan, A. (2009). Microstructural analysis of the ageing of pseudo-binary (Ti, Zr)Ni intermetallic compounds as negative electrodes of Ni-MH batteries. *Electrochimica Acta,* **54**, 2781–2789.

Halpert, G. (1984). Past developments and the future of nickel electrode cell technology. *Journal of Power Sources,* **12**, 177–192.

Hampson, N. A. and Mcneil, A. J. S. (1985). The electrochemistry of porous zinc V. The cycling behaviour of plain and polymer-bonded porous electrodes in koh solutions. *Journal of Power Sources,* **15**, 261–285.

Han, X., Xie, X., Xu, C., Zhou, D. and Ma, Y. (2003). Morphology and electrochemical performance of nano-scale nickel hydroxide prepared by supersonic coordination-precipitation method. *Optical Materials,* **23**, 465–470.

Han, X. J., Xu, P., Xu, C. Q., Zhao, L., Mo, Z. B. and Liu, T. (2005). Study of the effects of nanometer [beta]-Ni(OH)2 in nickel hydroxide electrodes. *Electrochimica Acta,* **50**, 2763–2769.

Hariprakash, B., Shukla, A. K. and Venugoplan, S. (2009). Secondary batteries – nickel systems | nickel-metal hydride: Overview. *In: Encyclopedia of Electrochemical Power Sources,* X00FC and Rgen, G. (eds.). Amsterdam: Elsevier.

Hassoun, J., Mulas, G., Panero, S. and Scrosati, B. (2007). Ternary Sn-Co-C Li-ion battery electrode material prepared by high energy ball milling. *Electrochemistry Communications,* **9**, 2075–2081.

He, G., Jiao, L. -F., Yuan, H. -T., Zhang, Y. -Y. and Wang, Y. -J. (2008). Preparation and electrochemical properties of MgNi-MB (M = Co, Ti) composite alloys. *Journal of Alloys and Compounds,* **450**, 375–379.

He, X., Wang, L., Li, W., Jiang, C. and Wan, C. (2006). Ytterbium coating of spherical Ni(OH)2 cathode materials for Ni-MH batteries at elevated temperature. *Journal of Power Sources,* **158**, 1480–1483.

Hills, S. (1965). Beneficial effect of lithiated electrolyte on iron battery electrodes. *Journal of the Electrochemical Society,* **112**, 1048–1049.

Hsu, L. -C. and Sheibley, D. W. (1982). Inexpensive cross-linked polymeric separators made from water-soluble polymers. *Journal of the Electrochemical Society,* **129**, 251–254.

Hu, W.-K. (1998). Effect of microstructure, composition and non-stoichiometry on electrochemical properties of low-Co rare-earth nickel hydrogen storage alloys. *Journal of Alloys and Compounds,* **279**, 295–300.

Hu, W. -K., Gao, X. -P., Noréus, D., Burchardt, T. and Nakstad, N. K. (2006). Evaluation of nano-crystal sized [alpha]-nickel hydroxide as an electrode material for alkaline rechargeable cells. *Journal of Power Sources,* **160**, 704–710.

Hu, W.-K. and Noréus, D. (2003). Alpha nickel hydroxides as lightweight nickel electrode materials for alkaline rechargeable cells. *Chemistry of Materials,* **15**, 974–978. 10.1021/cm021312z.

Ikoma, M., Hoshina, Y., Matsumoto, I. and Iwakura, C. (1996). Self-discharge mechanism of sealed-type nickel/metal-hydride battery. *Journal of the Electrochemical Society,* **143**, 1904–1907.

Ikoma, M., Komori, K., Kaida, S. and Iwakura, C. (1999). Effect of alkali-treatment of hydrogen storage alloy on the degradation of Ni/MH batteries. *Journal of Alloys and Compounds,* **284**, 92–98.

Iwakura, C., Asaoka, T., Yoneyama, H., Sakai, T., Oguro, K. and Ishikawa, H. (1988). Electrochemical characteristics of lani5 system hydrogen-absorbing alloys as negative electrode materials for nickel-hydrogen batteries. *Nippon Kagaku Kaishi*, Issue 8, 1482–1488.

Iwakura, C., Kajiya, Y., Yoneyama, H., Sakai, T., Oguro, K. and Ishikawa, H. (1989). Self-discharge mechanism of nickel-hydrogen batteries using metal hydride anodes. *Journal of the Electrochemical Society,* **136**, 1351–1355.

Iwakura, C., Nohara, S., Furukawa, N. and Inoue, H. (2002). The possible use of polymer gel electrolytes in nickel/metal hydride battery. *Solid State Ionics,* **148**, 487–492.

Jain, M., Elmore, A. L., Matthews, M. A. and Weidner, J. W. (1998). Thermodynamic considerations of the reversible potential for the nickel electrode. *Electrochimica Acta,* **43**, 2649–2660.

Jayalakshmi, M., Venugopal, N., Reddy, B. R. and Rao, M. M. (2005). Optimum conditions to prepare high yield, phase pure [alpha]-Ni(OH)2 nanoparticles by urea hydrolysis and electrochemical ageing in alkali solutions. *Journal of Power Sources,* **150**, 272–275.

Joubert, J. M., Latroche, M., Percheron-Guégan, A. and Bouet, J. (1996). Improvement of the electrochemical activity of Zr-Ni-Cr Laves phase hydride electrodes by secondary phase precipitation. *Journal of Alloys and Compounds,* **240**, 219–228.

Jungner, W. (1901). Sätt att pá elktrolytisk väg föstora ytan af sadana metaller, hvilkas syreföreningar äro kemiskt olösliga i alkaliska lösningar. Swed Pat, 15,567.

Kabanov, B. and Leikis, D. (1946). *Zhurnal Fizicheskoi Khimii,* **20**, 995–1003.

Kalaignan, G. P., Umaprakatheeswaran, C., Muralidharan, B., Gopalan, A. and Vasudevan, T. (1996). Electrochemical behaviour of addition agents impregnated in cadmium hydroxide electrodes for alkaline batteries. *Journal of Power Sources,* **58**, 29–34.

Kamath, P. V., Dixit, M., Indira, L., Shukla, A. K., Kumar, V. G. and Munichandraiah, N. (1994). Stabilized alpha-Ni(OH)[sub 2] as electrode material for alkaline secondary cells. *Journal of the Electrochemical Society,* **141**, 2956–2959.

Katz, M. H., Adler, T. C., Mclarnon, F. R. and Cairns, E. J. (1988). The effect of pulse charging on the cycle-life performance of zinc/nickel oxide cells. *Journal of Power Sources,* **22**, 77–95.

Kazarinov, I. A., Burashnikova, M. M. and Stepanov, A. N. (2001). Role of heterophase interactions between cadmium and nickel in cadmium electrode activation. *Russian Journal of Applied Chemistry,* **74**, 430–433. 10.1023/a:1012789426890.

Khaldi, C., Mathlouthi, H. and Lamloumi, J. (2009). A comparative study, of 1 M and 8 M KOH electrolyte concentrations, used in Ni-MH batteries. *Journal of Alloys and Compounds,* **469**, 464–471.

Kiani, M. A., Mousavi, M. F. and Ghasemi, S. (2010). Size effect investigation on battery performance: Comparison between micro- and nano-particles of [beta]-Ni(OH)2 as nickel battery cathode material. *Journal of Power Sources,* **195**, 5794–5800.

Kim, J. S., Paik, C. H., Cho, W. I., Cho, B. W., Yun, K. S. and Kim, S. J. (1998). Corrosion behaviour of Zr1-xTixV0.6Ni1.2M0.2 (M=Ni, Cr, Mn) AB2-type metal hydride alloys in alkaline solution. *Journal of Power Sources,* **75**, 1–8.

Latroche, M., Percheron-Guégan, A. and Chabre, Y. (1999). Influence of cobalt content in MmNi4.3-xMn0.3Al0.4Cox alloy (x=0.36 and 0.69) on its electrochemical behaviour studied by in situ neutron diffraction. *Journal of Alloys and Compounds,* **293–295**, 637–642.

Latroche, M., Percheron-Guégan, A., Chabre, Y., Bouet, J., Pannetier, J. and Ressouche, E. (1995). Intrinsic behaviour analysis of substituted LaNi5-type electrodes by means of in-situ neutron diffraction. *Journal of Alloys and Compounds,* **231**, 537–545.

Lewis, H., Jackson, P., Salkind, A., Danko, T. and Bell, R. (2001). Advanced membranes for alkaline primary and rechargeable alkaline cells with zinc anodes. *Journal of Power Sources,* **96**, 128–132.

Li, S. L., Wang, P., Chen, W., Luo, G., Chen, D. M. and Yang, K. (2010a). Effect of non-stoichiometry on hydrogen storage properties of La(Ni3.8Al1.0Mn0.2)x alloys. *International Journal of Hydrogen Energy,* **35**, 3537–3545.

Li, X., Song, Y., Wang, L., Xia, T. and Li, S. (2010b). Self-discharge mechanism of Ni-MH battery by using acrylic acid grafted polypropylene separator. *International Journal of Hydrogen Energy,* **35**, 3798–3801.

Li, Y., Han, S., Zhu, X. and Ding, H. (2010c). Effect of CuO addition on electrochemical properties of AB3-type alloy electrodes for nickel/metal hydride batteries. *Journal of Power Sources,* **195**, 380–383.

Li, Z. P., Liu, B. H., Hitaka, K. and Suda, S. (2002). Effects of surface structure of fluorinated AB2 alloys on their electrodes and battery performances. *Journal of Alloys and Compounds,* **330–332**, 776–781.

Lim, H. S. and Stadnick, S. J. (1989). Effect of precharge on nickel-hydrogen cell storage capacity. *Journal of Power Sources,* **27**, 69–79.

Lim, H. S. and Verzwyvelt, S. A. (1988). KOH concentration effect on the cycle life of nickel-hydrogen cells: III. Cycle life test. *Journal of Power Sources,* **22**, 213–220.

Linden, D. (1984). *Handbook of Batteries and Fuel Cells*, McGraw-Hill, New York.

Liu, C., Song, S., Li, Y. and Liu, A. (2008). Investigations on structure and proton diffusion coefficient of rare earth ion (Y3+/Nd3+) and aluminum codoped [alpha]-Ni(OH)2. *Journal of Rare Earths,* **26**, 594–597.

Liu, Y., Xu, L., Jiang, W., Li, G., Wei, W. and Guo, J. (2009). Effect of substituting Al for Co on the hydrogen-storage performance of La0.7Mg0.3Ni2.6AlxCo0.5-x (x = 0.0–0.3) alloys. *International Journal of Hydrogen Energy,* **34**, 2986–2991.

Lundquist Jr, J. T. (1983). Separators for nickel-zinc batteries. *Journal of Membrane Science,* **13**, 337–347.

Lv, J., Tu, J. P., Zhang, W. K., Wu, J. B., Wu, H. M. and Zhang, B. (2004). Effects of carbon nanotubes on the high-rate discharge properties of nickel/metal hydride batteries. *Journal of Power Sources,* **132**, 282–287.

Machida, N., Yamamoto, H., Asano, S. and Shigematsu, T. (2005). Preparation of amorphous 75L2S·xP2S3·(25-x)P2S5 (mol%) solid electrolytes by a high-energy ball-milling process and their application for an all-solid-state lithium battery. *Solid State Ionics,* **176**, 473–479.

Manzo, M. A. (1990). Nickel-hydrogen capacity loss on storage. *Journal of Power Sources,* **29**, 541–554.

Mao, Z. and White, R. E. (1991). A mathematical model of the self-discharge of a Ni-H[sub 2] battery. *Journal of the Electrochemical Society,* **138**, 3354–3361.

Markin, T. L. and Dell, R. M. (1981). Recent developments in nickel-oxide hydrogen batteries. *Journal of Electroanalytical Chemistry,* **118**, 217–228.

Mcbreen, J. (1990). Nickel oxide electrode: Structure and performance. *Modern Aspects of electrochemistry*, **21**, 29–67.

Mcbreen, J. and Gannon, E. (1983). The effect of additives on current distribution in pasted zinc electrodes. *Journal of the Electrochemical Society,* **130**, 1980–1982.

Mcdowall, J. (2003).*Memory Effect in Stationary Ni-Cd Batteries? Forget About It*! *Battcon*, The Battcon™ Stationary Battery Conference and Trade Show, Florida, Marco Island. 22–1.

Meli, F., Züttel, A. and Schlapbach, L. (1995). Electrochemical and surface properties of iron-containing AB5-type alloys. *Journal of Alloys and Compounds*, **231**, 639–644.

Michalowski, T. D. (1899). *Ger. Pat.*, 112, 351.

Micka, K. and Z Bransk, Z. (1987). Study of iron oxide electrodes in an alkaline electrolyte. *Journal of Power Sources*, **19**, 315–323.

Miller, L. (1986). An advanced nickel-cadmium battery cell design. *Journal of Power Sources*, **18**, 155–160.

Miller, L. (1987). Test summary for advanced hydrogen cycle nickel-cadmium cell. *Journal of Power Sources*, **21**, 339–342.

Montalenti, P. and Stangerup, P. (1977). Thermal simulation of NiCd batteries for spacecraft. *Journal of Power Sources*, **2**, 147–162.

Morishita, M., Ochiai, S., Kakeya, T., Ozaki, T., Kawabe, Y., Watada, M., Tanase, S. and Sakai, T. (2008). Structural analysis by synchrotron XRD and XAFS for manganese-substituted alpha- and beta-type nickel hydroxide electrode. *Journal of the Electrochemical Society*, **155**, A936–A944.

Moriwaki, Y., Gamo, T., Shintani, A. and Iwaki, T. (1989). Laves phase alloys as hydrogen storage electrodes for nickel hydrogen batteries. *Denki Kagaku*, **57**, 488–491.

Munshi, M. Z. A., Tseung, A. C. C. and Misale, D. (1988). The behaviour of polyvinyl alcohol at the planar Cd/Cd(OH)2 electrode interface. *Journal of Power Sources*, **23**, 341–350.

Munshi, M. Z. A., Tseung, A. C. C., Parker, J. and Dawson, J. L. (1985). Effect of an organic additive on the impedance of cadmium in alkaline solution. *Journal of Applied Electrochemistry*, **15**, 737–744. 10.1007/bf00620570.

Negeevich, V. M., Marchenko, G. P., Pavlova, L. R. and Sagoyan, L. N. (1990). Compactibility of active electrode mass based on cadmium oxide. *Powder Metallurgy and Metal Ceramics*, **29**, 860–864. 10.1007/bf00794016.

Nichols, J. T., Mclarnon, F. R. and Cairns, E. J. (1985). Zinc electrode cycle-life performance in alkaline electrolytes having reduced zinc species solubility. *Chemical Engineering Communications*, **37**, 355–379.

Notten, P. H. L. and Latroche, M. (2009). Secondary batteries – nickel systems | nickel-metal hydride: Metal hydrides. *In: Encyclopedia of Electrochemical Power Sources*, X00FC and Rgen, G. (eds.). Amsterdam: Elsevier.

Oliva, P., Leonardi, J., Laurent, J. F., Delmas, C., Braconnier, J. J., Figlarz, M., Fievet, F. and Guibert, A. D. (1982). Review of the structure and the electrochemistry of nickel hydroxides and oxy-hydroxides. *Journal of Power Sources*, **8**, 229–255.

Olurin, O. B., Wilkinson, D. S., Weatherly, G. C., Paserin, V. and Shu, J. (2003). Strength and ductility of as-plated and sintered CVD nickel foams. *Composites Science and Technology*, **63**, 2317–2329. 10.1016/s0266-3538(03)00265-3.

Orikasa, H., Karoji, J., Matsui, K. and Kyotani, T. (2007). Crystal formation and growth during the hydrothermal synthesis of [small beta]-Ni(OH)2 in one-dimensional nano space. *Dalton Transactions*, **14**, 3757–3762.

Oshitani, M., Takayama, T., Takashima, K. and Tsuji, S. (1986). A study on the swelling of a sintered nickel hydroxide electrode. *Journal of Applied Electrochemistry,* **16**, 403–412. 10.1007/bf01008851.

Ovshinsky, S. R., Fetcenko, M. A. and Ross, J. (1993). A nickel metal hydride battery for electric vehicles. *Science,* **260**, 176–181. 10.1126/science.260.5105.176.

Park, M.-S., Kang, Y.-M., Rajendran, S., Kwon, H.-S. and Lee, J.-Y. (2006). Si-Ni-Carbon composite synthesized using high energy mechanical milling for use as an anode in lithium ion batteries. *Materials Chemistry and Physics,* **100**, 496–502.

Peng, M. X. and Shen, X. Q. (2007). Template growth mechanism of spherical Ni(OH)(2). *Journal of Central South University of Technology,* **14**, 310–314. 10.1007/s11771-007-0061-9.

Pensabene, S. F. and Gould, J. W. (1976). Unwanted memory spooks nickel-cadmium cells. *IEEE Spectrum,* **33**, 32–36. http://adsabs.harvard.edu/abs/1976IEEES.

Periasamy, P., Ramesh Babu, B. and Venkatakrishna Iyer, S. (1996). Electrochemical behaviour of Teflon-bonded iron oxide electrodes in alkaline solutions. *Journal of Power Sources,* **63**, 79–85.

Petchjatuporn, P., Sirisuk, P., Khaehintung, N., Sunat, K., Wicheanchote, P. and Kiranon, W. (2008). Low cost RISC implementation of intelligent ultra fast charger for Ni-Cd battery. *Energy Conversion and Management,* **49**, 185–192.

Pistoia, G. (2009). Battery categories and types. *Battery Operated Devices and Systems.* Amsterdam: Elsevier.

Poa, S. P. and Lee, S. J. (1979). Experimental optimization of alkaline zinc-silver oxide primary cell with respect to the zinc electrode preparation and composition. *Journal of Applied Electrochemistry,* **9**, 307–313. 10.1007/bf01112484.

Pralong, V., Delahaye-Vidal, A., Chabre, Y., Beaudoin, B. and Tarascon, J. M. (2001). The outcome of cobalt in the nickel-cobalt oxyhydroxide electrodes of alkaline batteries. *Journal of Solid State Chemistry,* **162**, 270–281.

Provazi, K., Giz, M. J., Dall'Antonia, L. H. and Córdoba De Torresi, S. I. (2001). The effect of Cd, Co, and Zn as additives on nickel hydroxide opto-electrochemical behavior. *Journal of Power Sources,* **102**, 224–232.

Rahman, I. Z., Razeeb, K. M., Kamruzzaman, M. and Serantoni, M. (2004). Characterisation of electrodeposited nickel nanowires using NCA template. *Journal of Materials Processing Technology,* **153–154**, 811–815.

Raju, M., Ananth, M. V. and Vijayaraghavan, L. (2009). Influence of electroless coatings of Cu, Ni-P and Co-P on MmNi3.25Al0.35Mn0.25Co0.66 alloy used as anodes in Ni-MH batteries. *Journal of Alloys and Compounds,* **475**, 664–671.

Rashidi, R., Dincer, I., Naterer, G. F. and Berg, P. (2009). Performance evaluation of direct methanol fuel cells for portable applications. *Journal of Power Sources,* **187**, 509–516.

Ratnakumar, B. V., Timmerman, P. and Di Stefano, S. (1996). Simulation of temperature-compensated voltage limit curves for aerospace Ni-Cd batteries using a first principles' model. *Journal of Power Sources,* **63**, 157–165.

Reisner, D. E., Salkind, A. J., Strutt, P. R. and Xiao, T. D. (1997). Nickel hydroxide and other nanophase cathode materials for rechargeable batteries. *Journal of Power Sources,* **65**, 231–233.

Rongeat, C., Grosjean, M. H., Ruggeri, S., Dehmas, M., Bourlot, S., Marcotte, S. and Rou, L. (2006). Evaluation of different approaches for improving the cycle life of MgNi-based electrodes for Ni-MH batteries. *Journal of Power Sources,* **158**, 747–753.

Rongeat, C. and Rou, L. (2005). On the cycle life improvement of amorphous MgNi-based alloy for Ni-MH batteries. *Journal of Alloys and Compounds,* **404–406**, 679–681.

Rozentsveig, S. A. and Shcherbakova, Z. V. (1961). Effect of sulfur on the iron electrode in alkaline solution. *Zhurnal Fizicheskoi Khimii,* **35**, 2547–2552.

Rozentsveig, S. A., Uflyand, N. Y. and Shcherbakova, Z. V. (1962). Adsorption of sulfur on iron in alkali solutions. *Zhurnal Fizicheskoi Khimii,* **36**, 557–561.

Sakai, G., Miyazaki, M. and Kijima, T. (2010). Synthesis of β-$Ni(OH)_2$ hexagonal plates and electrochemical behavior as a positive electrode material. *Journal of the Electrochemical Society,* **157**, A932–A939.

Sakai, T., Oguro, K., Miyamura, H., Kuriyama, N., Kato, A., Ishikawa, H. and Iwakura, C. (1990). Some factors affecting the cycle lives of LaNi5-based alloy electrodes of hydrogen batteries. *Journal of the Less Common Metals,* **161**, 193–202.

Salvarezza, R. C., Videla, H. A. and Arv, A, A. J. (1982). The electrodissolution and passivation of mild steel in alkaline sulphide solutions. *Corrosion Science,* **22**, 815–829.

Sathyanarayana, S. (1985). Ideally rechargeable cadmium electrodes for alkaline storage batteries. *Journal of Applied Electrochemistry,* **15**, 453–458. 10.1007/bf00616001.

Sato, Y., Takeuchi, S. and Kobayakawa, K. (2001). Cause of the memory effect observed in alkaline secondary batteries using nickel electrode. *Journal of Power Sources,* **93**, 20–24.

Sawa, H., Ohta, M., Nakano, H. and Wakao, S. (1989). Effects of oxidation treatment of Ti-Zr-Ni hydride electrodes containing Zr_7Ni_{10} phase on their electrochemical properties. *Zeitschrift fur Physikalische Chemie N. F.,* **164**, 1527.

Shaoan, C., Anbao, Y., Hong, L., Jianqing, Z. and Chunan, C. (1998). Effects of barium and cobalt on electrochemical performance of nickel hydroxide with chemically co-precipitated zinc. *Journal of Power Sources,* **76**, 215–217.

Sheibley, D. W. and Manzo, M. A. (1980). Control of volume resistivity in inorganic-organic separators. *Journal of the Electrochemical Society,* **127**, 2392–2397.

Sheibley, D. W., Manzo, M. A. and Gonzalez-Sanabria, O. D. (1983). Cross-linked polyvinyl alcohol films as alkaline battery separators. *Journal of the Electrochemical Society,* **130**, 255–259.

Shivkumar, R., Paruthimal Kalaignan, G. and Vasudevan, T. (1995). Effect of additives on zinc electrodes in alkaline battery systems. *Journal of Power Sources,* **55**, 53–62.

Shu, K., Zhang, S., Lei, Y., L, G. and Wang, Q. (2003). Study on structure and electrochemical performance of melt-spun non-stoichiometry alloys Ml(NiCoMnTi)5+X. *International Journal of Hydrogen Energy,* **28**, 1101–1105.

Shukla, A. K. and Hariprakash, B. (2009a). Secondary batteries – nickel systems | electrodes: cadmium. *In: Encyclopedia of Electrochemical Power Sources,* X00FC and Rgen, G. (eds.). Amsterdam: Elsevier.

Shukla, A. K. and Hariprakash, B. (2009b). Secondary batteries – nickel systems | electrodes: Iron. *In: Encyclopedia of Electrochemical Power Sources,* X00FC and Rgen, G. (eds.). Amsterdam: Elsevier.

Shukla, A. K. and Hariprakash, B. (2009c). Secondary batteries – nickel systems | nickel-iron. *In: Encyclopedia of Electrochemical Power Sources,* X00FC and Rgen, G. (eds.). Amsterdam: Elsevier.

Shukla, A. K., Ravikumar, M. K. and Balasubramanian, T. S. (1994). Nickel/iron batteries. *Journal of Power Sources,* **51**, 29–36.

Shukla, A. K., Venugopalan, S. and Hariprakash, B. (2001). Nickel-based rechargeable batteries. *Journal of Power Sources,* **100**, 125–148.

Shukla, A. K., Venugopalan, S. and Hariprakash, B. (2009). Secondary batteries – nickel systems | nickel-cadmium: Overview. *In: Encyclopedia of Electrochemical Power Sources,* X00FC and Rgen, G. (eds.). Amsterdam: Elsevier.

Song, Q., Tang, Z., Guo, H. and Chan, S. L. I. (2002). Structural characteristics of nickel hydroxide synthesized by a chemical precipitation route under different pH values. *Journal of Power Sources,* **112**, 428–434.

Song, Q. S. and Chan, S. L. I. (2009). Nanostructured nickel hydroxides as electrode materials for nickel-based batteries. *In: Nano Materials for Energy Storage Applications,* Nalwa, H. S. (ed.). California, American Scientific Publishers.

Song, Q. S., Chiu, C. H. and Chan, S. L. I. (2006). Performance improvement of pasted nickel electrodes with an addition of ball-milled nickel hydroxide powder. *Electrochimica Acta,* **51**, 6548–6555.

Song, Q. S., Li, Y. Y. and Chan, S. L. I. (2005). Physical and electrochemical characteristics of nanostructured nickel hydroxide powder. *Journal of Applied Electrochemistry,* **35**, 157–162. 10.1007/s10800-004-6301-x.

Souza, C. A. C., Carlos, I. A., Lopes, M., Finazzi, G. A. and De Almeida, M. R. H. (2004). Self-discharge of Fe-Ni alkaline batteries. *Journal of Power Sources,* **132**, 288–290.

Stockel, J. F., Dunlop, J. D. and Betz, F. (1980). NTS-2 nickel-hydrogen battery performance. *Journal of Spacecraft and Rockets,* **17**, 31–34.

Stubicar, M., Blazina, Z., Tonejc, A., Stubicar, N. and Krumes, D. (2001). The effect of high energy ball milling on the crystal structure of GDNi5. *Physica B: Condensed Matter,* **304**, 304–308.

Tam, W. G. and Wainright, J. S. (2007). A microfabricated nickel-hydrogen battery using thick film printing techniques. *Journal of Power Sources,* **165**, 481–488. 10.1016)/j.jpowsour.2006.11.042.

Tamil Selvan, S., Nathira Begum, S., Chidambaram, V., Sabapathi, R. and Vasu, K. I. (1990). Effect of iron addition to the cadmium electrode. *Journal of Power Sources,* **32**, 55–62.

Taniguchi, A., Fujioka, N., Ikoma, M. and Ohta, A. (2001). Development of nickel/metal-hydride batteries for EVs and HEVs. *Journal of Power Sources,* **100**, 117–124.

Taucher-Mautner, W. and Kordesch, K. (2003). Influence of electrode composition on cycling performance of cylindrical nickel-zinc cells. *In: Batteries and Supercapacitors,* Nazri, G. A., Takeuchi, E., Koetz, R. and Scrosati, B. (eds.).

Taucher-Mautner, W. and Kordesch, K. (2004). Studies of pasted nickel electrodes to improve cylindrical nickel-zinc cells. *Journal of Power Sources,* **132**, 275–281.

Teplinskaya, T. K., Fedorova, N. N. and Rozentsveig, S. A. (1964). Nature of the product of the 2nd anodic process on the iron electrode of an alkaline battery. *Zhurnal Fizicheskoi Khimii,* **38**, 2176–2181.

Tessier, C., Guerlou-Demourgues, L., Faure, C., Demourgues, A. and Delmas, C. (2000). Structural study of zinc-substituted nickel hydroxides. *Journal of Materials Chemistry,* **10**, 1185–1193.

Thaller, L. H. and Zimmerman, A. H. (1996). Electrolyte management considerations in modern nickel/hydrogen and nickel/cadmium cell and battery designs. *Journal of Power Sources,* **63**, 53–61.

Thaller, L. H. and Zimmerman, A. H. (2003). Overview of the design, development, and application of nickel-hydrogen batteries. *The Aerospace Corporation, Los Angeles, California*, 44.

Unates, M. E., Folquer, M. E., Vilche, J. R. and Arvia, A. J. (1992). The influence of Foreign cations on the electrochemical behavior of the nickel hydroxide electrode. *Journal of the Electrochemical Society,* **139**, 2697–2704.

Van Mal, H. H., Buschow, K. H. J. and Kuijpers, F. A. (1973). Hydrogen absorption and magnetic properties of LaCo5xNi5-5x compounds. *Journal of the Less Common Metals,* **32**, 289–296.

Vermeiren, P., Adriansens, W., Moreels, J. P. and Leysen, R. (1998). Evaluation of the Zirfon® separator for use in alkaline water electrolysis and Ni-H2 batteries. *International Journal of Hydrogen Energy,* **23**, 321–324.

Vidotti, M., Salvador, R. P. and Córdoba De Torresi, S. I. (2009). Synthesis and characterization of stable Co and Cd doped nickel hydroxide nanoparticles for electrochemical applications. *Ultrasonics Sonochemistry,* **16**, 35–40.

Vijayamohanan, K., Balasubramanian, T. S. and Shukla, A. K. (1991). Rechargeable alkaline iron electrodes. *Journal of Power Sources,* **34**, 269–285.

Vijayamohanan, K., Shukia, A. K. and Sathyanarayana, S. (1990). Role of sulphide additives on the performance of alkaline iron electrodes. *Journal of Electroanalytical Chemistry,* **289**, 55–68.

Visintin, A., Anani, A., Srinivasan, S., Appleby, A. J. and Lim, H. S. (1995). Kinetic aspects of self-discharge of nickel-hydrogen batteries and methods for its prevention. *Journal of Applied Electrochemistry,* **25**, 833–840.

Wakao, S., Sawa, H. and Furukawa, J. (1991). Effects of partial substitution and anodic oxidation treatment of Zr-V-Ni alloys on electrochemical properties. *Journal of the Less Common Metals,* **172–174**, 1219–1226.

Wang, C., Marrero-Cruz, M., Soriaga, M. P., Serafini, D. and Srinivasan, S. (2002). Improvement in the cycle life of LaB5 metal hydride electrodes by addition of ZnO to alkaline electrolyte. *Electrochimica Acta,* **47**, 1069–1078.

Wang, C. Y., Sun, J., Liu, H. K., Dou, S. X., Macfarlace, D. and Forsyth, M. (2005). Potential application of solid electrolyte P11 OH in Ni/MH batteries. *Synthetic Metals,* **152**, 57–60. 10.1016/j.synthmet.2005.07.125.

Wang, S., Yang, Z. and Zeng, L. (2008). Study of calcium zincate synthesized by solid-phase synthesis method without strong alkali. *Materials Chemistry and Physics,* **112**, 603–606.

Wang, Y.-M. (1990). Effect of KOH concentration on the formation and decomposition kinetics of calcium zincate. *Journal of the Electrochemical Society,* **137**, 2800–2803.

Wang, Y. -M. and Wainwright, G. (1986). Formation and decomposition kinetic studies of calcium zincate in 20 w/o KOH. *Journal of the Electrochemical Society,* **133**, 1869–1872.

Wang, Z. M., Li, C. Y. V. and Chan, S. L. I. (2009). Effect of electrolyte on electrochemical characteristics of MmNi3.55Co0.72Al0.3Mn0.43 alloy electrode for hydrogen storage. *International Journal of Hydrogen Energy,* **34**, 5422–5428.

Wang, Z. M., Tsai, P.-J., Ip Chan, S. L., Zhou, H. Y. and Lin, K. S. (2010). Effects of electrolytes and temperature on high-rate discharge behavior of MmNi5-based hydrogen storage alloys. *International Journal of Hydrogen Energy,* **35**, 2033–2039.

Watanabe, K. -I., Koseki, M. and Kumagai, N. (1996). Effect of cobalt addition to nickel hydroxide as a positive material for rechargeable alkaline batteries. *Journal of Power Sources,* **58**, 23–28.

Watanabe, K. -I. and Kumagai, N. (1997). Electrochemical and thermodynamic studies of nickel electrodes in alkaline electrolytes. *Journal of Power Sources,* **66**, 121–127.

Watanabe, K., Kikuoka, T. and Kumagai, N. (1995). Physical and electrochemical characteristics of nickel hydroxide as a positive material for rechargeable alkaline batteries. *Journal of Applied Electrochemistry,* **25**, 219–226.

Wehrens-Dijksma, M. and Notten, P. H. L. (2006). Electrochemical quartz microbalance characterization of Ni(OH)2-based thin film electrodes. *Electrochimica Acta,* **51**, 3609–3621.

Willems, J. J. G. and Buschow, K. H. J. (1987). From permanent magnets to rechargeable hydride electrodes. *Journal of the Less Common Metals,* **129**, 13–30.

Wu, J. B., Tu, J. P., Han, T. A., Yang, Y. Z., Zhang, W. K. and Zhao, X. B. (2006). High-rate dischargeability enhancement of Ni/MH rechargeable batteries by addition of nanoscale CoO to positive electrodes. *Journal of Power Sources,* **156**, 667–672.

Wu, J. Z., Tu, J. P., Yuan, Y. F., Ma, M., Wang, X. L., Zhang, L., Li, R. L. and Zhang, J. (2009). Ag-modification improving the electrochemical performance of ZnO anode for Ni/Zn secondary batteries. *Journal of Alloys and Compounds,* **479**, 624–628.

Wu, M. S., Wu, H. R., Wang, Y. Y. and Wan, C. C. (2003). Effects of the stoichiometric ratio of aluminium and manganese on electrochemical properties of hydrogen storage alloys. *Journal of Applied Electrochemistry,* **33**, 619–625.

Yang, H., Zhang, H., Wang, X., Wang, J., Meng, X. and Zhou, Z. (2004). Calcium zincate synthesized by ballmilling as a negative material for secondary alkaline batteries. *Journal of the Electrochemical Society,* **151**, A2126–A2131.

Yang, J. L., Yuan, Y. F., Wu, H. M., Li, Y., Chen, Y. B. and Guo, S. Y. (2010). Preparation and electrochemical performances of ZnO nanowires as anode materials for Ni/Zn secondary battery. *Electrochimica Acta,* **55**, 7050–7054.

Yuan, A., Cheng, S., Zhang, J. and Cao, C. (1998). The influence of calcium compounds on the behaviour of the nickel electrode. *Journal of Power Sources,* **76**, 36–40.

Yuan, A., Cheng, S., Zhang, J. and Cao, C. (1999). Effects of metallic cobalt addition on the performance of pasted nickel electrodes. *Journal of Power Sources,* **77**, 178–182.

Yuasa, K. and Ikoma, M. (2006). Improvement of basic performances of a nickel-metal hydride battery. *Research on Chemical Intermediates,* **32**, 461–471. 10.1163/156856706777973727.

Yunchang, D., Hui, L., Jiongliang, Y. and Zhaorong, C. (1995). Effects of dopants on electrochemical performance of nickel cathodes. *Journal of Power Sources,* **56**, 115–119.

Züttel, A., Meli, F. and Schlapbach, L. (1994). Effects of pretreatment on the activation behavior of Zr(V0.25Ni0.75)2 metal hydride electrodes in alkaline solution. *Journal of Alloys and Compounds,* **209**, 99–105.

Zhang, C., Wang, J. M., Zhang, L., Zhang, J. Q. and Cao, C. N. (2001). Study of the performance of secondary alkaline pasted zinc electrodes. *Journal of Applied Electrochemistry,* **31**, 1049–1054. 10.1023/a:1017923924121.

Zhang, G., Fan, C., Pan, L., Wang, F., Wu, P., Qiu, H., Gu, Y. and Zhang, Y. (2005a). Magnetic and transport properties of magnetite thin films. *Journal of Magnetism and Magnetic Materials,* **293**, 737–745.

Zhang, H., Chen, Y., Zhu, Q., Zhang, G. and Chen, Y. (2008a). The effects of carbon nanotubes on the hydrogen storage performance of the alloy electrode for high-power Ni-MH batteries. *International Journal of Hydrogen Energy,* **33**, 6704–6709.

Zhang, L., Huang, H., Zhang, W. K., Gan, Y. P. and Wang, C. T. (2008b). Effects of conductive ceramic on the electrochemical performance of ZnO for Ni/Zn rechargeable battery. *Electrochimica Acta,* **53**, 5386–5390.

Zhang, W. K., Xia, X. H., Huang, H., Gan, Y. P., Wu, J. B. and Tu, J. P. (2008c). High-rate discharge properties of nickel hydroxide/carbon composite as positive electrode for Ni/MH batteries. *Journal of Power Sources,* **184**, 646–651.

Zhang, Z., Yang, J., Nuli, Y., Wang, B. and Xu, J. (2005b). CoPx synthesis and lithiation by ball-milling for anode materials of lithium ion cells. *Solid State Ionics,* **176**, 693–697.

Zhou, H. and Zhou, Z. (2005). Preparation, structure and electrochemical performances of nanosized cathode active material Ni(OH)2. *Solid State Ionics,* **176**, 1909–1914.

Zhu, W. -H., Ke, J. -J., Yu, H. -M. and Zhang, D. -J. (1995). A study of the electrochemistry of nickel hydroxide electrodes with various additives. *Journal of Power Sources,* **56**, 75–79.

Zimmerman, A. H. and Seaver, R. (1990). Cobalt segregation in nickel electrodes during nickel hydrogen cell storage. *Journal of the Electrochemical Society,* **137**, 2662–2667.

12

Redox flow batteries for medium- to large-scale energy storage

M. SKYLLAS-KAZACOS and C. MENICTAS,
University of New South Wales, Australia and
T. LIM, Ngee Ann Polytechnic, Singapore

DOI: 10.1533/9780857097378.3.398

Abstract: With the increasing integration of renewable energy sources into the electricity grids of many developed and developing countries, the need for energy storage has become a major priority for grid stabilisation. Flow batteries offer high energy efficiencies, very long cycle life and good cost structures for applications requiring more than 2 h of storage capacity. Of the flow battery technologies currently under development, the vanadium redox flow battery that was pioneered at the University of New South Wales (UNSW), Australia, has received the most attention, with more than 30 medium- to large-scale installations in Japan, Europe, USA and China demonstrating its benefits and features in a range of off-grid and grid-connected applications.

Key words: energy storage, redox flow batteries, vanadium redox flow battery, zinc-bromine battery, polysulphide-bromine battery.

12.1 Introduction

One of the major drawbacks of renewable energy technologies is their intermittent nature. In the case of solar energy, production occurs during sunlight hours, which is often out-of-phase with peak energy usage. Furthermore, intermittent cloud cover can create significant power output instability from photovoltaic panels, even on sunny days. Wind turbine output is equally unstable, and the irregular nature of wind power generation makes it impossible to meet the load power needs throughout the year. Hourly wind speed fluctuations add an additional supply continuity problem to the load that requires grid connection or other back-up power to ensure reliability. Integration of greater than 12–15% renewables to a grid will, however, lead to severe grid instability problems that can only be overcome by the incorporation of effective energy storage to maximise the utilisation of the renewable energy and to assist in load levelling. To date, several types of energy storage systems have been used successfully in demonstration and commercial applications. These include mechanical systems, such as pumped hydro, compressed air and flywheels, and electrochemical systems, such as fuel cells and batteries.

Pumped hydro storage currently comprises 3% of global power generation capacity and is the most common energy storage system available. Pumped hydro systems operate using two reservoirs of water, separated vertically, to generate and store electricity. The amount of storage capacity is determined by the size of the reservoirs and the power generated is dependent on the size of the turbines and the water flow rate. Accordingly, pumped hydro storage is generally used for large-scale applications, but is restricted by geography. Compressed air systems also offer gigawatt-scale power with several hours of storage capacity, but as with pumped hydro, these systems are limited by geology and geography to specific locations that may not be close to the renewable energy source or the load. Flywheels on the other hand are more flexible in terms of sizing and location, but their low energy density limits their use to power quality applications that require energy storage capacities of seconds to several minutes. While this would be adequate for levelling out the short-term instabilities from wind turbines and photovoltaic arrays, it is not generally considered viable for load shifting and load levelling of renewable energy sources that require several hours of storage capacity. On the other hand, electrochemical energy storage systems offer greater flexibility combined with the energy efficiency, performance and cost structures required to meet the needs of large-scale renewable energy storage and smart-grid applications (EPRI-DOE, 2003, 2004).

Of the electrochemical systems currently available, lead-acid batteries are capable of providing low cost energy storage and have been successfully implemented in back-up power systems and other small-scale applications. Despite their widespread use however, lead-acid batteries have relatively short life cycles under the deep discharge regimes required for most energy storage applications. A recent advance in lead-acid battery technology developed by the Australian Commonwealth Scientific and Industrial Research Organisation (CSIRO) is the UltraBattery, a hybrid energy storage device that integrates a supercapacitor with a lead-acid battery in one unit cell (CSIRO, 2011). The UltraBattery incorporates carbon plates at the negative electrode to act as supercapacitor hybrid electrodes, allowing fast charge acceptance compared with the conventional technology. This feature makes the lead-acid battery more suitable in hybrid vehicle applications, but could also be important for short-term wind turbine output power stabilisation. The cycle life of this hybrid system under deep discharge applications has yet to be demonstrated however, so its most likely application will be in hybrid vehicles and other applications requiring short-term energy storage capabilities.

Lithium ion batteries are being widely investigated for hybrid and electric vehicle applications, but are currently too expensive when compared to other storage systems (ESA, 2011). They do, however, have long life cycles, operating at close to 100% efficiency and have an energy density of approximately 300–400 kWh/m^3, making them ideally suited to the portable appliance market, particularly digital cameras. This high energy density enables the batteries to be low in weight, but leads to potential safety problems associated with the highly reactive lithium metal that could be formed at the negative electrode during accidental overcharge. To avoid potential explosions and fires therefore, lithium ion batteries require a protection circuit module (PCM) to prevent overcharge and discharge below the lower voltage limit. This also limits the charge and discharge current and renders the cell open circuit in the event of an abnormal condition (Balakrishnan *et al.*, 2006). Despite these limitations, a number of megawatt size Li-ion based demonstrations have recently been installed and tested in the USA (A123, 2011). Such systems would be capable of providing short-term power output stabilisation for wind turbines; however, for application in longer term storage of wind energy, the cost of Li-ion batteries need to be further reduced to compete with other options.

Another important candidate for large-scale energy storage is the sodium sulphur battery. The NaS battery consists of liquid (molten) sulphur at the positive electrode and liquid (molten) sodium at the negative electrode as active materials, separated by a solid beta alumina ceramic electrolyte. The operating temperature of the NaS battery is greater than 350°C, however, sophisticated methods of construction of the battery pack are required, together with auxiliary heating during charging and periods of non-use to prevent freezing of the sodium electrolyte and damage due to mechanical stresses. Despite their relatively high costs, NaS batteries have been installed at over 190 sites in Japan totalling more than 270 MW with stored energy suitable for 6 h daily peak shaving. The main limitation is the difficulty associated with the manufacture of the high-grade ceramic separators used in the devices and the potential safety hazards associated with the use of liquid metallic sodium and sulphur (Skyllas-Kazacos, 2010).

Unlike the lead-acid, Li-ion and NaS technologies that rely on solid state processes for their charge–discharge reactions that often lead to mechanical breakdown of the active material and life cycle limitations, flow batteries use solutions to store energy. Flow batteries therefore offer long cycle life under deep discharge operation, as well as the greatest flexibility in power to energy rating, since capacity is determined by the volume of electrolyte stored in external tanks.

Flow batteries, such as the zinc-bromine (Zn/Br) battery and vanadium redox battery (VRB), convert chemical energy into electricity by pumping

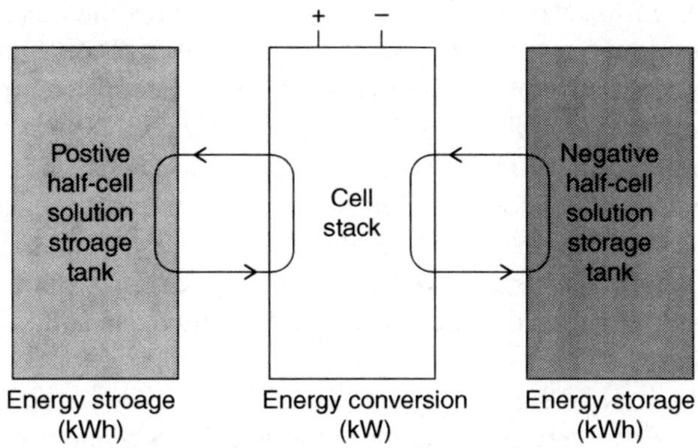

12.1 Flow cell schematic.

electrolytes from external storage tanks through a cell stack where electron transfer reactions take place on inert electrodes that usually comprise porous graphite felts that provide a high surface area for the electron transfer reactions. Electrolytes with differing oxidation states provide the electrochemical potential required to generate electricity. A number of reviews on flow battery research and development have been published in recent years and these cover a range of different flow battery chemistries and technologies (Skyllas-Kazacos *et al.*, 2011; Weber *et al.*, 2011)

Flow batteries differ from conventional batteries in that their active material is in the form of two redox couple solutions that are stored in external tanks and pumped through a stack of electrochemical cells where the solutions are separated by an ion-exchange membrane that prevents mixing of the two solutions (Fig. 12.1).

Flow cells fall into two categories. The first type is the metal/halide flow cell that involves the deposition of a metal at the negative electrode during charging. Examples are the Zn/Br battery and the zinc chloride (Zn/Cl) battery, which is one of the earliest types of flow batteries (Blevins, 1981) In both cases, capacity is determined by the quantity of zinc metal deposited at the negative electrode.

The second type of flow battery is the redox flow battery (RFB), which employs two fully soluble redox couple solutions in each half-cell. Unlike the Zn/Br and Zn/Cl flow batteries, the RFB has all reactants and products in the solution phase and no metals are plated on the electrodes during charging. The redox flow cell thus stores energy in the solutions, so that the capacity of the system is determined by the size of the electrolyte tanks, while the system power is determined by the size of the cell stacks. The redox flow cell is therefore more like a rechargeable fuel cell than a battery.

Initial work on redox batteries began in the 1970s with the development of the iron-chromium battery by NASA (Thaller, 1976; Swette and Jalan, 1984). Of the redox flow cell technologies that have been developed over the last 30 years, however, only the VRB invented at the UNSW in Australia has reached commercial fruition to date (Skyllas-Kazacos and Robbins, 1986); further developments in the electrolyte formulation have led to the Generation 2 (Skyllas-Kazacos, 2001) and Generation 3 vanadium (Li *et al.*, 2011) battery technologies that have higher energy densities than the original vanadium/sulphuric acid system (G1 VRB), but these are still at the development phase.

Each of the above RFB systems will be described in more detail later. The fundamental properties and performance characteristics of electrochemical cells as related to flow batteries will be first described below.

12.2 Electrochemical cells

When describing the performance of different types of electrochemical energy storage systems, it is important to understand the fundamentals of electrochemical cells and the factors that affect cell potential and energy efficiency. A brief outline of cell fundamentals is provided here, together with a description of the criteria used to assess the performance of electrochemical energy storage systems, in particular redox flow batteries.

12.2.1 Theoretical cell potential

The cell potential of an electrochemical cell is the potential difference occurring between the two electrodes of the cell, and arises due to the transfer of electrons through the external circuit of a cell that has not reached equilibrium. This is related to the Gibbs free energy change for the cell reaction given by:

$$\Delta G = -nE_{\text{cell}}F \tag{12.1}$$

where ΔG = Gibbs free energy change (J), n = number of electrons transferred per unit overall reaction (mol), E_{cell} = cell potential (V), F = Faraday constant (96 485 C mol^{-1}).

Furthermore, the cell potential is related to the composition of the reaction mixture via the Nernst Equation:

$$E_{\text{cell}} = E^{o}_{\text{cell}} - \frac{RT}{nF}\ln Q \tag{12.2}$$

where E_{cell} = cell potential (V), E^{o}_{cell} = standard cell potential (V), R = gas constant (8.314 J K^{-1} mol^{-1}), T = temperature (K), Q = reaction quotient = $\frac{[\text{Ox}]}{[\text{Red}]}$

12.2.2 Actual cell potential

Due to irreversible losses within a cell, the actual cell potential is lower than the theoretical cell potential described above. The losses arise from three sources, activation polarisation, concentration polarisation and ohmic polarisation, and result in an overall cell potential described by:

$$V_{cell} = E^{C} - E^{A} - \eta_{A} - \eta_{C} - iR_{cell} \quad [12.3]$$

where E^{C} = cathode potential (V), E^{A} = anode potential (V), η_{A} = activation overpotential for both cathode and anode reactions (V). This is the additional potential required to overcome the activation energy barrier for the electron transfer reactions. η_{C} = concentration overpotential at both the anode and cathode (V). This is the additional potential required to overcome the mass transport limitations at the electrode/electrolyte interface. i = current density (A cm^{-2}), R_{Cell} = ohmic resistance for a given electrode cross-sectional area (Ω cm^{2}).

Figure 12.2 provides a breakdown of the cell voltage components as a function of current density. Activation overvoltage is associated with the kinetics of the electrode reactions, and is a function of temperature and the properties of the electrode materials. Concentration overvoltage is associated with the mass transport processes at the electrolyte/electrolyte interface. This is affected by the electrolyte flow rate, particularly at high and low states of charge when the active material concentration in the electrolyte is low and the rate of transfer of the active ions to the electrode surface is unable to keep up with the transfer of electrons across the electrode/electrolyte interface. Ohmic resistance losses are associated with the resistances of the electrodes, membrane and current collectors, so material selection is critical in reducing these losses to achieve maximum efficiencies.

When using high surface area carbon or graphite felts, current density is reduced by several orders of magnitude, so activation overvoltage losses become insignificant. The cell will therefore operate in the region of ohmic polarisation and thus the potential can be more simply represented by:

$$V_{cell} = E^{C} - E^{A} - IR_{cell} \quad [12.4]$$

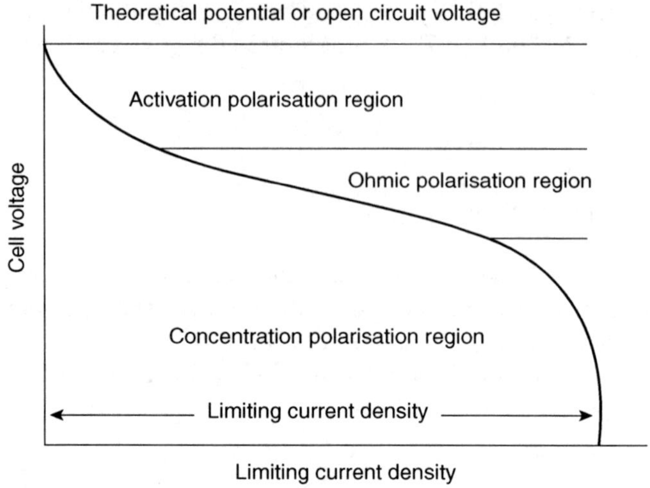

12.2 Actual cell potential as a function of current density.

At high current densities, mass transport processes are unable to supply the electroactive ions to the electrode surface at the rate needed to keep the reaction going and this gives rise to a limiting current. The electrode potential thus shifts to more negative or more positive values, leading to the breakdown of water and gassing side reactions that produce hydrogen and oxygen at the negative and positive electrodes respectively during charging. The magnitude of the limiting current is a function of the concentration of the electroactive ions in solution and this decreases at high states of charge (SoC) during charging, or at low SOCs during discharge. Increasing the rate of mass transport of the ions to the electrode surface, by either stirring or by pumping the solutions through each half-cell at a faster rate, will increase the limiting current, but concentration polarisation will lead to gassing side reactions if the current density exceeds the limiting current for the cell charging reactions at high SOCs. When this occurs, the cell can often go into 'overcharge', which is highly undesirable since this can potentially lead to the formation of hydrogen gas at the negative electrode. Where carbon electrodes are used in the cell, gassing side reactions during overcharge could also include the generation of CO_2 from the oxidation of the positive carbon electrode material. This can lead to a degradation of the positive electrode and result in reduced cycle life. To prevent overcharging of the cell, which can limit cell life and potentially produce hydrogen gas at the negative electrode, flow batteries are typically charged up to around 90–95% SOC and charging is ceased before the cell potential enters the range for water decomposition.

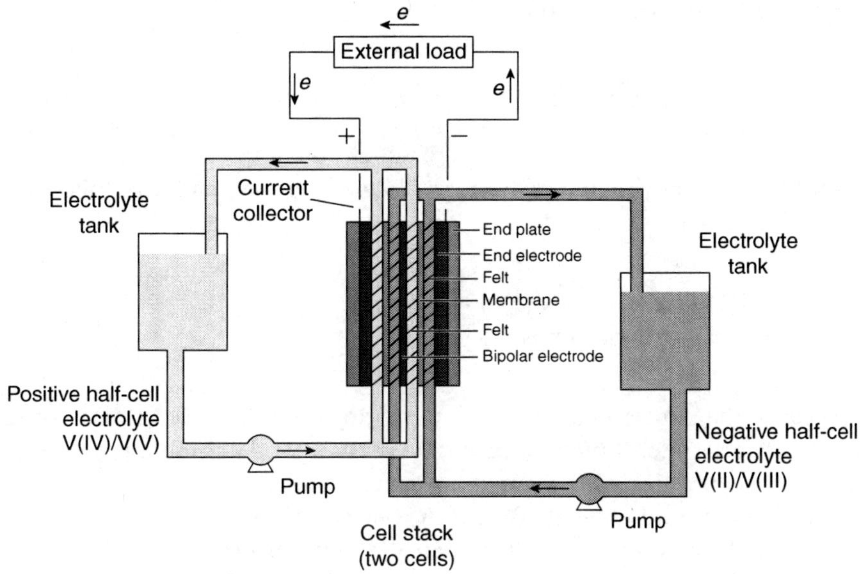

12.3 Typical bipolar stack arrangement.

Typically cells are connected in series and/or parallel in bipolar stacks in order to deliver the required current and voltage output. A bipolar stack configuration is illustrated in Fig. 12.3, showing series electrical connection of individual cells in the stack, with parallel hydraulic connection of the electrolyte channels.

12.2.3 State of charge

The SOC of the cell is a measure of the extent to which the half-cell reactions have taken place during charge and discharge cycles, and is proportional to the ratio of the electrode reactant and product concentrations in the electrolyte at any time. For most redox flow cell chemistries, changes in the oxidation state of the electroactive species in solution will give rise to colour changes that can often be used for the qualitative determination of SOC. As an example, in the case of the vanadium redox couples, the electrode reactions are given by (Skyllas-Kazacos and Kazacos, 2011):

Positive electrode reaction:

$$VO^{2+} + H_2O \underset{\text{discharge}}{\overset{\text{charge}}{\rightleftharpoons}} VO_2^+ + 2H^+ + e \qquad [12.5]$$

Negative electrode reaction:

$$V^{3+} + e^- \underset{\text{discharge}}{\overset{\text{charge}}{\rightleftarrows}} V^{2+} \qquad [12.6]$$

As the solution progresses through each oxidation state, the following colour changes are observed:

$$\begin{array}{l} V(II) \rightarrow V(III) \rightarrow V(IV) \rightarrow V(V) \\ \text{voilet} \rightarrow \text{green} \rightarrow \text{blue} \rightarrow \text{yellow} \end{array} \qquad [12.7]$$

Because the determination of colour is subjective, however, this method is not very accurate. A more appropriate method of determining SOC is to use the concentration of the vanadium species in the electrolyte.

The system is said to be at 100% SOC once all V(IV) has been converted to V(V) in the positive half-cell, and all V(III) has been converted to V(II) in the negative half-cell. Furthermore, when no V(II) or V(V) remains in the electrolyte at the end of discharge, the system is said to be at 0% SOC.

The relationship between SOC and vanadium ion concentrations can be expressed by the following equations:

$$[V(V)] = [V(II)] = V_T \cdot SOC/100 \qquad [12.8]$$

$$[V(IV)] = [V(III)] = V_T \cdot (1 - SOC/100) \qquad [12.9]$$

where V_T is the total vanadium ion concentration in each solution. By substituting Equations [12.8] and [12.9] into the Nernst Equation (Equation [12.2]) and assuming that the hydrogen ion concentration remains constant, it is possible to express the theoretical cell potential, or the cell open-circuit voltage (OCV), in terms of the SOC as shown below:

$$E_{\text{cell OC}} = E^0_{\text{cell OC}} + \frac{2RT}{nF} \ln \frac{SOC}{1 - SOC} \qquad [12.10]$$

where $E_{\text{cell OC}}$ = theoretical OCV of the cell, $E^0_{\text{cell OC}}$ = formal potential or theoretical OCV of the cell at 50% SOC for a fixed hydrogen ion concentration

The formal potential has been measured experimentally at 50% SOC for different electrolyte compositions, and this can range from 1.3 to 1.4

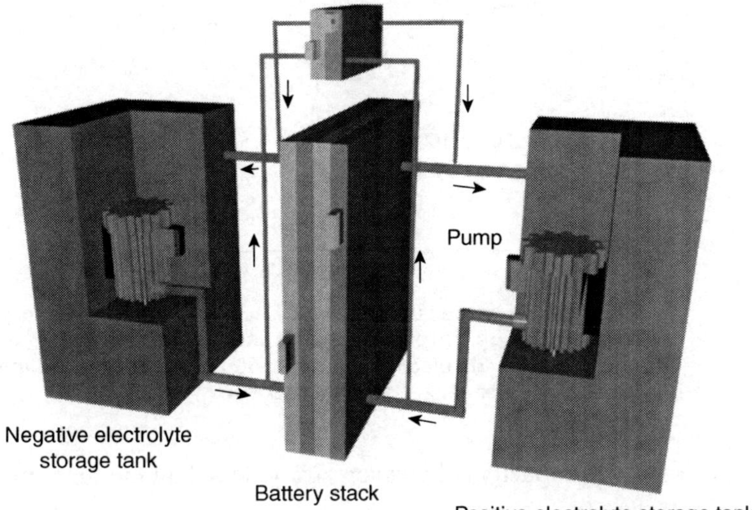

12.4 Flow battery system showing hydraulic arrangement of battery stack, OCV cell, electrolyte storage tanks and pumps.

V depending on the total vanadium and acid concentration. By continually measuring the potential difference between the two half-cell solutions (using an open-circuit cell), it is therefore possible to monitor the SOC of a redox flow cell, even under load – a feature unique to flow batteries. This is illustrated in Fig. 12.4, which shows the hydraulic connection of the battery stack, the open-circuit cell (also known as OCV cell), electrolyte storage tanks and pumps in a flow battery system.

12.2.4 Cell capacity

The theoretical cell capacity is given by Faraday's Law and is a function of the electrolyte volume and electroactive material concentration. At constant current, this is given by:

$$Q_T = It = m(nF) \tag{12.11}$$

where Q_T = capacity of the cell (A s), I = current passed through the cell (A), t = discharge time (s), m = moles of reactant required for complete cell discharge (mol), n = number of electrons transferred in the reaction (mol), F = Faraday constant (96 485 C mol^{-1}).

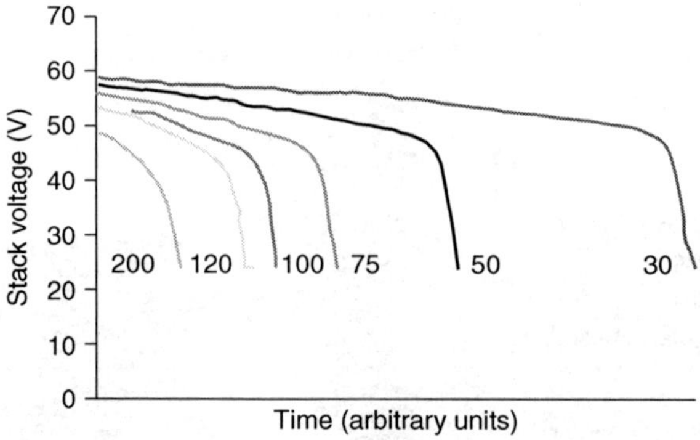

12.5 Typical discharge curves for a 5 kW vanadium flow battery at different discharge currents but same charge current of 30 A. Number of cells in stack = 38; electrode area = 1500 cm². (*Source*: Adapted from Skyllas-Kazacos *et al.*, 2010, Wiley and Sons.)

The actual cell capacity is, however, always less than the theoretical value, and this is due to:

- the SOC range used for charge–discharge cycling. For example, if the cell is only cycled between 5 and 95% SOC, the actual capacity will be 90% of the theoretical.
- polarisation losses that limit the operating SOC range at increased current densities for a given operating voltage range.

As with all electrochemical energy storage systems, the capacity of flow batteries will decrease with increasing current, but to a much smaller extent than with conventional batteries, which are severely limited by the rates of their solid state reactions (e.g., in the lead-acid battery, the charging and discharging processes involve the conversion of solid lead sulphate to spongy lead and lead dioxide at the negative and positive electrodes, respectively).

The relationship between capacity and current is illustrated in Fig. 12.5, which shows a series of typical discharge curves for a 38-cell VRB stack at different discharge currents. In this example, a charging current of 30 amps was used for all runs, allowing a high SOC to be reached in each case. During each subsequent discharge cycle, close to the maximum capacity could therefore be achieved. When charging at much higher currents, however, a much lower SOC would be reached for a set cut-off voltage, so subsequent discharge cycles would exhibit lower capacities.

As illustrated by Fig. 12.6, however, discharge capacity depends on the set cut-off voltage, and this is a function of the power electronics and inverters connected to the battery.

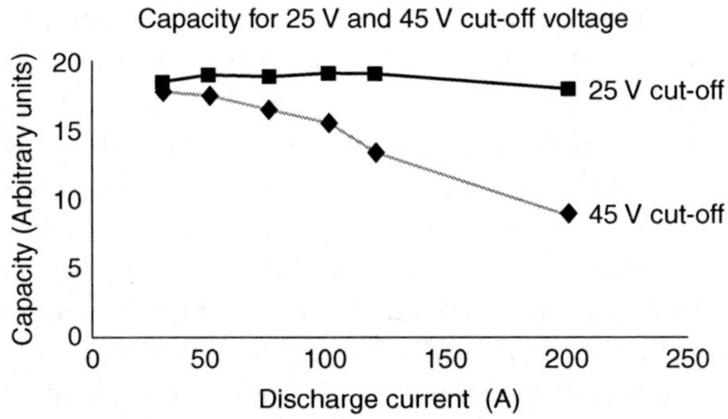

12.6 Capacity vs discharge current for different cut-off voltage limits corresponding to curves in Fig. 12.5.

An additional factor that can influence capacity is electrolyte flow rate. Operation at high flow rates will reduce concentration overvoltage losses and enable higher SOCs to be attained during charge, while greater active material utilisation will allow close to theoretical capacities to be achieved during discharge. Increased flow rate will however lead to greater pumping energy losses, so flowrate optimisation is required to maximise the overall energy efficiency of the system while also maximising available capacity. Typically, losses of 1–2% are associated with pumping energy requirements, depending on mode of operation.

12.2.5 Efficiencies

Coulombic efficiency

The total charge that is available during the discharge cycle relative to the number of coulombs used for the charge cycle is given by the coulombic efficiency. It is calculated using the following expression:

$$\eta_C = \frac{\oint I_{\text{dis}}\,\mathrm{d}t}{\oint I_{\text{ch}}\,\mathrm{d}t} \times 100\% \qquad [12.12]$$

where η_c = coulombic efficiency (%), I_{dis} = discharge current (A), I_{ch} = charge current (A), t = time (s).

Coulombic efficiency is maximised by minimising self-discharge processes and reducing the gassing side reactions during charging. Examples include self-discharge processes associated with the diffusion of the electroactive ions across the membrane, which can be prevented by using highly selective ion-exchange membranes, and voltage control, which can help prevent the gassing side reactions. Coulombic efficiency usually decreases at both high and low current densities. The low current density decrease is associated with greater degree of self-discharge across the membrane at the longer charging and discharging times, while the drop off at high current densities is caused by the greater rate of irreversible gassing side reactions during charging. Further coulombic efficiency losses (typically 1–2%) are associated with the shunt or bypass currents that flow through the common electrolyte manifolds in bipolar cell stacks. In the bipolar stack configuration the cells are interconnected hydraulically through common manifolds to provide parallel electrolyte flow through the individual cells in the stack. Shunt currents are produced by the voltage across the battery stack and flow via the electrolyte network. The magnitude of the shunt current increases with the number of cells in the stack and can also result in gas evolution and electrode degradation. Shunt currents will detract from the energy available from the system, so they need to be minimised by increasing the current path-length between adjacent cells. This typically involves the use of long, narrow electrolyte channels, but this will increase pumping pressure drop that also represents a parasitic loss in flow batteries. With good stack design and battery operation, however, these parasitic losses can be reduced to 2–3% of the total energy.

Voltage efficiency

Voltage efficiency measures the effects of cell polarisation or cell voltage losses. It is calculated via the following equation:

$$\eta_v = \frac{\oint V_{\mathrm{dis}}\,\mathrm{d}t}{\oint V_{\mathrm{ch}}\,\mathrm{d}t} \times 100\% \qquad [12.13]$$

where η_v = voltage efficiency (%), V_{dis} = discharge voltage (A), V_{ch} = charge voltage (A).

Various losses, including ohmic resistances, activation overpotential and concentration overpotential, will reduce the voltage efficiency. Voltage efficiency can be maximised by reducing the resistance of all cell components and using electrode materials with high electrical conductivity, good electroactivity and high surface area (Zhong *et al.*, 1993). In flow batteries, flow-through graphite felt electrodes provide the required high surface area for the electron transfer reactions, while pre-treatment has been shown to improve the wettability of the graphite felt materials and increase electrochemical activity by introducing surface functional groups that behave as active sites for the electron transfer reactions (Sun and Skyllas-Kazacos, 1992).

All cell losses increase with increasing current, so voltage efficiency decreases with increasing current density.

Energy efficiency

The overall energy efficiency of the cell is a measure of the amount of actual energy released on discharge relative to the amount of energy required to charge the cell. It is calculated via the following expression:

$$\eta_e = \frac{\oint I_{\text{dis}} V_{\text{dis}} \, \mathrm{d}t}{\oint I_{\text{ch}} V_{\text{ch}} \, \mathrm{d}t} \times 100\% \qquad [12.14]$$

where η_e = overall energy efficiency (%), I_{dis} = cell current during discharge (A), V_{dis} = cell voltage during discharge (V), I_{ch} = cell current during charge (A), V_{ch} = cell voltage during charge (V).

Energy efficiency can also be calculated from the product of the coulombic and voltage efficiencies as follows:

$$\eta_e = \eta_c \eta_v \qquad [12.15]$$

where η_e = energy efficiency (%), η_c = coulombic efficiency (%), η_v = voltage efficiency (%).

Energy efficiency is an indicator of the overall cell performance and will reflect the combined trends in coulombic and voltage efficiencies as a function of current density. This is illustrated by the example shown in Fig. 12.7.

The example illustrated in Fig. 12.7 shows the expected trends in coulombic and voltage efficiency with increasing current density. The combined effect on the overall energy efficiency is an optimal operating current density for maximum energy efficiency, although other factors including capital and maintenance costs could shift the optimal current density to higher values in practice.

12.2.6 Electrolyte flow rate

An RFB requires a minimum flow of electrolyte to ensure that the electroactive species is supplied to each half-cell at a rate that is sufficient to ensure that its concentration does not drop to zero before the electrolyte exits the cell. Each cell within a battery stack should ideally be supplied

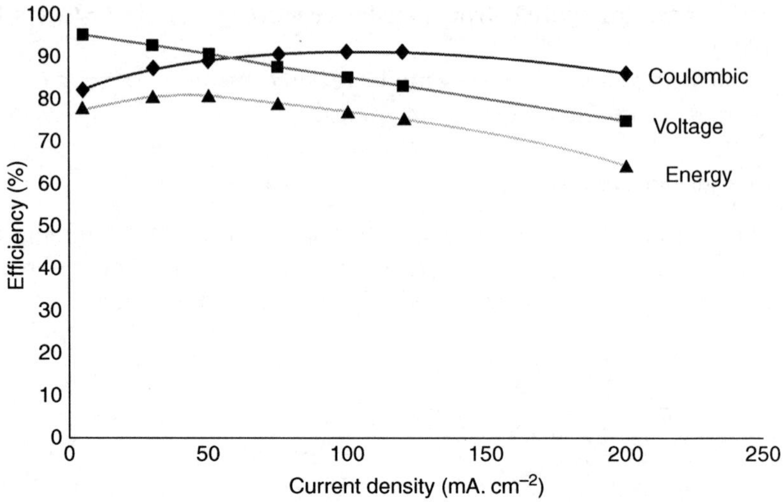

12.7 Typical trends in coulombic, voltage and energy efficiencies for a redox flow battery as a function of current density.

with the same flowrate; however, slight variations in electrolyte distribution may occur due to the pressure drop across the manifolds and non-uniform pressure drops through the porous graphite felt electrodes. The minimum theoretical flowrate, also known as the stoichiometric flow (F_{sf}), is a function of the reaction stoichiometry, the applied current and the cell SOC, and is given by:

$$F_{sf} = \frac{I}{N \times \text{SOC}} \tag{12.16}$$

where F_{sf} = stoichiometric flowrate (cm^3 min^{-1}), I = specified current (A), N = electrolyte capacity (A min cm^{-3}), SOC = state of charge (%).

The electrolyte capacity is a function of electrolyte concentration and can be determined via the following equation:

$$N = \frac{n \times M \times F}{1000\left(cm^3/L\right) \times 60\left(s/\text{min}\right)} \tag{12.17}$$

where N = electrolyte capacity (A min cm^{-3}), n = number of electrons per mole (mol^{-1}), M = concentration of electrolyte (mol L^{-1}), F = Faraday constant (96 485 C mol^{-1}).

The minimum flow rate required will thus vary with SOC of the system for a fixed electrolyte concentration. To minimise concentration overpotential losses, the actual flowrate used in a VRB is considerably higher than the stoichiometric flowrate and is a function of cell geometry and design.

Electrolyte flow is an important variable in flow battery design and operation. The electrolytes themselves act as a very efficient sink for the heat generated in the cell stack and can be integrated into a heat exchanger to provide good thermal management during high rate charge–discharge cycling. This eliminates problems of thermal runaway that is a major problem with non-flow batteries. Thermal modelling of VRB systems has been used as an effective tool for the design of efficient batteries and control strategies under a range of different operational regimes and climatic conditions (Ao *et al.*, 2012).

12.2.7 System integration

As with all types of energy storage systems, flow batteries must be integrated into the overall energy system to provide power to the end user. All batteries require direct current (DC) for charging and provide DC power at the terminals. On the other hand, grid supplied electricity is alternating current (AC), as are most consumer appliances and loads. When integrating any battery into an electricity grid that includes a range of power generation sources therefore, power conditioning equipment is required to match the voltage and currents between sources and loads. A typical system configuration for an off-grid power system incorporating wind, solar and storage is illustrated in Fig. 12.8.

In this example, electricity is generated by wind turbines, solar panels or a diesel generator, and is fed via a regulator or other form of power controller, into a battery system, to be stored for use when required by the load. The wind and solar sources usually produce unregulated, highly variable power flow, so regulator devices are used to stop the battery system from being overcharged, reducing any damage to the battery. Most household appliances use AC electricity, while batteries supply DC electricity. An inverter is therefore used to convert DC electrical energy into AC form. On the other hand, to charge the battery bank from the wind turbine or diesel generator, a converter is needed to convert the AC power to DC form. Solar and wind power generation alone are unlikely to be able to meet the load's power demand during periods of low wind speed or during high cloud cover and at night. An off-grid hybrid power system that uses solar and wind power turbine with a battery would be able to meet most of the load power demand if the battery could store sufficient energy to cover several days of possible cloud cover and low wind speeds. The cost of such a large battery would

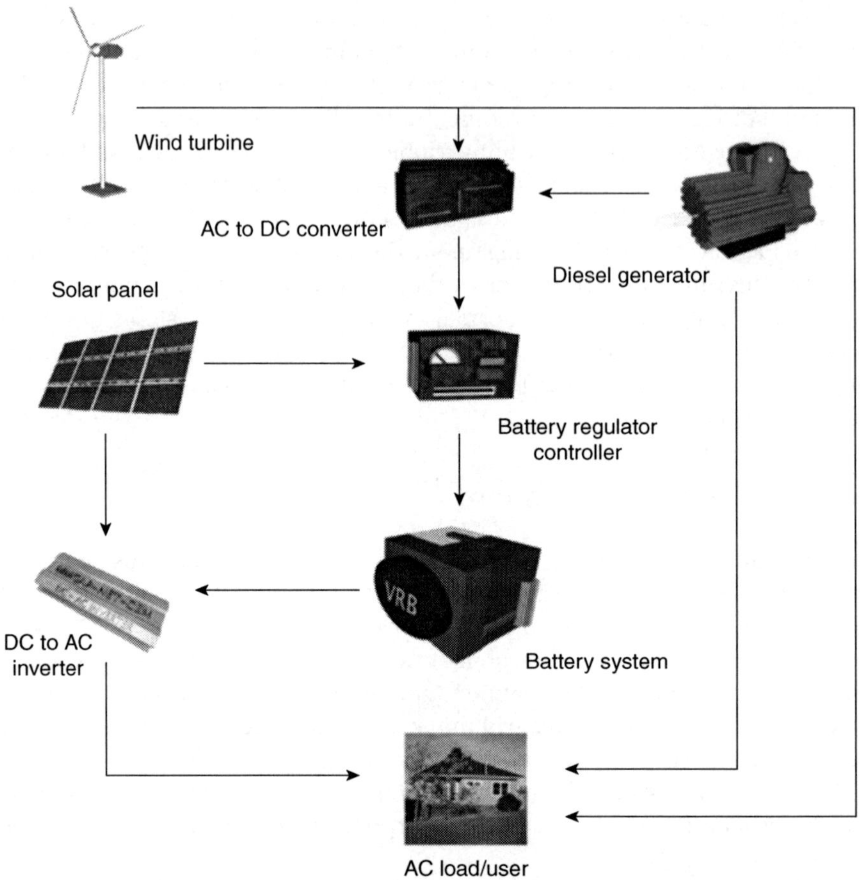

12.8 Typical off-grid power system configuration. (*Source*: Reproduced from Skyllas-Kazacos, M. 'Energy storage for stand-alone/hybrid systems: Electro-chemical Energy Storage Technologies' in Stand-alone and Hybrid Wind Systems: Technology, Energy Storage and Applications Editor: Professor J.K. Kaldellis, Woodhead Publishing, July 2010.)

be prohibitive, however, so that the best solution is to include a diesel generator to provide back-up power during the occasional periods when the battery SOC cannot be maintained during extended periods of continuous cloud cover and low wind speeds. The amount of storage and diesel back-up needed in an off-grid power system is site-dependent and is determined by the local solar irradiation, wind speeds and load profiles. A typical off-grid power system that utilises a high renewable energy fraction will typically require between 8 and 10 h of storage capacity to minimise diesel fuel consumption while providing reliable power to the consumer. For such large storage capacities, flow batteries offer the lowest cost per kWh capital and operating costs compared with other types of battery systems (Skyllas-Kazacos, 2010)

12.3 Flow battery chemistries

A number of different flow battery chemistries have been evaluated and developed since the 1970s. The chemistry, features and developmental status of the different types of flow battery technologies will be discussed below.

12.3.1 Vanadium redox batteries (VRBs)

The vanadium RFB is the most widely studied of all of the flow cell chemistries and will therefore be discussed in the most detail.

Features and benefits of the VRB

Several key properties give the VRB significant advantages over other forms of energy storage:

- Energy is stored in solution.
- The same electrolyte is used in both half-cells.
- The sources of energy and power are separated.

These features can have significant impacts on large-scale energy storage applications as well as smaller mobile uses.

Energy stored in solution

As the energy is stored in solution, no solid phase changes occur. This eliminates the possibility of short circuiting or shedding of the active material. The VRB can therefore be fully charged and discharged without the risk of damage, loss of battery life, or loss of capacity. The solution also has the ability to be stored for long periods with relatively no self-discharge occurring in the electrolyte tanks. Because electrolyte flows through the cell stack, the need for coolant and heat exchangers is often eliminated. These properties give the VRB a significant advantage over solid state energy storage systems such as the lead-acid battery.

Common electrolyte

By utilising the same electrolyte in both sides of the cell, additional benefits occur in terms of overall battery life, reduced maintenance costs and the overall economic costs. Problems occurring due to electrolyte cross-contamination are alleviated, which helps to reduce capital and maintenance costs, reducing the need for waste disposal and providing greater ease of operation.

Separation of energy/power rating

The VRB also separates the electrolyte from the cell stack, which means that capacity and power can be increased independently. The storage capacity can be increased by adding electrolyte solution, which also reduces the cost of the electricity supplied while the power can be increased through the addition of cell stacks. Cost per kWh drops dramatically for storage capacities greater than 4 h, which is particularly important in large-scale energy storage applications that require energy capacities of several hours. Figure 12.9 compares the cost/kWh of the G1 VRB with that for the lead-acid battery as a function of storage capacity.

In spite of a higher capital cost for the VRB compared with mass produced lead-acid batteries, its long cycle life and low maintenance and replacement costs make the cost of energy generated over a 25 year project life much lower in the case of the VRB for storage times greater than 3 h.

Other features include the ability to use underground electrolyte storage tanks, minimising the physical footprint of the system and reducing temperature fluctuations in extreme climates. Having separate electrolyte tanks

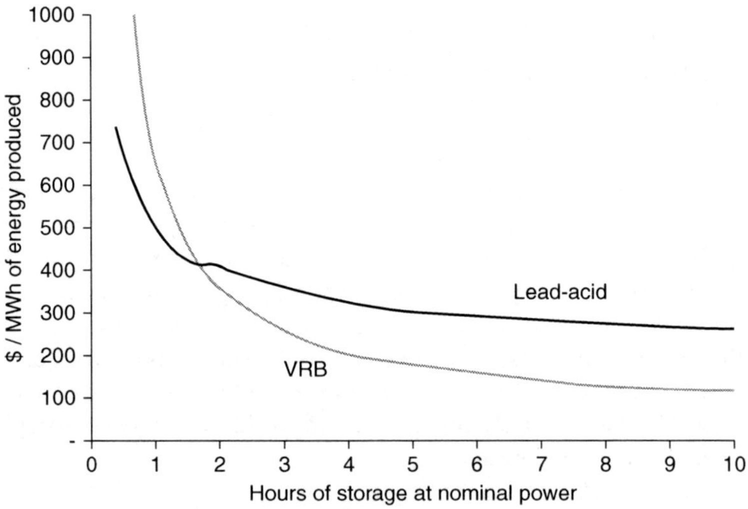

12.9 Cost per kWh comparison for lead-acid battery vs VRB for 5 kW stack costs and varying storage times. Figures shown are the cost of each kWh of energy produced over a 25 year project life. (VRB assumptions: V2O5 powder = $US 5/lb. Stack cost = $500/kW.) (*Source*: Adapted from Maria Skyllas-Kazacos, 2009, Elsevier).

also means that the risk of sudden energy release as a result of instantaneous mixing of solutions is eliminated, making the flow battery inherently safer than other types of batteries such as lithium. Flow batteries also offer great flexibility in design of both the cell stack and electrolyte tanks with a wide range of chemically resistant polymeric materials available for use. The critical requirement is that all materials in contact with the electrolyte must be capable of withstanding the highly oxidising conditions of the fully charge positive half-cell electrolyte. Materials such as polyethylene, polypropylene, Teflon, etc., have already been shown to possess excellent chemical and oxidation resistance to the highly oxidising V(V) solution in the fully charged positive half-cell electrolyte of the VRB, and these materials are usually used in the fabrication of the cell flow-frames, electrolyte tanks, pumps and plumbing components. Where MWh of storage capacity is required, however, it is possible to use concrete tanks with chemically resistant internal bladders or tank linings to provide strength and chemical resistance at a low cost.

Generation 1 VRB

The Generation 1 vanadium redox battery (G1 VRB) employs a solution of vanadium in sulphuric acid in both half-cells with the V^{2+}/V^{3+} redox couple operating in the negative half-cell and the VO^{2+}/VO_2^+ redox couple in the positive half-cell. The half-cell reactions are presented by Equations [12.5] and [12.6].

A detailed schematic diagram of the G1 VRB is given in Fig. 12.3. Thus, with an electrolyte concentration of 1 M, the overall standard cell potential is 1.26 V at 25°C. When using an electrolyte solution consisting of 2 M vanadium in 5 M sulphuric acid, however, the OCV observed is 1.4 V at 50% SOC and 1.6 V when fully charged (Skyllas-Kazacos *et al.*, 2011). At this concentration, the specific energy is in the range 25–30 Wh/kg and the G1 VRB has a temperature range of between approximately 10 and 40°C (Skyllas-Kazacos, 2010). This range is set by the saturation solubilities for the V(II) and/or V(III) ions at the lower limit and the thermal precipitation of V(V) at elevated temperature, as described by:

$$2VO^+_{2(aq)} + H_2O \rightarrow V_2O_{5(s)} + 2H^+ \qquad [12.18]$$

Increasing the acid concentration will therefore shift the above equilibrium to the left, increasing the solubility of V(V) ions. Increasing the sulphuric acid concentration will, however, reduce the solubility of the V(II), V(III) and V(IV) ions at low temperatures. An optimal sulphuric acid concentration therefore exists for the stabilisation of all of the oxidation states over the operating temperature range of the VRB. For a wider operating temperature range, a lower vanadium ion concentration is required, but this reduces the energy density of the electrolyte. Although this is of little significance in most stationary applications where there are no space or weight restrictions, energy density is of great importance in mobile applications. For this reason, the G1 VRB does not meet the energy density requirements for electric vehicles, despite their ability to be 'instantly' recharged by exchanging spent solutions at special refuelling stations.

Specific details on field trials of the G1 VRB are presented later.

Generation 2 vanadium bromide (G2 VRB)

Unlike the G1 VRB, which utilises a solution of vanadium in sulphuric acid in both sides of the cell, the G2 VRB employs a vanadium bromide/chloride mixed electrolyte in both half-cells. Since the bromide/polyhalide couple has a less positive potential than the V(IV)/V(V) couple, the bromide ions will preferentially oxidise at the positive electrode during charging. The positive half-cell thus utilises the $Br^-/ClBr_2$ or $Cl^-/BrCl_2^-$ redox couple while the negative half-cell utilises the same V^{2+}/V^{3+} redox couple reaction as the G1 VRB. The half-cell reactions for the charge and discharge processes are shown below (Skyllas-Kazacos, 2010):

Positive half-cell:

$$2Br^-_{(aq)} + Cl^-_{(aq)} \rightarrow ClBr^-_{2\,(aq)} + 2e^- \quad \text{(charge)} \qquad [12.19]$$

$$ClBr^-_{2\,(aq)} + 2e^- \rightarrow 2Br^-_{(aq)} + Cl^-_{(aq)} \quad \text{(discharge)} \qquad [12.20]$$

Negative half-cell:

$$V^{3+}_{(aq)} + e^- \rightarrow V^{2+}_{(aq)} \quad \text{(charge)} \qquad [12.21]$$

$$V^{2+}_{(aq)} \rightarrow V^{3+}_{(aq)} + e^- \quad \text{(discharge)} \qquad [12.22]$$

Using the same electrolyte in both half-cells, the G2 V/Br shares all of the benefits of the G1 VRB technology, particularly in that cross-contamination is eliminated, so that the solutions have an indefinite life. While the G1 VRB can utilise electrolyte concentrations of up to 2 M, however, the G2 VRB can use up to 4 M electrolytes, potentially doubling the amount of energy stored per litre or kg of solution compared with the G1 VRB. By increasing the energy density, the volume of electrolyte required to store a given amount of energy is reduced. This has significant advantages for energy storage, particularly for mobile applications.

Another major advantage is the increased temperature range at which the G2 VRB can successfully operate. While the G1 VRB has been shown to operate successfully between approximately 10–40°C, by eliminating the thermal precipitation reaction for V(V) and increasing the solubility limits for the other vanadium ions, the temperature range of the G2 VRB is expected to be between approximately 0–50°C. An increased temperature range allows the G2 VRB to operate successfully in geographical areas where the G1 VRB may not be effective, such as desert areas that have large diurnal temperature ranges.

The key issue that needs to be addressed to allow G2 V/Br system to become commercially viable, however, is the potential for bromine gas release. Extensive research at UNSW has successfully identified bromine complexing agents that can combine with the bromine to form a heavy oil that sinks to the bottom of the tank and prevents the release of any vapours. This is similar to the operation of the Zn/Br battery, which involved considerable early work on bromine complexes and will be described later. As with the Zn/Br battery electrolyte, however, the use of bromine complexing agents adds considerable cost to the electrolyte. Low cost alternatives therefore need to be developed before the G2 V/Br will become a commercial reality.

Generation 3 VRB (G3 VRB)

The term G3 VRB is used here to describe the most recent VRB chemistry developed by researchers at the Pacific Northwest National Laboratories (PNNL) in the USA (Li *et al.*, 2011). This chemistry involves the use of a mixed sulphuric acid/hydrochloric acid electrolyte and exhibits similar electrochemical reversibility of the vanadium redox couples to that observed for the sulphuric acid supporting electrolyte.

The half-cell reactions proposed for the VRB using the mixed sulphuric acid/hydrochloric acid supporting electrolyte are the same as in the G1 VRB in the case of the negative half-cell couple (Equation [12.6]). At temperatures below 20°C, the positive half-cell reactions are also the same as in the sulphuric acid electrolyte (Equation [12.5]). According to Li *et al.*, however, at temperatures above 20°C, the positive half-cell reaction becomes:

$$VO^{2+}{}_{(aq)} + Cl^{-}{}_{(aq)} + H_2O \underset{\text{discharge}}{\overset{\text{charge}}{\rightleftharpoons}} VO_2Cl_{(l)} + 2H^{+}{}_{(aq)} \qquad [12.23]$$

The interaction between the chloride and V(V) is thought to be responsible for the stabilisation of the V(V) ions against thermal precipitation

Table 12.1 Comparison of properties of G1, G2 and G3 VRB technologies

	G1	G2	G3
Electrolyte	V/sulphate in both half-cells	V/HBr/HCl solution in both half-cells	V/H_2SO_4/HCl in both half-cells
Negative couple	V^{3+}/V^{2+}	V^{3+}/V^{2+}	V^{3+}/V^{2+}
Positive couple	V(IV)/V(V)	Br/$ClBr_2^-$	V(IV)/V(V)
Maximum vanadium concentration	1.5–2 M	2.0–3.5 M	2.0–2.7 M
Specific energy	15–25 Wh/kg	25–50 Wh/kg	25–40 Wh/kg
Energy density	20–33 Wh/L	35–70 Wh/L	35–55 Wh/kg
Operating temperature range	10–40°C	0–50°C	0–50°C

at elevated temperatures, although the effect could in fact be due to the higher proton ion concentration used in the mixed acid electrolyte that effectively shifts the thermal precipitation equilibrium (Equation [12.18]) towards the left.

A comparison of the three VRB chemistries is given in Table 12.1.

The G3 VRB system can thus use electrolytes with vanadium ion concentrations up to 2.7 M, considerably higher than the G1 VRB electrolyte. While this is lower than the G2 VRB, it is free of bromine and the stability of the electrolytes can be maintained over an extended temperature range. As the same electrolyte is used in both half-cells, the G3 VRB shares all the benefits of the G2 V/Br and G1 VRB technologies, in that problems associated with cross-contamination of electrolytes and consequent irreversible capacity loss over continuous cycling is eliminated. As with the other vanadium battery technologies therefore, the electrolyte solutions have an indefinite life.

The expected operating temperature range of the G3 VRB is expected to be between 0 and 50°C, significantly greater that the G1 VRB with an operating range of 10–40°C. By slightly reducing the vanadium ion concentration or SOC operating range, it may be possible to further extend the G3 VRB operating range to approximately –5 to 50°C, thus making it suitable for areas that have significant daily and annual temperature variations.

PNNL researchers (Li *et al.*, 2011) also observed a small increase in the potential of the V(II)/V(III) and V(IV)/V(V) couples in the mixed sulphuric acid/hydrochloric acid electrolyte compared to the sulphuric acid supporting electrolyte. Initial battery tests have also indicated that the possibility of chlorine gas evolution during normal battery operation is very low. Further work is underway to scale up a G3 VRB system for early field trials.

Researchers at PNNL also conducted trials with 10 M HCl as the supporting electrolyte of an all-vanadium RFB (Kim *et al.*, 2011). The electrolytes at different SOCs were found to be stable with no precipitation evident after 15 days at vanadium concentrations of 2.3 M in 10 M chloride supporting electrolyte at a temperature range of –5°C to 40°C.

As mentioned previously, thermal precipitation is normally a problem for the pentavalent VO_2^+ species in sulphuric acid supporting electrolytes at concentrations above 2 M and temperatures higher than 40–50°C. Kim and co-workers (Kim *et al.*, 2011) found that the chloride solution V(V) showed no thermal precipitation in 10 M HCl at a vanadium concentration of 3 M and at a temperature up to 40°C. Again, the researchers proposed that this stabilisation is associated with the interaction between the V(V) and Cl^- ions; however, this phenomenon can also be explained by the very high proton concentration that effectively shifts the thermal precipitation equilibrium (Equation [12.18]).

During testing in a redox flow cell the researchers found that the current efficiency of the chloride supporting electrolyte system was approximately 96%, which is similar to that of the sulphuric acid supporting electrolyte. The energy efficiency was also similar between the two; however, the chloride system had an energy density approximately 30% higher.

One potential area of concern with the chloride system is that of chlorine evolution at the positive electrode during charging. This is not only potentially hazardous but also decreases current efficiency by reducing V(V) to V(IV) in the positive half-cell. The researchers found that at SOCs up to 80% and a temperature up to 50°C, however, there was no pressure build up in the positive half-cell.

These results are very encouraging, in that they demonstrate potential increases in energy density of the all-vanadium battery system that could expand the areas of application into both stationary and mobile systems, although further work is still needed to verify the results under a wide range of operating conditions.

Mixed V/Fe redox flow V cell

Also proposed by researchers at the PNNL in the USA, the mixed V/Fe redox battery employs a mixture of V and Fe ions in HCl or H_2SO_4/HCl acid mixtures in both half-cells. The half-cell reactions for the charge and discharge processes are shown below (Wang *et al.*, 2011):

Positive half-cell:

$$Fe^{2+} \underset{\text{discharge}}{\overset{\text{charge}}{\rightleftarrows}} Fe^{3+} + e \quad [12.24]$$

Negative half-cell:

$$V^{3+} + e \underset{\text{discharge}}{\overset{\text{charge}}{\rightleftarrows}} V^{2+} \qquad [12.25]$$

Initial studies attempted to use the Fe(II)/Fe(III) as the positive electrolyte and V(II)/V(III) couple in the negative electrolyte, giving a standard voltage of 1.02 V through the redox reaction. The studies showed a rapid capacity loss due to cross-contamination of the electrolytes. In order to overcome this limiting factor, mixed electrolytes were employed in both half-cells. By replacing the highly oxidising V(V) species with Fe(III) in the charged positive half-cell electrolyte, the researchers claim a possible cost reduction by using lower cost membranes that would otherwise disintegrate in V(V). This cell has a significantly lower cell voltage, however, (0.94 vs 1.4 V for the G1 VRB) and this, combined with the reduced solubility of the mixed electrolytes, significantly lowers the energy density of the V/Fe cell compared with other VRB systems.

Vanadium–oxygen redox fuel cell

As described previously the VRB employs vanadium electrolytes in both half-cells of the battery, but energy density is limited by the saturation solubilities of the vanadium ions in the positive and negative half-cell electrolytes at different temperature extremes. A very attractive way to substantially increase the energy density of the all-vanadium RFB is to eliminate the positive half-cell electrolyte and replace it with an air electrode to produce a hybrid vanadium–oxygen redox fuel cell (VOFC). This not only allows an increase in energy density by halving the electrolyte volumes, but also can provide substantial savings in production costs.

This concept was initially proposed by Kaneko and co-workers in 1992 and first evaluated by Menictas and Skyllas-Kazacos in 1997 at the University of NSW. A laboratory single-cell system, and later a multi-cell system, was evaluated, but early problems were identified with the stability of the membrane electrode assemblies caused by separation of the membrane due to swelling and expansion during hydration. Early problems were minimised, however, and a five-cell VOFC system was operated for a total of over 100 h without any deterioration in its performance (Skyllas-Kazacos and Menictas, 2011).

During discharge of the V/O_2 cell, the negative half-cell reaction is the same as in the VRB, but at the positive electrode, the reaction is:

$$O_2 + 4H^+ + 4e \rightarrow 4H_2O \qquad [12.26]$$

The standard redox potential for the oxygen reduction reaction is 1.23V, making the standard cell potential for the VOFC higher than the G1 VRB system. The very slow kinetics and mass transport limitation of the oxygen reduction reaction give rise to very high activation and concentration overvoltage losses, so in practice, cell voltage values of around 1 V are typically observed.

Previous studies by Skyllas-Kazacos and co-workers had shown that V(II)/V(III) concentrations as high as 4 M concentration could be achieved (Skyllas-Kazacos, 2003). The use of a 4 M vanadium chloride solution in the negative half-cell of a V/O_2 redox fuel cell has the capability therefore, of a four-fold increase in energy density compared with the G1 VRB system, making it practical for electric vehicle applications. Improvements in the catalytic activity of the oxygen diffusion electrode are, however, still needed, while further developments in the production of efficient membrane electrode assemblies are required to enhance the stability of the V/O_2 cell over several years of operation.

Case studies and field trials for generation 1 VRB systems

The UNSW Generation 1 VRB has been successfully implemented in more than 30 medium- to large-scale field trials around the world to demonstrate the benefits of the VRB in different applications and to enable researchers to gain information vital for the improvement of VRB technology.

The first field trial of a Generation 1 VRB system was in a demonstration solar powered house in Thailand in 1993 (Largent *et al.*, 1993). A 5 kW battery with 12 kWh of storage was installed to demonstrate an energy self-sufficient solar house that used the grid as back-up power instead of a diesel generator.

Under licence to the UNSW, Australia, the Japanese company Kashima-Kita Power Corporation installed a 200 kW VRB at one of the company's power plants in 1997 to demonstrate the load-levelling capabilities of the VRB. The capacity was 800 kWh and consisted of 23 000 L of 1.8 M electrolyte and eight 25 kW stacks. The VRB showed an overall energy efficiency of 80% at current densities between 80 and 100 $mA.cm^{-2}$ after 150 charge/discharge cycles. This application confirmed the technical viability of the VRB for large-scale energy storage, while demonstrating high energy efficiency, reduced maintenance and the ability to be upgraded easily compared to other storage systems (Skyllas-Kazacos, 2010).

A G1 VRB load-levelling battery was also installed by Canadian company VRB Power in March 2004 to help alleviate voltage regulation problems associated with increased electricity demand at the end of a rural transmission line in Utah, USA. A 250 kW battery with 140 000 L of electrolyte was installed with 8 h of storage in times of peak load. The capacity of the system is 2.3 MWh and, since the installation of the VRB, voltage disruptions have dropped by up to 90%, demonstrating the effectiveness of the VRB in load levelling (Prudent, 2011).

The largest G1 VRB installation to date has been a 4 MW/6 MWh system installed by Sumitomo Electric Industries at the Subaru Wind Farm on Hokkaido, Japan for wind energy storage and power stabilisation in 2005 (Tokudu *et al.*, 2000). The latter system was reported to give overall round trip energy efficiency of 80% with cycle life of over 270 000 cycles over 3 years of testing (Skyllas-Kazacos *et al.*, 2011).

In addition to the stationary storage applications, a vanadium battery powered electric golf cart was field tested at UNSW, using 40 L of 1.85 M vanadium electrolyte; a driving range of 17 km off-road was obtained, suggesting that the energy density of an optimised all-vanadium RFB could approach that of lead-acid, with the added advantage of rapid recharging by electrolyte replacement (Rychcik and Skyllas-Kazacos, 1988). Subsequent studies with a 3 M stabilised vanadium electrolyte gave a driving range of 31.5 km with partly filled electrolyte tanks, and showed that up to 54 km could be achieved if the tanks were filled to their maximum capacity (Skyllas-Kazacos *et al.*, 2011). Careful temperature control was, however, required to avoid vanadium precipitation at temperatures above 35 or below 15°C.

In 2010, the US Department of Energy funded the demonstration of a 1 MW/8MWh VRB for load-levelling trials at the Painesville Municipal Power Station in Ohio (DOE, 2010). The battery is to be used for load following to allow the coal-fired generators to operate at maximum capacity at all times and thereby minimise carbon dioxide emissions.

Since 2005, several companies have been established to manufacture VRB systems for a range of applications. These include Cellstrom in Austria and Prudent Energy and Rongke Power Co Ltd. in China. The most active of these companies is Prudent Energy, having acquired the original UNSW G1 VRB patents together with the Canadian VRB Power technology in 2008.

A 1 MWh vanadium RFB was installed in March 2011 as part of a wind and solar project at the National Wind Power Integration Research and Test Centre at Zhangbei, China (Prudent, 2001). The system was installed by Prudent Energy Services Corporation with funding from the China Electric Power Research Institute and was integrated to a 78 MW wind farm and a 640 kW of solar PV array. The 1MWh VRB comprises 175 kW modules and has a rated power of 500 kW with a peak power of 750 kW. The objective is to obtain control algorithms for the efficient integration of the vanadium RFB with wind and solar energy into the grid.

12.3.2 Iron/chromium redox flow battery (Fe/Cr)

The Fe/Cr system was the first RFB system to have been developed and evaluated for large-scale energy storage. After extensive screening of a large number of redox couples by NASA in the 1970s, the iron/chromium couple was selected on the basis of cost and availability (Swette and Jalan, 1984; Gahn *et al.*, 1985). In general, the system consisted of acidified solutions of chromium Cr(II)/Cr(III) and iron Fe(II)/Fe(III). The cell reactions of the iron/chromium system are as follows:

Positive electrode reaction:

$$Fe^{2+} \underset{\text{discharge}}{\overset{\text{charge}}{\rightleftarrows}} Fe^{3+} + e \qquad [12.27]$$

Negative electrode reaction:

$$Cr^{3+} + e^{-} \underset{\text{discharge}}{\overset{\text{charge}}{\rightleftarrows}} Cr^{2+} \qquad [12.28]$$

Initially the Fe/Cr cell used unmixed Fe and Cr reactants in the positive and negative half-cell electrolytes, respectively, but the solutions were subsequently mixed in order to address the issue of cross-mixing of the electrolytes across the membrane.

In premixed solutions, both the positive and negative electrolytes contained iron and chromium species as soluble salts in aqueous solutions of hydrochloric acid but only the Fe(II)/Fe(III) couple operates in the positive half-cell electrolyte with the Cr(II)/Cr(III) couple reacting at the negative electrode during charge–discharge cycling.

The first 1 kW prototype Fe-Cr systems was developed in 1980 by NASA (Johnson and Reid, 1985). Further work in Japan in the 1980s led to the development of a 10 kW Fe-Cr redox battery prototype with an 80% energy efficiency and 300 life cycles, as demonstrated by Kansai Electrical Power Co, Japan (Shimizu *et al.*, 1988). Operations involving the catholyte and anolyte circulation rates (in a 10 kW Fe-Cr RFB) to save energy, and a method of rebalancing, were developed by Mitsui Ltd., Japan (Nakamura, 1988).

Commercial development of the Fe-Cr battery was, however, abandoned because of problems of cross-contamination between anolyte and catholyte, poor energy efficiencies due to hydrogen evolution at the negative electrode, and fouling of the ion-exchange membranes. Another issue was the slow kinetics of the Cr(II)/Cr(III) couple that required the use of expensive gold catalysts. Hence, this system was largely ignored until the late 2000s, when it was reinvigorated by Deeya Energy (Deeya, 2011) that saw the Fe-Cr system as a potentially lower priced technology than the VRB.

12.3.3 Polysulfide bromine (PSB) flow battery

The polysulfide bromine (PSB) flow battery was first developed by Regenesis Technologies Ltd. in 1983. (Remick and Ang, 1984) This technology was later owned by Innogy Technologies Ltd. under the trademark 'Regenesis®', and Innogy Technologies spent considerable time further improving the technology, particularly on the stack design and fabrication, throughout 1990s.

The PSB flow battery employs sodium bromide electrolyte in the positive half-cell and sodium polysulfide electrolyte in the negative half-cell. During the charging cycle, bromide ions are oxidised to bromine and complexed to form tribromide ions at the positive electrode, while the polysulfide anion is reduced to sulphide ions at the negative electrode. The half-cell reactions for the charge and discharge processes are shown below (Ponce de Leon *et al.*, 2006, p. 720; Skyllas-Kazacos *et al.*, 2011):

Positive electrode reaction:

$$3Br^- \underset{\text{discharge}}{\overset{\text{charge}}{\rightleftarrows}} Br_3^- + 2e \qquad E^o = 1.09\,V \qquad [12.29]$$

Negative electrode reaction:

$$S_4^{-2} + 2e \underset{\text{discharge}}{\overset{\text{charge}}{\rightleftarrows}} 2S_2^{-2} \qquad E^o = -0.265\ V \qquad [12.30]$$

The two electrolytes are separated by a cation exchange membrane to prevent direct reaction between sulphide anions and the evolved bromine. The sodium ions pass the membrane during charging or discharging to complete the circuit.

The open-circuit potential of the polysulphide-bromine flow battery, is 1.5 V with energy efficiencies of 60–65% depending on operating conditions. The typical operating temperature range is between 20°C and 40°C. This flow battery attracted early attention because of the relative abundance and low cost of the electrolyte materials as well as the high solubilities in aqueous solutions, where up to 5 M NaBr and 2 M Na_2S have been tested in the PSB system (Ponce de Leon *et al.*, 2006, p. 721).

The PSB flow battery suffers several drawbacks, however, such as (Ponce de Leon *et al.*, 2006, p. 721):

- a cross-contamination problem, due to the use of two different electrolytes in each half-cell;
- electrolyte imbalances resulting from cross-contamination leading to change in electrolyte compositions over extended cycling;
- possibility of deposition of sulphur species on the membrane;
- the need to prevent $H_2S_{(g)}$ and $Br_2(g)$ formation.

Furthermore, unlike Zn/Br and G2 VRB system, the PSB flow battery does not use complexing agents to minimise the resulting bromine evolution at the positive electrode during charging, and this poses safety risks in operation (Skyllas-Kazacos *et. al.*, 2011).

Numerical modelling of the PSB system indicates that the performance of this flow battery is limited by mass transport over-potentials at the positive electrode during discharge (Scamman *et al.*, 2009). The model showed significant drift due to self-discharge and electro-osmotic effects. Therefore, careful management of the electrolytes is needed to ensure the reliability of the PSB flow battery operation. However, the complexity of electrolyte management will increase the maintenance cost significantly, thus reducing its cost competitiveness particularly for low- to medium-scale installations. Thus, the application of the PSB flow battery is restricted to MW large-scale operations where the contribution from electrolyte maintenance cost is minimal due to economy of scale.

Numerous efforts have been made to improve the energy efficiency of the PSB flow battery over the years and these are shown in Table 12.2 (Skyllas-Kazacos *et al.*, 2011). Despite these efforts, extensive research is still required to overcome the above technical challenges, high cost and safety concerns for commercialisation.

Early development of the polysulfide-bromide flow battery focused on large-scale applications for power ratings in excess of 5 MW. The construction of a 15 MW/120 MWh utility scale pilot plant was started in Little Barford, UK for arbitrage application; however, the project was abandoned in December 2003 due to funding issues. A separate mathematical model incorporating capital and operating costs to predict the technical and commercial performance of this pilot plant showed a net loss of 5.7 UK pence (US$0.09) per kWh at an optimum current density of 500 A/m^2 and energy efficiency of 64% (Scamman *et al.*, 2009b, p. 1237).

In 2004, Innogy Technologies granted an exclusive global licence to VRB Power Inc. for all intellectual property of the PSB system to commercialise this flow battery in systems ranging from 10 to 100 MW and 8–12 h storage duration. A second pilot plant was also planned in Mississippi at Tennessee Valley Authority (TVA) based on the same technology constructed at the Little Barford site (De Boer and Raadschelders, 2007, p. 5). However, this

Table 12.2 Research and development efforts in PSB redox flow battery

Year	Electrolyte	Membrane	Electrode materials	Comment
1984	1 M NaBr saturated with Br_2 and 2 M Na_2S	Nafion 125	Graphite and porous sulphide nickel electrodes	Single flow cell. Open-circuit potential at 50% charge = 1.5 V.
1999	5 M NaBr and 1.2 M Na_2S	Nafion 115	Activated carbon/polyolefin pressed electrodes	Monopolar flow cell. Cell voltage increased to 2.1 V during charging cycle for 30 min at 40 mA/cm^2.
2001	Br_3^-/Br^- and polysulphide/ sulphide	Sodium cation exchange (Du Pont)	Carbon-polyolefin composite bipolar	1 MW pilot scale facility at Innogy's Aberthaw power station, Cardiff, UK using 100 kW XL series stacks.
2004	1 M Br_2 dissolved in 2 M NaBr and 2 M Na_2S_2	Nafion	Negative electrode: Nickel catalyst supported on carbon. Positive electrode: Pt catalyst supported on carbon	Single flow cell. Power density = 0.64 W/cm^2. Voltage efficiency of 88.2% obtained at 0.1 A/cm^2 during charge and discharge cycle.
2005	4 M NaBr and mixture of 1.3 M Na_2S_4 and 1 M NaOH	Nafion 117	Nickel foam and carbon felt electrodes	Single flow cell. Internal ohmic resistance restricted energy efficiency to 77.2% at 40 mA/cm^2 and cell power density of 56 mW/cm^2.
2007	1 M NaBr and 0.5 M Na_2S_4 at pH 2	Nafion 115	Polyvinylidene-ifluoride (PVDF) and activated carbon composite on high density polyethylene (HDPE)/carbon core	Five-cell bipolar stack. Battery efficiency not determined.

Source: Skyllas-Kazacos *et al.*, 2011.

plant was also shelved as Innogy Technologies decided not to continue this joint venture. Since then, no further new large-scale demonstration of this flow battery has been reported.

12.3.4 Zinc-based flow batteries

The Zn/Br battery and Zn/Cl battery are two of the earliest types of flow batteries (Blevins, 1981). The working principle of zinc-based flow batteries is different from traditional redox flow batteries such as VRB and PSB systems. Zinc-based flow batteries involve the deposition of zinc at the negative electrode during charging so the total energy capacity is limited by the available electrode area for zinc deposition.

Zinc-bromine (Zn/Br) flow cell

The Zn/Br flow battery was first introduced more than 100 years ago; however, the poor performance and maintenance of the early Zn/Br batteries delayed commercialisation until Exxon and Gould developed a design for practical application in the mid 1970s and early 1980s to improve performance and maintenance.

Unlike the conventional redox flow cell technology, the working principle of the zinc-bromine flow cell is based on the deposition of zinc at the negative electrode and bromine evolution at the positive electrode. In contrast to conventional redox flow batteries that have capacity linked to the volume of the half-cell electrolytes, the energy capacity of the Zn/Br hybrid flow battery is thus limited by the amount of zinc deposited at the negative electrode, and this depends on the thickness and morphology of the metallic zinc layer.

The half-cell reactions for the charge and discharge processes are shown below:

Positive electrode reaction:

$$2\mathrm{Br}^- \underset{\text{discharge}}{\overset{\text{charge}}{\rightleftarrows}} \mathrm{Br}_2 + 2\mathrm{e} \quad E^o = 1.087\,\mathrm{V} \qquad [12.31]$$

Negative electrode reaction:

$$\mathrm{Zn}^{2+} + 2\mathrm{e} \underset{\text{discharge}}{\overset{\text{charge}}{\rightleftarrows}} \mathrm{Zn} \quad E^o = -0.763\,\mathrm{V} \qquad [12.32]$$

The schematic diagram of the Zn/Br hybrid flow battery is presented in Fig. 12.10 (Skyllas-Kazacos *et al.*, 2011).

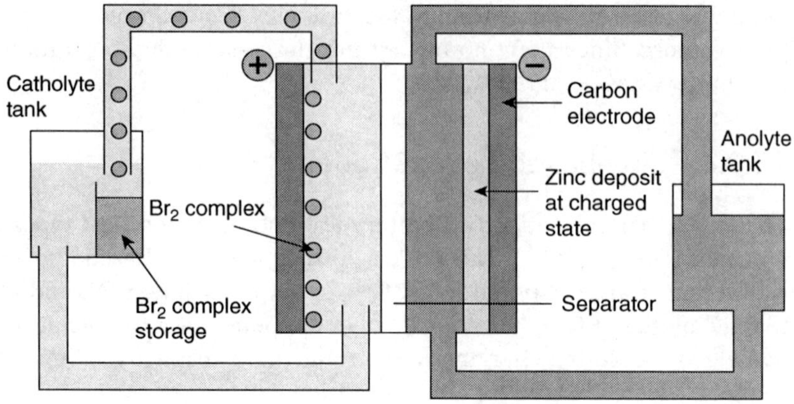

12.10 Schematic of zinc-bromine hybrid flow battery.

The zinc-bromine hybrid battery thus consists of anolyte and catholyte solutions containing aqueous $ZnBr_2$ and other salts stored in two external storage tanks and a power conversion cell. A small section of the catholyte tank is barricaded to form a compartment for collecting the evolved bromine. Both catholyte and anolyte are circulated between the power conversion cell and storage tanks by pumps. The early zinc-bromine hybrid flow battery suffered rapid self-discharge due to bromine being readily dissolved in the zinc bromide solution and diffusing to the negative electrode where it oxidises zinc according to the following reactions (Eustace, 1977):

$$Br_{2(aq)} + H_2O_{(l)} \rightarrow HBr_{(aq)} + HBrO_{(aq)} \qquad [12.33]$$

$$2HBr_{(aq)} + Zn_{(s)} \rightarrow ZnBr_{2(aq)} + H_{2(g)} \qquad [12.34]$$

To minimise this rapid self-discharge, the dissolved bromine concentration in the electrolyte solution at the positive half-cell should be kept as low as possible and both the positive and negative electrode should be separated by a micro-porous separator (Pavlov *et al.*, 1991, p. 29). The low dissolved bromine concentration in the aqueous phase can be achieved by adding bromine complexing agents or using propionitrile (PN) solvent (Singh *et al.*, 1983, p. 314).

The function of the bromine complexing agent is to reduce the concentration of bromine in the aqueous phase at the positive half-cell by extracting the bromine into a water immiscible organic phase that acts as bromine storage medium. The most commonly used bromine complexing agents are quaternary ammonium bromides (QBr) such as N-ethyl-N-methyl morpholinium bromide (MEM) and N-ethyl-N-methyl pyrrolidinium bromide

Table 12.3 Electrolyte compositions of zinc-bromine flow battery

Year	Research organisation	$ZnBr_2$ (M)	MEM (M)	MEP (M)	$ZnSO_4$ (M)	$ZnCl_2$ (M)	Dendrite inhibitor
1978	Exxon	3	0	1	0.2	0	0
1983	Meidensha	3	0.5	0.5	0	0	Sn^{2+}/Pb^{2+}
1985	Exxon	2	0.5	0.5	0	1	0
1989	Johnson Controls Inc. (JCI)	3	0.25	0.75	0	0	0
1995	ZBB	2.25	0	0.8	0	0.55	0
1999	SNL	2.25	0.8 (50/50)		0	0.55	0

Source: Poon, 2008.

(MEP) (Poon, 2008, p. 14). Each QBr can typically complex up to three molecules of bromine and can therefore bind a large amount of dissolved bromine, resulting in a very low bromine concentration in the aqueous phase and an almost 100-fold reduction in bromine gas evolution (Bajpal *et al.*, 1981, p. 3). Increasing the chloride concentration in addition to the bromine complexing agent can further reduce the aqueous bromine concentration, leading to improvements in energy efficiency and reduction in self-discharge (Kantner, 1985). Table 12.3 lists bromine complexing agents and chloride concentrations that have been explored by various research organisations (Poon, 2008).

Low bromine concentration in aqueous solution at the positive half-cell can also be obtained by adding PN solvent to the catholyte. In this case, the electrolyte compositions are different for both anolyte and catholyte, where the anolyte consists of 4 M $ZnBr_2$ and 3 M NaCl while the catholyte contains 2 M $ZnBr_2$ and 2 M Br_2 dissolved in water saturated PN (Singh *et. al.*, 1983). A synergetic effect was observed when QBrs were added to the PN electrolyte in the positive half-cell, resulting in a greater reduction in the aqueous bromine concentration than the addition of PN alone, and hence a higher energy efficiency (Cathro, 1988).

Zinc deposited on the negative electrode may form non-uniform dendrites and they can sometimes grow across to the positive electrode and cause shorting in the cells. To prevent this, it is important to maintain uniform zinc deposition. This can be done by circulation of both anolyte and catholyte solutions or by using smooth carbon plastic as negative electrodes. However, zinc dendrites may still form after extended cycling and cause channel blockage and short circuiting through the micro-porous separator. To overcome this problem, complete discharge is carried out periodically to completely removing all of the zinc deposit at the negative electrode. This is undesirable, however, since it disrupts normal operation of the energy storage and power delivery system (Skyllas-Kazacos *et al.*, 2011). Another drawback of the Zn/Br hybrid flow battery is that the faster kinetics of the Zn^{2+}/Zn couple compared with the Br_2/Br^- couple causes polarisation and eventual battery failure at high current operation. This problem can be overcome by the use of high surface area carbon electrodes at the positive electrode; however, the active surface area of the carbon will eventually decrease and oxidation of the carbon coating will occur (Ponce de Leon *et al.*, 2006, p. 726).

The theoretical cell potential of the Zn/Br hybrid battery is 1.85 V; however, due to the use of a complexing agent to prevent bromine gas evolution, the actual cell potential is lower, at 1.76 V. The theoretical specific energy density is 429 Wh/kg (Symons and Butler, 2001), while only about 65–75 Wh/kg (Clarke *et al.*, 2004 cited in Skyllas-Kazacos *et al.*, 2011) is achieved in practice. The high specific energy density is attributed to the very negative potential of the Zn^{2+}/Zn couple resulting in a high cell voltage. This is the main reason why the zinc-bromine hybrid flow battery still attracts wide interests despite suffering from high self-discharge, dendritic zinc deposition, high cost of electrodes, materials corrosion, unsatisfactory energy efficiency and low cycle life.

Other features of this hybrid flow battery include (Skyllas-Kazacos *et al.*, 2011):

- 100% depth of discharge,
- ambient temperature operation,
- high coulombic and voltage efficiencies, reported at 90% and 85%, respectively,
- abundant and inexpensive zinc materials,
- moderate energy capacity ranging from 50 to 400 kWh.

Applications of the Zn/Br flow battery are mainly in electric vehicles and load management. Johnson Controls (JCI) tested the feasibility of the Zn/Br flow battery for electric vehicle application using eight cells employing 3 M $ZnBr_2$, 0.25 M MEM and 0.75 M MEP electrolyte to power Ford ETX-II van (Bolstad and Miles, 1989 cited in Poon, 2008). A sustained 35 kW power with energy efficiency of 69.4% and specific energy density of 67.3 Wh/kg were achieved. The energy density was increased to 71.9 Wh/kg when 3 M NH_4Cl was added into the electrolyte; however, the energy efficiency was found to drop to 62.7%.

The feasibility of Zn/Br flow battery for load management application was tested by various research groups with varying electrolyte compositions, membrane, cell designs, capacity, etc. Table 12.4 shows the performance of Zn/Br flow battery systems for load management application by various research groups (Poon, 2008).

Despite improvements in Zn/Br flow battery design, the issues of bromine crossover and dendritic zinc deposition cannot be completely eliminated,

Table 12.4 Performance of zinc-bromine flow battery for load management application

	Exxon	Exxon	JCI Z Design	JCI V Design	ZBB V Design	MITI
Electrolyte	A	A	B	B	C	–
Membrane	M1	M1	M1	M1	M1	M2
Capacity (kWh)	20	30	20	20	50	50
Bipolar cell	80–100	124	78	78	60	30
Stack	2	2	2	2	3	24 × 2 Series
Electrode area (cm^2)	–	–	1170	1270	2500	1600
Coulombic efficiency (%)	86.7	81.8	80	85–90	–	92.4
Energy efficiency (%)	67.4	52.7	60	70–75	77	81.1

Notes:
A: 2 M $ZnBr_2$, 0.5 M $ZnCl_2$, 0.5 M MEM and 0.5 M MEP.
B: 3 M $ZnBr_2$, 0.25 M MEM and 0.75 M MEP.
C: 2.25 M $ZnBr_2$, 0.5 M $ZnCl_2$ and 0.8 M QBr (MEM:MEP= 50:50).
M1: Microporous polyethylene (MPPE) separator.
M2: Perforated polyolefin based membrane or ion-exchange membrane.
Source: Poon, 2008.

leading to low energy efficiency, high self-discharges and low cycle life (Ponce de Leon *et al.*, 2006, p. 726). At present, only three companies are commercially involved with the Zn/Br technology, these being RedFlow in Australia, and ZBB Energy and Premium Power in the USA.

Zinc-chlorine (Zn/Cl) flow cell

The zinc-chlorine flow cell was one of the earliest flow batteries developed by Charles Renard in 1884 to power his airship 'La France' around the same time as the Zn/Br flow cell, which first appeared in 1885 (Blevins, 1981). The working principle of the zinc-chlorine flow cell is similar to Zn/Br, which is based on the deposition of zinc at the negative electrode and chlorine formation at the positive electrode as shown in the following half-cell reactions:

Positive electrode reaction:

$$2Cl^- \underset{\text{discharge}}{\overset{\text{charge}}{\rightleftarrows}} Cl_2 + 2e \quad E^o = 1.357 \text{ V} \qquad [12.35]$$

Negative electrode reaction:

$$Zn^{2+} + 2e \underset{\text{discharge}}{\overset{\text{charge}}{\rightleftarrows}} Zn \quad E^\circ = -0.763 \text{ V} \qquad [12.36]$$

The theoretical cell potential of the zinc-chlorine battery is 2.12 V, which is higher than Zn/Br flow cell. Furthermore, as there is no complexing agent used to prevent chlorine evolution, the actual cell potential has the same value as the theoretical cell potential. The theoretical specific energy density is 465 Wh/kg, but in practice, only 60–80 Wh/kg is achieved with 96% depth of discharge (Pavlov *et al.*, 2011).

However, the chemistry of the zinc-chlorine flow cell is different from Zn/Br due to the difference in chemical and physical properties of chlorine and bromine. At room temperature, chlorine is gaseous while bromine is a reddish-brown liquid, and this leads to more complicated designs for the zinc-chlorine flow cell to prevent chlorine evolution even at room temperature. Chlorine is also slightly soluble in Zn/Cl solutions and will therefore evolve as chlorine gas during charging, which would affect the chloride ion balance in the electrolyte solutions. In order to prevent chlorine gas escaping, special design is required to collect and store the chlorine gas to be fed back into the electrolyte solutions and be reduced to chloride ions at the electrode. Two methods of chlorine gas collection and storage have been evaluated. The first method uses a compression technique and stores the gas as liquid chlorine at 70–80 psig. When current is generated by the flow cell, the stored liquid chlorine is mixed with the electrolyte solution and transported to the electrode (Pavlov *et al.*, 2011). The second method involves converting chlorine gas into chlorine hydrate $Cl_2.8H_2O$, a yellow ice-like slurry at temperatures below 9.6°C. (Bellows and Grimes, 1984)

The zinc-chlorine flow battery consists of an electrochemical stack, Zn/Cl electrolyte solution with added potassium chloride to improve electrolyte conductivity as well as a chlorine circulating loop and electrolyte cooling system. The electrolyte in the electrochemical stack is circulated using the pump. During charging, the chlorine gas is removed from the cell via the stack and mixes with the electrolyte solution, which has been cooled by a refrigeration system to below 9.6°C to form yellow ice-like chlorine hydrate, which is then pumped into the hydrate reservoir. During discharge, the cooled hydrate is passed through a heat exchanger to break down the chlorine hydrate; the chlorine rich stream is then pumped into the anode to be reduced to chloride (McBreen, 1984). Zinc is oxidised to zinc ions and reacts with chloride ions to form Zn/Cl at the negative electrode.

Due to the low solubility of chlorine in the electrolyte (~0.03 M), the chlorine electrode has to be made of a porous graphite material to enable chlorine containing solution to flow through the pores of the electrode allowing chloride ion formation. Since the chlorine solubility is low, a separator is not required for this flow cell battery.

Like the Zn/Br flow cell, the zinc-chlorine system also suffers from rapid self-discharge and dendrite formation. The rate of self-discharge can be reduced by circulating the electrolyte solution, while the dendrite formation can be overcome by completely stripping off the zinc deposit after each discharge prior to charging. However, this will disrupt the overall energy storage and power delivery operation. The acidic nature of the electrolyte solution used in the zinc-chlorine system also results in zinc oxidation and the release of hydrogen gas, which poses an explosion hazard in the chlorine storage reservoir. Furthermore, the graphite materials will be slowly oxidised to CO_2 during charging, due to the high electrode potential, particularly at high pH (McBreen, 1984).

Besides the above electrochemical problems, the engineering issues surrounding the zinc-chlorine flow cell are more challenging than for Zn/Br, as they require the integration of the cell stack with the chlorine storage system. As chlorine capture and storage require a cooling system to prevent chlorine gas escaping the electrolyte solutions, the cell design for the zinc-chlorine flow cell is much more complicated than that for the Zn/Br flow cell. The net energy efficiency is thus expected to be lower, as some of the energy will be used to supply auxiliary equipment such as the gas and electrolyte pumps, cooling system and a hydrogen recombination system.

For these reasons (dendrite formation, sensitivity to impurities, pH management, complex cell design involving cell stack, hydrate formation, system engineering and integration, as well as safety), zinc-chlorine has failed to attract attention for further development despite having higher energy density than the Zn/Br flow cell.

An attempt to design and develop a zinc-chlorine battery for energy storage was carried out by Furukawa Electric Co Ltd. in Japan in 1980s, where the company developed 1, 10 and 50 kW modules. However, the company decided to terminate the project after critical analysis of the technological, performance and economical parameters following the test result at the Government Industrial Research Institute in Osaka showing that this flow battery poses serious safety and environmental hazards (Pavlov *et al.*, 1991).

12.3.5 Lithium-based flow battery

The lithium-air battery holds great promise, due to its outstanding specific capacity of 3842 mAh/g as anode material. The lithium-air battery works by combining lithium ion with oxygen from the air to form lithium oxide at the positive electrode during discharge. A recent novel flow cell concept involving lithium is proposed by Chiang *et al.* (2009). They proposed to use typical intercalation electrode materials as active anodes and cathode materials. These active materials are prepared by suspending lithium-based compounds in a liquid electrolyte to form a semi-solid suspension. These two different suspensions are then pumped into and out of a reaction chamber, which is separated by a thin porous membrane (Abraham and Jiang, 1996). The system is named a semi-solid flow cell (SSFC). They claimed a higher energy density compared with conventional aqueous flow cells due to the higher concentration of active materials in the solid component of the liquid suspension (Duduta *et al.*, 2011). The estimated energy density using established lithium intercalation compounds is around 130–250 Wh/kg in the optimised SSFC system, thus paving the way for widespread adoption in electric vehicles. This novel concept demonstrates that slurry type active materials can be used to store energy, and hence this will open a new field of research to find better cathode and anode active materials as well as electrolytes within the semi-solid flow system to lower the cost to enable wider applications.

12.3.6 Other flow batteries chemistries

Besides the afore mentioned flow batteries, other RFB chemistries are currently being developed, including all-chromium, Tiron-Pb, Methylimidazolium iron chloride molten salt, actinide-based RFB, organic-based redox flow batteries and the new lithium-based flow battery.

Like the VRB, the all-chromium system involves only a single element to avoid cross-contamination of the electrolyte (Bae *et al.*, 2002). The tiron-Pb cell employs 0.25 M tiron in 3 M H_2SO_4 as the catholyte and a lead electrode as the anolyte. The theoretical cell potential is 1.10 V.

Actinide-based flow cells involve the use of excess depleted actinides arising from nuclear power activities as a potential redox couple for energy storage (Shiokawa *et al.*, 2000). Among the four actinide elements typically found in nuclear power generation, namely uranium (U), neptunium (Np), plutonium (Pu) and americium (Am), only neptunium couples and uranium couples appear promising in terms of good energy efficiency based on theoretical calculation; however, further investigation will be required to evaluate their safety and environmental impact before they can be commercialised.

The use of organic solvents for redox flow batteries is also attractive, due to their higher cell potentials. The solubility of many redox couples can be increased in organic solvents, leading to a likely increase in energy density of the redox flow batteries. However, organic solvents are more expensive, possibly less conductive, and more environmentally unfriendly than the aqueous system. Therefore, it is important to assess the performance and environmental compatibility of these organic electrolytes for flow battery applications.

Recent studies on ruthenium, chromium and vanadium acetylacetonate show that the energy efficiency of these organic-based redox flow batteries is very low, at about 5–10%, despite their higher open-circuit potentials than aqueous VRB (Skyllas-Kazacos *et al.*, 2011). Although chromium and vanadium acetylacetonate systems showed higher open-circuit potentials (3.4 and 2.2 V, respectively) than the aqueous VRB system (1.6 V), the low efficiency and the high cost of organic electrolytes have limited the development of these organic electrolytes for redox flow batteries applications. Further research into more cost effective solvents with acceptable environmental and chemical compatibility is needed in order to make them commercially competitive with aqueous-based flow batteries.

12.4 Conclusion

As the number of grid-connected renewable energy sources increases around the world, their inability to supply constant and predictable power has seen the resurgence of interest in the development of advanced energy storage systems. In particular, redox flow batteries have received considerable attention due to their ability to store large amounts of electricity in an efficient and cost effective manner.

Redox flow batteries possess many attractive features that make them ideally suited to large-scale energy storage in both off-grid and grid-connected applications that require more than 2 h of energy storage capacity. The independent sizing of power and energy capabilities provides great flexibility and allows systems to be designed to suit the specific needs of each application. Energy storage capacity can therefore be readily increased by simply using bigger electrolyte tanks and adding more electrolyte. The incremental cost of additional energy storage capacity is therefore determined only by the cost of extra electrolyte so the cost per kWh of generated energy decreases dramatically with increased storage time. Cost estimates have in fact shown that for storage capacities in excess of 2 h, the G1 VRB can deliver energy at less than half the cost of an equivalent lead-acid battery, which is the cheapest battery currently available commercially. Overall energy efficiencies of up to 80% have been demonstrated in large demonstration systems along with more than 200 000 charge–discharge cycles, which cannot be matched by any other battery technology currently under development. With further developments on improved hydraulic sealing, material selection and cost effective components with suitable longevity, increased commercial implementation of flow batteries is expected (Skyllas-Kazacos *et al.*, 2011; Weber *et al.*, 2011). The main challenge for flow battery developers, however, is to attract sufficient investment funding that will enable the required production volumes needed to achieve the necessary cost structure for the large emerging markets in renewable energy storage and smart-grid applications.

A unique feature of flow batteries for electric vehicle applications is their ability to be both electrically recharged and mechanically refuelled by electrolyte exchange. Their energy density relative to lithium ion and other battery technologies has, however, precluded their application in electric vehicles to date. With recent advances in the G2 and G3 vanadium battery technologies and the new lithium semi-solid flow battery developments, mobile applications in refuelable electric cars and buses may become viable in the not-too-distant future.

12.5 References

A123 (2012), Website: http://www.a123systems.com/ (Accessed 9 December 2012).

Abraham K.M. and Jiang Z. (1996), A polymer electrolyte-based rechargeable lithium/oxygen battery, *Electrochem. Soc.* **143**, 1–5.

Bae C.H., Roberts E.P.L and Dryfe R.A.W. (2002), Chromium redox couples for application to redox flow batteries, *Electrochimica Acta* **48**, 279–287.

Bajpal S.N. (1981), Vapour pressure of bromine-quarternary ammonium salt complexes for zinc-bromine battery application, *J. Chem. Eng. Data* **26**(1), 2–4.

Balakrishnan P.G., Ramesh R. and Kumar T.P. (2006), Review: Safety mechanisms in lithium ion batteries, *J. Power Sources* **155,** 401–414.

Bellows J. and Grimes P.G. (1984), *Zinc/Halogen Battery, in Power Sources for Electric Vehicles* (B.D. McNicol, D.A.J. Rand, eds.), *Elsevier,* Amsterdam, p. 621.

Blevins C.M. (1981), Life-testing of 1.7 kW h zinc-chloride battery system: Cycles 1-1000, *J. Power Sources* **7**, 121–132.

Bolstad J.J. and Miles, R.C. (1989), Development of the zinc/bromine battery at Johnson Controls Inc. *Institute of Electrical and Electronics Engineers (IEEE), Energy Conversion Engineering Conference* 1989 (IECEC-1989), Proceedings of the 24th Intersociety, August 6–11 1989, Washington D.C., USA.

Cathro K.J. (1988), Performance of zinc/bromine cells having a propionitrile electrolyte, *J. Power Sources* **23**, 365–383.

Chiang Y., Carter W.C., Ho B. and Duduta M. (2009), Massachusetts Institute of Technology, High Energy Density Redox Flow Device, US Provisional Patent No 61/287,180.

Clarke R., Dougherty B., Mohanta S. and Harrison S. (2004), Abstract 520, Joint International Meeting: 206th Meeting of the Electrochemical Society/2004 Fall Meeting of the Electrochemical Society of Japan, Honolulu, Hawaii, 3–8 October 2004.

Commonwealth Scientific and Industrial Research Organisation website (CSIRO) (2011): http://www.csiro.au/science/Ultra-Battery (Accessed 12 September 2011).

De Boer P. and Raadschelders J. (2007), *Flow Batteries: Briefing Paper*. Arnhem: Leonardo Energy. Available at: <http://www.leonardo-energy.org/webfm_send/164> (Accessed 6 July 2007).

Deeya Energy (2011), website: http://www.deeyaenergy.com/product/ (Accessed 19 January 2011).

DOE News Release (2010): US Department of Energy Smart Grid Program Award for Demonstration of V-Fuel Vanadium Battery Technology in the USA, 8 July 2010, URL: <http://www.energy.gov/news2009/8305.htm> (Accessed 15 December 2010).

Duduta M., Ho B., Wood V.C., Limthongkul P., Brunini V.E., Carter W.C. and Chiang Y. (2011), Semi-solid lithium rechargeable flow battery, *Adv. Energy Mater.* **1**, 511–516.

Engineering Conference 1989 (IECEC-1989), Proceedings of the 24th Intersociety, Washington D.C., USA 6–11 August 1989.

EPRI-DOE (2004), Energy Storage for Grid Connected Wind Generation Applications, EPRI, DOE Technical Update, Palo Alto, CA, Washington DC, 1008703, December.

EPRI-DOE (2003), *Handbook of Energy Storage for Transmission and Distribution Applications*, EPRI, DOE, Palo Alto, CA, Washington DC.

ESA (2011), Electricity Storage Association website: http://www.electricitystorage.org/technology/storage_technologies/technology_comparison (Accessed 9 December 2011).

Eustace D.J. (1977), Exxon Research and Engineering Co., *Metal Halogen Electrochemical Cell.* US Patent 4,064,324.

Gahn N.H. Hagedorn J.A. Johnson (1985), Cycling performance of the iron chromium redox energy storage system, NASA TM-87034, NASA, Dept. of Energy, US.

Johnson D. and Reid M. (1985), Chemical and electrochemical behavior of the Cr(III)/Cr(II) half-cell in the Iron-Chromium redox energy storage system, *J. Electrochem. Soc* **132,** 1058.

Kantner E. (1985), Exxon Research and Engineering Co., *Zinc-Bromine Battery with Improved Electrolyte.* US Patent 4,491,625.

Kim S., Vijayakumar M., Wang W., Zhang J., Chen B., Nie Z., Chen F., Hu J., Li L. and Yang Z. (2011), Chloride supporting electrolytes for all-vanadium redox flow batteries, *Phys. Chem. Chem. Phys.* **13**, 18186–18193. DOI: 10.1039/c1cp22638j.

Largent R. L., Skyllas-Kazacos M. and Chieng, J. (1993), Improved PV system performance using vanadium batteries, in: *Proceedings of the IEEE 23rd Photovoltaic Specialists Conference*, Louisville, Kentucky, May.

Li L., Kim S., Wang W., Vijayakumar M., Nie Z., Chen B., Zhang J., Xia G., Hu J., Graff G., Liu J. and Yang Z. (2011), A stable vanadium redox-flow battery with high energy density for large-scale energy storage, *Adv. Energy Materials* **1,** 394–400. DOI: 10.1002/aenm.201100008.

Maria Skyllas-Kazacos (2009), 'Secondary batteries: Redox flow battery – vanadium redox' in *Encyclopedia of Electrochemical Power Sources*, J. Garche, P. Moseley, Z. Ogumi, D. Rand and B. Scrosati, Eds, 444–453, Elsevier, Amsterdam.

McBreen J. (1984), Rechargeable zinc batteries, *J. Electroanal. Chem.* **168**, 415–432.

Nakamura Y. (1988), Operating method for redox flow type cell, Japanese Patent, 63150863.

Pavlov D., Papazov G. and Gerganska M. (1991), *Battery Energy Storage Systems.* Sofia: UNESCO Regional Office for Science and Technology for Europe (ROSTE). Available at: < http://unesdoc.unesco.org/images/0009/000916/091670eo.pdf> (Accessed 4 December 2011).

Ponce de Leon C., Frias-Ferrer A. and Gonzalez-Garcia J. (2006), Redox flow cells for energy conversion, *J. Power Sources* **160**, 713–732.

Poon G. (2008), Bromine Complexing Agents for use in Vanadium Bromide (V/Br) Redox Flow Cell, MSc., The University of New South Wales.

Prudent Energy (2011)– case study: VRB Technology in Japan, URL http://www.pdenergy.com/pdfs/casestudy_japan.pdf

Prudent Press Release (2001), http://www.pdenergy.com/press_030211_cepri.html

Remick R.J. and Ang P.G.P. (1984), Institute of Gas Technology, Electrically rechargeable anionically active reduction-oxidation electrical storage-supply system. US Patent 4,485,154.

Rychcik M. and Skyllas-Kazacos M. (1988), Characteristics of new all-vanadium redox flow battery, *J. Power Sources* **22,** 59.

Scamman D.P., Reade G.W. and Roberts E.P.L (2009), Numerical modelling of a bromide-polysulfide redox flow battery, Part 1: Modelling approach and validation for a pilot scale system, *J. Power Sources* **189**, 1220–1230.

Scamman D.P., Reade G.W. and Roberts E.P.L (2009), Numerical modelling of a bromide-polysulfide redox flow battery, Part 2: Evaluation for a utility scale system, *J. Power Sources* **189**, 1231–1239.

Shimizu M., Mori N., Kuno M., Mizunami, K. and Shigematsu T. (1988), Development of a redox flow battery, *Proc. Electrochem Soc* **88**, 249.

Shiokawa Y., Yamana H. and Moriyama H. (2000), An application of actinide elements for a redox flow battery, *J. Nuclear Sci. Technol.* **37**(3), 253–256.

Singh P., White K. and Parker A.J. (1983). Application of non-aqueous solvents to batteries: Part 1. Physicochemical properties of propionitrile/water two phase solvent relevant to zinc-bromine battery, *J. Power Sources*, **10**, 309–318.

Skyllas-Kazacos M. and Robins R. (1986), All-Vanadium Redox Battery. US Patent 4,786,567.

Skyllas-Kazacos (2003), Novel vanadium chloride/polyhalide redox flow battery by Maria Skyllas-Kazacos, *J. Power Sources* **24**, 299–302.

Skyllas-Kazacos M. Kazacos G., Poon G. and Verseema H. (2010), Recent advances with UNSW vanadium-based redox flow batteries, *Int. J. Energy Res.* (Energy Storage Special Issue), **34**, 182–189.

Skyllas-Kazacos M. (2010), 'Energy storage for stand-alone/hybrid systems: Electrochemical Energy Storage Technologies' in *Stand-alone and Hybrid Wind Systems: Technology, Energy Storage and Applications,* J.K. Kaldellis (ed.), Woodhead Publishing, July 2010.

Skyllas-Kazacos M, Chakrabarti M.H., Hajimolana S.A., Mjalli F.S. and Saleem D. (2011), Progress in flow battery research and development, *J. Electrochem. Soc.* **158**(8), R55–R79.

Skyllas-Kazacos M. and Kazacos M. (2011), State of charge monitoring methods for vanadium redox flow battery control, *J. Power Sources* **196**, 8822–8827.

Skyllas-Kazacos M. and Menictas C. (2011), Performance of vanadium/oxygen redox fuel cell, *J. Appl. Electrochem.* **41**,1223–1232. DOI 10.1007/s10800-011-0342-8.

Sun B. and Skyllas-Kazacos M. (1992), Modification of graphite electrode materials for vanadium redox flow battery application – I. Thermal treatment, *J. Electrochimica Acta* **37**(7), 1253–1260.

Sun B. and Skyllas-Kazacos M. (1992), Modification of graphite electrode materials for vanadium redox flow battery application – II. Acid treatments, *J. Electrochimica Acta* **37**(13), 2459–2465.

Swette L. and Jalan V. (1984), Development of electrodes for the NASA iron/chromium redox system and factors affecting their performance, NASA CR-174724, DOE/NASA/0262-1.

Symons P.C. and Butler P.C. (2001), *Introduction to Advanced Batteries for Emerging Applications*. Albuquerque, NM: Sandia National Laboratories. Available at:<http://prod.sandia.gov/techlib/access-control.cgi/2001/012022p.pdf> (Accessed 2 December 2011).

Tang A., Ting S., Bao J. and Skyllas-Kazacos M. (2012), Thermal modelling and simulation of the all-vanadium redox flow battery, *J. Power Sources* **203**, 165–176. Available on-line at http://dx.doi.org/10.1016/j.jpowsour.2011.11.079

Thaller L.H. (1976), Electrically rechargeable redox flow cell, US Patent **3**,996.

Tokuda N, Kanno T., Hara T., Shigematsu T., Tsutsui Y., Ikeuchi A., Itou T. and Kumamoto T. (2000), Development of a redox flow battery system, *SEI Tech. Rev.* **50,** 88.

Wang W., Kim S., Chen B., Nie Z., Zhang J., Xia G., Li L. and Yang Z. (2011), 'A new redox flow battery using Fe/V redox couples in chloride supporting electrolyte', *Energy Environ. Sci.* **4**, 4068. DOI: 10.1039/c0ee00765j.

Weber A.Z., Mench M.W., Meyers J.P., Ross P.N., Gostik J.T. and Liu Q. (2011), Redox flow batteries: A review, *J. Appl. Electrochem.* **41**, 1137–1164.

Yang Z., Zhang J., Kintner-Meyer M.C.W., Lu X., Choi D., Lemmon J.P. and Liu J. (2011), Electrochemical energy storage for green grid, *Chem. Rev.* **111**(5), 3577–3613.

Zhong S., Padeste C., Kazacos M. and Skyllas-Kazacos M. (1993), Physical chemical and electrochemical properties comparison for rayon and PAN based graphite felt electrodes, *J. Power Sources* **45**, 29–41.

13
Superconducting magnetic energy storage (SMES) systems

P. TIXADOR, Grenoble INP/Institut Néel – G2E lab, France

DOI: 10.1533/9780857097378.3.442

Abstract: Superconducting magnetic energy storage (SMES) is one of the few direct electric energy storage systems. Its specific energy is limited by mechanical considerations to a moderate value (10 kJ/kg), but its specific power density can be high, with excellent energy transfer efficiency. This makes SMES promising for high-power and short-time applications. So far SMES uses Nb-Ti conductors, due to their low cost. Apart from the cost, YBCO appears very attractive due to its current capacity, mechanical performance and 'high' operating temperature. Since the first studies in the 1970s, successful demonstrations of SMES have been carried out but the high cost has prevented commercialization.

Key words: SMES, storage devices, large-scale superconductivity, magnet.

Note: This chapter is a revised and updated version of Chapter 9 'Superconducting magnetic energy storage (SMES) systems' by P. Tixador, originally published in *High temperature superconductors (HTS) for energy applications*, ed. Z. Melhem, Woodhead Publishing Limited, 2012, ISBN: 978-0-85709-012-6.

13.1 Introduction

An inductance, which carries a current, is characterized by an energy of magnetic origin. To store this energy the inductance should be short-circuited. However, the inductance must be without any loss, then superconducting so that this energy is not quickly dissipated by Joule effect. A short-circuited superconducting magnet stores energy in magnetic form, thanks to the flow of a persistent direct current (DC). The current really remains constant due to the zero DC resistance of the superconductor (except in the joints). The current decay time is the ratio of the coil's inductance to the total resistance in the circuit. File and Mills performed measurements of the persistent current decay and determined decay time constants of 10^5 years.[1] The equivalent resistivity is around 10^{-25} Ωm. The resistivity of copper at room temperature is 1.7 10^{-8} Ωm. Thus, the decay time for a copper coil at room temperature of the same dimensions and inductance would be less than 0.1 ms. Superconductors are thus indispensable for magnetic energy storage systems, except for very

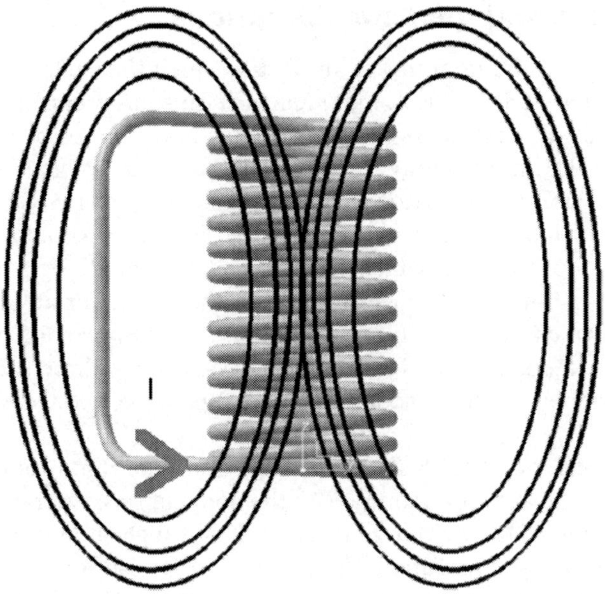

13.1 SMES concept (short-circuited SC winding with magnetic field lines).

short storage durations (lower than 1 s). This storage system is known as SMES.[2,3] This rather simple concept was proposed by M. Férrier in 1969.[4]

The magnetic stored energy (W_{mag}) is determined by the coil's self-inductance (L) and its current (I) or, equivalently, by the magnetic flux density and field integrated over all space (Fig. 13.1):

$$W_{mag} = \frac{1}{2}LI^2 = \frac{1}{2}\iiint_{space} BH\,\mathrm{d}x\,\mathrm{d}y\,\mathrm{d}z \qquad [13.1]$$

Ferromagnetic materials cannot store a great deal of energy since their magnetic field is close to zero. However, they increase the magnetic flux density in air that is adjacent to them. This tends to increase the total magnetic energy. Unfortunately, the weight of the ferromagnetic component increases the total weight so that the specific energy (energy/mass) increases with their use. As a result, they are nearly never used.

When the short circuit is opened, the stored energy is transferred in part or totally to a load. The current of the coil decreases via a negative voltage, while a positive voltage charges or energizes the magnet by increasing the current (Equation [13.2]):

$$V = L\frac{\mathrm{d}I}{\mathrm{d}t} \qquad [13.2]$$

V is the voltage across the magnet.

13.2 Current and load considerations

The current is the state quantity of an SMES, which thus acts as a current source. Nevertheless, SMES is not an ideal current source since the current decreases with its discharge (Fig. 13.2). The current decreases by a factor of 2 when 73% of the initial energy is recovered.

SMES is the 'dual' of a capacitor, which acts as a voltage source (Table 13.1). The loads in which a capacitor or a SMES are discharged should be different. For a capacitor the load, which fixes the current, should be a resistance or an inductance, but not a capacitor. On the other hand, for an SMES system, the load, which fixes the voltage, should be a resistance or a capacitor, but not an inductance. A SMES system cannot be used to directly charge or discharge a magnet. An intermediate voltage type power conversion stage is required.

Even if the load for a SMES should be capacitor type, it may have a small inductance (L_{load}) given, for example, by the connections. Energy will then be lost at the time of the discharge in the switch, which should then simultaneously withstand voltage and current. The current cannot vary instantaneously in the load inductance and is zero before the SMES discharges. The lost energy is expressed by:[5]

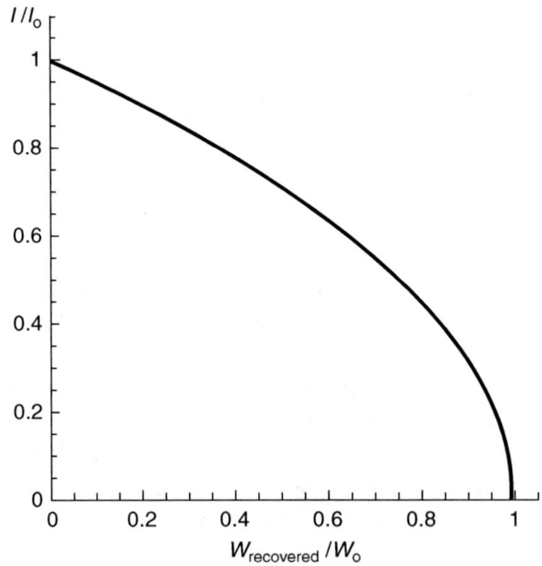

13.2 Current vs the energy recovered in a SMES $(W_o = LI_o^2/2)$.

Table 13.1 Comparison of capacitor and coil for energy storage (time constant for resistive load)

	Storage mode		Discharge mode	Time constant	Source
SMES	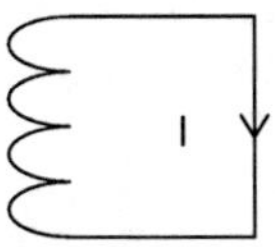 Short circuit $V = 0$	$W = \frac{1}{2} L I^2$	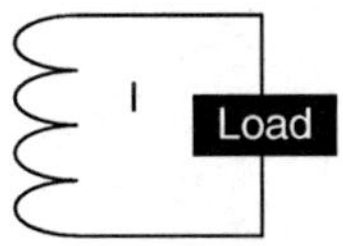	$\frac{L}{R}$	Current
Capacitor	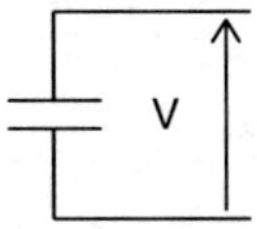 Open-circuit $I = 0$	$W = \frac{1}{2} C V^2$	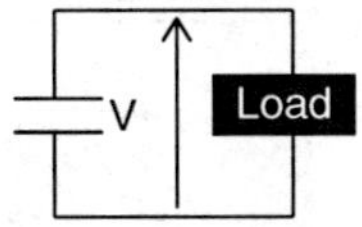	RC	Voltage

$$\Delta W = \frac{L_{\text{load}}}{L_{\text{SMES}} + L_{\text{load}}} \frac{1}{2} L_{\text{SMES}} I_{\text{o}}^2 \qquad [13.3]$$

The load inductance should be much lower than the SMES inductance to lower the energy loss and the stresses on the switch.

The resistance value of the load affects the discharge of magnets and capacitors in opposite ways. A low-resistance load slowly discharges an SMES (thus a long discharge time (L/R)) and rapidly discharges a capacitor (short discharge time (RC)).

In storage mode SMES does not offer any danger in terms of voltages since the voltage is zero, whereas dielectric capacitors for fast discharges often require high voltages in the storage mode.

13.3 SMES systems

Figure 13.3 shows a schematic drawing of an SMES system connected to a load. Consisting of five major components or subsystems as shown:

- A superconducting magnet ① with its mechanical supporting structure and current leads (electric connections between the SC magnet and the room temperature circuit).
- A cryogenic system ② (cryostat, vacuum pumps, cryocooler, etc.).
- A power conditioning system ③ (interface between the superconducting magnet and the load).

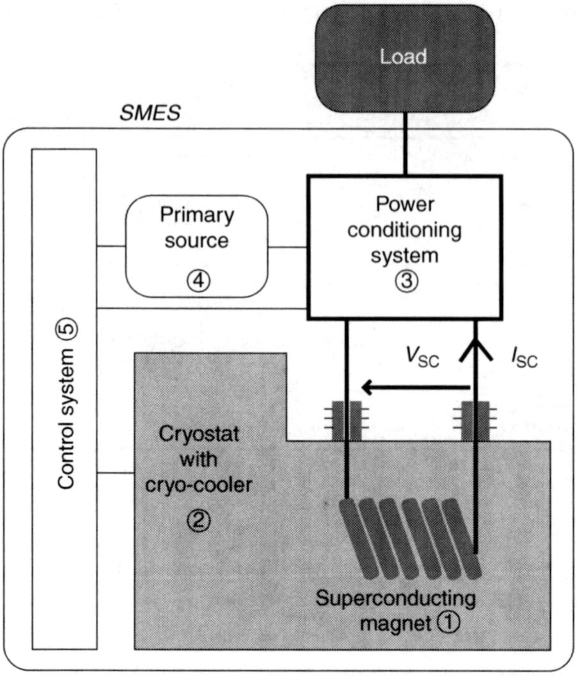

13.3 SMES system with its components. See text for explanation.

- A primary source ④ to energize the superconducting magnet.
- A control and management system ⑤ (electronics, cryogenics, magnet protection, etc.).

When the load is directly connected to the SMES (Fig. 13.4), the power conditioning system is only a switch. However, the current is not controllable and decreases as the magnet discharges (Fig. 13.2).

In Fig. 13.5, the SMES is connected to the electric grid with which it exchanges energy. The SMES is directly energized by the grid. The SMES may supply energy at a time of voltage sags or power outages. It acts then as an uninterruptible power source (UPS) for sensitive loads. The SMES may also damp low-frequency power oscillations via an energy exchange, for example by receiving energy from the grid and then delivering energy to the grid, thus stabilizing the grid.

When it acts in that fashion, the power conditioning system is a rectifier/inverter. This power electronic circuit is required to convert the DC quantities of the superconducting magnet to alternating current (AC) ones and vice versa, since the very large majority of grids are AC.

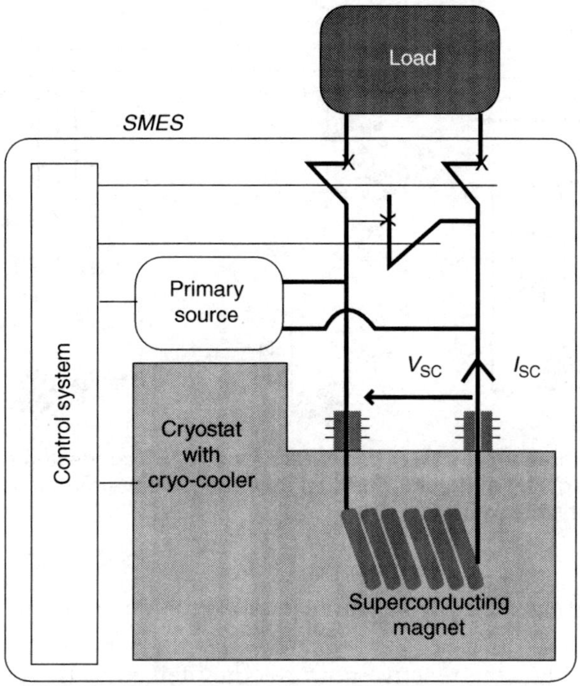

13.4 SMES directly connected to a load.

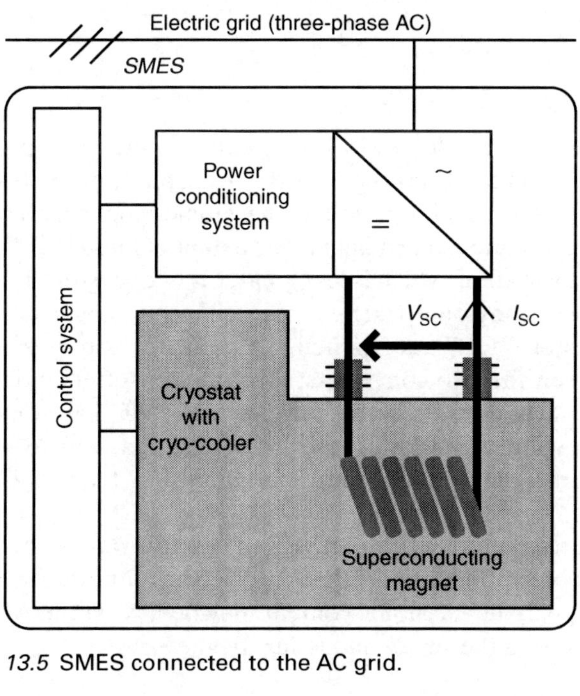

13.5 SMES connected to the AC grid.

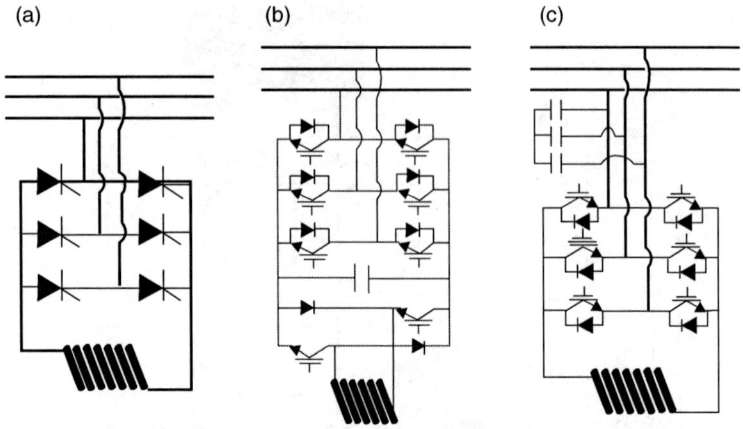

13.6 The three main power converters for a SMES connected to a three phase AC grid.[6] (a) thyristor bridge, (b) voltage converter and chopper, (c) current converter.

The power converter mainly has three possible topologies (Fig. 13.6):[6]

- A thyristor bridge: the reactive power is then linked to the exchanged active power and cannot be independently controlled:

$$P = \frac{3\sqrt{2}}{\pi} UI_{sc} \cos\alpha \;\; ;Q = \frac{3\sqrt{2}}{\pi} UI_{sc} \sin\alpha = P \; tg\alpha \qquad [13.4]$$

  U is the voltage amplitude (RMS value) between the three phases and I_{sc} is the current in the SC magnet, α is the firing angle of the thyristors. This converter induces a lot of voltage harmonics and causes some AC losses in the superconducting magnet. The extent of these losses depends on the current variation, which is less greater is energy stored.
- A voltage source converter in series with a chopper using insulated gate bipolar transistor (IGBT) components: a capacitor is required for the DC bus between the two converters; this topology offers independent control of the active and reactive powers, a low AC harmonic distortion, and low voltage ripples across the SC magnet lead to some AC losses. Of course, only the chopper is required if a DC bus is already available.
- A current source converter: the active and reactive powers can also be independently controlled; the SMES is directly connected to the DC side (no chopper); the harmonic content may be low on the AC side, as are the AC losses in the superconducting magnet.

In all three cases a transformer, not represented in Fig. 13.6, is generally inserted between the grid and the converter. The cost of the power converter is a significant part of the SMES system cost.

13.4 SMES limitations

SMES is an emerging energy storage technology, which has to be compared with other alternatives. The main characteristics of an energy storage device are:

- Specific energy (volume–mass).
- Specific power (volume–mass)/discharge duration.
- Investment and running costs (kWh–kW).
- Energy conversion efficiency, losses in storage mode.
- Operational security/failure characteristics.
- Lifetime/cycle number.
- State of the technology (demonstration/emerging/mature).
- Environmental considerations.

13.4.1 Comparisons with other energy storage methods

The importance of the different items listed above depends on the application. Nevertheless, the energy and the power densities are fundamental. Energy storage systems are classically compared using the Ragone chart, which plots the specific power vs the specific energy (Fig. 13.7). SMES is in terms of energy density between conventional capacitors and batteries, comparable to supercapacitors/ultracapacitors but with higher power density capabilities. For SMES, the grey zone indicates the presently attained values. The black zone covers theoretically possible ranges, but requires much more research and development.

The power characterizes how quickly the energy can be released. The mean power is the ratio of the energy to the discharging time. Figure 13.8 plots the ranges of power and discharge time for different storage technologies. Short discharge times correspond to power quality applications, and long discharge times to energy management applications.[7]

CAES (Compressed Air Energy Storage) uses underground reservoirs (salt cavern, old hard rock mine, etc.), to pressurize large volumes of air and then release the stored volume to recover the energy. Pumped hydro storage (two water reservoirs at different elevations) and CAES are the only available technologies for bulk energy storage systems to level the consumption, for example.

The Ragone plot (Fig. 13.7) shows that SMES is more a power source than an energy source such as a battery. This is why SMES is suitable for high

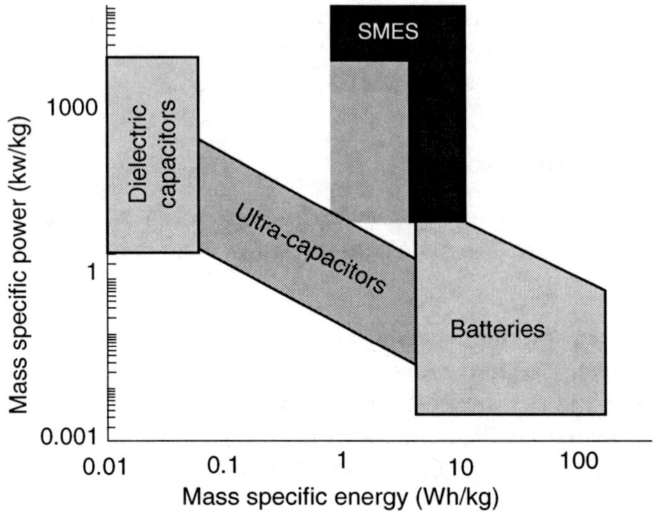

13.7 Ragone plot for SMES, batteries and capacitors.

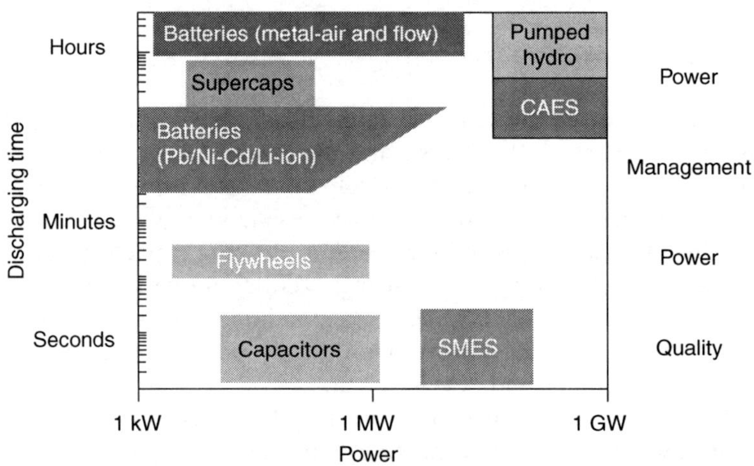

13.8 Discharging time *vs* power for various energy-storing devices[7] (http://electricitystorage.org/).

power, up to 100 MW, and short duration (Fig. 13.8), under a few seconds. Of course, complementary energy sources can be associated such as an SMES for short durations and batteries for long durations to cover a broader range.

It is important to compare the specific energy of storage systems to classical sources of energy (Table 13.2) to see that their energy performances are very low. One kWh is a lot of energy: it is the kinetic energy in a 10 ton truck at a speed of 100 km/h.

Table 13.2 Comparison of the specific energy of various systems

Source of energy	Energy density (kWh/kg)
Coal	8
Wood	4
Oil	10–12
Natural gas	10–14
Enriched uranium	600 000
Water fall (1000 m)	0.003
Pb/Li-ion battery	0.03/0.2
SMES	0.003
Dielectric capacitor	0.00015

13.4.2 Specific energy limitation

Although the attainable magnetic flux density limits the energy per unit volume given by Equation [13.1] $(B^2/2\mu_0)$, the real limit of the energy stored in a SMES is mechanical in nature in addition to the critical current in the superconductor. The virial theorem[8] clearly shows that the magnetic energy is related to the structural mass. For one dimension stresses, the relation between the masses of the mechanical structure in tension and in compression (M_{tension}, $M_{\text{compression}}$) and the stored energy (W_{mag}) is:

$$W_{\text{mag}} = \frac{\sigma}{d}\left(M_{\text{tension}} - M_{\text{compression}}\right) \qquad [13.5]$$

σ is the working stress (assumed to be the same for tension and compression) and d is the density of the structural material.

The total structure mass becomes:

$$M_{\text{structure}} = M_{\text{tension}} + M_{\text{compression}} = 2M_{\text{compression}} + W_{\text{mag}}\frac{d}{\sigma} \qquad [13.6]$$

The ultimate limit is when the full mechanical structure only works in tension:

$$M_{\min} = W_{\text{mag}}\frac{d}{\sigma} \qquad [13.7]$$

This ultimate limit is theoretically approached for force-balanced or force-reduced coils[9,10] but they remain very complex for the winding (design and construction). 'Force-free' configurations in which the current flows always parallel to the magnetic field were proposed.[11] However, they are physically impossible. The intent of the proposers was to devise a geometry that would cancel the Lorentz force. In principle, cancellation might be possible in some infinite structures, which are inherently impossible as well. Thus, expression [13.7] gives the absolute minimum for real structures. In the fundamental formation of energy and force, the arguments for magnetic systems apply equally for any mechanical system, including kinetic energy storage (flywheels) and compressed air.

Worked example

Since the virial theorem is very important for SMES, let us consider an example to find the expression again. A simple case is the thin infinite solenoid with a thickness e, small compared to the inner radius (R). The maximum hoop stress in such a solenoid considering that turns act dependently is:

$$\sigma = \frac{JB_{\max}R}{2}\;(B_{\max} = \mu_0 Je) \qquad [13.8]$$

Energy is only stored inside a thin infinite solenoid:

$$W_{\text{mag}} = \frac{1}{2\mu_0}B^2_{\max}\,\pi R^2 h \qquad [13.9]$$

h is the unit height of the solenoid.

The magnetic energy per solenoid volume is simply expressed by the hoop stress:

$$W_{\text{mag}} = \frac{\sigma}{2}\text{Vol}_{\text{solenoid}} \qquad [13.10]$$

This expression does not take into account the compressive force, which must be accommodated by having additional structure that is not in hoop tension[12] and the exact relation for an infinite thin solenoid becomes, with the compressive contribution:

$$W_{\text{mag}} = \frac{\sigma}{3}\text{Vol}_{\text{solenoid}} \qquad [13.11]$$

This is one third of the ultimate virial limit.

More generally, the minimum superconducting volume may be expressed by:

$$Vol_{SC}^{min} = C_1 \frac{W_{mag}}{\sigma} \qquad [13.12]$$

where C_1 depends on the magnet geometry ($C_1 \geq 1$). For a thin wall infinite solenoid it is 3. References 12 and 13 give this coefficient for solenoids. It reaches the minimum value (one, ultimate virial limit) for a thin, very short solenoid (one turn). On the other hand, the energy is very low since the available space for the conductor tends towards zero. This optimal theoretical geometry is then not used.

The expression [13.12] is the same as for the rotating part of a flywheel. The magnetic energy is then the kinetic one and the electromagnetic stresses are replaced by the stresses due to the centrifugal forces. C_1 depends on the rotor geometry.

Assuming a reasonable working stress of 100 MPa, expression [13.7] gives for a magnet with steel structure working in full tension (Equation [13.7]) a maximum value of 12.5 kJ/kg (3.5 Wh/kg) for the stored energy per unit mass. The working stress of 100 MPa may be increased somewhat, but the mass specific energy will still be limited to some tens of kJ/kg.

The compact muon solenoid (CMS)[14] magnet of the LHC collider almost reaches this value for its cold mass (2.6 GJ/225 tons or 11 kJ/kg). The record specific energy reaches 13.4 kJ/kg[15] for a thin Nb-Ti solenoid for particle astrophysics.

Implications

The mechanical design of an SMES is a challenging issue; the magnet conductor must be designed to contain tremendous stresses and deformations without degradation of superconducting properties. The only method used to contain the large Lorentz forces of an SMES is the self-supported/cold concept.[2] The cold structure of the magnet supports the magnetic forces. The functions have to be combined to optimize the specific energy: the conductor should provide both current transport and mechanical support by adequate strength. Nevertheless, as the stored energy increases, additional structure is often required to contain the forces, which increase with size.

Another approach is to transfer the forces outside the magnet to an ambient temperature support. For example, the magnet could be installed in an underground cavern or in reinforced surface trenches. Such a solution might be possible for huge energies, higher than GWh.[2,16]

13.4.3 Superconducting volume

A relationship between the superconducting volume and the stored energy is:[17]

$$Vol_{SC} = C_2 \frac{W_{mag}^{3/2}}{J_{ov}\left(\mu_0 B\right)^{1/3}} \quad [13.13]$$

C_2 mainly depends on the magnet geometry. J_{ov} is the average current density in the magnet and B is the magnetic flux density. In this expression the mechanical stress, one of the major limitations for SMES, does not appear and varies with the size. However, Equation [13.13] shows important points. Contrary to many other energy storage systems, SMES active volume does not increase linearly with stored energy. SMES is then more interesting in terms of quantity of superconducting (then cost) when its energy is high. That point is reinforced by the cryogenics, which penalizes above all small magnets in terms of additional volume and weight. On the other hand, the amplitude of the magnetic flux density does not play a predominant part in reducing the superconducting volume. The average current density is much more important. It is limited by the performance of the superconductor, but above all by protection issues (see Section 13.5.4).

13.4.4 Volume energy density

The expression of the stress in a solenoid (Equation [13.8]) shows that at a given stress two main options are possible:

- High magnetic flux density and small radius/volume.
- Low magnetic flux density and large radius/volume.

Both magnets will have similar weights for the same magnetic energy. With low magnetic flux densities, the operating temperature may be much higher, due to the $J_c(B, T)$ characteristic of superconductors (see Fig. 13.15).

Table 13.3 illustrates these two options for 5 MJ SMES with similar stresses.[18]

13.4.5 Design examples

We have designed simple solenoids (internal and external radii (R_i,R_e) height (h)) for SMESs. The current density (J_{ov}) is the same in the solenoid cross-section ((R_e – R_i) h). The dimensions (R_i, R_e and h) have been optimized using a genetic algorithm to maximize the specific mass energy (then

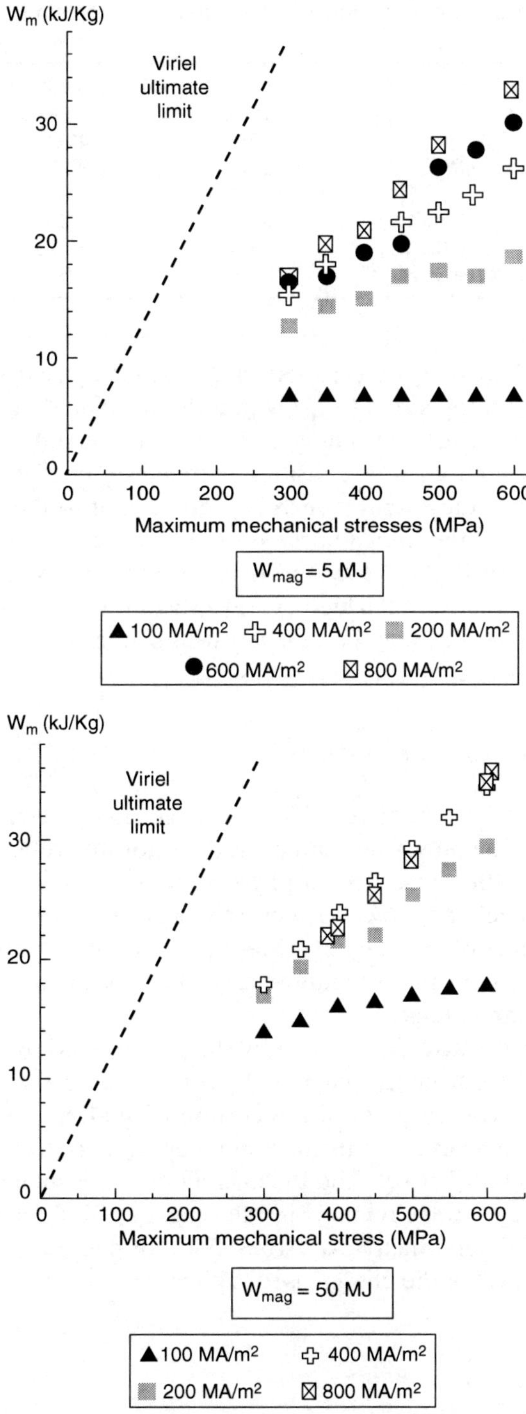

13.9 Mass specific energy vs maximum stress for different current densities and for two energies.

Table 13.3 Two superconducting solenoid magnets for similar energy and stresses

Internal/external radius (m)	0.1684/0.1884	0.6393/0.6459
Height (m)	1.3906	1.283
Current density (MA/m²)	400	400
Max hoop stress (MPa)	370	399
Energy (MJ)	4.82	4.99
Max magnetic flux density (T)	9.76	3.09
Energy mass density (kJ/kg)	19.35	18.26
Energy volume density (MJ/m³)	31.1	2.97

the minimum solenoid cross-section) for given energy (W_{mag}), current density (J_{ov}) and maximum hoop stress (σ_{max}) within the winding.[19] Figure 13.9 shows the results for two different energies. They well illustrate the limitations of SMES. For high current densities (depending on the energy), mechanics limits the specific energy, which is proportional to the stress in accordance with the virial theorem. On the other hand the current density becomes the limitation if it is too low and high specific energy cannot be reached. The figures also show that a high energy is favourable for an SMES. Higher mass specific energies may be reached with similar constraints on the superconductor (current density and stresses).

13.4.6 Specific power limitation

If a SC magnet does not store, relative to its size, a huge energy, it can release it in a very short time. The power per unit mass does not theoretically have a limit, but often has little sense from a physical point of view. In practical applications this level can, however, be extremely high (100 MW/kg). The power is the product of the magnet current (I_{sc}) and the terminal voltage (V_{sc}) across it. High powers thus require large currents and an excellent electric isolation for high voltages.

The maximum power is also determined by the power conditioning system. As energy is removed from the magnet, the current decreases since the energy is proportional to the square of the current (Equation [13.1]). The rated power cannot then be fixed for the maximum operating current I_{SC}^{max}. It refers typically to about half of I_{SC}^{max}, and the magnet operates then between $I_{SC}^{max}/2$ and I_{SC}^{max}. Energy always remains in the magnet ($W_o/4$ for $I_{SC}^{max}/2$). This is similar to a flywheel, which never stops. For a discharge at constant power (P), the fraction (f) of the energy used is given by:

$$f = 1 - \left(\frac{P}{V_{SC}^{max} I_{SC}^{max}} \right)^2 \qquad [13.14]$$

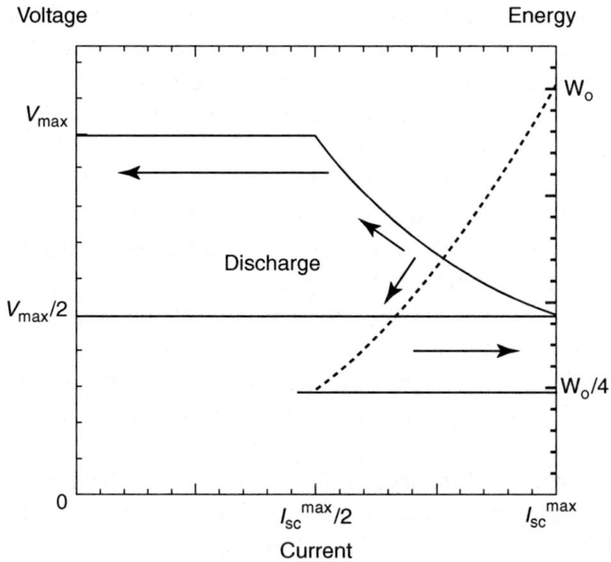

13.10 Voltage and energy evolutions vs current during a discharge at constant power.

V_{SC}^{max} is the maximum voltage across the magnet terminal.

Figure 13.10 gives the voltage and energy vs current for a discharge at constant power from I_{SC}^{max} to $I_{SC}^{max}/2$. The voltage has a maximum value at the end of the discharge to compensate the current reduction.

However, an SMES can be totally discharged but not at constant power. For ultrafast discharging times, under 100 μs, the series and derivative capacities of the coil should be taken into account.

13.4.7 Energy transfer efficiency

During current changes (charge and the discharge) some energy is lost due to the AC losses in the superconducting coil and to eddy current losses in the cryostat. These two contributions can be kept to a very low level (some % of the stored energy) thanks to a suitable design of a low-AC-loss superconducting conductor and of the cryostat. Therefore, the SMESs show excellent energy conversion efficiencies, greater than 95%. This value is very high compared to other storage systems (batteries 70–90%, pumped hydro up to 70%). This inherently high energy efficiency is due to the absence of energy conversion to and from another form, mechanical or chemical. For the same reason, the capacitors also show a high energy conversion factor of 90–95%.

Though this efficiency is important to adjust the SMES to the load energies, it is not the real energy efficiency of the system, and may be irrelevant depending on the application. The energy required by the cryogenic system should be then taken into account and it degrades the high transfer efficiency when the cycle number is low.

Charging the magnet must be less rapid than its discharge. The operating margins indeed decrease during charging, while they increase during discharge. The temperature margin is the difference between the operating (T_o) and current sharing (T_{cs}) temperatures. At T_{cs} the superconductor starts to be dissipative and T_{cs} decreases with the transport current. T_{cs} reaches T_o for the critical current at T_o and T_c for no current. During charges and discharges the operating temperature increases due to the AC losses, but T_{cs} increases or decreases during discharge or charge, respectively

The number of charge-discharge cycles can be very high since it is mainly limited by the mechanical fatigue of the support structure. A SMES may be then used for repetitive and rapid bidirectional exchanges of power with the load.

13.4.8 Summary and main SMES applications

In summary, mass specific energy is limited by mechanical stresses to values of tens of kJ/kg (0.003 kWh/kg), much lower than batteries or other energy sources (Fig. 13.7, Table 13.2). But the energy can be very quickly, and with little loss, released so that the SMES is a short time/pulse energy source. It is basically a transient/pulse power source, not an energy source. The charge can be fast and the number of cycles very high. The SMES acts as a current source and the load should suit. Without moving parts except in the refrigeration system, SMES requires only very light maintenance. It is an environmentally friendly device.

The characteristics of SMESs make them a promising candidate for pulse power sources in different fields, such as some flexible AC transmission systems (FACTS), the rail electromagnetic launchers[20,21] or catapults (aircraft launch), magnetic forming (use of electromagnetic forces to form a metal),[22] and possibly other. The rail electromagnetic launcher uses a pair of conductive rails with a moving part with sliding/rolling contacts in between. Supplied with huge currents, the velocities may reach several km/s. Surpassing the velocities of conventional guns, a possible application of the electromagnetic launcher is military (rail gun). It also could be used to launch small payloads into suborbital or orbital altitudes.[23] A FACTS is a device that enhances the operation (security, capacity and flexibility) of power grids. FACTS cover various grid applications and SMESs suit those

Table 13.4 Some SMES characteristics for different purposes

	SMES plant[2]	SMES/ETM[24]	5 MVA SMES[25]
Energy	5250 MWh (18.9 TJ)	20.4 MWh (73 GJ)	7.3 MJ
Power	1000 MW	400 MW	5 MW
Magnet diameter	1000 m	129 m	0.648 m (4 pole)
Magnet height	19 m	7.5 m	0.7 m configuration)
Current	200 kA	200 kA	2657 A
Superconductor	Nb-Ti	Nb-Ti	Nb-Ti
Operating temperature	1.8 K	1.8 K	4 K
Status	Only design	Abandoned	Used for voltage dips

characterized by a high power and a moderate energy such as transient stability, voltage sag smoothing and flicker mitigation.

Table 13.4 gives some characteristics of three SMESs intended for different power and energy ranges. Only the smallest was constructed and operated.

13.5 Superconducting magnets

The heart of the SMES system, the SC magnet, must be designed to:

- minimize the SC volume for a given magnetic energy, taking into account the mechanical stresses,
- ensure proper cooling and mechanical support of the electromagnetic forces.

The magnet must fulfil the specified electromagnetic signature, for example the position of the 0.5 mT line. It must be protected in case of a quench, which should be avoided by proper design. Nevertheless, quench is always a possible event, which must not be allowed to degrade the magnet.

13.5.1 Magnet topologies

There are two main magnet topologies: solenoidal and toroidal (Fig. 13.11).

A solenoid has a simple structure and its electromagnetic forces are easier to handle than for a toroid; a toroid is subjected to a net large radial force towards the central axis, in addition to the transverse and longitudinal forces. Quench of a toroidal coil is problematic, since it creates an imbalance in the force distribution. The main advantage of a toroid is its naturally low stray field, since the field is contained only within the magnet bore. Its disadvantage is that it stores only about half of the energy per unit conductor stored by the non-shielded solenoid. Nevertheless, the conductor quantity

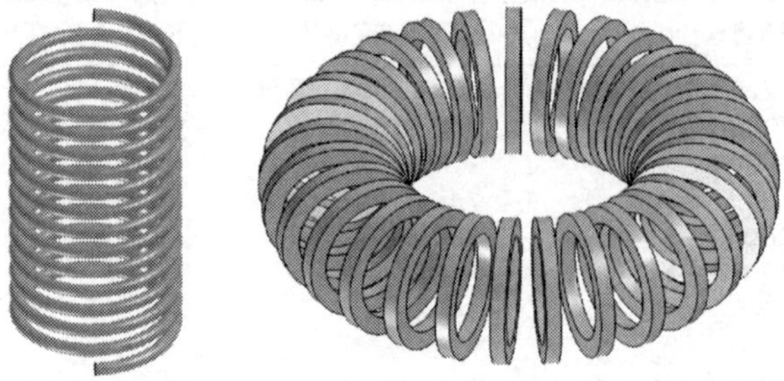

13.11 The two basic magnet topologies for an SMES: (a) solenoidal and (b) toroidal.

per unit stored energy becomes nearly the same when compared to an actively shielded solenoid. Active shielding uses compensated coils around the main magnet in order to cancel the magnetic field outside.

A hexagonal arrangement of solenoids with opposite polarity for adjacent coils (Fig. 13.12), can be used to reduce the stray field of solenoids.[26] Moreover this topology offers a modular design with elementary 'small' solenoids and the production of several identical coils leads to cost savings. However, for fixed physical dimensions, the geometry with alternating polarity stores much less energy than the geometry with all coils having the same polarity.

The energy per unit conductor volume is optimized in a non-shielded solenoid with a diameter to height ratio of about 5 but without stress constraint.

13.5.2 Cryogenics

To remove dissipation (P_{cold}) at low temperature (T_{cold}) power must be provided at room temperature (300 K). The minimum power (P_{min}) is given by the Carnot formula:

$$\frac{P_{min}}{P_{cold}} = \frac{300 - T_{cold}}{T_{cold}} \qquad [13.15]$$

Figure 13.13 shows that the minimum power required to remove 1 W at low temperature (Equation [13.15]) decreases by a factor of 5.3 at 20 K and 14.8 at 50 K when compared to 4 K operation. A higher operating temperature only slightly reduces the cost of the cryostat, but it reduces very

13.12 Schematic drawing of hexagonal arrangement of solenoids for stray field reduction.[26]

significantly the cost of the cryocooler, in a way even more pronounced than the Carnot efficiency. To illustrate that, Fig. 13.13 also shows the cold power capacity from a 7 kW compressor, which approximately fixes the cryocooler cost. The inverse of this cold capacity gives then an approximate cost per watt (Fig. 13.13), which drastically decreases when the operation temperature increases (Fig. 13.13).

However, in large systems, the cryogenic cost is only a fraction (even if not very small) of the total cost. Therefore, a higher operating temperature cannot bring a very strong reduction of the system total cost and thus a tremendous modification of the SMES competitiveness. Nevertheless, operation at higher temperatures brings a lot of advantages for the magnet itself (Section 13.5.3).

13.5.3 Magnet conductor

The superconducting conductor for the magnet winding must meet several requirements:

- High engineering (overall, i.e., the superconductor and normal matrix/shunt/substrate) current densities in large magnetic fields.

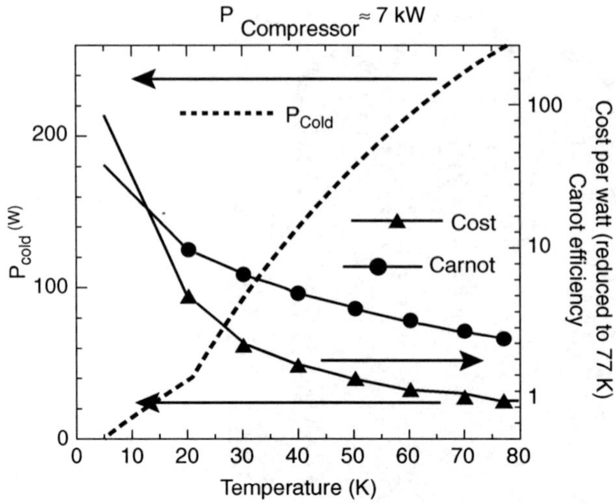

13.13 Cold capacity from a 7 kW compressor, cost per cold power reduced to the 77 K value and Carnot efficiency.

- Support of mechanical stresses/deformations to integrate the mechanical function.
- Low cost, since the real present SMES bottleneck remains its cost.
- Operating temperature as high as possible.

The cable-in-conduit conductor (CICC)[27,28] combines for example several functions: the external sheath may contain most of the Lorentz forces, and is used as cooling fluid vessel.

Low temperature superconductors (LTS), Nb-Ti

At present, only Nb-Ti conductors meet the first three requirements, but their operating temperature is unfortunately low, at or slightly above 4.2 K, the liquid helium temperature. 4 K cryogenics is perfectly mastered, but remains very expensive in terms of capital investment cost and to a much lesser extent operational costs.

Nb-Ti magnets benefit from steady advances in cryogenic cooling. Gradual improvements in large cryocoolers extend the maintenance cycle and reduce the refrigeration electrical load. The introduction of high-critical-temperature (high temperature superconducting (HTS)) current leads was another important improvement. These have significantly reduced the related losses (1/10 at 4 K, 1/3 total at 300 K), a large contributor towards the total loss.[29]

With LTS, for reasons of stability, protection and stress containment, the engineering current density decreases with energy.[30] For energies above 1 MJ

(0.3 kWh), the overall current density should be lower than 100 MA/m^2. The current density is no more limited by the SC current capacity.

High temperature superconductor (HTS)

HTS materials offer very interesting opportunities for SMES. They offer the possibility to operate at higher temperatures, which reduces the cryogenic operating cost and above all the cryocooler cost (Fig. 13.13). The reduced refrigeration power then translates into improved efficiency.

The operation at higher temperatures brings a very significant enhancement of the material specific heat (Fig. 13.14) (factors of 80 and 1000 between 4, 20 and 50 K, respectively), then a much more stable operation, so less sensitive to external perturbations. The minimum quench energy (MQE) characterizes the energy density deposited in the magnet required to quench it, then its stability. In adiabatic conditions, the expression of the MQE per unit volume may be estimated by:[31]

$$\mathrm{MQE} = \mathrm{H}(T_{cs}) - H(T_o) \approx c_p(T_o)[T_{cs} - T_o]\left|1 - \frac{I_{SC}}{I_c}\right| \qquad [13.16]$$

The large increase in the specific heat (Fig. 13.14), in addition to the larger temperature margin ($T_{cs} - T_o$) enhances the MQE greatly.

The constraints relative to stability are much lower compared to those of LTS, and higher overall current densities may be foreseen. Since SMES systems are in general characterized by high power, constraints linked to protection are not a limitation (see Section 13.5.4).

Nevertheless, the increased specific heat is not favourable for magnet protection.[32] The zone, which has lost its SC state, propagates very slowly, and may experience high temperature excursions and thus be damaged. Quench detection becomes more difficult: the voltage to be detected is proportional to the quenched zone. Improved detection methods have been proposed.[33]

The YBCO-coated conductor or second generation (2G) HTS wire is particularly promising for SMES. Their current capacity under high magnetic flux densities keeps high values, even when temperature increases (Fig. 13.15), exceeding 1000 MA/m^2 at 20 K under 13 T. The 2G wires continuously experience remarkable improvements and advances; these values will increase still further in the future.

The greater the mechanical stresses in a SMES system, the lower is the magnet weight (Equation [13.7]). To minimize the weight, the conductor should combine the function of current transport and mechanical support. This is the reason for the great interest in IBAD 2G tapes for SMES, since they withstand more than 800 MPa without any degradation.

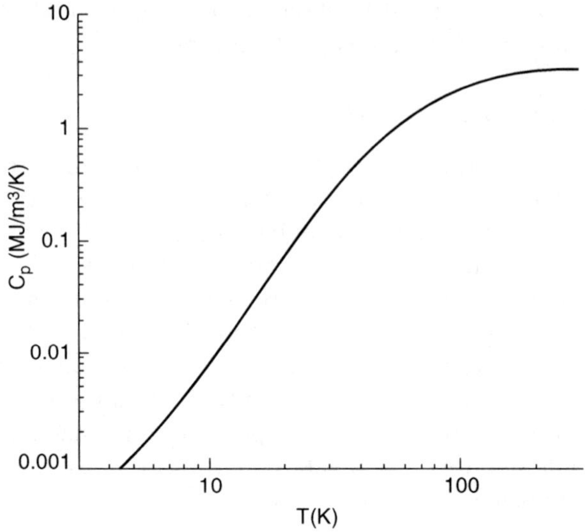

13.14 Typical specific heat vs temperature (copper example).

Nevertheless, the 2G wires are anisotropic thin tapes. The magnet designer should reduce the transverse magnetic flux density, which significantly affects the current capacity (Fig. 13.15). High current cables from 2G wires remain an issue, due to their geometry, but some means to address this are under investigation, such as Roebel bars.[34]

Though 2G HTS wires offer exciting perspectives, their present cost is the real bottleneck, but they are still in development stage with reduced production. They are today at least 20 times more expensive than Nb-Ti, and Nb-Ti SMES systems are already too expensive. The cost of HTS will decrease a lot in the future, so HTS SMES offers great promise.

MgB_2 wires, another area of research, do not have the potential for a very significant breakthrough, even at low conductor cost. The operating temperature increase is not significant enough compared to Nb-Ti (10–15 K instead of 4 K). The current capacities at higher temperatures decrease rapidly with magnetic flux density.

13.5.4 Magnet protection

To protect a magnet is to design it such that the hot spot temperature after a quench does not reach values that may damage the magnet. The hot spot temperature is overestimated from the thermal adiabatic equation:

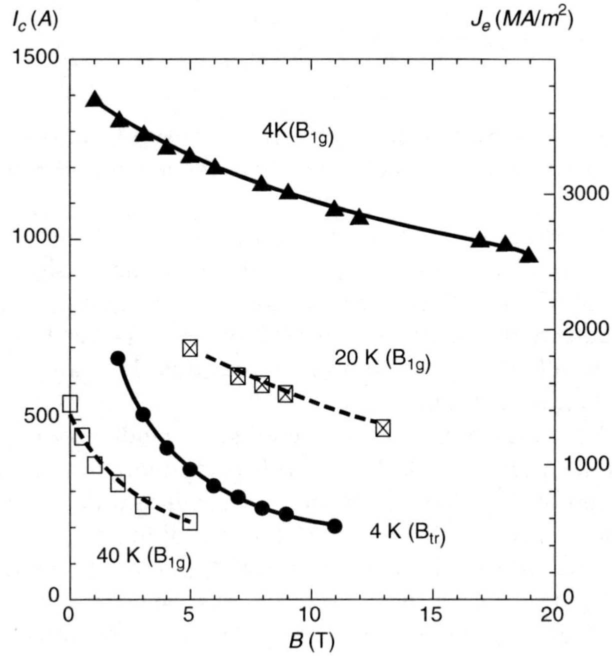

13.15 Overall tape current densities at various temperatures *vs* longitudinal magnetic field except at 4.2 K (transverse characteristic as well) data from CEA and CERN.

$$\rho(T)\, j(t)^2 = c_p(T)\frac{\mathrm{d}T}{\mathrm{d}t} \Rightarrow j(t)^2\mathrm{d}t = \frac{c_p(T)}{\rho(T)}\mathrm{d}T \quad [13.17]$$

The time integration of the two members gives the hot spot temperature ($T_{\max}$):

$$\int_0^\infty j(t)^2\mathrm{d}t = \int_{T_o}^{T_{\max}} \frac{c_p(T)}{\rho(T)}\mathrm{d}T = F(T_{\max}) \quad [13.18]$$

The classical technique is to discharge the magnet into an external discharge resistance (R_d) when the quench is detected (t_{det} after the real quench). The magnet resistance is supposed negligible compared to the discharge resistance. The time integration of the square of the current is then very simple, and the equation can be rewritten introducing the relevant quantities:

$$F(T_{\max}) = J_o^2 \left| t_{\det} + \frac{W_o}{I_o V_{\max}} \right| = J_o^2 \left| t_{\det} + \frac{W_o}{P_{\max}} \right| \quad [13.19]$$

W_o is the energy for the SC current I_o or current density J_o. $V_{\max}$ is the maximum voltage across the SC magnet, which fixes the maximum power ($P_{\max} = V_{\max} I_o$).

This equation shows that a magnet becomes more difficult to protect when the current density and/or the energy increases. The sensitivity of the quench detection, which defines the detection delay ($t_{\det}$), plays an important part. But detection is difficult with HTS magnets. Nevertheless, since SMESs are above all attractive as power sources, their low ratio $W_o/P_{\max}$ is favourable for their protection.

Let us consider some orders of magnitude. Superconductors experience no damage for temperatures of 400 K with some margin. Some YBCO tapes have indeed reached 720 K without any significant degradation[35] in fault current limiters (FCL). For magnets, thermal differential contraction stresses have to be studied. Taking 400 K for $T_{\max}$ and 100 ms for $t_{\det}$ as well as for $W_o/P_{\max}$ the overall current density must be lower than 450 MA/m^2 for the SCS4050 SuperPower tape (http://www.superpower-inc.com/). Reducing the tape resistivity (thicker copper stabilizer) may increase this already rather high value. Such high current densities are only required for small SMESs. For SMESs above 50 MJ Fig. 13.9 has shown that overall current densities of the order of 200 MA/m^2 are enough to design solenoid magnets with mass specific energy of the order of 26 kJ/kg (two times today's record value).

13.6 Applications of SMES

Table 13.5 summarizes the main applications of SMESs.

13.6.1 SMES for the power grid

Load levelling

The idea of the SMES was first conceived in 1970; the motivation was to level the diurnal load in the French electricity network.[4] The energies required (thousands of MWh) led to huge magnets (1 km in diameter, see Table 13.1) with a lot of realization difficulties. Furthermore, SMES is not the best solution in this case, due to the long cycles (hours). Pumped hydroelectric offers an alternative which is generally chosen. Of course, the need for two nearby water reservoirs with significant altitude difference limits the available sites for this technology.

Table 13.5 Main applications of SMES

Application	UPS	FACTS	Load levelling	Pulse power source
Use	Voltage-power quality and security for sensitive-critical loads	Active (and reactive) power exchanges with the grid to improve its operation (stability, capacity enhancement, etc.)	Pulsed loads with large stored energies (pulsed magnets, etc.)	Devices with high power / short time requirements: • Electromagnetic launcher • Magnetic forming

On the other hand SMES is very interesting to level some pulse loads with short cycles. Some power supplies use an energy tank to significantly smooth the power consumption of pulse loads. A typical example is the proton synchrotron (PS) circular accelerator at CERN. This accelerator operates in pulse mode. The operational magnetic flux density is typically reached in 0.7 s, maintained during 0.3 s and returned to zero in 0.7 s. Six to eight millions of such cycles are performed each year. The maximum dissipative power of this resistive accelerator is 10 MW, but 50 MW is required at the end of the magnet charge, whereas the power supply should absorb a large amount of power to lower the magnetic flux density. It is very interesting, even indispensable, to have an energy tank to provide energy during the magnetic field to raise and to recover energy during the magnetic field fall. The first PS power supplies used rotating machines as storage device, and now capacitors are used. This last power supply is called power converter for the PS (POPS main magnets[36]). An SMES system is of interest to replace it, due to the mass and volume of the capacitors.

Flexible AC transmission system (FACTS)

With energies of some to tens of MJ, SMES can be used in the electricity networks as a FACTS[37] device. An SMES system operating in part as a FACTS device was the first superconducting application installed in a real power grid. In 1976, the Los Alamos Scientific Laboratory began the development of a 30 MJ SMES for use on the Bonneville Power Authority's Tacoma substation[38] to damp low-frequency power oscillations. This SMES operated in real grid conditions for a very short time with only a few hundred energy transfers during a test, evaluation and commissioning period. The main operational issue was the refrigerator, which had been damaged during shipping from the manufacturer. No problems occurred with the Nb-Ti magnet and the cryostat. An alternative solution for power flow variations was found while the 30 MJ coil was under construction. That approach was to replace some of the rectifier diodes on the northern terminus of the existing DC link between the Pacific Northwest and Southern California with SCRs. That solution successfully damped the oscillations. As a result, the LANL-BPA SMES operation was stopped. However, this experience was the first successful and conclusive demonstration and field test of a large SMES operation in a real power grid.

More recently, in 2000, the American Superconductor company installed six SMES units at key points in the grid in northern Wisconsin, USA, to enhance its stability.[39] This grid was experiencing voltage instability problems with large momentary voltage depressions, which could have led to grid collapse. The six SMESs at different key locations in the grid injected real and reactive powers into the grid to boost the voltage and they increased the power transmission capabilities by 15%. Each SMES could provide continuously 2.8 MVAR (5.6 MVAR during 1 s) and 2 MW during a short duration. These SMESs were packaged in standard trailers for easy and rapid deployment. The commissioning of a 345 kV line some years later solved the voltage instability problems, and these SMES units were disconnected.

But electric grids require much more reactive than active power to enhance their operation. The active power requirement is rather rare. An SMES is only required for active power in FACTS. For reactive power, a FACTS using capacitors with a relatively low stored energy is sufficient: D-VAR for example.[40] Such a FACTS meets most of the grid requirements.

Bridging power

The development of renewable energies for electricity grids poses the problem of storage, since they often are intermittent. If SMES cannot solve the issue of the management of energy or load levelling, it may be an attractive solution for short durations, up to seconds for power quality or bridging power. SMES may assure the continuity of service before the starting of other storage technologies better suitable for long durations. So, associated with other storage devices, an SMES may make it possible to cover the full range of durations.

13.6.2 SMES for local power conditioning: uninterruptible power supplies (UPS)

Several SMESs have been operated in UPS with power ratings in the order of MW. These are used locally for critical loads requiring ultra-clean power for sensitive processing (such as semiconductor chip fabrication facilities, for example) or military and research laboratory applications.

The superconducting magnet replaces the batteries classically used. These SMESs have been mainly provided by American Superconductor.[41] They have led to a considerable amount of test experience. From 2000, this company had accumulated more than 35 unit-years of operation.

One of the first systems was installed in 1993 for an ammonia production furnace. Another 1.4 MVA/2.4 MJ SMES was installed at the Brookhaven National Laboratory (USA) to offer high quality power for a synchrotron radiation source.[42] It provides power during voltage sags or momentary interruptions to avoid beam loss. Owens Corning's extrusion and production lines in North Carolina have been protected by SMES from voltage sags.[43] Likewise, in South Africa, an SMES has protected a paper machine against 72 dips in 11 months.[44]

In Japan several SMESs have been built. One objective is the protection of sensitive load against voltage dips. In 2003 a 5 MW–7 MJ SMES was fabricated using Nb-Ti solenoid in a four-pole configuration.[25] It has compen-

13.16 10 MJ Chubu electric SMES in operation at Kameyama Factory, Sharp corporation.

sated voltage dips in a liquid crystal manufacturing factory. Another SMES system with a rating of 10 MVA was built in 2005 (Fig. 13.16).

A national Japanese programme is being carried out on SMES for load fluctuation compensations. They have initiated the development of high-T_c SMES and tested a 1 MVA SMES using a Bi-2212 PIT wire, but operating at 4 K.[45] Due to the advances and performances of YBCO-coated conductors, SMES projects are based on these conductors now.

13.6.3 SMES for pulse power sources

Another SMES class are pulse power sources for dedicated applications with very high power demand over short time. In the 1980s, a large SMES development programme was carried out in United States under the Strategic Defense Initiative (SDI).[46,47] The main objective was a power source for the Free Electron Laser, but utility applications were studied as well. As intermediate stage, the SMES engineering test model (ETM) aimed to design and build a 20–30 MWh (72–108 GJ) – 400/1000 MW magnet (see Table 13.1, middle column). Two teams (Bechtel and Ebasco) were in competition in the design phase and they proposed rather different solutions. One used a 60 kA Al stabilized Nb-Ti conductor with a helium bath cooling. The second was based on a 200 kA Nb-Ti CICC. It was developed and tested up to 303 kA under 5 T at 1.8 K, which still is a world record in terms of current capacity. Abandoning the SDI resulted in the termination of this SMES project.

Pulse power SMES sources using toroidal coil magnets have been also studied in Russia.[48,49]

The Délégation Générale pour l'Armement (DGA), the French institution in charge of research for army, has launched in the first decade of this century a programme to develop an HTS SMES for rail electromagnetic launchers/guns[50,51] with Institute Néel and G2Elab among others. Since this application requires a short current pulse, SMES is the ideal power source with a very simple implementation (Fig. 13.3). Today power capacitors are used but a dissipative conversion circuit is required to transform the voltage source (capacitor) into a current source necessary for the rail gun. The huge currents (100–1000 kA) required by a rail gun remain a challenging issue for an SMES. Using an SMES, the current can be lowered compared to a capacitor-based source since the supplied current can be more constant during the shot.[5] The XRAM concept (MARX spelt backwards) may be used to increase the discharge current: the magnet is made with several coils charged in series and discharged in parallel. Figure 13.17 shows the successful tests carried out with Institut Saint Louis (ISL) on the XRAM generator.[52]

This DGA programme has led inter alia to an HTS SMES (Fig. 13.18) designed for 800 kJ (425 kJ achieved) based on a BiSrCaCuO conductor from Nexans. It operates at 20 K using only conduction cooling from

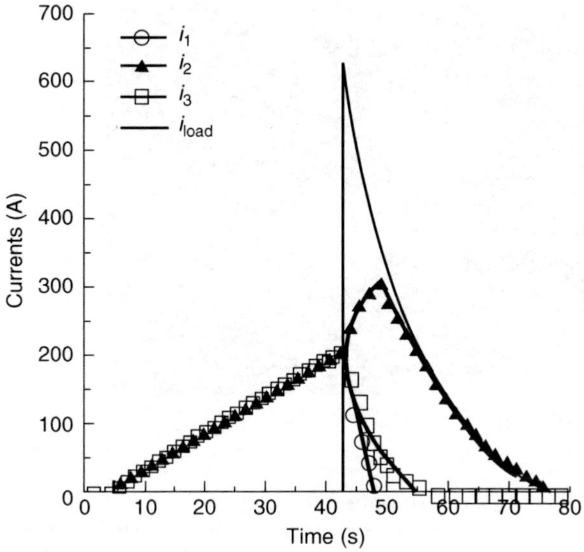

13.17 XRAM generator based on a SMES with three coils, charge in series, discharge in parallel.

(a)

13.18 (a, b) 800 kJ 20 K BSCCO conducting cooling SMES built in Grenoble.

13.18 Continued

Table 13.6 Main HTS SMES throughout the world

Organization	Country	Year	Data	SC	Application
Chubu[45]	Japan	2004	1 MVA, 1 MJ	Bi-2212	Voltage stability
CAS[53]	China	2007	0.5 MVA, 1 MJ	Bi-2223	Power quality
KERI[54]	Korea	2007	0.6 MJ	Bi-2223	Power, voltage quality
DGA/CNRS[50]	France	2007	0.8 MJ	Bi-2212	Pulse application/ rail gun
KERI[54]	Korea	2011	2.5 MJ	YBCO	Power quality
Chubu[55] M-PACC project	Japan		100 MVA, 2.4 GJ	YBCO	Power system control
SuperPower, ABB, BNL, TcSUH[56]	USA	2011–2013	20 kW, 3.4 MJ	YBCO	Load levelling

cryocoolers for user friendly cryogenics. Taking into account the severe military environment, HTS suits much better than LTS. The purpose was to qualify several technological solutions on a representative level and to acquire an essential operational experience feedback. It has been followed by a reflection phase with ISL to optimize the complete system. The use of YBCO has been analysed and show opportunities to enhance the mass specific energies (Fig. 13.9) so the interest of SMESs.

13.6.4 Summary

In summary, several SMES systems have proved their operational capabilities for short-term (seconds) power at MW scale. They are commercially available and the field test experience is very large in the US and Japan. Nevertheless, the number of SMES systems sold remains very low. The major reason is the high initial cost coupled with the competition with more mature technologies. Moreover, SMES addresses niche applications having high active power and short time demands.

R&D mainly now concerns HTS SMES (Table 13.6).

13.7 Conclusion

SMES is particularly suitable for power sources of short duration, because the power density is much higher than the stored energy density. SMES is thus an excellent solution for applications such as pulse/transient power sources especially current sources, UPS or FACTS for power grids. A number of SMES units have been installed and operated successfully during many years demonstrating their very satisfying performances. The bottleneck for widespread commercialization of SMES remains the high capital cost.

The deregulation of the electricity market and the requirements to enhance the power capacities of the present grids bring the opportunity for FACTS or bridging power using SMES.

The need for pulse power sources for emerging applications such as electromagnetic launchers also offers very good opportunities for SMES since they offer a high performance, well-adapted solution.

Several characteristics of HTSs are very attractive for SMESs, especially YBCO-coated conductors, for which the following characteristics are significant: current capacity under large magnetic flux densities at temperatures much higher than 4 K, mechanical strength. But these very favourable and promising properties will be used only if these conductors can be produced at low cost.

13.8 Acknowledgements

This work was performed under a DGA programme and an ANR project. The author warmly thanks A. Badel for fruitful discussions. The author is particularly grateful to W. Hassenzahl for his contribution to this chapter, in particular his extremely careful reading and the interesting exchanges with the author.

13.9 References

1. J. File and R.G. Mills, 'Observation of persistent current in a superconducting solenoid', *Physical Review Letters*, vol. **10**, 93–96 (1963).
2. W. Hassenzahl, 'Superconducting magnetic energy storage', *IEEE Transactions on Magnetics*, vol. **25**, 750–758 (1989).
3. C. A. Luongo, 'Superconducting storage systems: an overview', *IEEE Transactions on Magnetics*, vol. **32**, 2214–2223 (1996).

4. M. Ferrier, 'Stockage d'énergie dans un enroulement supraconducteur', *Low temperature and Electric Power*, Pargamon Press, 425–432 (1970).
5. A. Badel, 'High critical temperature SMES as pulse source', PhD, Grenoble INP, 2010.
6. M. H. Ali, B. Wu and R. A. Dougal, 'An overview of SMES applications in power and energy systems', *IEEE Transactions on Sustainable Energy*, vol. **1**, 38–47 (2010).
7. J. Kondoh, I. Ishii, H. Yamaguchi, A. Murata, K. Otani, K. Sakuta, N. Higuchi, S. Sekine and M. Kamimoto, 'Electrical energy storage systems for energy networks', *Energy Conversion and Management*, vol. **41**, 1863–1874 (2000).
8. F. C. Moon, 'The virial theorem and scaling laws for superconducting magnet systems', *Journal Applied Physics*, vol. **53**, 9112–9121 (1982).
9. S. Nomura, N. Watanabe, C. Suzuki, H. Ajikawa, M. Uyama, S. Kajita, Y. Ohata, H. Tsutsui, S. Tsuji-Iio and R. Shimada, 'Advanced configuration of superconducting magnetic energy storage', *Energy*, vol. **30**, 2115–2127 (2005).
10. J. L. Smith, 'Field analysis for a SMES magnet with radial force balance', *IEEE Transactions on Applied Superconductivity*, vol. **5**, 357–360 (2005).
11. G. E. Marsh, 'Force-free magnetic field: solutions, topology and application', *World Scientific*, ISBN: 978-981-02-2497-4.
12. Y. M. Eyssa, 'Design of single layer superconductive energy storage magnets', *Journal Physics D: Applied Physics*, vol. **13**, 1719–1726 (1980).
13. A. A. Kuznetsov, *Soviet Physics Technical Physics*, vol. **6**, pp. 472–475 (December 1961).
14. D. Campi, B. Curé, A. Gaddi, H. Gerwing, A. Hervé, V. Klyuklin, G. Maire, G. Perinic, P. Bredy, P. Fazilleau, F. Kircher, B. Levésy, P. Fabbricatore, S. Farinon and M. Greco, 'Commissioning of the CMS magnet', *IEEE Transactions on Applied Superconductivity*, vol. **17**, 1185–1190 (2007).
15. A. Yamamoto, Y. Makida, H. Yamaoka, H. Ohmiya, K. Tanaka, T. Haruyama, T. Yoshida, K. Yoshimura, S. Matsuda, K. Kikuchi, Y. Ootani and S. Mizumaki, 'A Thin Superconducting Solenoid Magnet for Particle Astrophysics', *IEEE Transactions on Applied Superconductivity*, vol. **12**, 438–441 (2002).
16. C. Rix, C. Luongo, W. Bingham, A. Bulc, K. Cooke, D. Lieurance, K. Partain and S. Peck, 'A self-supporting superconducting magnetic energy system (SMES) concept', *Cryogenics*, vol. **34**, No. 1, 737–740 (1994).
17. W. V. Hassenzahl, 'A comparison of the conductor requirements for energy storage devices made with ideal coil geometries', *IEEE Transactions on Magnetics*, vol. **25**, 1799–1802 (1989).
18. P. Tixador, N. T. Nguyen, J.M. Rey, T. Lecrevisse, V. Reinbold, C. Trophime, X. Chaud, F. Debray, S. Semperger, M. Devaux and C. Pes, 'SMES optimization for high energy densities', *IEEE Transactions on Applied Superconductivity*, vol. **22**, 5700704 (2012).
19. B. Vincent, P. Tixador, T. Lecrevisse, J.-M. Rey, X. Chaud and Y. Miyoshi, 'HTS magnets: opportunities and issues for SMES', *IEEE Transactions on Applied Superconductivity*, vol. **23**, 5700805 (2013).
20. H. D. Fair, 'Electric launch science and technology in the United States', *IEEE Transactions on Magnetics*, vol. **39**, 11–17 (2003).
21. P. Lehmann, 'Overview of the electric launch activities at the French-German Research Institute of Saint-Louis (ISL)', *IEEE Transactions on Magnetics*, vol. **39**, 24–28 (2003).

22. T. E. Motoasca, H. Blok, M. D. Verweij and P. M. van den Berg, 'Electromagnetic forming by distributed forces in magnetic and non magnetic materials', *IEEE Transactions on Magnetics*, vol. **40**, 3319–3330 (2004).
23. P. Lehmann, O. Bozic, H. Grobusch and J. Behrens, 'Electromagnetic Railgun technology for the deployment of small sub-/orbital payloads', *IEEE Transactions on Magnetics*, vol. **43**, 480–485 (2007).
24. R. J. Loyd, T. E. Walsh and E. R. Kimmy, 'Key design selections for the 20.4 MWh SMES/ETM', *IEEE Transactions on Magnetics*, vol. **27**, 1712–1715 (1991).
25. S. Nagaya, N. Hirano, M. Kondo, T. Tanaka, H. Nakabayashi, K. Shikimachi, S. Hanai, J. Inagaki, S. Ioka and S. Kawashima, 'Development and performance results of 5 MVA SMES for bridging instantaneous voltage dips', *IEEE Transactions on Applied Superconductivity*, vol. **14**, 699–704 (2004).
26. W. Weck, P. Ehrhart, A. Müller and G. Reiner, 'Superconducting inductive pulsed power supply for electromagnetic launchers: design aspects and experimental investigation of laboratory set-up', *IEEE Transactions on Magnetics*, vol. **33**, pp. 524–527 (1997).
27. M. O. Hoenig, Y. Iwasa and D. B. Montgomery, 'Supercritical-helium cooled' bundle conductors' and their application to large superconducting magnets', *Proceedings of the 5th International Conference on Magnet Technology*, Roma, Italy, 21–25 April 1975, 519–524 (1975).
28. C. A. Luongo, K. D. Partain, J. R. Miller, G. E. Miller, M. Heiberger and A. Langhorn, 'Quench initiation and propagation study (QUIPS) for the SMES-CIC', *Cryogenics*, vol. **31**, No. 1, 611–614 (1994).
29. J. R. Hull, 'High-temperature superconducting current leads', *IEEE Transactions on Applied Superconductivity*, vol. **3**, 869–875 (1993).
30. M. N. Wilson, 'Stabilization, protection and current density: some general observations and speculations', *Cryogenics*, vol. **31**, 449–503 (1991).
31. L. Dresner, '*Stability of Superconductor*', New York: Plenum Press, ISBN 0-306-45030-5.
32. T. Effio, U. P. Trociewitz, X. Wang and J. Schwartz, 'Quench induced degradation in $Bi_2Sr_2CaCu_2O_{8+x}$ tape conductors at 4.2 K', *Superconductor Science and Technology,* vol. **21**, 045010, p. 10 (2008).
33. A. Badel, P. Tixador, G. Simiand and O. Exchaw, 'Quench detection system for twin coils HTS SMES', *Cryogenics*, vol. **50**, 674–681 (2010).
34. W. Goldacker, A. Frank, A. Kudymow, R. Heller, A. Kling, S. Terzieva and C. Schmidt, ' Status of high transport current ROEBEL assembled coated conductor cables', *Superconductor Science and Technology*, vol. **22**, 034003, p.10 (2009).
35. M Schwarz, C. Schacherer, K-P Weiss and A Jung, 'Thermodynamic behaviour of a coated conductor for currents above T_c', *Superconductor Science and Technology*, vol. **21**, 054008, p. 4 (2008).
36. C. Fahrni, A. Rufer, F. Bordry and J. P. Burnet, A novel 60 MW pulsed power system based on capacitive energy storage for particle accelerators, *Proceedings of the Twelfth European Conference on Power Electronics and Applications*, Aalborg, Denmark, 2–5 September (2007).
37. X. D. Xue, K. W. E. Cheng and D. Sutanto, 'A study of the status and future of superconducting magnetic energy storage in power systems', *Superconductor Science and Technology,* vol. **19**, R31–R39 (2006).

38. H. J. Boenig and J. F. Hauer, 'Commissioning tests of the Bonneville power administration 30 MJ superconducting magnetic energy storage unit, *IEEE Transactions on Power Apparatus and Systems*, vol. **104**, 302–312 (1985), and W. V. Hassenzahl private communication.
39. T.R. Abel, 'D-SMES for Wisconsin', *Modern Power Systems*, vol. **19**, No. 10, 28–29 (1999).
40. H. G. Sarmiento, G. Pampin and J. Diaz de Leon, 'FACTS solutions for voltage stability problems in a large metropolitan area', *Power System Conference and Exposition*, IEEE PES, New York, 10–13 October 2004, 275–282 (2004).
41. American Superconductor, http://www.amsuper.com/.
42. J. Cerulli, 'Operational experience with a SMES device at Brookhaven National Laboratory, New York', *IEEE Power Engineering Society Winter Meeting*, New York, 31 January–4 February 1999, 1247–1252 (1999).
43. J. Cerulli, G. Melotte and S. Peele, 'Operational experience with a superconducting magnetic energy storage device at Owens Corning Vinyl Operations, Fair Bluff, North Carolina', *IEEE Power Engineering Society Summer Meeting*, Edmonton, Canada, 18–22 July 1999, 524–528 (1999).
44. R. Schottler and R. G. Coney, 'Commercial application experiences with SMES ', *The IEE Power Engineering Journal*, vol. **13**, 149–152 (1999).
45. K. Shikimachi, H. Moriguchi, N. Hirano, S. Nagaya, T. Ito, J. Inagaki, S. Hanai, M. Takahashi and T. Kurusu, 'Development of MJ-Class HTS SMES system for bridging instantaneous voltage dips', *IEEE Transactions on Applied Superconductivity*, vol. **15**, 1931–1934 (2005).
46. R. L. Verga, 'SMES and other large-scale SDI cryogenic application programs', *Advances in Cryogenic Engineering*, vol. **35**, Plenum Press, 555–564 (1990).
47. G. W. Ullrich, 'Summary of the DNA SMES development program', *IEEE Transactions on Applied Superconductivity*, vol. **5**, 416–421 (1995).
48. E. P. Polulyakh, L. A. Plotnikova, V. A. Afanas'ev, M. I. Kharinov, A. K. Kondratenko, Yu. Klimenko and S. I. Novikov, 'Development of toroidal superconducting magnetic energy storages (SMES) for high-current pulsed power supplies', *Advances in Cryogenic Engineering*, vol. **48A**, Plenum Publ., 721–727 (2000).
49. E. Yu. Klimenko and E. P. Polulyakh, 'Closed Flux Winding for SMES', *Advances in Cryogenic Engineering*, vol. **47A**, Plenum Publ., 323–328 (2002).
50. P. Tixador, B. Bellin, M. Deleglise, J. C. Vallier, C. E. Bruzek, S. Pavard and J. M. Saugrain, 'Design of a 800 kJ HTS SMES', *IEEE Transactions on Applied Superconductivity*, vol. **17**, 1707–1710 (2007).
51. A. Badel, P. Tixador, K. Berger and M. Deleglise, 'Design and preliminary tests of a twin coil HTS SMES for pulse power operation', *Superconductor Science and Technology*, vol. **24**, 055010 (2011).
52. P. Dedié, V. Brommer, A. Badel and P. Tixador, 'Three-Stage Superconducting XRAM Generator', *IEEE Transactions on Dielectrics and Electrical Insulation*, vol. **18**, 1189–1193 (2011).
53. X. Liye, W. Zikai, D. Shaotao, Z. Jinye, Z. Dong, G. Zhiyuan, S. Naihao, Z. Fengyuan, X. Xi and L. Liangzhen, 'Fabrication and tests of a 1 MJ HTS magnet for SMES', *IEEE Transactions on Applied Superconductivity*, vol. **18**, 770–773 (2008).

54. K. C. Seong, 'An introduction of HTS-SMES project in Korea', *IEEE/CSC & ESAS European Superconductivity News Forum (ESNF)*, No. 7, January 2009.
55. K. Shikimachi, N. Hirano, S. Nagaya, H. Kawashima, K. Higashikawa and T. Nakamura, 'System coordination of 2 GJ class YBCO SMES for power system control', *IEEE Transactions on Applied Superconductivity*, vol. **19**, 2012–2018 (2009).
56. Q. Li, 'Long term superconducting magnetic energy storage (SMES) for GRIDS, air and space applications', Presented at ASC Conference, Portland, 7–12 October (2012).

Index

Lightning Source UK Ltd.
Milton Keynes UK
UKOW03n0858081013

218677UK00001B/51/P

9 781845 697846